Biomedical Sciences

Biomedical Sciences

Essential Laboratory Medicine

Edited by

Ray K. Iles
Anglia Ruskin University, UK

Suzanne M. Docherty
Queen Elizabeth II Hospital, Welwyn Garden City, UK

A John Wiley & Sons, Ltd., Publication

This edition first published 2012 © 2012 by John Wiley & Sons, Ltd

Wiley-Blackwell is an imprint of John Wiley & Sons, formed by the merger of Wiley's global Scientific, Technical and Medical business with Blackwell Publishing.

Registered office: John Wiley & Sons, Ltd, The Atrium, Southern Gate, Chichester, West Sussex, PO19 8SQ, UK

Editorial offices: 9600 Garsington Road, Oxford, OX4 2DQ, UK
The Atrium, Southern Gate, Chichester, West Sussex, PO19 8SQ, UK
111 River Street, Hoboken, NJ 07030-5774, USA

For details of our global editorial offices, for customer services and for information about how to apply for permission to reuse the copyright material in this book please see our website at www.wiley.com/wiley-blackwell.

The right of the author to be identified as the author of this work has been asserted in accordance with the UK Copyright, Designs and Patents Act 1988.

Library of Congress Cataloging-in-Publication Data

Biomedical sciences : essential laboratory medicine / Raymond Iles and Suzanne Docherty.
    p. ; cm.
  Includes bibliographical references and index.
  ISBN 978-0-470-99775-8 (cloth) – ISBN 978-0-470-99774-1 (pbk.)
  1. Diagnosis, Laboratory–Textbooks. 2. Medical laboratory technology–Textbooks. I. Iles, Raymond. II. Docherty, Suzanne.
  [DNLM: 1. Pathology, Clinical–methods. 2. Laboratory Techniques and Procedures. QY 4]
  RB37.B56 2011
  616.07′5–dc23

2011019935

A catalogue record for this book is available from the British Library.

This book is published in the following electronic formats: ePDF 9781119950929; ePub 9781119962410; Mobi 9781119962427

Set in 10/12pt, Minion by Thomson Digital, Noida, India

First Impression 2012

The editors would like to dedicate this book to the memory of Marion Docherty.

Contents

List of Contributors

Dr Stephen A. Butler, B.Sc., Ph.D.
Centre for Investigative and Diagnostic
Oncology
Middlesex University
The Burroughs
London, UK

Dr Iona Collins, MBBS, BMedSci (Hons.),
FRCS (Orth.)
Consultant in Spinal Surgery
Morriston Hospital
Swansea
Wales, UK

Dr Suzanne M. Docherty, MBBS,
BMedSci (Hons.), Ph.D.
Core Medical Trainee
Queen Elizabeth II Hospital
Welwyn Garden City
Hertfordshire, UK

Catherine S. Fontinelle, B.Sc. Hons., ARCS,
FIBMS, CSci
Senior Biomedical Scientist
Microbiology Department, St Helier Hospital
Carshalton,
Surrey, UK

Dr Sarah J Furrows, MBBS, M.Sc., MRCP,
FRCPath.
Consultant Microbiologist and Infection
Control Doctor
Kingston Hospital
Kingston-upon-Thames
Surrey, UK

Professor Ray K. Iles, B.Sc., M.Sc., Ph.D.,
CBiol, FSB, FRSC
Faculty of Health and Social Care
Anglia Ruskin University
Bishop Hall Lane
Chelmsford, UK

Dr David Ricketts, CSci, DBMS, FIBMS
Department of Clinical Biochemistry
North Middlesex University Hospital
London, UK

Professor Ivan M. Roitt, D.Sc., FRS, Hon. FRCP
Director, Centre for Investigative and Diagnostic
Oncology
School of Health and Social Science
Middlesex University
The Burroughs
London, UK

Dr Christopher M. Stonard, MA, MB, BChir,
FRCPath
Consultant Histopathologist
Department of Histopathology and Cytopathology
Chesterfield Royal Hospital NHS Foundation Trust
Calow
Chesterfield, UK

Jennifer H Stonard, B.Sc., LIBMS
Biomedical Scientist
Department of Histopathology and Cytopathology
Northern General Hospital
Sheffield Teaching Hospitals NHS Foundation Trust
Herries Road
Sheffield
South Yorkshire, UK

Preface

The practice of clinical medicine and the diagnosis and management of human disease becomes ever more complex with each year that passes and our knowledge of the molecular basis of pathology expands seemingly exponentially. There is thus an ever greater need for well-trained, highly skilled biomedical scientists — the professionals who perform the vital laboratory tests and investigations that underpin the diagnosis of disorders and the evaluation of the effectiveness of treatment.

With this textbook on *Biomedical Sciences*, we set out to create a comprehensive — yet focused — resource that students can use at all levels of their study and career progression in biomedical science. After an overview of the anatomy and physiology of major organ systems, individual chapters cover those aspects of science that are relevant to the clinical laboratory: pathophysiology; clinical cell biology and genetics; cellular pathology; clinical chemistry; medical microbiology; clinical immunology; haematology and transfusion science, and then concludes with a chapter on professional practice. The book includes contributions from a number of registered Biomedical Scientists which greatly enhances its clinical relevance and interest as well as giving a sense of what happens in the real world, and at the bench in the working clinical laboratory.

We hope this textbook helps to take you successfully into a fulfilling career in biomedical science or an allied profession that you enjoy as much as the various contributors have to date.

R.K.I and S.M.D

Chapter 1

Anatomy and physiology of major organ systems

Professor Ray K. Iles, B.Sc., M.Sc., Ph.D., CBiol, FSB, FRSC,
Dr Iona Collins, BMedSci, MBBS, FRCS and
Dr Suzanne M. Docherty, BmedSci, MBBS, Ph.D.

No area of medical science is truly self-contained; all systems interact, so as we study our chosen speciality we have to put this in a holistic context of human biology. This is as true for the clinical laboratory specialist as for any other medical professional. This introductory chapter is not aimed to be a comprehensive text on anatomy and physiology as there are numerous extremely good volumes published on this subject. However, the reader may wish to dip into these explanatory notes as a refresher or source of direction for further study. After all, students of clinical biomedical science will find they have to read around our specific substantive chapters on haematology, clinical chemistry, microbiology and especially histopathology if they do not have a grasp of anatomical systems.

1.1 The skeletal system

The obvious functions of the skeleton are to provide support, leverage and movement and protection of organs, for example the skull protects the brain, the rib cage the lungs, heart, liver and kidneys, and the pelvis the bladder. In addition, the skeletal system is a storage site for calcium and phosphate minerals and lipids (yellow marrow) and critically a site for the production of blood cells (red bone marrow).

The characteristics of bone are that they are very lightweight yet very strong — resistant to tensile and compressive forces. Interestingly, healthiness (bone density) depends on continuous stressing or loading (i.e. activity). Bones are characterized by their shape (Figure 1.1) into long bones, short bones, flat bones and irregular bones.

1.1.1 The anatomical structure of a bone

Best exemplified by long bones, the bone itself is subdivided by internal and external structures. The bone is covered by a layer of cartilage called the periosteum underneath which is a layer of dense compacted calcified compact bone: however, beneath this layer can either be a hollow chamber (medullary

Biomedical Sciences: Essential Laboratory Medicine, First Edition. Edited by Ray K. Iles and Suzanne M. Docherty.
© 2012 John Wiley & Sons, Ltd. Published 2012 by John Wiley & Sons, Ltd.

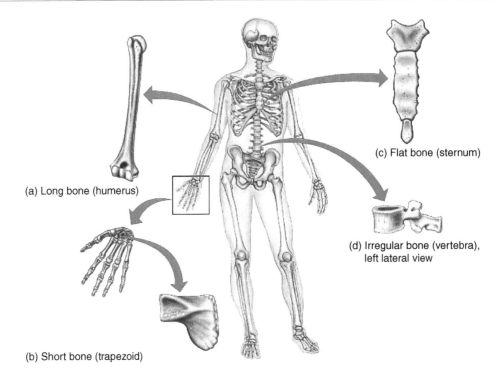

(a) Long bone (humerus)

(b) Short bone (trapezoid)

(c) Flat bone (sternum)

(d) Irregular bone (vertebra), left lateral view

Figure 1.1 The human skeleton and the four bone categories which are shape descriptors. *Essentials of Human Anatomy & Physiology*, 9th Edition, Marieb, 2008 © Pearson Education Inc.

cavity) filled with the specialist tissue of the bone marrow or a spongy bone of small cavities. The spongy bone is always found at the end structures of articulating long bones and is a region of continued bone turnover lying above a line of active bone cells called the epiphyseal line. This spongy bone region is called the epiphysis, whilst the bone marrow dominant region between the two epiphyseal lines is termed the diaphysis where highly active bone turnover (remodelling) does not continuously occur (Figure 1.2).

Bone is derived from connective tissue and there are two types of connective tissue in the skeletal system — calcified bone and cartilage. Cartilage tissue forms a covering of articular surfaces, ligaments and tendons, as well as sheaths around bone (periosteum).

Bone tissue is calcium phosphate ($Ca_3(PO_4)$) crystals embedded in a collagen matrix peppered with bone cells. Thus bone is 60% minerals and collagen and 40% water where the collagen enables bones to resist tensile forces (i.e. are elastic) and minerals which enable bones to resist compressive forces, but this does makes them brittle.

Bone (osseous tissue) is, however, living tissue and therefore has an abundant blood and nerve supply:

periosteal arteries supply the periosteum (see Figure 1.3 (a)); nutrient arteries enter through nutrient foramen supplies compact bone of the diaphysis and red marrow (see Figure 1.3(b)) and metaphyseal and epiphyseal arteries supply the red marrow and bone tissue of epiphyses (see Figure 1.3(a)).

1.1.2 Spongy bone and compact bone

Bone tissue is of two types — spongy and compact. Spongy bone forms 'struts' and 'braces' with spaces in between. Spaces contain bone marrow allowing production and storage of blood cells (red marrow) and the looser structure allows the bone to withstand compressive forces. Compact bone makes up the outer walls of bones, it appears smooth and homogenous and always covers spongy bone. Denser and stronger than spongy bone, compact bone gives bones their rigidity. Spongy and compact bone are biochemically similar, but are arranged differently. In compact bone the structural unit is the osteon (see Figure 1.4).

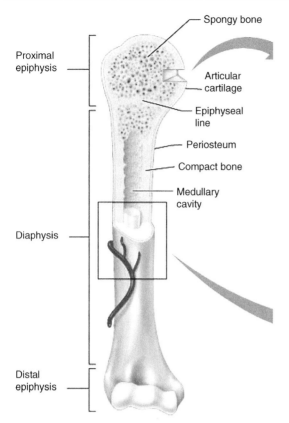

Figure 1.2   Structural components of the long bone. *Essentials of Human Anatomy & Physiology*, 9th Edition, Marieb, 2008 © Pearson Education Inc.

1.1.3   Osteocytes – mature bone cells

There are two types of bone cell:

- **Osteoblasts** – bone forming cells.
- **Osteoclasts** – bone destroying cells.

In the formation of new bone osteoblasts cover hyaline cartilage with bone matrix. Enclosed cartilage is digested away leaving the medullary cavity. Growth in width and length continues by the laying down of new bone matrix by osteoblasts. Remodelling to ensure the correct shape is effected by osteoclasts (bone-destroying cells). In mature bones osteoblast activity decreases whilst oesteoclast remodeling activity is maintained. However, bone remodelling requires both oesteocytes. Triggered in response to multiple signals stress on bones means that there is considerable normal 'turnover' – bone is a dynamic and active tissue; for example, the distal femur is fully remodelled every 4 months.

Osteoclasts carve out small tunnels and osteoblasts rebuild osteons: osteoclasts form a leak-proof seal around cell edges and then secrete enzymes and acids beneath themselves. The resultant digestion of the bone matrix releases calcium and phosphorus into interstitial fluid. Osteoblasts take over bone rebuilding,

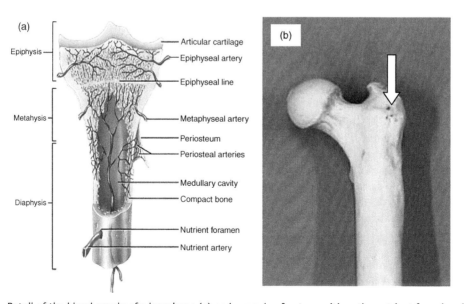

Figure 1.3   Detail of the blood supply of a long bone (a) and example of entry position, the nutrient foramina, is indicated in (b). *Essentials of Human Anatomy & Physiology*, 9th Edition, Marieb, 2008 © Pearson Education Inc.

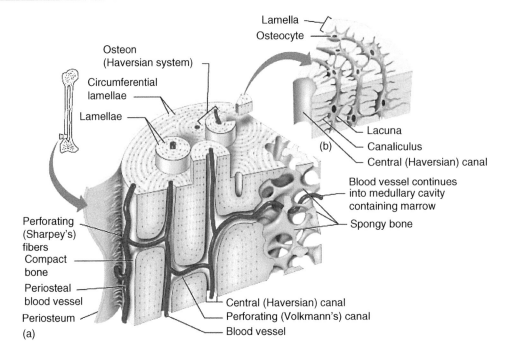

Figure 1.4   Microanatomy of the bone. *Essentials of Human Anatomy & Physiology*, 9th Edition, Marieb, 2008 © Pearson Education Inc.

continually redistributing bone matrix along lines of mechanical stress.

1.1.4   How bones grow

Bone growth only occurs in those young enough to still have an active, unfused epiphyseal plate (roughly < aged 16–19). The epiphyseal plates fuse earlier in females than in males – generally, females have stopped growing by around the age of 16, while for males this is around 18 to 19 (see Figure 1.5).

Cartilage cells are produced by mitosis on the epiphyseal side of plates (ends of bones) – this is continuous with articular cartilage at the end of the bone. Cartilage cells are destroyed and replaced by bone on the diaphyseal side of plates (middle of long bone)and a zone of resting cartilage anchors the growth plate to the bone. The epiphyseal plate is at the top of Figure 1.5, and this is where new cartilage cells are being created by mitosis. As they are 'pushed away' from the epiphyseal plate by new cartilage cells being created 'behind' them, osteoblasts lay down a calcium phosphate matrix in and around the cartilage cells, ossifying the area. This gradually takes on the structure of bone. The epiphyseal plate cartilage is continuous with the articular cartilage at the end of the bone, and new cartilage (and bone formation) is occurring in both areas rather than strictly just at the epiphyseal plate. Furthermore, the bone has to be remodelled as it increases in length, or the whole bone would be as wide as the epiphysis – but what you actually need is a narrower diaphysis (shaft) in the middle of the bone. The thick articular cartilage, at either end of the bone, is continuous with the thin (but tough) periosteum around the outside of the rest of the bone. Periosteum has a rich blood supply which is important when you consider bones grow not only in length but in width.

Periosteal cells (from membrane around the bone) differentiate into osteoblasts and form bony ridges and then a tunnel around a periosteal blood vessel. Concentric lamellae fill in the tunnel to form an osteon (see Figure 1.6). Blood vessels around the outside wall of the bone, on the periosteum, are 'walled in' as periosteal cells convert into osteoblasts and build new bone around them. This is why cortical bone is composed of osteons.

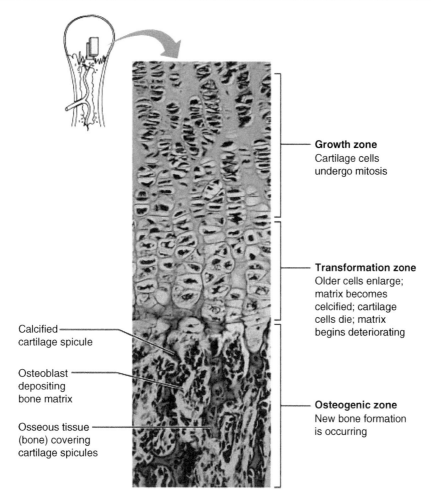

Growth zone
Cartilage cells
undergo mitosis

Transformation zone
Older cells enlarge;
matrix becomes
celcified; cartilage
cells die; matrix
begins deteriorating

Calcified
cartilage spicule

Osteoblast
depositing
bone matrix

Osseous tissue
(bone) covering
cartilage spicules

Osteogenic zone
New bone formation
is occurring

Figure 1.5   Histological appearance of epiphyseal plate. *Essentials of Human Anatomy & Physiology*, 9th Edition, Marieb, 2008 © Pearson Education Inc.

1.1.5   Endocrine regulation and nutritional requirement of bone growth

Several hormones are involved in endocrine control of bone growth: growth hormone, thyroid hormone, insulin and calcitonin. Before puberty growth hormone is the most important hormone involved in regulating bone growth. The metabolic hormones, thyroid hormones and insulin are involved in modulating the activity of growth hormone and ensuring proper proportions in the skeleton. Together these maintain the normal activity at the epiphyseal plate until the time of puberty. At puberty the increase in sex hormone production results in an acceleration of bone growth. These hormones promote the differences in the shape of the skeleton associated with males and females such as density and shape such as a flatter and wider pelvis in females. However, in both sexes the rate of ossification starts to outpace the rate of cartilage formation at the epiphyseal plates. Eventually the plates ossify and bone growth stops when the individual reaches sexual and physical maturity.

For adequate bone growth good nutrition is also required as are adequate levels of minerals and vitamins: calcium and phosphorus, vitamin D for bone formation, vitamin C for collagen formation and vitamins K and B_{12} for protein synthesis.

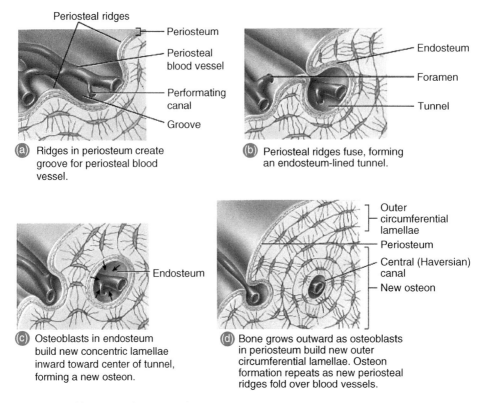

Periosteal ridges

Periosteum

Periosteal
blood vessel

Performating
canal

Groove

(a) Ridges in periosteum create
groove for periosteal blood
vessel.

Endosteum

Foramen

Tunnel

(b) Periosteal ridges fuse, forming
an endosteum-lined tunnel.

Endosteum

(c) Osteoblasts in endosteum
build new concentric lamellae
inward toward center of tunnel,
forming a new osteon.

Outer
circumferential
lamellae

Periosteum

Central (Haversian)
canal

New osteon

(d) Bone grows outward as osteoblasts
in periosteum build new outer
circumferential lamellae. Osteon
formation repeats as new periosteal
ridges fold over blood vessels.

Figure 1.6 Appositional bone growth

1.1.6 The role of bone as a mineral store

A critical mineral which bones are involved in regulating is calcium as its ion concentrations in plasma must be very carefully controlled. Calcium homeostasis is affected by a negative feedback system involving the action of two primary hormones; calcitonin, produced from parafollicular cells of the thyroid gland in the neck and parathyroid hormone (PTH, also called parathormone) produced by the parathyroid glands (which lie on top of the thyroid gland). Responding to a fall in plasma calcium ions, released PTH, among other effects, induces the release of calcium by bone, whilst a rise in plasma calcium results in calcitonin which has the opposite effects, one of which is to promote increased deposition of calcium in bone.

1.2 The digestive system

This section aims to give an overview of the anatomy of the digestive system, identifying the major organs of the alimentary canal and the accessory digestive organs. In particular, the structure and function of the following organs and accessory organs of the alimentary canal are briefly described (see Figure 1.7):

- the oral cavity, pharynx and oesophagus;
- the stomach;
- the small intestine;
- the liver and gallbladder;
- the pancreas;
- the large intestine.

In so doing, it is possible to outline the major processes occurring during digestive system activity and give an overview of digestion and absorption.

1.2.1 Nutrition and absorption

The overall function of the digestive tract is to process not only the macronutrients (carbohydrates, proteins and fats) but also vitamins and minerals. Vitamins are complex organic substances essential for health,

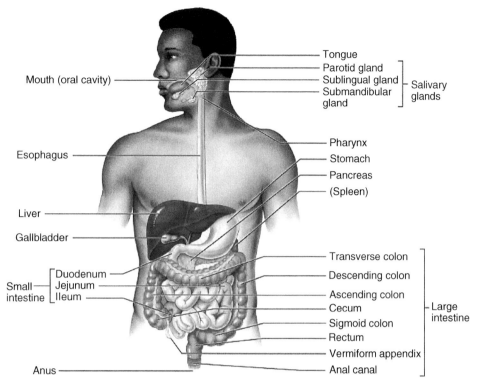

Figure 1.7   Overall anatomy of the digestive system. *Essentials of Human Anatomy & Physiology*, 9th Edition, Marieb, 2008 © Pearson Education Inc.

required in very small amounts (mg or µg per day) but most cannot be made by the body. They function as cofactors in enzyme activity, antioxidants to deal with free radicals generated during metabolism, and even as prohormone (i.e. vitamin D).

Minerals are inorganic compounds required by the body, like vitamins, for a variety of functions but often as cofactors or the reactive centres of functional proteins. Some minerals are needed in larger amounts than others, for example calcium, phosphorus, magnesium, sodium, potassium and chloride. Others are required in smaller quantities and are sometimes called trace minerals, for example iron, zinc, iodine, fluoride, selenium and copper. However, despite being required in smaller amounts, trace minerals are no less important than other minerals.

In order to extract macro- and micronutrients from food stuffs the digestive system must bring about ingestion, digestion (mechanical and chemical), enable movement through the digestive tract, facilitate absorption of nutrients and finally defaecation of the nondigestible elements and some waste products.

1.2.2   Ingestion

The oral cavity is a far more complex mechanism than just a set of teeth. You unconsciously analyse food when you put it in your mouth to check it isn't too large a chunk to sensibly chew, that it doesn't contain very hard bits, and that it isn't in some way mouldy or otherwise unpleasant. Only then do you start chewing properly and contemplating swallowing it. Thus the oral cavity analyses the food, mechanically processes (chews to smaller pieces), lubricates (saliva) and starts the process of chemical digestion via the enzymes secreted as part of saliva (see Figure 1.8).

After chewing we swallow but there are two phases:

- buccal phase (voluntary);
- pharyngeal phase (involuntary).

1.2.2.1   Pharynx and oesophagus

During the pharyngeal phase, the airways have to be shut off by the **epiglottis** to prevent food from going

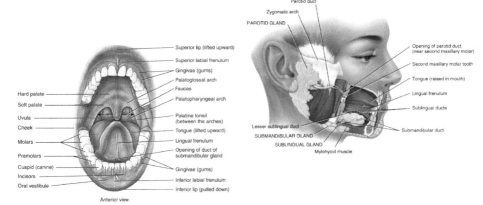

Figure 1.8   Structures and exocrine glands of the oral cavity. From Tortora and Derrickson, *Principles of Anatomy and Physiology*, Twelfth Edition, 2009, reproduced by permission of John Wiley & Sons Inc.

down the air passages/windpipe (see Figure 1.9). Babies don't have quite the same set up, and this allows them to breathe while drinking milk. Peristalsis carries food in one direction only — down, so you can eat and drink standing on your head if you want to; animals such as horses effectively do this by eating with their heads lower than the level of their stomach.

1.2.3   The stomach

Lying in the upper part of the abdominal cavity, this sac or balloon like stomach occupies a volume of 50 mL empty, but expands to 4 L when full. The different orientations of muscle layers in the stomach allow it to contract in different directions to maximize the effectiveness of mechanically breaking down food. The folds (rugae) increase the surface area for maximum absorption (see Figure 1.10). It is also important to note that there is a cardiac sphincter between the oesophagus and stomach, and a pyloric sphincter between the stomach and duodenum — sometimes the pyloric sphincter is malformed (this predominantly affects baby boys) and will not open, which causes projectile vomiting and failure to thrive until it is surgically corrected. At the other end the stomach sits below the diaphragm, but sometimes part of the stomach is squeezed up through the diaphragm, resulting in heartburn and reflux as acid enters the oesophagus.

1.2.3.1   Stomach mucosal lining

The gastric mucosa contain three predominant differentiated cell types: parietal cells which secrete hydrochloric acid and intrinsic factors facilitating the absorption of vitamin B12; chief cells which secrete pepsinogen (inactive form of pepsin) — which is activated by HCl and begins the digestion of protein; and mucous cells. The stomach secretes a thick mucus to protect itself from its own hydrochloric acid (see Figure 1.11).

1.2.3.2   The gastric digestive process

Swallowed food collects in the upper storage area. Starch (complex carbohydrate) continues to be digested until the mass has been mixed with gastric juice. Small portions of mashed food are pushed into the digesting area of the stomach where acid in gastric juice unwinds (denatures) the proteins and the enzyme pepsin breaks up the chains

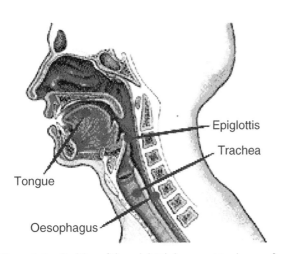

Figure 1.9   Position of the epiglottis in respect to closure of the trachea

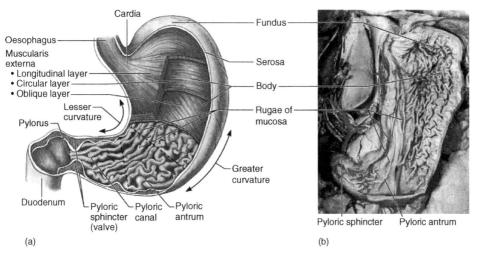

Figure 1.10   (a) Anatomy and (b) cross-sectional appearance of the stomach. *Essentials of Human Anatomy & Physiology*, 9th Edition, Marieb, 2008 © Pearson Education Inc.

of amino acids. This all forms a thick liquid mass called chyme which moves on and enters the small intestine. Fat forms a separate layer on the top.

1.2.4   The small intestine

The small intestine consists of three distinct anatomical regions: the duodenum, pyloric sphincter to jejunum; jejunum, duodenum to ileum; and ileum, jejunum to large intestine. The small intestine is where most nutrients are absorbed, and it is all about surface area maximization (see Figure 1.12).

The mucosal folds of the small intestine are covered in villi (Figure 1.12(a) and (b)), and each villus in turn is lined with columnar cells that have a brush border (Figure 1.12(c)), all to give a large surface area for

absorption. Note too that each villus has a rich blood supply to help with this too (see Figure 1.12(c)).

1.2.5   Liver and gall bladder

Positioned below the diaphragm and protected by the lower half of the rib cage, the liver is divided into a right and left lobe by the round ligament. The gall bladder nestles into it from underneath and in real life this is a dark green colour and really stands out. Among other functions the liver produces bile. Bile contains bile acids, which assist with the absorption of fats and fat-soluble vitamins in the small intestine. Many waste products, including bilirubin, are eliminated from the body by secretion into bile and elimination in faeces. Adult humans produce 400 to 800 mL of bile per day.

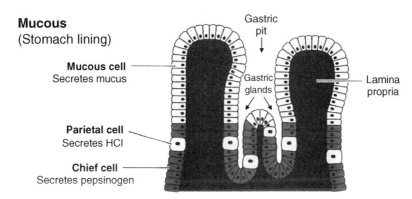

Figure 1.11   Microanatomy of the stomach lining

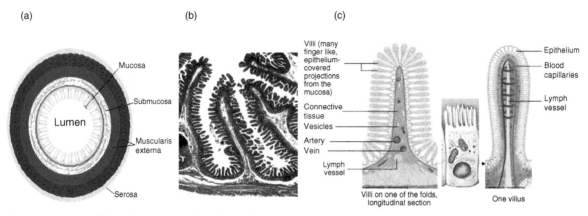

Figure 1.12   Small intestine: cross-sectional and microanatomy

Further modification of bile occurs in that organ. The gall bladder stores and concentrates bile during the fasting state. Typically, bile is concentrated fivefold in the gall bladder by absorption of water and electrolytes – virtually all of the organic molecules are retained. The liver drains bile out towards the gall bladder in the bile duct, and further down this is joined by secretions from the pancreas to form a common bile duct, which secretes a mixture of bile and pancreatic juices into the duodenum as food passes through (see Figure 1.13). The bile duct can become obstructed by small gallstones (and other things, like a tumour in the head of the pancreas), which causes jaundice and is described in several of the following chapters.

1.2.6   The pancreas

The pancreas is both an exocrine and endocrine gland. Its endocrine function is fulfilled by the pancreatic islets cells which secrete insulin and glucagon, whilst its digestive system exocrine function is to produce and secrete digestive enzymes: trypsin and chymotrypsin which break proteins into peptides (short chains of amino acids); pancreatic lipase which digests triglycerides into a monoglyceride and two free fatty acids; amylase which hydrolyses starch to maltose (a glucose–glucose disaccharide) and others such as nucleases (ribonuclease, deoxyribonuclease) and those that digest fibrous tissues (e.g. gelatinase and elastase). The pancreas is a highly sensitive organ and can become inflamed (pancreatitis) – this is caused by pancreatic enzymes from damaged pancreatic cells leaking into pancreatic tissue and digesting it.

1.2.7   Small intestine and associated organs and digestion

It must be remembered that the small intestine is a major site in the digestive process, the pancreas liver

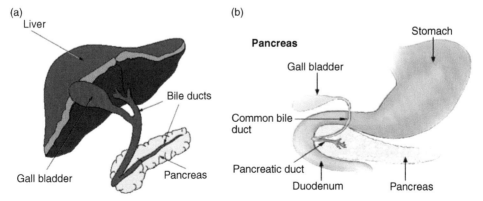

Figure 1.13   (a) Diagrammatic representation of the relative positional anatomy of the liver, gall bladder and pancreas, and (b) in relationship to the stomach and small intestine

and gall bladder all work in concert with the absorption function of the mucosal folds and villi cells found here. Thus, chyme squirts into the duodenum from the stomach and peristaltic movement pushes the chyme along and mixes with secretions for chemical digestion. In particular pancreatic juice and bile help to digest carbohydrate, lipids and proteins. All the while the macro- and microanatomy of the small intestine optimizes absorption and facilitates the transport of nutrient across the mucosal barrier and into the blood stream. Most digested food is absorbed in the small intestine so there is a rich and complex net of blood and lymphatic channels around and leading to and from the small intestine as it winds backwards and forwards.

1.2.8 The large intestine

The residual chyme moves from the small intestine via the Illeocecal valve (which prevents the contraction from these larger vessels forcing waste back into the small intestine) into the first pouch or haustra of the large intestine — the caecum. Herbivores have a large caecum and appendix that contain symbiotic bacteria that synthesize the enzyme cellulase, allowing them to digest plants cell walls, these pass through us as fibre. The human appendix is roughly the size of the little finger, but in some people it is relatively long and thin (with a small diameter that is more likely to block, possibly resulting in appendicitis).

The movement of this residual digestive chyme through the large intestine is slow and rather laborious in the mechanical mechanism that operates: pouches (or haustra) fill to capacity, when stretched they contract and force the contents into the next haustra (and section of the colon). During this slow passage water is absorbed or reabsorbed, and vitamins and minerals are absorbed along with it. As water is absorbed the residual chyme is dehydrated and compacted to form faeces. Mass peristalsis forces the contents into the rectum for the storage of faecal material prior to defecation (see Figure 1.14).

The average passage time of undigested food residues through the human gut is about 50 h in men and 57 h in women, but ranges from well under 20 to over 100 h. It also changes from one day to the next. However, about 80–90% of the entire transit time of food in the body is spent in the colon, so it needs to be large and have a good capacity. Thus, movement through the digestive tract varies dramatically section

per section: Oesophageal peristalsis is fast with a transit time of about 3 s; time in the stomach is about 1–3 h; small intestine digestion and absorption is 2–6 h, whilst 12–48 h is spent in the large intestine prior to defecation.

1.3 The cardiovascular system

The function of the cardiovascular system is as a transport system of the body carrying:

- respiratory gases;
- nutrients;
- hormones and other material to and from the body tissues.

The fluid component of this system — blood — is a complex of specialized cells and solution of salts (electrolytes) and soluble proteins. At the centre of the cardiovascular system is the heart to which structurally distinct vessels — arteries — carry blood away, and equally structurally distinct vessels — veins — carry blood back to the heart.

However, the cardiovascular system has two divisions: pulmonary and systemic (see Figure 1.15). In the pulmonary division, blood flows from the right ventricle of the heart to alveolar capillaries of the lungs and back to the left atrium of the heart. In the systemic division the left ventricle pumps blood to the rest of the body and all other body capillaries, and the blood returns to the heart's right atrium. Hence there is an asymmetry in muscle mass between the two ventricles. In addition the two divisions have two different profiles with respect to the transport of respiratory gases; pulmonary arteries are low in O_2 high in CO_2 whilst the arteries of the systemic division are high in O_2 low in CO_2 (and the pulmonary–systemic veins vice versa). Capillaries, minute blood vessels found throughout tissues, connect the small arteries to the small veins. Exchange of respiratory gases and nutrients with the tissues occurs across the walls of the capillaries.

1.3.1 The heart

The heart is a complex structure of four chambers, powerful muscles, specialized valves that contract and

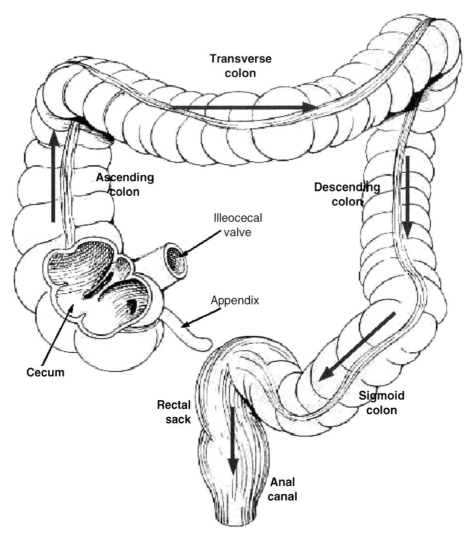

Figure 1.14   Diagrammatic representation of the structure of the large intestine

open/close in a coordinated manner, regulated by its own specialist sensory and responsive neurological system (see Figure 1.16). The entire organ is surrounded by a protective barrier called the pericardium. It is in fact a protective sac surrounding the heart consisting of an outer tissue layer called the parietal pericardium, a proteinacious (pericardial) fluid and a heart wall contacting tissue, the visceral pericardium also referred to as the epicardium.

The heart wall consists of two tissue layers the inner endocardium which is contiguous with blood vessel endothelium and the myocardium of specialist cardiac muscle. The heart has two structural classes of chambers: receiving chambers or atria (singular, atrium) and pumping chambers or ventricles. The right ventricle pumps for the pulmonary circulation, the left ventricle pumps for the systemic circulation. The 'Great Vessels' of the heart are the aorta and pulmonary trunk. Heart valves ensure the one-way flow of blood through the heart and there are two types: semilunar valves (pulmonary semilunar and aortic semilunar) lead from the ventricles and prevent back flow from pulmonary and systemic vasculature. The atrio-ventricular (AV) valves are the tricuspid − right atrium into the right ventricle; and bicuspid (mitral) − left atrium to the left ventricle. If a section is cut through the heart at the atrial ventricular boundary through the four valves a structural skeleton of (four) fibrous rings can be seen. These tough fibrous rings provide rigidity to prevent the dilation of valves and provide a point of

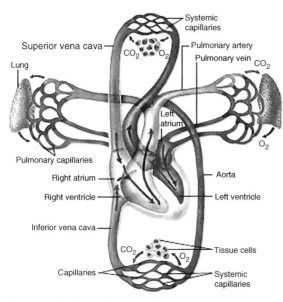

Figure 1.15   The systemic and pulmonary systems

attachment for valves. This fibrous skeleton also electrically isolates the atria from the ventricles. The AV bundle ('bundle of His') is the only electrical connection between the atria and the ventricles (see Figure 1.17).

The origin of heartbeat is located in a sinoatrial (SA) node of the heart, where a group of specialized cells continuously generates an electrical impulse. The SA node generates such impulses about $100-120$ times per min at rest. However, in a healthy individual the resting heart rate (HR) would never be that high. This is due to continuous control of the autonomic nervous system (ANS) over the output of SA node activity, which net regulatory effect gives real HR. In a healthy subject at rest it is ranging between 50 and 70 beats per min.

The electrical impulse of the sino-atrial (SA), stimulated by blood flow, first induces the muscle tissue of the atrial chamber to contract. The electrical impulse travels to the atrio-ventricular node and synchronizes with this tissue's inherent but weaker electrical pulsivity. This combined and synchronized electrical pulse travels down the conductive fibres (bundle of His, bundle branches)of the noncontractive muscular cardiac septum (i.e. this tissue does not contract in response to this electrical signal) to the Purkinje fibres which originate at the base of the ventricle muscle walls and travel up towards the atrioventricular boundary. The result is that the signal

is delayed, atria muscles are relaxing, but the impulses then induce waves of contraction of the ventricles from the bottom up. This efficiently empties the heart ventricles — like squeezing a toothpaste tube from the bottom and not the middle, whilst the atrium refill. The order of impulse spreading all over the heart muscle through specialized pathways creates synchronized heart muscle contraction between both atriums (first) and then the ventricles which contracts in a wave starting from the bottom of the heart to the top of the ventricles.

1.3.2   The vasculature

The blood vessels are the **arteries**, **arterioles**, **capillaries**, **venules** and **veins** and all blood vessels are lined with specialist cells of the endothelium (see Figure 1.18). The arteries which carry blood away from the heart are subject to the highest blood pressure and located deep within tissues. Subject to much lower pressures, veins return blood to the heart.

1.3.2.1   Structure of arteries and arterioles

Arteries consist of three tissue layers: **tunica interna**, **tunica media** and **tunica externa**. However, there are two types of artery: elastic arteries, which contain elastic fibres in the tunica media and interna, which are the largest. Muscular arteries have little elasticity and abundant smooth muscle in the tunica media.

Arterioles are less than 1 mm in diameter and consist of endothelium and smooth muscle. It is the ability of arteries to contract by virtue of the dense smooth muscle layers that allows these vessels to regulate blood pressure in a general and locality specific manner. Indeed the **metarterioles** regulate the flow of blood into capillaries (see Figure 1.19).

Capillaries are the sites of exchange, they are very thin and permeable, allowing exchange between blood and tissue cells in systemic capillaries and the exchange between blood and air in pulmonary capillaries.

1.3.2.2   Structure of veins and venules

Veins are thinner than arteries, of a much larger diameter and located both deep and superficially

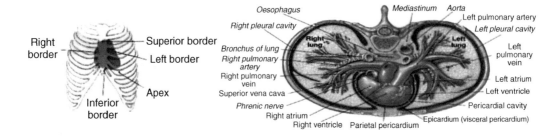

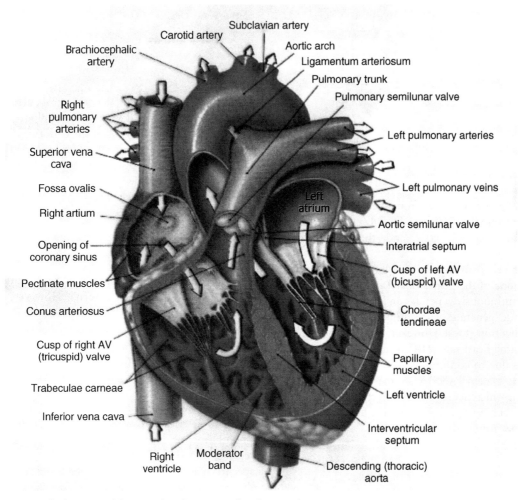

Figure 1.16   The heart, position associated organs and major vessels

within tissue. A key difference is that veins have valves. Since the blood in veins is under much lower pressure after a forward flow pressure beat from the left ventricle, the blood could flow backwards again. The valves prevent this backwards flow and veins within muscles are squeezed by external contraction of muscle tissue mass as a result of movement (and general muscle tone) to help return blood.

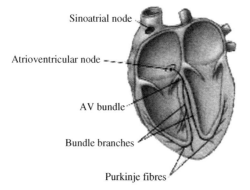

Figure 1.17   Neurological conduction system of the heart

1.3.3   *Blood distribution*

Blood does not spend its time equally between veins and arteries. Indeed, as demonstrated in Table 1.1, most of the time blood is in the systemic venous system.

The reason is that exchange is not just one way — from the oxygen rich systemic arteries to tissues but huge exchange occurs in the hepatic portal system where food and metabolites are absorbed (see Figure 1.20). Similarly, coming away from bone (marrow) in the exiting veins are new blood cells.

1.3.3.1   The heart's blood supply

The highly active muscle and neurological tissue of the heart has its own surrounding network of capillaries fed by the coronary circulation, some of the major vessel of which are the right coronary artery, the left coronary artery, the circumflex, the anterior interventricular (also LAD) and the coronary sinus.

1.3.3.2   Blood flow

Defined as the volume of blood flowing through a vessel, an organ, or the entire circulation in a given period blood flow is measured in mL per minute. Equivalent to cardiac output (CO), considering the entire vascular system this is relatively constant when at rest. However, it varies widely through individual organs, according to immediate needs.

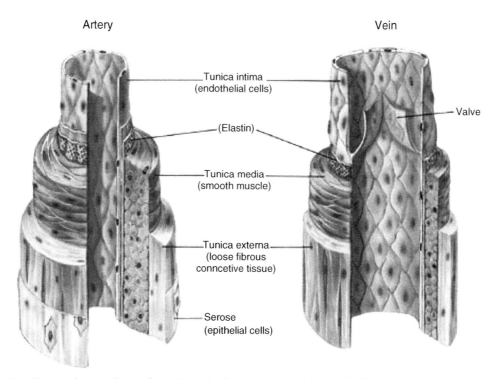

Figure 1.18   Structural comparisons of arteries and veins. *Human Physiology*, 4th Edition, Fox, 1993 © William C. Brown

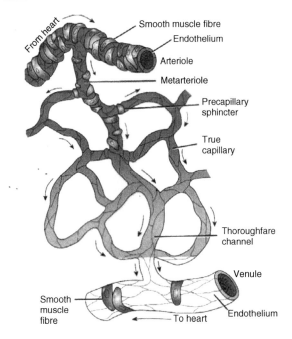

Figure 1.19   Structural representations of arteriole capillary venule 'mesh'

Table 1.1   Percentage distribution of blood in the cardio-vasculature

Systemic venous system	**64%**
Systemic arterial system	13%
Heart	7%
Systemic capillaries	7%
Pulmonary venous system	4%
Pulmonary arterial system	3%
Pulmonary capillaries	2%

Resistance — opposition to flow — is the measure of the amount of friction blood encounters as it passes through vessels. Resistance is more significant in the systemic circulation and is referred to as peripheral resistance (PR).

The three important sources of resistance are blood viscosity — thickness or 'stickiness' of the blood, total blood vessel length — the longer the vessel, the greater the resistance encountered — and blood vessel diameter. Changes in vessel diameter are frequent and significantly alter peripheral resistance. Resistance varies inversely with the fourth power of vessel radius

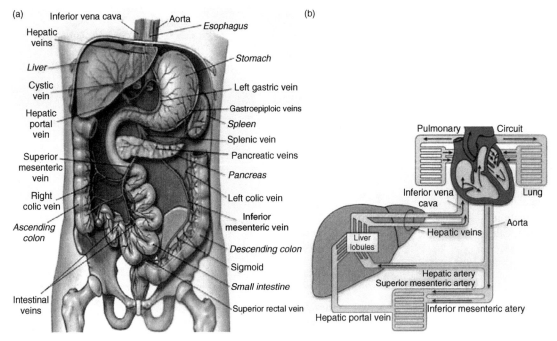

Figure 1.20   (a) Diagrammatic anatomical and (b) functional representations of the hepatic portal circulations demonstrating the dominance of the venous vessels

(one-half the diameter), for example, if the radius is doubled, the resistance is 1/16 as much.

Blood flow (F) is directly proportional to the difference in blood pressure (ΔP) between two points in the circulation. If ΔP increases, blood flow speeds up; if ΔP decreases, blood flow declines. Blood flow is inversely proportional to resistance (R), If resistance (R) increases, blood flow decreases. Resistance is more important than difference in blood pressure in influencing local tissue blood pressure.

The pumping action of the heart generates blood flow through the vessels along a pressure gradient — always moving from higher- to lower-pressure areas — and pressure results when flow is opposed by resistance. Blood pressure is defined as the force per unit area exerted on the wall of a blood vessel by its contained blood and is expressed in millimetres of mercury (mm Hg).

Thus, as the blood vessels get generally wider we find that systemic blood pressure Is highest in the aorta, declines throughout the length of the pathway and is 0 mm Hg in the right atrium. The steepest change in blood pressure occurs in the arterioles.

The fact that there is really no pressure left by the time the blood is returning to the heart means that the body relies on the 'sucking' effect of the diaphragm lowering to cause inhalation (i.e. creating negative pressure in the thoracic cavity) within the thoracic cavity to help draw venous blood back up towards the heart.

1.4   The urinary system

The urinary system comprises the kidneys, ureter(s), bladder and urethra (Figure 1.21). The Functions of the urinary system are:

- to excrete organic waste;
- to regulate blood volume;
- to regulate blood pressure;
- to regulate ion concentrations (sodium, potassium, chloride, calcium, etc.);
- to maintain blood pH at physiological range (7.35–7.45).

The kidneys lie retroperitoneal (behind the peritoneal cavity), so they are separated from the abdominal organs that lie in front of them, in the peritoneal cavity. Weighing approximately 150 g, the distinctively shaped kidney (approximately 12 cm (long) × 6 cm

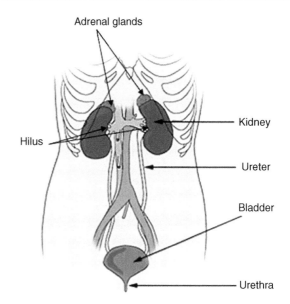

Figure 1.21   Gross anatomy of the urinary system

(wide) × 3 cm (thickness)) lies approximately at vertebral level T12–L3 with the upper parts protected by the 11th and 12th ribs. The organs themselves have three layers of protective tissue: renal capsule, adipose capsule and renal fascia. The capsule is a tough connective tissue membrane around the kidney which is quite hard to peel off it. The kidney is also covered in a very thick layer of fat which gives it considerable protection against trauma.

The female urethra is considerably shorter than the male urethra (meaning that females suffer from far more urinary tract infections than males — bacteria have only a short distance to travel to get into the bladder in women, and the rectum is very close to the urethra in females too). The right kidney is lower than the left kidney, which means it gets more kidney infections (a shorter urethra for bacteria to travel along from the bladder) and suffers more from trauma, as it has less protection from the ribcage (see Figure 1.22).

1.4.1   The kidneys

The kidney internal structure consists of a cortex which produces urine and a series of collecting ducts that take the urine from the kidney via a connective tube (ureter) to the major reservoir for eventual excretion (urination from the bladder). In its cross-sectional anatomy the kidney has a ureteric interface cavernous

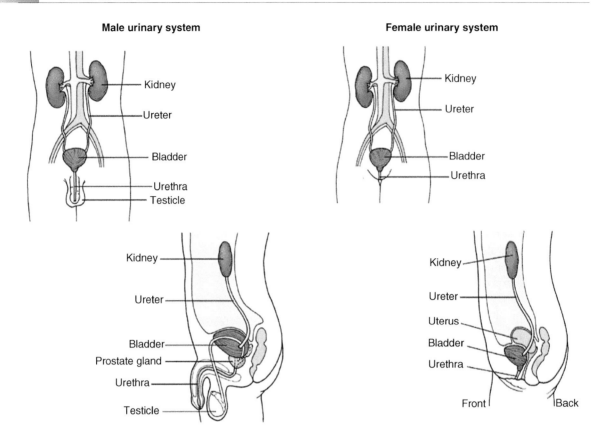

Figure 1.22 Comparative anatomy of the female and male urinary system

region called the hilus (where kidney stones often occur) to which a larger and larger urine draining region (termed calyces) empties. The minor calyces are composed of aggregated collecting ducts – these in turn aggregate into major calyces, which finally become the hilus and ureter. The renal artery supplies the kidney – the kidney is highly metabolically active and needs a good supply of oxygen and glucose. Relative to its small size, it actually uses about a quarter of the arterial blood supply in the body. This also allows it to be highly efficient in terms of filtration of large volumes of plasma and production of urine. The renal vein drains deoxygenated, filtered blood from the kidney (see Figure 1.23).

1.4.1.1 The nephron

The physiological unit of the kidney is the nephron which anatomically straddles the renal cortex with a descending loop into the renal medulla. The blood vessels in the cortex of a kidney reduce from larger vessels, to lots of small round 'tufts' of capillaries each one of these is one glomerulus, the knot of capillaries within a Bowman's capsule (see Figure 1.24).

Blood vessels wrap around the whole nephron, and after the efferent arteriole leaves the Bowman's capsule it continues and is wrapped around the proximal convoluted tubule, loop of Henle, and distal convoluted tubule. This means that water, ions, amino acids, drugs and so on can easily move between the nephron and the bloodstream along the length of the nephron.

Pressure within the glomerulus is kept very high by the diameter of the outgoing arteriole being narrower than the ingoing arteriole. In the Bowman's capsule, fluid is forced out of the glomerulus and into the capsule from the bloodstream. Blood within the glomerulus is under very high pressure, and the capillaries here are very leaky, so a lot of the fluid component of blood is forced straight out into the capsule, and a

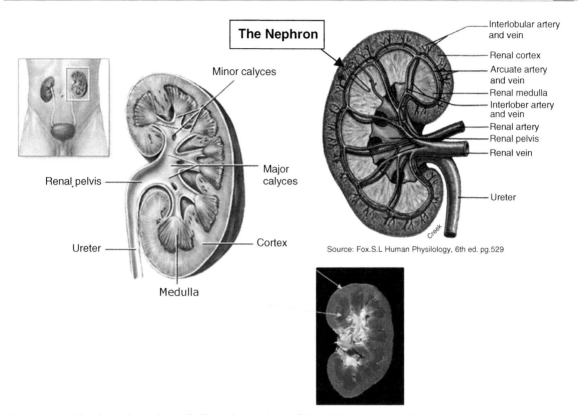

The Nephron

Minor calyces

Renal pelvis

Ureter

Major calyces

Cortex

Medulla

Interlobular artery and vein

Renal cortex

Arcuate artery and vein

Renal medulla

Interlober artery and vein

Renal artery

Renal pelvis

Renal vein

Ureter

Source: Fox.S.L Human Physilology, 6th ed. pg.529

Figure 1.23   Blood supply and metabolic active regions of the kidney. *Human Physiology*, 4th Edition, Fox, 1993 © William C. Brown

large volume of fluid literally pours into the proximal tubule. Proteins greater than 10 kD are generally too large to be forced out of blood (urine should be essentially protein-free in the healthy individual — that is at very very low levels compared to the blood), and the cells of blood are obviously much too large to get between the gaps in the capillaries too (urine should not contain red or white blood cells in the healthy).

1.4.1.2   Glomerular filtration rate (GFR)

This is the volume of blood filtered per unit time by all glomeruli combined, approximately 125 mL per minute (or 7.5 L/h). However 7.5 L of filtrate per hour is entering the nephrons of your kidneys, but how much urine are you actually producing? It should be around 60 mL/h, which tells you that over 99% of the water in the filtrate alone is being reabsorbed. This is fortunate as your bladder capacity is only about 500 mL, and you would otherwise need to empty it 15 times per hour.

GFR varies directly with glomerular blood pressure which, in turn, is determined by systemic blood pressure. This can vary dramatically due the environment or activity; however, renal autoregulation regulates the diameter of incoming arterioles to keep blood flow within normal limits and maintain GFR. If you lose a lot of your blood volume, epinephrine/adrenalin is produced by the sympathetic nervous system, and causes blood pressure to rise (by increasing heart rate and stroke volume as well as via arteriole constriction). However, the kidneys are also partially masters of their own incoming blood pressure via production of renin, which increases blood pressure via the renin–angiotensin system. The cells that produce renins, the juxtaglomerular cells, are located between the glomerulus and the distal convoluted tubule of the same nephron.

1.4.1.3   Reabsorbtion

Nutrients (glucose and amino acids) which pass into the filtrate at the glomerulus are normally all reabsorbed

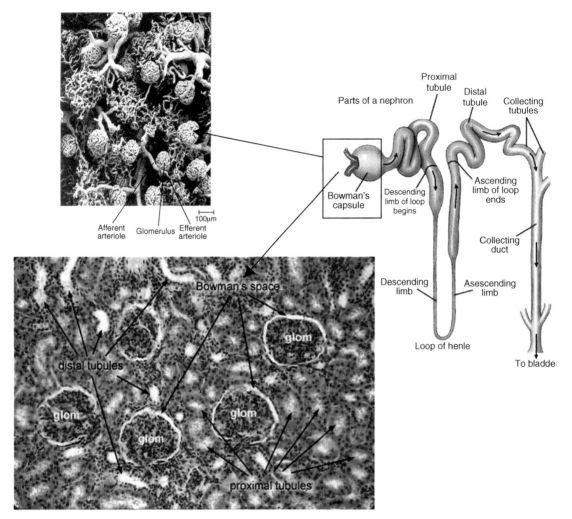

Figure 1.24   The nephron and its microanatomy and histology

again. This reabsorption occurs selectively at specific regions of the proximal and distal tubules and the ascending and descending regions of the loop of henle (see Figure 1.25) We generally need to expend energy (i.e. use ATP) to reabsorb all of the small molecules that left the bloodstream in the leaky glomerulus and entered the nephron.

Sodium (Na) is actively reabsorbed due to the action of aldosterone (produced by the adrenal gland) by activating specific Na transporters in the cells of the proximal tubules. This increased local blood level of Na causes osmotic uptake of water from the loop of henle. ADH, antidiuretic hormone, is produced in the pituitary gland in response to dehydration. It causes more water to be reabsorbed from the nephron and put back into the bloodstream, that is it makes it easier for osmotic reabsorption.

1.4.1.4   Tubular secretion

Most of the processes of tubular secretion again involve energy expenditure. These processes tend to happen further down the nephron than tubular reabsorption, around the distal convoluted tubule area (see Figure 1.25). It is a process in which substances move into the distal and collecting tubules from the blood:

- disposing of substances not already in the filtrate (e.g., certain drugs such as penicillin);

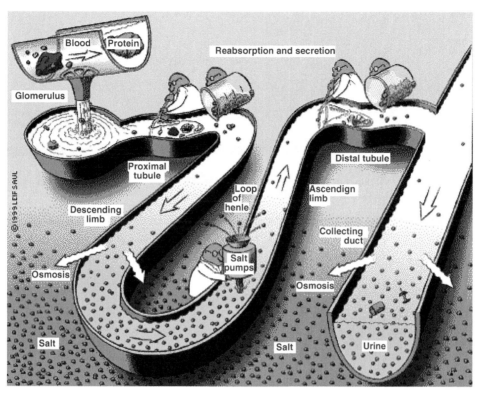

Figure 1.25   Reabsorption and secretion regions of the nephron's post glomerulus draining tubules

- eliminating undesirable substances (e.g., waste products like urea);
- ridding the body of excess potassium ions; and
- controlling blood pH by H^+ secretion.

1.5 Respiratory system

Each lung is divided into lobes. The right lung, which has three lobes, is slightly larger than the left, which has two. The lungs are housed in the chest/thoracic cavity, and covered by a protective membrane — the pleura (see Figure 1.26). The diaphragm, the primary muscle involved in respiration, separates the lungs from the abdominal cavity. To breathe in (inhale), the diaphragm tightens and flattens and the rib cage rises increasing the volume of the thoracic cavity. This creates a decrease in pressure, a partial vacuum, sucking the air into your lungs. When the diaphragm relaxes to its original shape and the ribs lower, the thoracic cavity volume decreases again and forces out the inspired air.

There are two modes of breathing:

Quiet breathing is where inspiration is active: the diaphragm, external intercostals muscles are in-

volved. However, expiration is passive, that is, no muscles are involved and no energy is expended — elastic rebound of the rib cage and diaphragm reduces the thoracic cavity volume expelling the air.

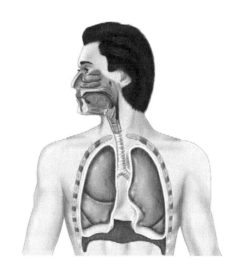

Figure 1.26   Basic anatomy of the lungs. *Essentials of Human Anatomy & Physiology*, 9th Edition, Marieb, 2008 © Pearson Education Inc.

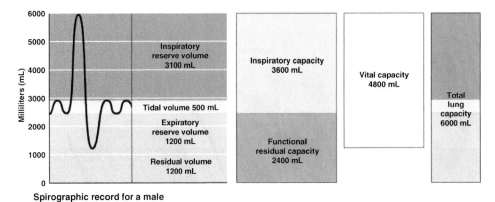

Figure 1.27 Respiratory volumes and capacities. *Essentials of Human Anatomy & Physiology*, 9th Edition, Marieb, 2008 © Pearson Education Inc.

In *forced* breathing both inhalation and exhalation are active processes: during inspiration the sternocleidomastoid, scalene and pectoralis minor muscles contract whilst during exhalation the internal intercostals and abdominal muscles contract reducing the thoracic cavity more than with elastic rebound.

1.5.1 Lung function measures

The two different breathing modes give different lung function measures (see Figure 1.27):

- tidal volume (TV) – air that moves into and out of the lungs with each breath (approximately 500 mL);
- inspiratory reserve volume (IRV) – air that can be inspired forcibly beyond the tidal volume (2100–3200 mL);
- expiratory reserve volume (ERV) – air that can be evacuated from the lungs after a tidal expiration (1000–1200 mL); and
- residual volume (RV) – air left in the lungs after strenuous expiration (1200 mL).

This leads to different measures of capacity:

- inspiratory capacity (IC) – total amount of air that can be inspired after a tidal expiration (IRV + TV);
- functional residual capacity (FRC) – amount of air remaining in the lungs after a tidal expiration (RV + ERV);
- vital capacity (VC) – the total amount of exchangeable air (TV + IRV + ERV); and
- total lung capacity (TLC) – sum of all lung volumes (approximately 6000 mL in males).

1.5.2 Respiration

In the exchange of gases – alveoli to blood occurs at the alveoli of the lungs – gas is exchanged between the air in the alveoli and the blood by the process of diffusion. In this process surfactant covers the internal surface of the alveoli air sac (see Figure 1.28). The surfactant is important in decreasing surface tension, increasing pulmonary compliance (reducing the effort needed to expand the lungs) and reducing the tendency for alveoli to collapse.

1.5.3 Oxygen and carbon dioxide exchange

Molecular oxygen is carried in the blood: some is bound to haemoglobin (Hb) within red blood cells and some is dissolved in blood plasma. Each Hb molecule binds four oxygen molecules in a rapid and reversible process. The haemoglobin–oxygen combination is called oxyhaemoglobin (HbO_2) and haemoglobin that has released oxygen is called reduced haemoglobin (HHb).

Carbon dioxide is transported in the blood in three forms: dissolved in plasma, about 7–10%; chemically bound to haemoglobin, about 20% is carried in RBCs as carbaminohaemoglobin; and as bicarbonate ions in plasma, about 70% is transported as bicarbonate (HCO_3^-).

Carbon dioxide must be released from the bicarbonate form before it can diffuse out of the blood into the alveoli. Bicarbonate ions combine with hydrogen ions to form carbonic acid and carbonic acid splits to

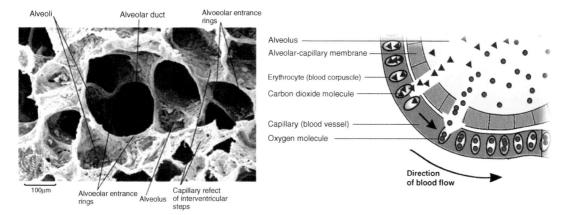

Figure 1.28   The alveoli and gaseous exchange

form water and carbon dioxide. The carbonic acid-—bicarbonate buffer system resists blood pH changes. If hydrogen ion concentrations in blood begin to rise, excess H^+ is removed by combining with HCO_3^- and if hydrogen ion concentrations begin to drop, carbonic acid dissociates, releasing H^+ ions.

1.6   The nervous system

The central nervous system (CNS) consists anatomically of the brain, where high reasoned thought processes occur which influence autonomic nervous responses (of the brain stem), and the spinal cord which affects these neural control signals and the five senses:

- sight;
- sound;
- smell;
- taste; and
- touch.

With such fundamentally important functions the nervous system, at is most critical parts, is protected by the skull and spinal cord (see Figure 1.29).

1.6.1   Central nervous system

The brain and spinal cord make up the central nervous system (CNS). The brain is made of two

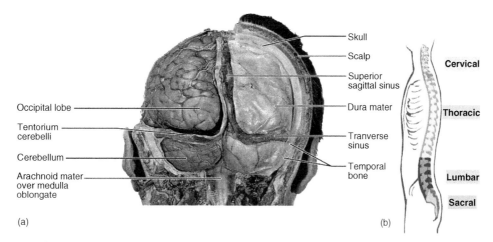

(a)   (b)

Figure 1.29   (a) Protection for the brain in the head/skull, and (b) and spinal cord by the various vertebrae. *Essentials of Human Anatomy & Physiology*, 9th Edition, Marieb, 2008 © Pearson Education Inc.

hemispheres, connected by a bridging connection of nerves called the corpus callosum. These cerebral hemispheres are the newest addition to the human brain in evolutionary terms, so it is called the neocortex. The older part of the brain is the hindbrain. The hindbrain, which is at the back of the brain, includes the cerebellum, which controls balance and the brainstem, which controls the most basic functions of the body, including breathing, blood pressure and heart rate.

Messages are sent from − and received by − the brain by a direct continuation of the spinal cord. The spinal cord has a two-tone colour scheme of grey and white when viewed in cross section with the naked eye (see Figure 1.30(a)). The grey matter looks like a butterfly with outstretched wings, sitting in the middle of the surrounding white matter. The grey matter has nerves which act at a specific site of the body, for example the grey matter at the level of the fourth lumbar vertebra is involved with knee sensation and function, whereas the grey matter at the level of the 10th thoracic vertebra is involved with sensation and muscle control around the umbilicus.

Information is packaged in bundles, or *tracts* within the cord, so that damage to a specific part of the cord creates a predictable functional deficit (Figure 1.30(b) shows the main tracts in the spinal cord − but there are many more).

The corticospinal tracts are nerves which go from the motor cortex of the brain to the muscles which obey the brain and cause a movement. The spinothalamic tracts convey information to the sensory cortex that a specific part of the body has just experienced a sharp pain, a change in temperature or a firm touch. The posterior columns convey information to the brain about where the body currently lies in space, for example standing, raising one arm and so on. This information helps to control balance. The posterior columns also tell the brain if part of the body is experiencing light touch.

The brain is also composed of different centres, which perform specific roles. For example the pituitary gland and hypothalamus are parts of the brain which release hormones (Figure 1.30(c)).

1.6.2   Autonomic nervous system

What cannot be overlooked are the autonomic nervous pathways which reside in the brain stem and control breathing, heart rate, eye responses and even swallowing (see Figure 1.31 and Table 1.2).

1.6.3   Peripheral nervous system

All nerves outside the brain and spinal cord constitute the peripheral nervous system (PNS). The autonomic nervous system is part of the peripheral nervous system and this set of nerves helps to regulate blood pressure, heart rate, crying and sexual function. The autonomic system is subdivided into the sympathetic and parasympathetic nervous systems. In general, the sympathetic system makes us feel excited, angry or frightened − it is part of the 'fight or flight' impulse. On the other hand, the parasympathetic system calms us down, lowering our blood pressure and slowing down our heart rate.

Peripheral nerves/neurons all have a similar structure. They consist of an axon, with a cell body which communicates with other neurons via lots of branches termed *dendrites.*

Signals are transmitted by electrical and chemical (neurotransmitter) means. The electrical signal is created by sodium and potassium ions moving across a membrane, called an *action potential,* which is able to travel considerable distances without losing amplitude, because the nerve cells are lagged with a fatty substance called myelin which acts as an insulating material. The myelin is made by Schwann cells. When one nerve needs to communicate with another nerve, they do so by their dendrites, which have tiny gaps called synapses where the electrical impulse jumps from one cell to the next. Nerve impulses only travel in one direction along a nerve, from the cell body to the terminal endings (Figure 1.32)

There are several different neurotransmitters, including acetyl choline, noradenaline, adrenaline, dopamine, GABA, serotonin and substance P. Different neurotransmitters tend to be found in special areas, for example GABA is associated with brain signals, ACH with muscle and dopamine with sexual function.

When a motor nerve stimulates a skeletal or cardiac muscle to contract, the motor cortex of the brain relays the signal along neurons in the corticospinal tract, exit along the anterior horn and the action potential stimulates the release of acetylcholine (ACh) at the motor end plate. The ACh alters calcium channels in the muscle fibres and this causes structural changes within the muscle fibres. The cells in skeletal and cardiac

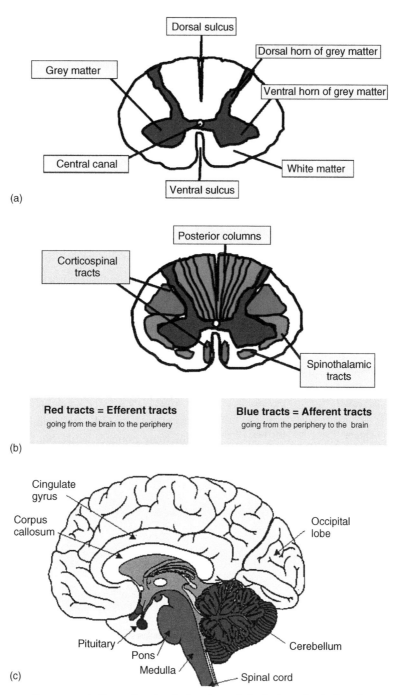

Figure 1.30   Diagrammatic representations of anatomical regions of the brain and spinal cord

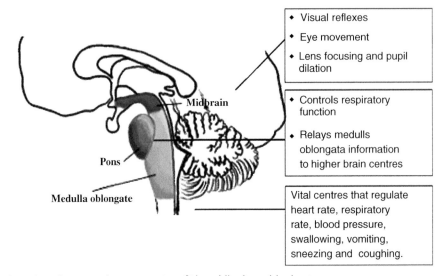

	• Visual reflexes
	• Eye movement
	• Lens focusing and pupil dilation
	• Controls respiratory function
	• Relays medulls oblongata information to higher brain centres
	Vital centres that regulate heart rate, respiratory rate, blood pressure, swallowing, vomiting, sneezing and coughing.

Figure 1.31   Location of autonomic nerve centre of the midbrain and brain stem

Table 1.2   Actions of the parasympathetic and sympathetic systems

	Sympathetic (fight and flight)	Parasympathetic (rest and digest)
Pupils	Constrict	Dilate
Heart rate	Increase	Decrease
Blood supply to stomach	Decrease	Increase
Blood supply to brain	Increase	Decrease
Sexual function	Erection	Orgasm

muscle cells contain sarcomeres with protein filaments called actin and myosin. When a muscle contracts, the myosin proteins, which are found on the thick filaments, ratchet along the thin filaments which contain actin proteins, resulting in increased overlap of the thick and thin filaments. This results in smaller H zones and shorter sarcomeres, as demonstrated in Figure 1.33.

When a muscle contracts it moves its attachments. The muscle is attached to bone at each end by tendons, which are noncontractile. So, for example, the psoas muscle is attached to bone at one end at the top of the femur (specifically on the lesser trochanter) and on the

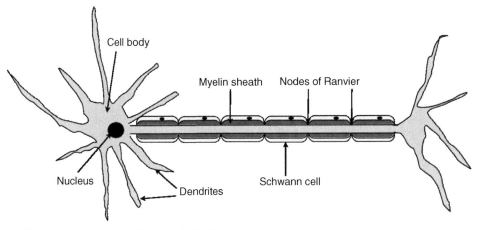

Figure 1.32   Diagrammatic representation of a typical peripheral nerve structure

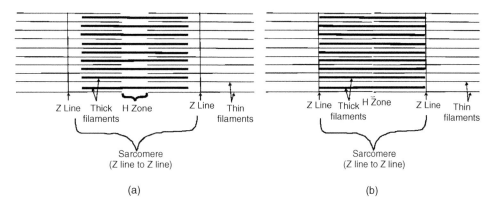

Figure 1.33   (a) Diagrammatic representation of the physiological anatomy of muscle filaments and (b) the changes in positioning upon contraction

other end to the lumbar spine (specifically, the transverse processes of L1 to L5). The main joint between the femur and the spine is the hip joint. As the psoas muscle runs in front of the hip joint, the psoas muscle flexes the hip joint. This results in either the femur being flexed while the torso is still, the torso flexing while the femur is still, or both the femur and torso flexing together. In order to control which movement results from psoas contraction, other muscles contract either with or against the psoas. So, if you want to perform a sit-up from a lying down position, you will want your legs to remain on the floor, while you flex your hips and raise your torso from the ground. Psoas will contract to flex the hip joint, supported by iliacus which runs next to it, the anterior abdominal wall will contract (these muscles attach to the pelvis, so won't influence leg movement) and the hamstrings, which run behind the hips joint, from the pelvis to the tibia will contract to extend the hip joint — thus keeping the femurs from moving. As you begin to appreciate here, a seemingly simple movement involves the coordination of several different muscles working together and against each other to achieve the desired effect — the muscles receive the messages by motor nerves which travel along the front and sides of the spinal cord (corticospinal tracts) from the motor cortex in the brain (see Figure 1.34). The movement will result in the back leaving the floor and this change in sensation, as well as change of position in space is conveyed as nerve impulses along sensory and proprioceptive fibres in the posterior columns of the spinal cord to the sensory cortex and other areas of the brain. The change in position in space is supported by information gleaned from the semicircular canals in the middle ear, because the head position has also moved from a horizontal to vertical position.

1.6.4   The five senses

1.6.4.1   Sight

The eye receives light waves and converts this information into electricity by things called photoreceptors cells. The electrical energy is transmitted to the visual cortex in the brain by the optic nerve. Whereas they eye works like a camera, capturing data as an inverted image, the visual cortex interprets the electrical energy into correctly-orientated images. The eye is globe-shaped and divided into two compartments. The front compartment is called the anterior chamber and contains the lens with its attachments. The watery fluid contained in the anterior chamber is called the aqueous humour and the gelatinous fluid in the posterior chamber is called the vitreous humour (Figure 1.35).

The lens' shape is maintained by muscles (ciliary muscles) which make the lens thinner or thicker to alter focus. The pupil size is controlled by the iris, which is the structure giving us our eye colour (blue, green or brown). Once the light has hit the retina at the back of the eye, two specialized types of cells convert the light energy to electrical energy. The rods are colour-sensitive cells and contain photoreceptors. The rods are found in the highest numbers at the fovea. This is a single spot on the retina, which collects the most

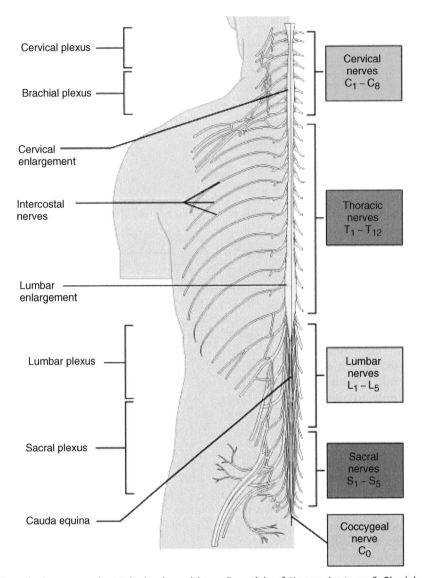

Figure 1.34 The spinal nerves and vertebral exit positions. *Essentials of Human Anatomy & Physiology*, 9th Edition, Marieb, 2008 © Pearson Education Inc.

information. The cones are cells which differentiate between light and dark and are the cells which give us our night vision, where we see in black and white. The optic nerve conveys the information received by the retina onwards to the brain. The optic nerve attaches to the retina by a disc of nerve endings called the optic disc. There are no rods or cones on the optic disc so that any light which lands on the optic disc is not recognized; this is the blind spot.

To find our blind spot, fix your eyes on an object in front of you. Take a small coloured object and move it around your visual field. You will find that the object vanishes at one consistent point with the right eye, and a different point with the left eye. Our brains usually

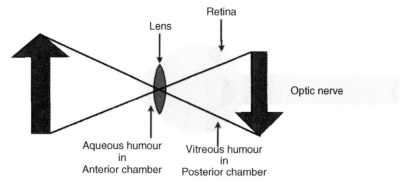

Figure 1.35   Diagrammatic representations of the function components of the eye

ignore the blind spot. The next part of the pathway is very interesting from a clinical viewpoint, as it is sometimes possible to diagnose which part of the brain is damaged when a person loses part of their vision (see Figures 1.36 and 1.37).

The left visual cortex receives information from the outer(lateral) half of the same-sided (=ipsilateral) left eye and the inner (medial) half of the opposite (=contralateral) right eye. The right visual cortex receives information from the ipsilateral visual field's lateral half and the contralateral medial half's visual field. Tracts B and C cross, or decussate at the optic chiasm, which is just above, or superior to, the pituitary gland.

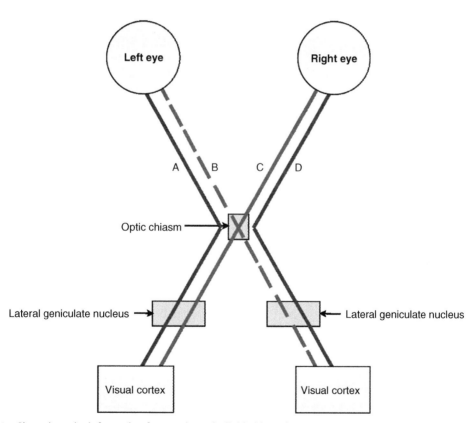

Figure 1.36   Shows how the information from each eye is divided into three tracts per eye, six in total; tracts A and B come from the left eye and tracts C and D come from the right eye

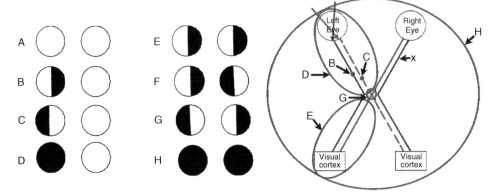

Figure 1.37   Illustration of visual perception as a result of damage to the optic nerve pathways: A = normal vision; B = suspiciously discrete lesion damaging exactly one half of tract B; C = same suspicion as for Mr B, with tract C; D = damaged left optic nerve; E = left-sided brain damage beyond the optic chiasm – termed homonymous hemianopia – may have lost normal vision as a result of a stroke, or a severe head injury; F = two identical discrete lesions at B and X, very unlikely; G = bitemporal hemianopia, a visual defect which may come on slowly and gradually worsen, commonly a pituitary tumour, pressing upwards and against the optic chiasm; H = damage to both optic nerves, an enormous pituitary tumour compressing all optic tracts, or massive posterior brain damage, affecting both visual cortices (cortical blindness)

1.6.4.2   Sound

Hearing and balance are closely associated, as both rely on structures inside the inner ear. Sound is received by the ear and processed via three compartments into electrical signals, which are then relayed to the brain. The ear is divided simply into the outer ear, middle ear and inner ear (see Figure 1.38). The outer ear is the part that we see, that is, the ear, or pinna, itself. The hole in the ear is the external auditory meatus (='the hole outside that listens', Latin again) and sound waves are transmitted to the ear drum, which is the beginning of the middle ear. The ear drum is called the tympanic membrane. It is called a drum, rather than a membrane, or wall, due to its function. The tissue is stretched like the skin on a drum and, as sound waves strike the drum, it vibrates according to the frequency (pitch) and amplitude (volume) of the sound waves received. The three smallest bones in the body are attached in series to the other side of the ear drum. These are called the malleus (hammer), incus (anvil) and stapes (stirrup). The malleus attaches to the ear drum and the stapes attaches to the oval window, which is a smaller version of the ear drum. The incus links the two bones together. The oval window is the partition between the middle ear and the inner ear. The ossicles of the middle ear act as a dampening system for sound waves which hit the ear drum. When the sound wave amplitude is very large, the ear drum vibration is large too. A negative feedback system comes into play, whereby a muscle attached to the malleus, called tensor tympani, stretches the tympanic membrane, thus reducing the size of the vibrations' amplitude. The second dampening system occurs at the other end of the ossicle chain, by a muscle attached to the stapes, called stapedius. The vibrations from the stapedius are transmitted to the cochlea via the oval window and transferred to electrical energy, which is conveyed to the auditory cortex in the brain via the auditory nerve. Next to the cochlea lie the semicircular canals, which are organized in x,y,z planes to help maintain balance.

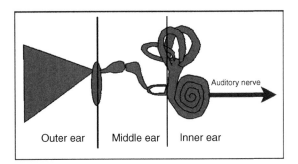

Figure 1.38   Diagrammatic representation of the functional components of the ear

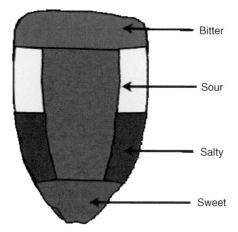

Figure 1.39 Diagrammatic representations of sensory regions of the tongue

1.6.4.3 Taste

This sense is mostly about taste buds on the tongue. There are a few taste buds in the walls of the mouth and the upper oesophagus as well. Different parts of the tongue are better at detecting different types of taste. There are four main tastes (see Figure 1.39).

The chemicals which make up different tastes are converted into electrical activity via the taste buds and then transmitted to the brain via the vagus nerve (10th cranial nerve) and the glossopharyngeal nerve (ninth cranial nerve).

1.6.4.4 Smell

This is another sense that relies on chemical stimuli which are converted to electrical energy. A small surface area around the size of a postage stamp in each nose is responsible for smell. The receptors are called olfactory receptor cells and the olfactory nerves travel up the nose to the cribiform plate at the front of the brain. The olfactory nerves (first cranial nerve) regenerate continually throughout life and for this reason they are the focus of lots of research into nerve regeneration following spinal cord injuries and degenerative nerve diseases.

1.6.4.5 Touch

The fifth sense is touch. Whereas the other special senses have nerves which communicate directly with the brain, via cranial nerves, touch — or sensation — uses mainly nerves which transmit information to the brain via the spinal cord. These nerves are called

spinal nerves and they belong to the peripheral nervous system.

Touch can be classified into light touch, vibration, pain and temperature perception. These different types of touch are managed by different types of nerve, which travel along different parts of the spinal cord to the brain.

1.7 The endocrine system

It is appropriate that the endocrine system should be considered in juxtaposition with the nervous systems as there are many similarities and contrasts in their functionality as control systems of the body: both act as communication networks; both use chemical substances as mediators between cells and both cause actions at a distance. In cellular origins they are closely related and in the adrenal medulla and hypothalamus specialized 'neurons' secrete 'hormones' into the circulation. However, whilst neurological signal control is a fast response, short-lived and specific effect,

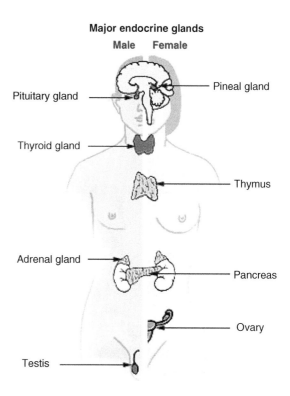

Figure 1.40 The anatomical positioning of the major endocrine glands

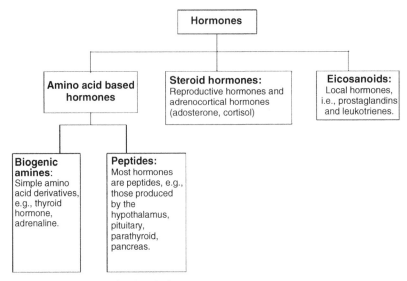

Figure 1.41   Classification of hormones by chemical structure type

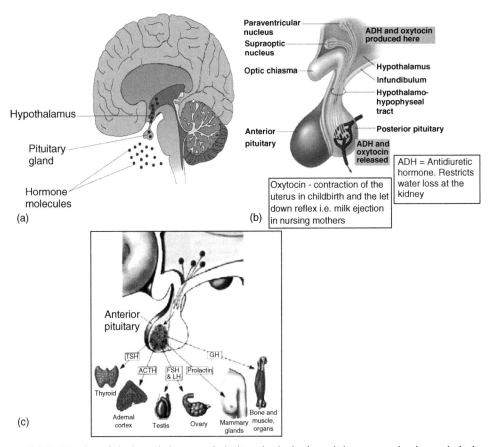

Figure 1.42   (a) Positioning of the hypothalamus and pituitary in the brain and the neuroendocrine and pituitary peptide hormone systems ((b) and (c) respectively) controlled

endocrine control is a slower response to stimuli but responses last for longer and the responses are more generalized.

The major endocrine glands are illustrated in Figure 1.40 and the hormones produced by these glands are chemical messengers carried in the bloodstream to distant target cells where they cause specific changes in function.

1.7.1 Main functions of hormones

The main function of the hormones is (Figure 1.41):

- to promote the development of physical, sexual and mental characteristics (e.g. the sex steroid hormones);
- to promote the adjustments of important bodily functions (e.g. thyroid hormone and adaptation to cold);
- to keep certain physiological parameters constant (i.e. maintain homeostasis, e.g. insulin and blood sugar).

The function of various endocrine organs systems is described in Chapter 2 and others in Chapter 7; however, the hypothalamus and pituitary are the master controllers of the endocrine system.

1.7.2 The hypothalamus and pituitary glands

Positioned at the base of the brain (see Figure 1.42 (a)) the hypothalamus is a direct link between the central nervous system and endocrine signalling. Neuroendocrine cells directly secrete hormone into the circulation in the anterior pituitary (see Figure 1.42(b)) or induce specialist cells of the posterior pituitary to produce more complex peptide hormones. Nevertheless the effects and system controlled by these hormones are broad and widely spread throughout the body. The anterior pituitary gland is often referred to as the 'master' endocrine organ, but it too has a 'master'. The hypothalamus through the secretion of releasing and inhibiting hormones controls all the systemic hormones released (see Figure 1.42(c)).The hypothalamus through sympathetic activation has a direct neural control of the adrenal medulla (which produces adrenaline and nor adrenaline) by controlling the release of adrenocorticotrophic hormone — ACTH — by specialist corticotroph cells of the anterior pituitary. The posterior pituitary is not really an endocrine gland as it simply stores and releases two hormones — oxytocin and ADH — but these are actually made and secreted by the main body of neuroendocrine cells found in the hypothalamus (see Figure 1.42(b)).

Bibliography

Marieb, E.N. (2008) *Essentials of Human Anatomy and Physiology*, 9th edn, Pearson Education.

Pocock, G. and Richards, P. (2009) *The Human Body: An Introduction for the Biomedical and Health Sciences*, 1st edn, Oxford University Press.

Ross and Wilson (2010) *Anatomy and Physiology in Health and Illness*, 10th edn (eds A. Waughand A. Grant) Churchill Livingstone.

Chapter 2
Pathophysiology

Dr Suzanne M. Docherty, BMedSci, MBBS, Ph.D.

2.1 Pathophysiology: a definition

Pathophysiology is the study of medicine which deals with any disturbances of body function caused by disease. This can result in alterations to the mechanics and/or biochemistry of the tissue concerned, with an enormous variety of consequences, but also of scope for investigation. It is thus a very large field of study. In considering the pathological conditions that can affect the individual, students can also usefully consider how these occur within the entire population, and how the cause of outbreaks of certain conditions can be studied. In this chapter we encompass an introduction to the fascinating field of epidemiology, from its historical roots and important work in working out the root cause of important ailments affecting large numbers of people, through to modern epidemiology. We will also include an overview of pharmacology to consider how the common classes of drugs work in treating human pathology.

2.2 Introduction to epidemiology

Epidemiology looks at the occurrence of diseases and accidents in different groups of people, and attempts to find out why disease has occurred. This is an integral part of the work of Public Health organizations — discovering what is at the root of morbidity and mortality, and putting strategies in place to address the causes and thus improve the health of a population. The population at risk of a particular problem or illness — the target population — can sometimes comprise an entire nation of people, but in most epidemiological studies it will be a narrower section of society. From the target population, subjects for study should be identified; this may be according to where they live (geographical), where they are a patient or in care (special care), what job they do (occupational — this can include schoolchildren!), or a disease they have (diagnostic). Subjects should be chosen at random whenever feasible to prevent the introduction of bias

Biomedical Sciences: Essential Laboratory Medicine, First Edition. Edited by Ray K. Iles and Suzanne M. Docherty.
© 2012 John Wiley & Sons, Ltd. Published 2012 by John Wiley & Sons, Ltd.

to an epidemiological study. However, this is not always possible, as a breakdown of the types of epidemiological study and their design will show. Finally, once a study population has been defined and studied, comparisons are made to allow conclusions to be drawn.

To undertake studies on a disease, diagnostic criteria for it have to be clearly established to define what is a '**case**' of a particular condition and to enable division of the study group into those who have been affected by it and those who are unaffected. For example, for the purposes of an epidemiological study on anaemia in adult females, anaemia may be defined as a haemoglobin level of <11.5 g/dL. Case definitions need to be as clear and standardized as is practicable.

Once cases have been identified in a population, the incidence and prevalence of the condition can be worked out in relation to the target population. **Incidence is defined as the rate at which new cases occur in a defined population over a set period of time**, that is

$$\frac{\text{Number of cases}}{(\text{Population at risk}) \times (\text{Time over which cases were identified})}$$

though it should be noted that the population should be relatively constant!

'**Prevalence' is defined as the proportion of a population who are cases at any one time**; for example, a questionnaire might identify that 1% of all farmers in Worcestershire had a broken limb at one time point. However, examining the prevalence of a condition over a longer time ('period prevalence') will often give a better measure of the true prevalence of a condition.

'**Mortality' is defined as the incidence of death from a disease.**

Once the incidence of a condition is known in a specific population, comparisons can be made with other populations who have higher or lower incidences of the same defined condition to try to identify factors that put the higher incidence population 'at risk', or that are acting as protective factors in a population at lower risk. However, investigators need to guard against confounding factors in studies, which are differences between individuals in the target population that work to increase or decrease the risk of developing a particular condition independently of the factor under investigation. The biggest confounding factors in most epidemiological studies are age and sex; for example, two populations might have drastically different incidences of heart disease, but further examination reveals that one population has more females (who have lower incidence of heart disease than males), or that one population has a lower mean age than the other (since increasing age is associated with an increased incidence of heart disease). To reduce the risk of confounding factors compromising epidemiological investigations, population standardization is often undertaken, weighted averages of age and sex disease occurrence rates are used, or the number of cases observed is compared to that of a reference population. Mathematical modelling techniques can also be used to adjust data for risk factors, protective factors and confounding factors.

Bias can also be problematic for epidemiological studies. Bias may be introduced at the stage of selecting subjects to study, in that people selected may not be truly representative of the target population. Where questionnaires are used to survey a population, the people who respond may be different in some way to those who fail to respond, and this can bias the study. 'Recall bias' is also important when subjects are questioned about previous exposure to a potential risk factor for a condition — those with the condition are more likely to recall exposure than those who do not have the condition. Bias can never be completely eliminated from epidemiological studies, but an awareness of potential sources of bias can influence study design to minimize it.

2.2.1 Observational epidemiological studies

2.2.1.1 Longitudinal studies

In longitudinal studies, subjects are selected from within a target population and followed. The numbers involved and timescale used vary widely; it may be a few people followed over weeks, or whole populations over decades, or even their lifetime. Most will be examining the effect of exposure to known or hypothesized risk factors on developing (or dying from) a particular disease, and usually a group of 'exposed' subjects will be compared to a group of 'unexposed' subjects for any differences in the incidence of the disease. Confounding factors can be factored in, often by matching up individual exposed subjects (cases) with a control unexposed subject of the same age and gender (and possibly for similarity in other variables too, e.g. social class or ethnicity) and following both sets of subjects over the course of the study.

The disadvantages of longitudinal studies are numerous. By studying subjects over long time periods, many will be lost to follow up (e.g. from migrating overseas), and studies can be very complex to administer as large numbers of subjects are involved. Exposure to some hypothesized risk factors may be extremely rare, and thus require large numbers of subjects to be included to have sufficient numbers of exposed subjects. There may also be a long time delay between exposure and the manifestation of disease (e.g. it is often several decades between exposure to asbestos and the development of the lung tumour mesothelioma).

2.2.1.2 Case-control studies

A different approach to epidemiological studies that may be preferable to the wide-ranging longitudinal studies is the case-control study. In a case-control study, people with a particular condition (cases) are identified, and their exposure to one or more theoretical risk factors is established retrospectively, for example by questioning about areas lived in, diet, or industrial exposure in different jobs. This is compared to the exposure to the same risk factors in matched controls who do not have the condition. Cases and controls can be matched up according to exposure to potential confounding factors, improving the validity of the study, and the results statistically adjusted for confounding variables too. Once the exposure of both cases and controls to a particular risk factor is known, an odds ratio can be calculated for the likelihood of developing that condition after exposure.

2.2.2 Experimental epidemiological studies

Longitudinal and case-control studies are observational; investigators study people as they are, and confounding factors can affect the results in ways that investigators have little control over. Experimental studies can be set up instead, in which investigators control exposure to possible risk or protective factors for a condition, and observe the outcome for experimental subjects.

2.2.2.1 Randomized controlled trials

The 'gold standard' of epidemiological experimental study is the randomized controlled trial. Subjects who are representative of the target population and meet predefined entry criteria are recruited, and their consent for the experiment obtained. The number of subjects required can be calculated in advance based on what will give the study statistical 'power', that is, the ability to show a real change. The subjects are divided into two groups; one will receive a particular treatment/exposure to a possible risk factor (subjects), while the other will not (controls). In an ideal situation a randomized controlled trial will be 'blinded' so that subjects are randomly assigned to one group or the other, and will not know which group they are in — this can be achieved using computer-generated random numbers. Controls can be given a 'placebo' treatment if appropriate, for example in a vaccine study case subjects will receive the vaccine, while control subjects will receive a placebo injection that has all other components of the vaccine except the active compounds. Better still is the 'double blind' trial, in which the investigator is also unaware of which group subjects have been assigned to, since this eliminates the potential for reporting bias of the results by an investigator who is expecting a particular outcome. However, arrangements must exist for rapidly 'breaking the code' if a problem arises. The trial continues until a set end point, which may be weeks, months or years later, and the outcomes from the case and control subjects are compared and statistically analysed to evaluate the intervention.

Some randomized controlled trials are designed to be 'crossover' trials where feasible, such that the case subjects and control subjects cross over in the middle of the trial, with the cases then becoming controls, and vice versa. This enables a treatment or exposure to be evaluated in twice the number of people as a noncrossover trial. However, the crossover trial is only suited to interventions that have a short 'half-life', for example it could not be used for a trial on a vaccine in which the effects will last for years — the cases could not go on to act as controls if they have received the vaccine and still have the relevant antibodies!

2.2.3 Epidemiology in outbreaks and epidemics

Outbreaks of disease are still of international importance, as recent global pandemics of SARS and Swine Flu demonstrated. One of the first, and best known, epidemiological studies was that undertaken on cholera by the physician John Snow in London, in which he investigated the source of a cholera outbreak centred in Soho in 1854. Snow's work occurred in the days before knowledge that disease could be transmitted by microorganisms, when it was commonly believed that such diseases were transmitted by breathing 'foul air'. However, Snow worked out that the cholera outbreak centred around one water pump in Broad Street by studying the patterns of disease among the residents, and enquiring about which pump had been used by those infected with cholera. Snow later had the pump handle removed from the pump, which was found to be contaminated by a cesspit in close proximity. He also performed a statistical analysis to link the number of cases of cholera with the quality of the water supply. Snow became known as the 'father of epidemiology'.

Since Snow's day, epidemiology has been used to investigate the causes of numerous outbreaks of illness, from the more mundane local clusters of food poisoning to instances of serious illness caused by environmental contamination (such as the aluminium contamination of the drinking water supply at Camelford in 1988), right up to the identification of the cause of AIDS and its likely mode of transmission in the early 1980s. However, defining what is a 'case' of a particular condition remains important in the epidemiological study of outbreaks of disease, as 'pseudoepidemics' are a possibility where a case definition is not well established. The population at risk also needs to be identified before hypotheses about the outbreak are developed, refined and tested. Once the source of an outbreak is clear, control and preventative measures can be put in place to halt the epidemic.

2.3   Introduction to pharmacology

Pharmacology is the study of the science of drugs and their affect on the human body. Drugs are not necessarily simply those substances prescribed by a doctor, or bought from the chemist — a number of everyday substances such as alcohol and tobacco are also drugs, as are the herbal remedies that some people choose to take. All of these substances contain active compounds that can provoke a change in the body's physiology via a variety of mechanisms outlined in this section. (Note that homeopathic 'remedies' fall outside of this definition, since they are composed of water that is so diluted since initial exposure to an active compound

that is effectively only water once again, and numerous studies suggest homeopathic remedies simply exert a 'placebo' effect.) Before considering the pathological conditions that can arise in the human body and their treatment, it is useful to have an understanding in how pharmacological treatments actually achieve their effect, so that pathophysiology and treatment can both be better understood.

2.3.1   Mechanisms of drug action

Drugs bring about changes in the body in one of five general ways:

- action on receptors on cell membranes (including membranes of organelles within the cell) — this is the most important conceptually in pharmacology;
- blocking the action of specific enzymes, for example, angiotensin converting enzyme inhibitors, used to treat high blood pressure (see Section 2.8.9);
- inhibiting cell transport mechanisms — for example, the drug probenecid inhibits transport of organic anions in the renal tubule, and is often used along with penicillin to prevent penicillin being rapidly excreted by the kidneys;
- action on invading organisms — antibiotics are all examples of this class of drug, as are antiviral, antifungal and antihelminth drugs;
- nonspecific interactions — these are usually simple chemical reactions, for example, using antacid drugs (alkali compounds) to neutralize gastric acid.

2.3.2   Drugs and receptors

To bind to a receptor and produce an effect, a drug needs to have a very specific structure. In general, drugs need to have a very similar structure to the natural compound (e.g. hormone) that usually binds to a receptor, and drugs generally need a minimum three-point attachment at a receptor to have an effect. Drug binding to receptors can utilize any of the chemical bonds the reader should already be familiar with; hydrophobic interactions between nonpolar groups on drug and receptor to minimize water contact, ionic interactions between the charged groups on drug molecules and receptor molecules, or occasionally via covalent bonds. Since covalent bonds are extremely strong and difficult to reverse, few drugs bind to receptors using them except in situations where there is toxicity.

One of the most important concepts in considering the action of drugs at receptors is that of the agonist and antagonist. An agonist drug is a drug that causes a stimulation-type response at a receptor; that is, it mimics the effect of the natural molecule that binds the receptor. A partial agonist does not fit the receptor as well as the natural molecule that binds there, so it causes a weaker response. An antagonist is a drug that blocks or dampens the normal response at a receptor, as it only partially fits the receptor, that is, it can block it and prevent the natural molecule (e.g. neurotransmitter) or an agonist drug binding there, but it cannot activate the receptor.

Many antagonists work by competition with the natural agonist, so the relative concentrations of the agonist molecule and antagonist drug are important — at low concentrations of antagonist, little blockade of the receptor will be achieved, while at high concentrations of antagonist compared to natural agonist, good levels of blockade of the receptor will be achieved (Figure 2.1).

2.3.2.1 The adrenergic receptor – the receptor of the sympathetic system

The sympathetic system (the body's 'fight or flight' system, triggered in times of stress) uses receptors that are activated by adrenalin, and these are known as adrenergic receptors. The effects of adrenalin at these receptors are to prepare the body to take rapid action in a threatening situation, for example causing the heart to pump faster, the bronchi to dilate to let maximal oxygen reach the lungs, the arterioles to constrict and increase blood pressure, to name the major effects (see Chapter 1 to revise all the effects of activation of the sympathetic system). There are different types of adrenergic receptors found on different tissues of the body; the heart has β1 adrenergic receptors, the bronchioles of the lungs have β2 adrenergic receptors, and the blood vessels have mainly α adrenergic receptors. Adrenalin acts on all types of adrenergic receptor, but the subtle differences in these receptors allow drugs to be made which will selectively act on one type of adrenergic receptor. Salbutamol is a drug used to treat asthma, a condition in which the bronchioles are constricted and breathing becomes difficult. Salbutamol is an agonist drug that acts at β2 receptors; it activates them, causing dilation of the bronchioles. However, it will also activate the similar β1 receptors on the heart to a degree, meaning that a rapid heart rate is a common side effect of salbutamol.

Beta blocker drugs have also been developed for people with high blood pressure. These drugs are antagonists for β receptors in the heart, blocking them so that adrenalin is unable to bind to them, and thus

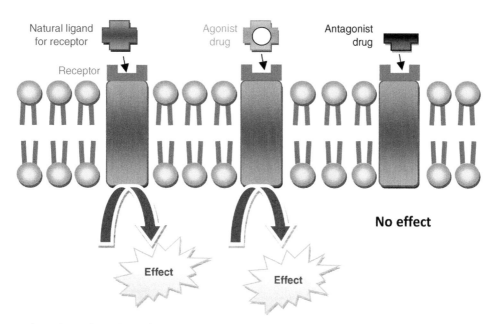

Figure 2.1   The actions of agonist and antagonist drugs at a receptor

causing the heart to pump less hard, with subsequent reduction of blood pressure. The first beta blocker drug, propranolol, was not 'selective' and blocked both β1 receptors in the heart and also β2 receptors in bronchioles, which had the unfortunate side effect of triggering asthma in individuals who were susceptible. Later beta blockers such as atenolol were said to be 'cardioselective', meaning that they predominantly blocked only β1 receptors and were less likely to trigger asthma; however, it remains good practice not to give beta blocker drugs to patients with asthma.

2.3.2.2 The cholinergic receptor

The parasympathetic system (our normal day-to-day operating system) uses acetylcholine as its neurotransmitter, and thus its receptors are known as cholinergic receptors. Binding of acetylcholine to cholinergic receptors of the parasympathetic system causes responses such as a lowering of heart rate, vasodilation, bronchoconstriction, with increased gastric motility and saliva production. Drugs which block cholinergic receptors (i.e. are antagonists of cholinergic receptors) are known as anticholinergic drugs, and can be used in nausea and vomiting to decrease gastric motility. An example is atropine, which is used to increase heart rate by the blockade of cholinergic receptors. Common side effects of anticholinergic drugs are a dry mouth, rapid heart rate, blurred vision and urinary retention — these

are largely predictable if you know the normal effects of the parasympathetic system on the body!

2.3.3 Enzyme inhibition by drugs

Some drugs act by inhibiting the action of cellular enzymes. Good examples include nonsteroidal anti-inflammatory drugs (NSAIDs) such as aspirin and ibuprofen, which inhibit the action of the enzyme cyclo-oxygenase, and block the production of prostaglandins from arachidonic acid, thus decreasing inflammation. However, prostaglandins also decrease acid production and increase mucus production in the stomach, so one side effect of NSAIDs in blocking prostaglandin formation is to trigger gastritis in a significant proportion of those who take NSAIDs for any length of time.

Another example of enzyme inhibition by drugs is angiotensin converting enzyme inhibitors (ACE inhibitors) such as lisinopril, which inhibit the angiotensin converting enzyme and ultimately prevent the formation of the potent vasoconstrictor angiotensin II, thus keeping blood pressure down. The powerful antacid drug omeprazole is also an enzyme inhibitor; it binds to the proton pump enzyme in parietal cells in the stomach to prevent the formation of H^+ ions, and thus ultimately the formation of HCl for gastric acid (Figure 2.2).

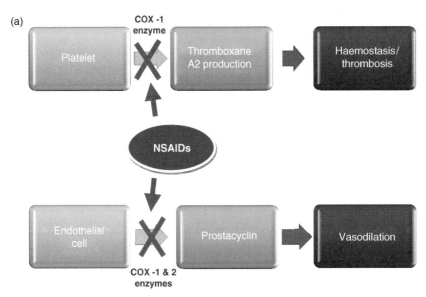

Figure 2.2   Enzyme inhibition pathways of (a) NSAIDs, (b) ACE inhibitors and (c) omeprazole

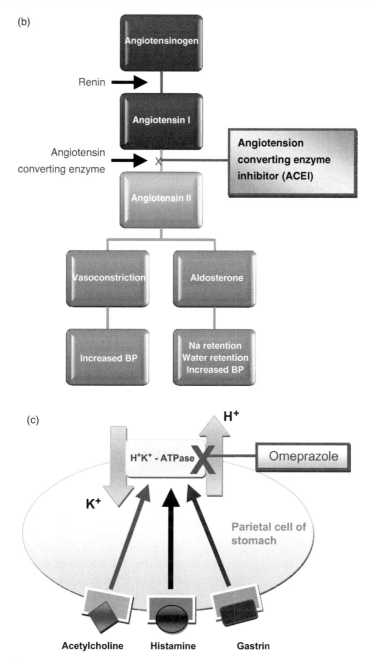

(b)

(c)

Figure 2.2   (*Continued*)

2.3.4 Drugs that affect transport within a cell

Some drugs act by preventing the movement of ions across cell membranes. A good example of this class of drug is calcium channel blockers such as nifedipine; these prevent the movement of calcium ions across the cell membrane in the smooth muscle of blood vessels, preventing contraction of blood vessels and thus maintain vasodilation and lower blood pressure. They are

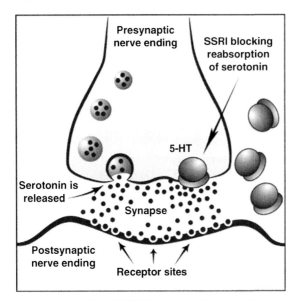

Figure 2.3   Action of SSRI drugs as antidepressants

useful in some patients with hypertension, and in coronary artery disease. Many drugs that act within the central nervous system do so by interfering with the movement of neurotransmitter molecules — specific serotonin reuptake inhibitor antidepressants (SSRIs) such as fluoxetine act by preventing the reuptake of the neurotransmitter serotonin into the presynaptic nerve ending, and prolong the action of serotonin (which in simple terms is a 'happiness' neurotransmitter!) by doing this (Figure 2.3).

2.3.5   Antibiotics

The mode of action of antibiotic drugs — encompassing antibacterial, antiviral and antifungal agents — is considered in detail in Chapter 6.

2.3.6   Drugs with nonspecific actions

Some drugs have nonspecific actions; that is they are not interacting with receptors or enzymes. Antacid drugs such as sodium bicarbonate act to neutralize stomach acid in the straightforward 'acid + base = salt + water' reaction, and emollient drugs can be applied to the skin to hold in moisture. General anaesthetic vapours such as halothane act to reduce the function of the CNS, but it is as yet unclear exactly how they achieve this effect.

2.3.7   Adverse drug reactions

An adverse drug reaction is defined as any unwanted effect resulting from a drug's use in treatment. Adverse drug reactions are common, and constitute the most common cause of iatrogenic injury in hospital (i.e. injuries caused by medical treatment). Many are caused by errors in prescribing or administering a drug, and most are preventable. Unlike toxic drug effects — which occur at doses above the usual therapeutic range of a drug — adverse drug reactions are unpleasant, unwanted effects that occur at doses normally intended for diagnosis, therapy or prophylaxis. The two major types of adverse drug reaction are Type A and Type B, summarized in Table 2.1.

There are other types of adverse drug reaction in addition to Type A and Type B, though these are the most important in clinical practice. Type C adverse drug reactions result from chronic effects of a drug, such as colon dysfunction due to chronic use of laxatives. Type D adverse drug reactions are delayed effects of drugs that occur years (sometimes decades) after treatment, for example the cancers that can result from the chemotherapy given for a different initial malignancy, for example leukaemia some years/decades after the use of alkylating chemotherapy for an initial lymphoma. Type E adverse drug reactions are 'end-of-treatment' effects that occur when drugs are stopped, such as seizures after withdrawal of the anti-epileptic drug phenytoin.

One of the best ways of preventing adverse drug reactions is via monitoring of drug levels for those drugs that have a narrow therapeutic index, that is those drugs where there is a narrow window for the drug levels in serum to give a therapeutic effect rather than a toxic or subtherapeutic effect (Table 2.2).

The antibiotic drug gentamicin is an example of a drug with a narrow therapeutic index in which the levels of drug in serum are monitored daily in

Table 2.1   Type A and Type B adverse drug reactions

	Type A	Type B
Definition	'Augmented' adverse drug reaction: predictable, common and dose-related	'Bizarre' adverse drug reaction: unpredictable, rare, and not dose-related
Morbidity and mortality	High morbidity, low mortality	High mortality, low morbidity
Preventable?	Most are avoidable	Most are unavoidable
Examples	Excessive pharmacological effect, e.g. respiratory depression from morphine, heart block with digoxin, gastric ulceration from aspirin, low blood sugar with insulin	Allergic reactions, e.g. penicillin allergy
	Rebound responses, e.g. rebound hypertension after clonidine withdrawal	Genetically determined effects, e.g. autosomal recessive trait for 'slow acetylation' of some drugs in the liver means they are more slowly metabolized, putting slow acetylators at risk of side effects. One example is the drug-induced form of systemic lupus erythematosus that occurs with the antihypertensive drug hydralazine in slow acetylators.
	Crossreaction with other receptors, e.g. palpitations from salbutamol	Idiosyncratic effects of drugs, e.g. swelling around the eyes from taking proton pump inhibitor drug omeprazole, DVT after taking oral contraceptive pill

patients who are prescribed it. If levels of gentamicin increase and climb above the therapeutic range, damage to the nerves in the ear that control hearing and balance can result, along with damage to the kidneys. See also Section 2.3.7 for further details of drug monitoring.

Having considered how the drugs used to treat human disease attain their desired effects (and how the safe use of drugs can best be achieved), we can now move on to look at the diseases of major body systems; understanding their diagnosis, investigation and management.

Table 2.2   Drug with a wide therapeutic index (left); drug with a narrow therapeutic index (right)

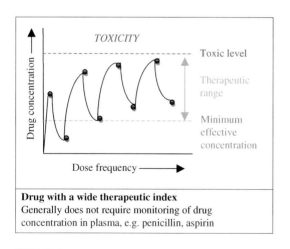

Drug with a wide therapeutic index
Generally does not require monitoring of drug concentration in plasma, e.g. penicillin, aspirin

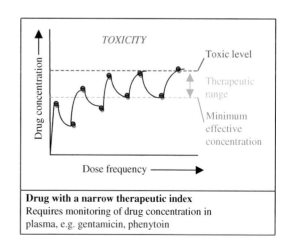

Drug with a narrow therapeutic index
Requires monitoring of drug concentration in plasma, e.g. gentamicin, phenytoin

2.4 Gastroenterology

2.4.1 Symptoms of GI disease

The gastrointestinal (GI) tract is some 5 m long in an adult human, and symptoms of gastrointestinal disease vary according to which part of the GI tract is affected. Problems in the mouth, throat and oesophagus can lead to dysphagia, a difficulty in swallowing. Heartburn — a burning sensation felt behind the sternum — may be experienced where acid reflux from the stomach into the oesophagus occurs. Vomiting may occur due to stimulation of vomiting centres in the medulla of the brain, or diarrhoea (passing increased volumes of loose stool) may be experienced. Steatorrhoea is the passage of pale bulky stools, and occurs where fat absorption is impaired. Constipation usually means the infrequent passage of hard stools — often with difficulty — but may be interpreted differently by different people.

The gastrointestinal system comprises several components; a bolus of food passes through the mouth and throat into the oesophagus, then the stomach, after which components will travel on through the small intestine and large intestine. Each of these aspects of the gastrointestinal tract can give rise to their own unique symptoms and presentations of disease — these are considered in turn below.

2.4.2 Oesophagus

The oesophagus conducts boluses of masticated (chewed) food from the mouth into the stomach through the oesophageal sphincter. Problems of the oesophagus can lead to difficulty with swallowing — 'dysphagia' — or 'heartburn' (a burning sensation experienced behind the sternum) from reflux of gastric acid up into the oesophagus.

2.4.2.1 Acid reflux and Barrett's oesophagus

A degree of reflux of gastric contents into the oesophagus is normal, but repeated, lengthly episodes of gastric acid in contact with the oesophageal mucosa can cause the symptoms of heartburn. This is worsened by bending down or lying flat, by alcohol and smoking, but can be relieved by the action of simple antacids in neutralizing gastric acid. H2 receptor antagonists such as ranitidine can also be used to decrease the production of gastric acid, or proton pump inhibitors such as omeprazole to block the proton pump producing H^+ ions in the stomach.

The normal squamous epithelium of the oesophagus may undergo metaplasia to a columnar epithelium following long-term exposure to gastric acid; this is known as Barrett's oesophagus, and may be identified from biopsies taken from the oesophagus during endoscopy undertaken for investigation of longstanding heartburn, or dysphagia (Figure 2.4).

Barrett's oesophagus may undergo malignant transformation and give rise to adenocarcinoma of the oesophagus in a proportion of patients with it, such that surveillance for malignant change is often undertaken once Barrett's oesophagus is identified. However, most tumours of the oesophagus are squamous cell carcinomas that have arisen in the usual squamous epithelium of the oesophagus. Oesophageal carcinoma

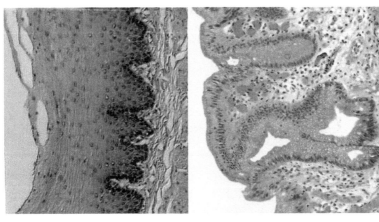

Figure 2.4 Normal squamous epithelium of the oesophagus (left); Barrett's oesophagus (right)

causes progressive dysphagia, starting with difficulty in swallowing solids, then difficulty with swallowing liquids too. A barium swallow X-ray is used to investigate the oesophagus initially, in which barium (which is opaque on X-ray films) is swallowed, and X-rays are taken as it travels through the oesophagus. A narrowing of the oesophagus is seen where there is a tumour (Figure 2.5).

An endoscopy is performed to visualize any oesophageal stricture, and biopsies taken to confirm the diagnosis. CT and MRI scanning can also provide useful information for staging a tumour. Treatment is with surgery to remove a tumour if possible, but in most patients will comprise symptomatic treatment to relieve obstruction (e.g. insertion of a stent to keep the oesophagus open).

2.4.3   The stomach

2.4.3.1   Gastritis

Gastritis is inflammation of the stomach lining, with an inflammatory cell infiltrate seen in the gastric mucosa microscopically. Breaks in the mucosa may be present and these can bleed, sometimes extensively (Figure 2.6).

Most gastritis is caused by chronic infection with *Helicobacter pylori*, a Gram-negative, urease-producing bacterium that produces an acute gastritis that rapidly becomes chronic. Use of nonsteroidal antiinflammatory drugs such as aspirin and ibuprofen, which block prostaglandin production in the gastric mucosa and remove the protective effect of these from the action of gastric acid, is also a common cause. However, alcohol, infections such as the herpes simplex virus, stress and burns can all lead to gastritis. The symptoms and signs of gastritis include heartburn, vomiting and haemorrhage from the gastric mucosa, and an endoscopy (with or without biopsies) may be performed if symptoms merit it to establish a diagnosis, along with specific testing for *Helicobacter*

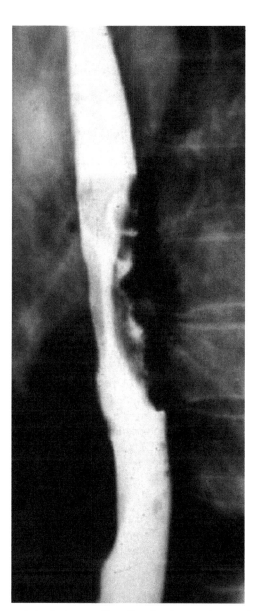

Figure 2.5   Barium swallow film showing a stricture where there is an oesophageal tumour narrowing the lumen. Source: emedicine.medscape.com

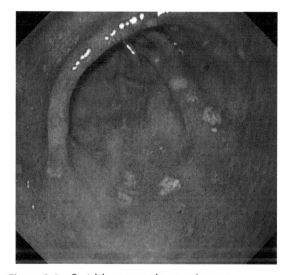

Figure 2.6   Gastritis as seen via an endoscope

infection, such as antibody serology. Treatment of gastritis involves removing identified causes, for example stopping ibuprofen, or eradicating any *Helicobacter* infection with a multidrug regime comprising an antibiotic and a proton pump inhibitor drug. Proton pump inhibitor drugs or H2 receptor antagonists are used in most patients with gastritis to decrease gastric acid production and thus damage to the inflamed mucosa.

2.4.3.2   Peptic ulcers

Peptic ulcers occur where the epithelium of an area of mucosa is eroded; most occur in the stomach (gastric ulcers) or duodenum (duodenal ulcers), and around 75% are associated with *Helicobacter pylori* infection. Peptic ulcers cause pain in the upper abdomen which is usually relieved by antacids, sometimes with nausea and heartburn. Occasionally, peptic ulcers can perforate right through the wall of the stomach or duodenum, or haemorrhage. A gastroscopy will allow visualization of an ulcer and biopsy of areas of interest, and treatment comprises eradicating any *Helicobacter pylori* and suppressing gastric acid, for example with a proton pump inhibitor drug such as omeprazole (Figure 2.7).

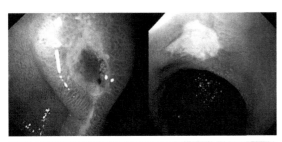

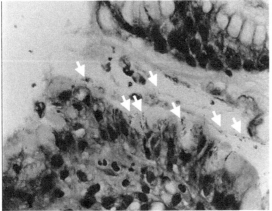

Figure 2.7   Large duodenal (left) and gastric (right) ulcers seen on endoscopy; histology of a gastric ulcer showing the seagull-shaped *Helicobacter* organisms (arrows) on the surface

2.4.3.3   Gastrointestinal bleeding

Bleeding into the gastrointestinal tract can be divided into upper GI bleeding (originating in the oesophagus, stomach or small intestine), and lower GI bleeding (from the colon). Upper GI bleeding is usually due to chronic peptic ulceration, but other common causes are varicose veins in the oesophagus, tears from repeated vomiting (Mallory–Weiss tears), or from tumours in any part of the upper GI tract. Upper GI bleeding causes haematemesis – or vomiting blood, which looks dark due to the action of gastric acid – and melaena, the passing of dark, tarry stools as blood from the upper GI tract is altered during its passage to the rectum. The bleeding usually settles spontaneously, but may require a blood transfusion and an urgent endoscopy to establish the site of bleeding and enable specific treatment of the underlying cause, for example using proton pump inhibitor drugs for peptic ulcers, or coagulation of actively bleeding ulcers by laser.

Bleeding from the lower GI tract is common, and the great majority of cases are due to haemorrhoids (also known as piles) around the anus. Massive bleeding is fortunately rare, and where it does occur is usually due to diverticular disease (see Section 2.4.5) or ischaemia of the colon, where its blood supply is disrupted. However, tumours of the colon, inflammatory bowel disease, polyps and congenital malformation of blood vessels in the colon can all also cause lower GI bleeding. The symptoms are passage of bright red blood from the rectum, sometimes with abdominal pain depending on the cause. Rectal examination and techniques such as a colonoscopy can be used to determine the site of bleeding, though angiography or the use of Technitium-labelled red cells may be required to look for malformed, bleeding blood vessels in the colon. Treatment is of the underlying cause, for example surgical resection of any tumour, with red cell transfusion as needed.

2.4.3.4   Gastric cancer

The incidence of gastric cancer is declining in the UK relative to other cancers, and it is currently the eighth most common cancer in men. Most cases are related to chronic infection with *Helicobacter pylori*, though smoking, alcohol, a diet low in fruit and vegetables and high in salt are also implicated. Most cases are adenocarcinomas, and may be localized ulcerated tumours ('intestinal type') or more extensive tumours ('diffuse type'). Gastric cancer causes very similar symptoms to peptic ulceration, that is upper abdominal pain, along with weight loss, nausea and loss of appetite. Metastatic spread occurs in advanced disease to the liver and peritoneum, with symptoms of ascites and an enlarged liver. A barium meal or gastroscopy with biopsies are used to diagnose gastric cancer, with CT scanning to stage the tumour (i.e. see if and how far it has spread). Treatment is with surgical resection where possible, though overall survival is poor.

2.4.4   Small intestine

The small intestine's major function is nutrient absorption, and to optimize this function its surface area is massively increased by villi and microvilli. Disturbance of small intestinal function causes diarrhoea, abdominal pain, weight loss and steatorrhoea (pale stools that contain fat), and specific deficiencies of nutrients. The commonest causes of malabsorption from the small intestine in the UK are coeliac disease and Crohn's disease.

2.4.4.1   Coeliac disease and malabsorption

Coeliac disease is an immunological condition in which antibodies are formed against the gliadin component of gluten, a protein found in wheat. Further ingestion of gluten once this has occurred causes chronic inflammation in the small intestine, with loss of villi (villous atrophy), and thus a massively decreased absorptive capacity. It is known to run in families, and is strongly associated with inheritance of particular HLA genes: HLA-A1, -B8, -DR3, -DR7 and -DQW2. Coeliac disease diagnosis is often delayed as symptoms are common, nonspecific ones such as tiredness, diarrhoea, anaemia and abdominal pain; the peak ages of incidence are in infancy, and age 30–40 years. The diagnosis is made by finding antigliadin and antiendomysial antibodies in serum, followed by biopsy of the jejunum via endoscopy and confirmation of loss of villi on histology (Figure 2.8). Treatment of coeliac disease is with a lifelong gluten-free diet. There is an increased risk of intestinal lymphoma and small intestinal cancer in coeliac disease.

2.4.4.2   Carcinoid tumours of the small intestine

Carcinoid tumours are rare tumours that produce serotonin (also known as 5-hydroxytryptamine, or 5-HT), and are most usually found in the terminal

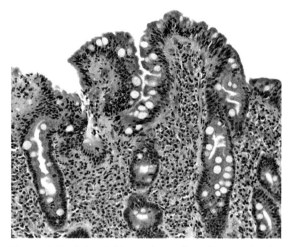

Figure 2.8   Jejunal biopsy in coeliac disease showing villous atrophy

ileum, appendix or rectum. If they metastasize to the liver, serotonin produced by the metastatic carcinoid deposit is able to avoid metabolism and pass directly into the hepatic vein and then into the systemic circulation. This causes 'carcinoid syndrome' of flushing, wheezing, diarrhoea, abdominal pain and fibrosis of cardiac valves on the right side of the heart. It is diagnosed by measuring levels above the normal range of a metabolite of serotonin, 5-hydroxyindoleacetic acid (5-HIAA) in a 24 h urine collection, and liver ultrasound will reveal carcinoid metastases. Treatment is via surgical resection of carcinoid tumour, and the use of serotonin antagonist drugs such as octreotide to improve symptoms (Figure 2.9).

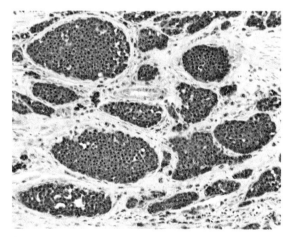

Figure 2.9   A carcinoid tumour seen in islands in the appendix (tumour cells stained using antibodies against chromogranin, a marker they express)

Table 2.3   Comparison of the features of Crohn's disease and ulcerative colitis

Crohn's disease	Ulcerative colitis
More common in smokers	Less common in smokers
Incidence the same for men and women	More common in women
Affects any part of the gastrointestinal tract	Affects just the colon
'Skip' lesions seen: affected segments separated by normal tissue	Lesions are continuous
Cobblestone appearance of mucosa, with fissures and deep ulcers	Red, friable mucosa that bleeds easily
Inflammation extends throughout the entire wall	Inflammation confined to the mucosa
Many granulomata seen	No granulomata. Crypt abscesses seen and depletion of goblet cells

2.4.5   The colon

2.4.5.1   Inflammatory bowel disease

Inflammatory bowel disease (IBD) is increasingly common in the Western world, with a peak incidence at around 20–40 years. It is known to be associated with the inheritance of HLA-B27, and men and women are equally affected. The cause is currently unknown, but there are various hypotheses around the role of infection with mycobacteria and viruses such as the measles virus, and autoimmune processes. There are two main forms of IBD: Crohn's disease and ulcerative colitis, and the features of both are compared in Table 2.3.

The symptoms of Crohn's disease depend on which part of the GI tract is involved, but commonly include diarrhoea, weight loss and abdominal pain. Ulcerative colitis is confined to the colon, and presents with bloody diarrhoea — in a severe attack there is also fever, rapid heart rate, a raised ESR, anaemia and a low serum albumin. Some patients develop 'extraintestinal manifestations' such as arthritis of a single joint, specific skin rashes (e.g. erythema nodosum and pyoderma gangrenosum), liver cirrhosis or renal stones. X-rays with barium contrast may be used to assist with the diagnosis, and will often show ulceration and narrowing in the affected segments. A colonoscopy is used to obtain biopsies for histological diagnosis, which is useful to distinguish between Crohn's disease and ulcerative colitis. Not every case can be given an exact type of IBD, and such cases are designated 'indeterminate colitis'.

The treatment of IBD uses preparatations of 5-acetylsalicylic acid such as mesalazine to induce remission, often alongside corticosteroids (either may be used orally, or via enemas/suppositories). Some patients will require immunosuppressive treatment with drugs such as azathioprine or cyclosporin in addition, and others benefit from the use of an 'elemental diet' of liquid formulations of carbohydrates, fat and amino acids. Surgery may be required to remove affected segments of the gut in patients in whom drug therapy fails, or where there are complications such as abscess formation or perforation of the inflamed gut.

2.4.5.2   Diverticular disease

Diverticular disease is common; around 50% of people over age 50 are affected, and they are associated with a low fibre diet. Diverticulae are outpouchings of mucosa that push through the weak areas of the muscular wall of the colon, forming pockets in which faecal material can become trapped. This may result in diverticulitis, or inflammation of diverticulae with abdominal pain, nausea, fever, though most cases of simple diverticulosis are asymptomatic (Figure 2.10). Treatment of diverticulitis is with resting the bowel (i.e. keeping a patient nil by mouth), and giving intravenous fluids and antibiotics. Surgical intervention is occasionally required in cases that do not settle.

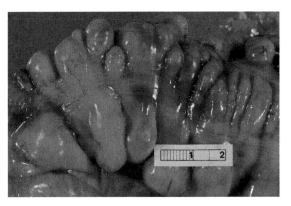

Figure 2.10   Diverticulae in the colon

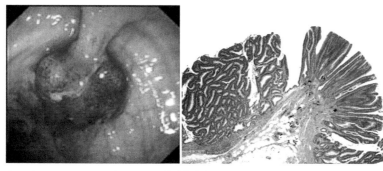

Figure 2.11   Colonic polyp as seen on colonoscopy (left); histology of a colonic adenoma (right)

2.4.5.3   Polyps

Colon polyps are areas of colonic mucosa that are raised above the mucosal surface. They are almost always benign adenomas, and are usually entirely asymptomatic and discovered incidentally (e.g. during a colonoscopy) since they are found in around 10% of the population. However, patients who inherit the autosomal dominant gene for familial polyposis coli will develop thousands of colonic polyps, with an extremely high risk of malignant transformation due to the sheer number of polyps present – in such patients resection of the colon should be performed early to eliminate this risk (Figure 2.11).

2.4.5.4   Colorectal cancer

Colorectal cancer is one of the more common malignancies; it is currently the third most frequent cancer in the UK, and its incidence increases with increasing age. It is associated with a low fibre, high fat diet, and both ulcerative colitis and familial polyposis coli predispose to it. The most common genetic alterations in sporadic bowel cancers are mutations that activate an oncogene known as KRAS, along with the mutation or loss of the tumour suppressor genes APC, SMAD4 and TP53. Most tumours occur in the rectosigmoid area of the colon and spread directly through the bowel wall, and into the surrounding lymphatics, possibly metastasizing to the liver. More than 90% are adenocarcinomas. Symptoms and signs of colorectal cancer include changes in bowel habit, rectal bleeding and abdominal pain, and investigation commonly uses colonoscopy to visualize any masses, and acquire biopsies. X-ray studies with barium contrast may also be used. Treatment is via surgical resection of a tumour where feasible, with rejoining of the two ends of the bowel where possible (where this is not possible, the

Table 2.4   Staging of colorectal cancer

Dukes' Stage	Features
A	Tumour confined to the bowel wall
B	Tumour extending through the bowel wall
C	Regional lymph nodes involved
D	Distant metastases

proximal end of the colon will be brought out onto the abdominal wall in a 'colostomy'). Duke's classification is used to 'stage' colon cancer from histological appearances and information from scans, as shown in Table 2.4 (Figure 2.12 and Figure 2.13).

Following surgical resection of a malignant colorectal tumour, further treatment with chemotherapy or radiotherapy may also be advised, depending on the stage of the tumour, and the patient's age and general health. Some patients may be suitable for therapy with monoclonal antibodies (e.g. Cetuximab) against proteins encoded by mutated oncogenes that give malignant cells a survival advantage, and this improves survival for those whose tumours express the relevant antigen.

2.4.5.5   Appendicitis and peritonitis

Acute appendicitis is common, and is caused by obstruction of the appendix with a small impacted piece of faecal material, triggering inflammation which can lead on to gangrene and perforation of the appendix if untreated. It can be seen in all age groups, and causes acute abdominal pain that begins centrally and moves to the right side, along with loss of appetite and fever. Blood tests show a raised white cell count, and a scan may show the inflamed appendix. Treatment is with surgical removal of the appendix, often done via a laparoscope ('keyhole surgery'). However, the appendix

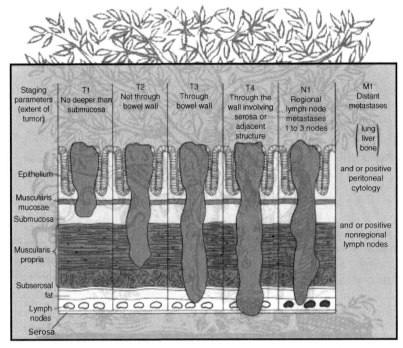

Figure 2.12   Staging of colorectal cancer — layers of the colon wall and the meaning of Dukes' staging. From Abelhoff *et al*, *Abelhoff's Clinical Oncology, Fourth Edition, 2008© Elsevier*

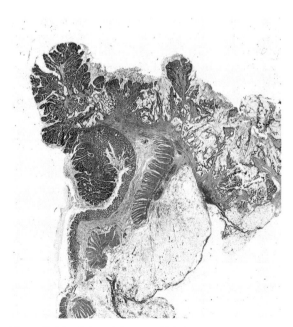

Figure 2.13   Colorectal cancer histology

may perforate and cause a local abscess in the abdomen, or generalized peritonitis.

Peritonitis is a serious condition that follows rupture of an abdominal viscus (e.g. appendix, or perforated gastric ulcer). It causes sudden abdominal pain with shock, and an erect chest X-ray will show a line of air under the diaphragm. Treatment is that of the underlying condition, but peritonitis has a high mortality rate.

2.5   Liver, biliary tract and pancreatic disease

2.5.1   *Symptoms and diagnosis of liver, biliary tract and pancreatic disease*

The presentation of liver disease can vary according to whether patients have acute or chronic liver disease.

Acute liver disease typically gives rapid onset of symptoms such as nausea, tiredness and loss of appetite, with jaundice developing later. Chronic liver disease often has no symptoms, and patients often present at a late stage of the disease with a variety of symptoms such as confusion, drowsiness, collections of fluid in the abdominal cavity (ascites), gastrointestinal bleeding, or jaundice (with which many patients experience an unbearable itching of the skin — pruritis — where bile salts are deposited). Serum for liver function tests (LFTs) is often sent to the clinical biochemistry laboratory to investigate potential liver disease, and serum levels of bilirubin (total and conjugated), serum albumin, and liver enzymes alkaline phosphatase (ALP), γ-glutamyl transpeptidase (GGT), alanine transaminase (ALT), aspartate transaminase (AST) are measured (see Table 2.5 for the functions and diagnostic use of these). Clotting tests — specifically, measurement of prothrombin time — are also often useful to assess liver function as clotting factors are made by the liver and have a short half-life, and thus deranged blood clotting can also be a useful indicator of liver disease.

Ultrasound examination of the liver following confirmation of jaundice with LFTs will show a dilated bile duct if there is obstruction to the biliary tree, and serum viral markers for the hepatitis viruses will be undertaken too in case they are the cause for jaundice. In the presence of a dilated biliary tree, an endoscopic examination called endoscopic retrograde cholangio-pancreatography (ERCP) may be performed in which an endoscope is passed through the stomach and into the duodenum to the opening of the common bile duct (CBD) to inject dye into the biliary tree and visualize it on X-ray. Biopsies can be taken of any obstructing masses to arrive at a diagnosis, gallstones can be removed if obstructing the CBD, or a stent might be inserted into a narrowed bile duct to hold it open.

Table 2.5   Liver enzymes and their significance

Liver function test component		Significance
Bilirubin — total		Bilirubin is derived from breakdown of haem, which in turn comes from breakdown of aging or defective red blood cells, and the liver process and excretes this. Bilirubin is taken up into hepatocytes and conjugated with glucoronic acid to make it water soluble, after which it can be excreted in bile. Elevated levels of total bilirubin cause jaundice, with bilirubin deposited in body tissues causing the characteristic yellowing of skin and mucous membranes.
Bilirubin — conjugated		Assessment of levels of conjugated bilirubin helps to narrow the possible causes of jaundice; if conjugated bilirubin is elevated as well as total bilirubin, conjugation is proceeding normally but the liver is unable to excrete bile effectively. If unconjugated bilirubin is normal but total bilirubin is raised, this indicates a surplus of conjugated bilirubin and indicates a problem in the bilirubin pathway before this point.
Serum albumin	Alb	Albumin is made in the liver and is the major protein found in serum. Levels are decreased in chronic liver disease, some forms of renal disease where it is lost into urine, and in malnutrition. Its half-life is relatively long (20 days), making it a less sensitive marker of hepatic disease compared to clotting factor measurement.
Alkaline phosphatase	ALP	ALP is an enzyme present in the biliary ducts, and levels are increased in biliary duct obstruction (e.g. by gallstones or pancreatic cancer) and by tumours present in the liver. However, levels of ALP are also elevated in patients with Paget's disease of bone, in hyperthyroidism, and in growing children.
γ-glutamyl transpeptidase	GGT	GGT is a more sensitive indicator of obstruction to the outflow of bile than ALP is and will show small elevations even in mild, early liver disease, but is also elevated with chronic alcohol abuse and when patients are on drugs that induce liver enzymes (e.g. the anticonvulsant drug phenytoin).
Alanine transaminase	ALT	ALT is found within hepatocytes, and acute damage to hepatic cells causes it to leak into serum causing dramatic rises (often measured in multiple of the upper limit of normal, or ULN) e.g. in acute viral hepatitis infection.
Aspartate transaminase	AST	Like ALT, AST is found in hepatocytes and will similarly rise rapidly in acute liver injury. However, it is also found in other cells of the body (e.g. red blood cells, the heart and skeletal muscle); rises in AST are thus seen in myocardial infarction and damage to skeletal muscles.

2.5.1.1 Jaundice

Jaundice refers to the yellow discolouration of skin and eyes where serum bilirubin levels are greater than approximately twice the upper limit of the normal range (Figure 2.14).

Bilirubin is mostly formed from the breakdown of red blood cells, and is carried in blood bound to albumin to be conjugated to glucuronic acid in the liver. Conjugated bilirubin is soluble and can be excreted from the liver in bile, where it undergoes conversion to urobilinogen in the terminal ileum of the small intestine. Some urobilinogen is excreted in faeces as stercobilinogen — which give faeces their characteristic brown colour — and some is reabsorbed and excreted by the kidneys (giving urine its characteristic yellow colour). Causes of jaundice are divided into prehepatic causes (i.e. haemolysis), hepatic causes (e.g. liver disease) and posthepatic causes (obstruction to the outflow of bile).

Prehepatic jaundice is caused by increased production of bilirubin where excessive numbers of red blood cells are being destroyed, for example in autoimmune haemolytic anaemia, in red cell membrane defects such as hereditary spherocytosis, in malaria infection, thalassaemia and sickle cell anaemia, and in glucose-6-phosphate dehydrogenase deficiency. Investigations will show haemolytic anaemia along with an increase in unconjugated bilirubin and normal liver enzymes.

Hepatic jaundice is caused by ineffective bilirubin conjugation, for example in liver disease such as cirrhosis, viral or drug-induced hepatitis, liver tumours (primary or secondary), and where the liver is immature in premature and newborn babies. Inherited disorders of bilirubin metabolism can also cause hepatic jaundice. The commonest example is Gilbert's syndrome, which affects around 5% of the UK population and is usually asymptomatic, with a wide range of defects possible in the bilirubin metabolism pathway. It is characterized by an isolated increase in serum bilirubin, with normal liver enzymes and no indication of liver disease.

Posthepatic jaundice is due to obstruction of the outflow of bile from the liver, where there are gallstones in the common bile duct, inflammation along the bile duct (cholangitis), or with compression of the common bile duct by tumours of the biliary tree or in the head of pancreas.

2.5.2 Liver disease

2.5.2.1 Hepatitis

Hepatitis is acute (lasting less than 6 months) or chronic (lasting more than 6 months) inflammation of the liver, with the presence of inflammatory cells; that is, neutrophils or lymphocytes in the tissues, and liver cell necrosis. This causes a rise in AST levels in serum. The liver is large and inflamed, and hepatitis sometimes causes jaundice. It is most often caused by infection with a hepatitis virus, though it can also be caused by the Epstein—Barr virus, cytomegalovirus, autoimmune disease, drugs (paracetamol overdose, or the anaesthetic gas halothane), pregnancy and alcohol. Complete recovery is usual. However, hepatitis can induce massive necrosis of hepatocytes and is occasionally fatal.

In hepatitis, a marked rise in serum AST is seen, and an increased bilirubin where there is jaundice. Full blood count will show an increase in white cells, and serum markers for the hepatitis viruses will be positive where one is the causative agent, for example anti-Hepatitis A Virus (anti-HAV) antibodies. Where autoimmune disease is the cause for hepatitis, circulating autoantibodies will be present, for example antinuclear and antismooth muscle antibodies), with an increased bilirubin, AST and ALT.

Treatment of hepatitis is by treating the cause where possible, immunosuppressant drug treatment for autoimmune disease, interferon-α for Hepatitis B and C, stopping alcohol where this is the likely cause of hepatitis, or withdrawal of drugs known to cause hepatitis. Immunization is available for Hepatitis A and Hepatitis B and this should be given in at-risk groups.

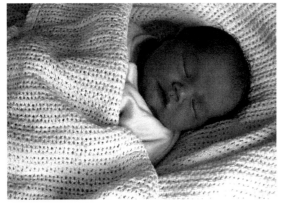

Figure 2.14 Jaundice in a newborn Caucasian baby suffering from ABO incompatibility (see Chapter 8)

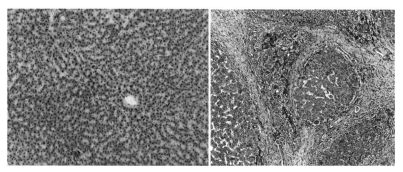

Figure 2.15   Normal liver (left) stained with H&E; cirrhotic liver (right) stained with Masson's trichrome, showing scar tissue replacing normal liver tissue

2.5.2.2   Cirrhosis

Liver cirrhosis results where there has been necrosis of liver cells, and this is replaced by scar tissue (fibrosis). This causes deterioration of liver function and changes in liver architecture (Figure 2.15).

Symptoms and signs likely to be found with cirrhosis include jaundice, an enlarged liver and spleen, spider naevi (see Figure 2.16), ascites (free fluid in the peritoneal cavity) and dilated veins over the abdomen. In the final stages of liver cirrhosis patients become confused and disorientated as toxins usually removed by the liver build to such a level that they begin to cross the blood—brain barrier.

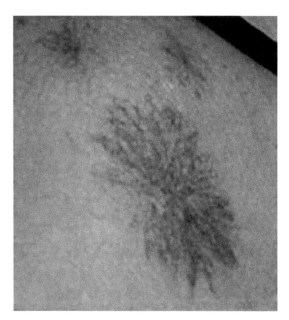

Figure 2.16   Spider naevi on skin

The commonest cause of liver cirrhosis is alcohol in the developed world, but worldwide the commonest cause is Hepatitis B infection. Other causative agents include Hepatitis C, cystic fibrosis, haemochromatosis (a genetic condition in which excess iron is absorbed in the gut and deposited in tissues, including the liver), biliary cirrhosis, autoimmune chronic active hepatitis, and α1-antitrypsin deficiency. The diagnosis may be suspected from elevations of AST and ALT in liver function tests, and a low serum albumin, with a prolonged prothrombin time in clotting tests. Further testing for Hepatitis B and C serology should be done, along with autoantibodies to look for an autoimmune basis for the disease, serum iron, ferritin and total iron binding capacity for haemochromatosis, and serum α1-antitrypsin levels. However, liver cirrhosis is a histological diagnosis and can only be made from biopsy of the liver, with demonstration of the altered architecture typical of cirrhosis.

Cirrhosis is irreversible, though progression can be halted by treating the cause if possible, for example abstaining from alcohol. In end-stage liver cirrhosis, liver transplantation is often advised, and referral to a specialist centre made.

2.5.2.3   Liver tumours

Most liver tumours are metastatic from other areas of the body rather than primary liver tumours. The most common sites to metastasize to the liver are breast, lung and gastrointestinal tract. Primary liver tumours may be benign or malignant, and of malignant tumours, hepatocellular carcinoma (HCC) is the commonest worldwide (though it is relatively rare in developed countries).

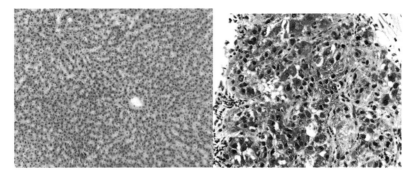

Figure 2.17   Normal liver histology (left); hepatocellular carcinoma histology (right)

2.5.2.4   Hepatocellular carcinoma

Hepatitis B infection is the biggest single predisposing factor to HCC, and liver cirrhosis from any cause also increases the likelihood of HCC developing. Patients with HCC experience loss of appetite, fevers, ascites, weight loss and pain around the area of the liver. Serum levels of α-fetoprotein (a normal antigen produced in the fetus that is reexpressed by HCC cells) are raised, and CT scan or ultrasound of the liver will show a mass. Liver biopsy and histology of the mass will confirm the diagnosis. Surgical excision of the tumour may prolong survival, but in most cases HCC is fatal within months (Figure 2.17).

2.5.3   Disorders of the biliary tract and pancreas

2.5.3.1   Gallstones

Around 20% of the UK population have gallstones. They are more frequently found in women than men, and the old adage was that they mostly occurred in people who were, 'fair, fat, female, fertile and 40'; risk factors for developing gallstones do include obesity and use of the contraceptive pill. Gallstones are either

composed of cholesterol (80% of gallstones are cholesterol stones), or pigment stones composed of bilirubin compounds that are found in people with chronic haemolysis. Most people with gallstones will have no symptoms, though a gallstone can become trapped in the entrance to the gall bladder or in the bile duct and cause severe pain, nausea and jaundice, with fever — this is acute cholecystitis. Ultrasound of the gall bladder will show it to contain stones and be distended. Removal of the gall bladder (cholecystectomy) should be performed some days to months after cholecystitis (Figure 2.18).

2.5.3.2   Pancreatitis

Pancreatitis is inflammation of the pancreas, and causes severe abdominal pain thought to be due to necrosis of pancreatic cells due to autodigestion from release of proteolytic enzymes that should normally be confined to lysosomes within pancreatic cells. Common causes include gallstones, alcohol, hyperlipidaemia and steroid drugs. Pancreatitis may be acute or chronic, and patients with pancreatitis experience severe pain in the upper abdomen, with nausea and vomiting. An elevated level of serum amylase — a digestive enzyme normally produced by the pancreas

Figure 2.18   Gall bladder containing predominantly bile pigment gallstones (left); paler cholesterol gallstones (right)

— assists with the diagnosis, and CT or ultrasound scans show an enlarged pancreas. Treatment uses analgesia and resting the gut by keeping patients nil by mouth, with a nasogastric tube to remove gastric secretions. However, acute pancreatitis can be severe, and may trigger multiorgan failure and even death in up to 50% of the most serious cases. Patients who recover may also go on to have further attacks of pancreatitis, particularly where their causative factor (e.g. alcohol) continues.

2.5.3.3 Pancreatic cancer

Pancreatic cancer is increasing in incidence in developing countries, and is now the 11th most common cancer in the UK, but the fourth cause of cancer deaths. It is more frequent in older people (two thirds of all cases are diagnosed in people over the age of 70) and occurs more often in men than in women. Risk factors for developing pancreatic cancer include smoking, alcohol, diabetes, pancreatitis, being overweight, and a family history of pancreatic cancer. Most cases affect the head of the pancreas, and cause a painless jaundice as the tumour presses on the bile duct and obstructs it (Figure 2.19).

Abdominal pain is common, and diabetes may ensue as functional pancreatic tissue is lost to tumour. The diagnosis is made with CT, ultrasound scanning, and endoscopic retrograde cholangiopancreatography (ERCP), in which an endoscope is passed through the stomach and into the duodenum, and dye injected into the bile duct, with biopsies taken of the mass if it is accessible. Most cases are adenocarcinomas, and many express the tumour marker CA 19-9, which can be monitored during treatment. In most cases, surgical excision of pancreatic cancer is not possible, and treatment with chemotherapy and radiotherapy is recommended where patients are fit enough. A stent is often placed in the compressed bile duct to hold it open, and this relieves jaundice and associated symptoms (e.g. severe itching) for most patients. Pancreatic cancer sadly has a very low 5 year survival rate; most cases are diagnosed late as the early stages cause no symptoms, and by this stage most cases are inoperable.

2.5.3.4 Endocrine tumours of the pancreas

Some tumours that arise in the pancreas are derived from hormone-producing cells, and usually secrete one hormone that gives rise to specific symptoms,

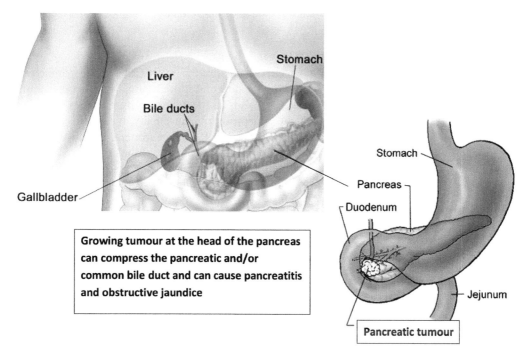

Figure 2.19   Cancer of the head of the pancreas, and its relationship to the bile duct

and can be measured to assist with diagnosis. Gastrinomas occur in G cells of the pancreas and produce large quantities of gastrin, causing development of large gastric ulcers. Vipomas secrete vasoactive intestinal peptide, which produces watery diarrhoea. Glucagonomas produce glucagon, and are derived from the α cells of the pancreas – this produces diabetes mellitus and a rash. Treatment of these endocrine tumours is with surgical removal of the tumour where possible, and treatment of the symptoms, for example the proton pump inhibitor drug omeprazole for gastrinomas.

2.6 Rheumatology

2.6.1 Symptoms of rheumatological disease

Problems with the musculoskeletal system are extremely common in the general population, and account for a large proportion of consultations with general practitioners. Patients with rheumatological problems may experience pain in one or more joints (arthralgia), which may arise from the joint itself, or from tissues surrounding the joint (e.g. tendons, ligaments or the joint capsule). The term arthritis is used to describe an abnormal joint: one which is swollen, deformed and may contain extra fluid – effusion – in the joint. However, it is important to realize that the underlying defect in patients with rheumatological disease is an autoimmune condition in which the immune system is erroneously attacking different tissues, and rheumatological conditions can thus manifest themselves in any organ of the body. This is outlined in the most common presentation patterns for different rheumatological disease below.

2.6.2 Arthritis

Arthritis can be subdivided into osteoarthritis and rheumatoid arthritis.

2.6.2.1 Osteoarthritis

People with osteoarthritis experience pain in joints which is increased by movement, and relieved by resting, often with increased stiffness in joints first thing in the morning. It is caused by a gradual destruction of the articular cartilage that normally lines a joint surface (i.e. the cartilage would normally prevent one bone from grinding along the surface of another bone in a joint). The underlying bone is exposed and becomes thickened, with cysts and an increased proportion of blood vessels developing. As new cartilage growth is attempted and calcifies, small outgrowths of bone called 'osteophytes' appear at the articular surface, causing further pain. The distal interphalangeal joints – the joints at the end of fingers – are most commonly affected. X-rays will show a narrowed joint space and the presence of osteophytes, with sclerosis (thickened bone) and cysts in the bones. Blood tests are typically normal. Treatment consists of advising overweight patients to lose weight, and using nonsteroidal antiinflammatory drugs (e.g. aspirin or ibuprofen) to control symptoms. Some patients will require surgery to replace badly damaged joints.

Osteoarthritis is found more often in women than men, and occurs more commonly with age: 50% of those older than 60 have osteoarthritic changes on X-ray, but not all of these will have symptoms. It may be secondary to previous trauma to, or congenital malformation of, a joint, but most cases are primary osteoarthritis, and biochemical abnormalities in cartilage may underlie this.

2.6.2.2 Rheumatoid arthritis

Rheumatoid arthritis (RA) causes symmetrical pain, swelling and stiffness, usually in the small joints of the hands and feet first. As it progresses, it leads to joint deformity and instability, and spreads to other joints of the body. RA differs from osteoarthritis in the pattern of joints involved, as shown in Figure 2.20, and in the involvement of other parts of the body other than joints in the illness.

RA tends to occur at a younger age than osteoarthritis (with a peak in the fourth decade of life), and affects women three times more than men, in common with other autoimmune conditions. It is a disorder of the synovial membrane that lines a joint cavity, in which an autoimmune process causes its infiltration by macrophages, lymphocytes and plasma cells. The inflamed synovium grows over the articular cartilage surface in a thick 'pannus', destroying the articular cartilage and underlying bone. Apart from the joint involvement in rheumatoid arthritis, subcutaneous

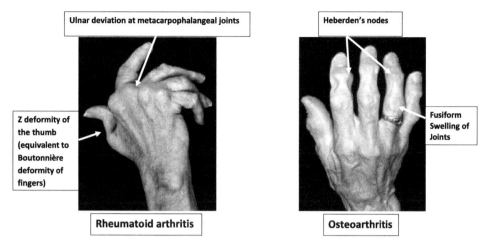

Figure 2.20 Patterns of joint involvement in the hands in rheumatoid arthritis (left) and osteoarthritis (right)

nodules appear in the skin too — these contain granulomata, and are also found in the lungs and heart. Muscle wasting may also be seen, along with fever, fatigue, swollen lymph nodes, anaemia, pleural effusions and pericarditis: RA is a multisystem disorder, and the immune system can turn against several tissues as part of the disease process. Blood tests will often show a normocytic, normochromic anaemia (the 'anaemia of chronic disease'), an ESR and CRP that are raised in proportion to the disease activity, and rheumatoid factor is positive in around 80% of patients. A newer test used in RA is anticitrullinated cyclic peptide (antiCCP) antibody titre measurement — a high titre antiCCP is highly sensitive for RA, and antiCCP antibodies can be found several years before RA becomes clinically apparent. X-rays of affected joints will show a narrowed joint space, bone erosion and cysts in bones at the joint surface.

RA can be difficult to manage and requires the input of several specialist medical teams. Nonsteroidal antiinflammatory drugs are often used to manage pain and swelling, but drugs to modify the disease process ('disease-modifying antirheumatic drugs', DMARDs, such as methotrexate or sulphasalazine) should be started as soon as possible to prevent joint erosion, deformity and loss of function ensuing as RA progresses. These drugs can have serious side-effects including bone marrow failure, and the use must be monitored as appropriate for the drug, for example with regular full blood counts for patients on methotrexate. Corticosteroids can also be used for RA flares,

either orally or injected into affected joints. The latest treatments for RA for patients who have not been successfully treated with at least two different DMARDs use injections of antiTNF antibodies (e.g. with adalimumab or etanercept), often with dramatic improvement.

2.6.2.3 Infective arthritis

Fortunately, infections of joints are uncommon, but when they do occur, joint destruction may ensue within hours. Bacteria, viruses and fungi can all cause infective arthritis. Septic arthritis is said to occur when joints are infected with pus-producing organisms, particularly with *Staphylococcus aureus*, which reaches joints following local spread from infected tissues, from trauma, or from spread via the bloodstream from a distant site. Tuberculous arthritis may occur in a small proportion of patients with TB, and usually affects the spine; meningococcal arthritis can occur in meningococcal septicaemia, or gonococcal arthritis in patients with genital or oral infections with gonorrhoea. Septic arthritis leads to a red, hot, swollen and painful joint with or without fever. Fluid is aspirated from the joint to arrive at the diagnosis, and in septic arthritis there is pus in the joint fluid ($>50\,000 \times 10^6$/L white blood cells). Gram staining can be performed on the fluid to illustrate the presence of bacteria, and culture of the fluid will identify the organism. Blood cultures may similarly identify the infective organism. Immediate treatment with appropriate intravenous antibiotics should be started to minimize joint destruction.

2.6.3 Systemic lupus erythematosus

Systemic lupus erythematosus (SLE) is the commonest form of connective tissue disorders that cause arthritis, along with immune complex deposition in tissues, and vasculitis (inflammation of blood vessels). It is a disorder of multiple body systems and can present in different ways in different patients; it mainly affects younger females, and is seen far more frequently in people of Afro-Carribean descent. It is believed to be largely genetically inherited, but female hormones and some viral infections (particularly with the Epstein–Barr virus) have also been implicated. SLE is characterized by the development of antinuclear antibodies, and may be associated with a defect in apoptosis that causes intracellular components to be presented to lymphocytes as if they are foreign antigens, triggering the development of antibodies against components of 'self' cells. The presentation of SLE often involves an assymetrical distribution of painful joints which changes to different joints ('migratory arthralgia'), sometimes with fever, fatigue and depression. Some patients develop a characteristic 'butterfly' facial rash. Pleurisy, renal disease and involvement of the heart (with aortic valve lesions, or inflammation of the pericardium and/or endocardium) are also common in SLE, but other body systems can also be involved. A full blood count often shows a normochromic, normocytic anaemia (commonly with decreased neutrophil and platelet counts) with raised ESR, and the presence of antinuclear antibodies is found. Futher antibodies against intracellular materials such as double-stranded DNA (dsDNA) and phospholipids may also be identified in patients with SLE. Biopsies of affected tissues will show vasculitis and the presence of granulomata.

Treatment of SLE depends on its severity, and the organs affected. Nonsteroidal antiinflammatory drugs may be used for mild joint pain, but most patients will require treatment with corticosteroids, and some with disease-modifying or immunosuppressant drugs such as hydroxychloroquine or azathioprine, often alongside corticosteroids.

2.6.4 Crystal deposition diseases

2.6.4.1 Gout

Gout is the commonest crystal deposition disease, and occurs when abnormal uric acid metabolism leads to deposition of sodium urate crystals in joints, soft tissues and the urinary tract. It is 10 times more common in men than in women. The underlying problem is hyperuricaemia, or increased levels of uric acid, due to its overproduction, or decreased renal excretion. The latter is more commonly responsible, and may be seen in chronic renal disease, with high blood pressure, with alcoholism or with the use of thiazide diuretic drugs. Overproduction of uric acid may occur as a result of increased purine production, since urate is a product of purine metabolism; purines are increased in myeloproliferative diseases such as leukaemia or polycythaemia rubra vera, psoriasis or carcinoma. Acute gout causes a sudden onset of severe pain, redness and swelling in a joint, usually the joint in the big toe. It may affect multiple joints, and crystals may also be deposited in cartilage, most notably around the ear. Examination of aspirated synovial fluid from infected joints shows the long, pointed urate crystals, and treatment in the short term is with nonsteroidal antiinflammatory drugs, injection of corticosteroids into affected joints. In the longer term, patients who are overweight should be advised to lose weight, alcohol consumption should be decreased; drugs to inhibit purine breakdown (e.g. allopurinol) or drugs to promote the renal excretion of uric acid (e. g. probenecid) can also be used.

2.6.4.2 Pseudogout – pyrophosphate arthropathy

Calcium pyrophosphate crystals can also be deposited in tissues, mainly in articular cartilage of the knee, causing pseudogout. The cause is unknown, but it tends to occur in older people with equal incidence in men and women. It may follow primary hypoparathyroidism, gout, haemochromatosis or hypothyroidism. In pseudogout, microscopy of aspirated synovial fluid shows small, rectangular crystals of calcium pyrophosphate (which show birefringence under polarized light – see Chapter 4). Treatment is with joint aspiration, nonsteroidal antiinflammatory drugs and corticosteroids.

2.6.5   Bone disease

2.6.5.1   Osteomalacia

Osteomalacia occurs due to a lack of sufficient calcium and phosphate mineralization of bone following closure of epiphyseal plates (where this occurs prior to closure of epiphyseal plates, rickets — bowing of the bones of the lower leg — is seen instead). It is often due to vitamin D deficiency, from poor diet $+/-$ lack of sun exposure. However, it may be due to renal disease that results in poor production of active vitamin D from its precursors, from malabsorption of vitamin D from the diet (e.g. in coeliac disease) or from hepatic failure. Osteomalacia leads to bone fractures and muscle and bone pain. Low levels of calcium and phosphate are seen in serum, along with an increased level of alkaline phosphatase, and X-rays identify the poorly mineralized bones. Bone biopsy and histology will give a definitive diagnosis, and treatment is with oral vitamin D supplementation.

2.6.5.2   Osteoporosis

Osteoporosis differs from osteomalacia as it is a deficiency of all components of bone, that is both matrix and minerals. It is more common in women (particularly after the menopause, and with other causes of oestrogen deficiency such as ovary removal), and associated with smoking, lack of exercise, people with a low BMI, excess alcohol consumption, corticosteroid use, rheumatoid arthritis, renal failure, and where there is a family history of osteoporosis. It leads to an increased risk of fractures, usually of the vertebrae, wrist and neck of femur. Blood tests (including calcium, phosphate and alkaline phosphatase levels) are normal, but bone densitometry scanning will identify 'thinned' osteoporotic bones. Prevention is the ideal in people at risk; oestrogen therapy will reduce bone loss following the menopause, with attention to sufficient calcium intake. Bisphosphonate drugs reduce bone loss and may be prescribed alongside calcium supplements.

2.6.5.3   Paget's disease

Paget's disease is characterized by overproduction of weak, pathological new bone following overabundant osteoclast and osteoblast activity. Viruses may play a role in its causation, since viral inclusions can be seen in the osteoclasts of Paget's disease, and it is predomi-
nantly a disease of the elderly. Paget's disease mainly affects the pelvis, skull, tibia, spine and femur, leading to bone pain and deformities (e.g. bowed legs, or misshapen skull), sometimes with fractures and, rarely, bone tumours. A significantly raised alkaline phosphatase level may be found and X-rays of affected bones show areas of bone lysis and thickening. Paget's disease is treated, once symptomatic, with bisphosphonate drugs and calcitonin to decrease bone turnover.

2.7   Urinary tract disease

2.7.1   Symptoms and investigation of urinary tract disease

Disorders of the renal tract commonly cause problems or changes to the pattern of urination (e.g. dysuria, or painful urination, polyuria, frequent urination, nocturia, needing to urinate at night, anuria, passing no urine, or haematuria, passing urine with blood in it). Pain may be experienced from the kidney or bladder regions, and renal disease may cause systemic symptoms and signs such as hypertension, fatigue, an increased serum creatinine and urea, or protein in urine (proteinuria).

Renal disease is most commonly investigated initially by simple dipstick testing of urine and comparison of the colour of each test strip with a chart. This enables rapid assessment of the presence or absence of a wide range of analytes in urine, such as blood, protein, glucose, ketones, bilirubin and white blood cells, and the estimation of the specific gravity of urine and its pH (Figure 2.21).

The presence of glucose in urine may indicate diabetes mellitus, while bilirubin signifies jaundice. White cells imply a urinary tract infection, and urine can be cultured to identify any organisms present. Protein detected in urine (proteinuria) is sometimes found in small amounts during episodes of fever and after exercise, but significant proteinuria seen on repeated urine samples should be investigated further by organizing a collection of all urine voided over 24 h for measurement of the total amount of protein excreted in one day. More than 2 g of protein in urine in 24 h implies renal disease. Haematuria may be macroscopic (blood in urine is visible with the naked eye) or microscopic (urine looks normal, but dipstick testing reveals the presence of haemoglobin). If microscopic

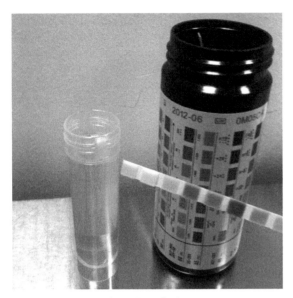

Figure 2.21 Dipstick testing of urine

haematuria is detected, microscopy of urine should be undertaken to look for red cells, and for red cell 'casts' (which indicates bleeding within the glomeruli of the nephron). Further testing might include looking at the cells present in urine — urine cytology — and radiographic studies of the urinary tract. A cystoscopy could also be performed, in which a flexible camera is inserted into the urethra to visualize the bladder, and biopsies of any abnormal tissue identified can be taken.

2.7.2 Urinary tract infection

Urinary tract infection (UTI) is common in females, who have a far shorter urethra than males, allowing bacteria a quicker route into the bladder where they can multiply. Many bacteria that cause UTI are derived from the bowel, and the proximity of the urethra and anus in the female particularly predisposes them to UTI. The commonest organism cultured is *Escherichia coli*, but skin organisms such as *Staphylococcus saprophyticus* and *Staphylococcus epidermidis* also account for some 20% of UTIs in women. UTIs acquired in hospital are often caused by *Klebsiella aerogenes* or *Enterococci* species; catheterization of the bladder, bladder stones and obstruction of the urinary tract all make UTI more likely. Tuberculosis can occasionally cause UTI, usually in immigrant populations.

Patients with a UTI experience urination that is painful (dysuria) and frequent, and may have haematuria. Pain over the region of the bladder is usually experienced, and patients may complain that their urine smells offensive. If infection ascends from the bladder up the ureter(s) to the kidney, pyelonephritis results, and patients tend to become systemically unwell with fever and loin pain in addition to symptoms of lower UTI. Patients with UTI should have urine cultured, and finding >100 000 of one bacterial species per ml of urine is diagnostic. Appropriate oral antibiotics that are renally excreted to concentrate in the bladder are prescribed, such as trimethoprim or nitrofurantoin. In patients with pyelonephritis, intravenous antibiotics may be required.

2.7.3 Autoimmune disease and the kidney

2.7.3.1 Glomerulonephritis

Glomerulonephritis is a group of disorders in which the immune system causes damage to the glomerulus of the renal nephron. Circulating antigens (e.g. DNA in systemic lupus erythematosus) become trapped in the glomerulus and antibody—antigen complexes begin to form, or antigen—antibody complexes present in the circulation may be deposited in the glomerulus. In many patients, the causative antigen is never discovered, though some may be produced by drugs (e.g. penicillamine), bacteria (e.g. *Streptococcus viridans*), viruses (e.g. mumps) and fungi, as well as by innate antigens. In other patients, glomerulonephritis is caused by the presence of antiglomerular basement membrane (antiGBM) antibodies that react against the glomerular basement membrane to directly damage it. Once immune complexes are present in the glomerulus, they trigger inflammation and the appearances seen under the light microscope after renal biopsy are used to classify the resulting glomerulonephritis, which may affect the entire kidney ('diffuse' glomerulonephritis), be limited to part of the kidney ('focal' glomerulonephritis), or be seen to affect certain parts or all of the glomeruli ('segmental' or 'global' glomerulonephritis).

Glomerulonephritis may present in different ways, with asymptomatic proteinuria and haematuria right up to renal failure. Some patients develop 'acute nephritic syndrome' of oedema and hypertension along with proteinuria and haematuria, or 'nephrotic syndrome' in which there is heavy proteinuria, a fall in plasma albumin and oedema. It is treated according to the clinical picture; in some patients, a full recovery is achieved with supportive measures and careful assessment of fluid balance with daily weighing and measurement of fluid input and output. Hypertension is treated with salt restriction, diuretics and vasodilators, and in some patients, immunosuppressant drugs such as azathioprine or prednisolone may be required. Renal failure is managed as detailed in Section 2.7.7.

2.7.3.2   Tubulointerstitial nephritis

Tubulointerstitial nephritis is inflammation of the supporting cells around the renal tubules, with tubular damage, and is usually due to hypersensitivity to drugs such as penicillin, ibuprofen, furosemide, phenytoin and rifampicin. It causes acute renal failure along with fever, a rash, an increase in eosinophils in the blood, and joint pain, and renal biopsy will demonstrate eosinophil infiltration amid the interstitial cells. Treatment is by stopping the offending drug and managing the renal failure (see Section 2.7.7), often alongside corticosteroid drugs, and tubulointerstitial nephritis will often resolve well.

2.7.4   Hypertension and the kidney

Kidney disease can cause dramatic hypertension, but equally hypertension can trigger renal disease, and distinguishing the correct order of events is not always straightforward. Essential hypertension — hypertension with no identifiable cause — leads to thickening of the walls of renal blood vessels and a shrinkage in kidney size, often to the detriment of renal function. Very sudden, severe hypertension can trigger necrosis and fibrin deposition in the glomerular arterioles, causing renal damage with subsequent worsening hypertension and further renal damage.

Secondary hypertension is due to bilateral renal disease in 80% of cases, and glomerulonephritis, chronic pyelonephritis, polycystic kidney disease and renal artery stenosis account for most of this. Hypertension is triggered by activation of the renin—angiotensin—aldosterone system and retention of salt and water in diseased kidneys. Some cases of secondary hypertension may also occur due to unilateral kidney disease in the case of renal artery stenosis, where a poor blood supply from a constricted renal artery triggers an increased renin release from that kidney, with a subsequent rise in angiotensin II, and thus blood pressure. Renal artery stenosis can be demonstrated using arteriography of renal vessels with injected contrast, and is usually treated using antihypertensive drugs (except ACE inhibitors, which are unsafe in renal artery stenosis). Some may be treated surgically by dilating the narrowed renal artery.

2.7.5   Renal stones

Renal stones are relatively common in the UK population, and affect males twice as often as females. Most (80%) comprise calcium oxalate and calcium phosphate compounds in patients with unexplained high levels of calcium secretion in urine (idiopathic hypercalciuria), though some comprise magnesium ammonium phosphate (often following urinary tract infection with organisms that produce urease, such as *Proteus* and *Klebsiella* species), and others are formed from uric acid or cystine. Stones in the urinary tract can cause no symptoms, but can cause excruciating loin or bladder pain (depending on their location), haematuria, and infections and obstruction of the urinary tract. They are often visible on a plain X-ray of the abdomen, and further blood tests can be performed to pinpoint causative factors such as serum calcium, uric acid and bicarbonate levels, with chemical analysis of any stone that is passed. A 24 h urine collection can be analysed for calcium, oxalate and uric acid output. Treatment is with strong analgesia. Stones of <5 mm diameter will often pass spontaneously, while larger stones can be broken down with shock waves (lithotripsy) or removed via cystoscopy or open surgery. Any underlying conditions that predispose to the development of renal stones should be treated, and patients encouraged to drink enough to stay well hydrated.

2.7.6   Cystic renal disease

The presence of one or more renal cysts is extremely common in the general population, becoming more so with increasing age in any one person, and is usually of no significance. However, adult polycystic disease (APCD) is an autosomal dominant genetic condition in which both kidneys develop numerous cysts that compress renal tissue and cause progressive loss of renal function in enlarging kidneys. This causes loin pain, haematuria in some, and development of hypertension and rising urea levels in serum. Some also develop cysts in other organs, and Berry aneurysms in the brain (see Section 2.8.10). Renal ultrasound confirms the diagnosis, and treatment is with control of hypertension. Most patients with APCD will eventually require dialysis or kidney transplantation, and other first-degree relatives should be screened for the disease.

2.7.7   Renal failure

Renal failure is characterized by a fall in glomerular filtration rate (GFR, see Chapter 7) and a failure to maintain renal excretion of waste products. This is seen alongside variable levels of failure of erythropoietin secretion, lack of vitamin D hydroxylation, problems with acid–base balance, salt and water balance, and blood pressure. Acute renal failure is a rapid decline in renal function over days to weeks, and is often reversible. It may be caused by 'prerenal' problems that lead to poor perfusion of the kidneys with blood (e.g. haemorrhage, burns, diabetic ketoacidosis, sepsis, myocardial infarction, renal artery obstruction), renal problems (e.g. glomerulonephritis, malignant hypertension, interstitial nephritis, nephrotoxins such as the antibiotic gentamicin) or 'postrenal' problems (obstruction to urine flow anywhere along the pathway). Acute renal failure is often asymptomatic at the outset, and then polyuria or oliguria (large or small urine volumes produced, respectively) is commonly seen, with weakness, nausea, vomiting and confusion as urea levels rise in serum. Symptoms of anaemia (pallor, fainting, shortness of breath) may be experienced if erythropoietin excretion deteriorates.

Investigation of acute renal failure aims to give an accurate assessment of renal function at that point in time, and also to identify the underlying cause quickly to attempt to reverse it. A full history and examination is extremely important. A full blood count, urine dipstick testing and urine microscopy are performed, along with measurement of urinary electrolytes, then imaging of the kidneys with ultrasound and/or CT scanning. Renal biopsy may be undertaken if the cause is still unexplained and the kidneys are of normal dimensions. Other investigations that may be done in some patients include

serum autoantibodies, serology for Hepatitis B and C, and HIV, blood cultures and serum protein electrophoresis (to exclude multiple myeloma – see Chapter 8). Acute renal failure should be treated by reversing or removing identified problems, for example restoring fluid volume after haemorrhage, stopping nephrotoxic drugs, treating hypertension or relieving obstructions to urine flow. Fluid balance must be carefully monitored and urine volume, serum urea and electrolytes measured, and the potential need for dialysis considered. Patients with severe hyperkalaemia – a serum $K^+ > 6.5\,\text{mmol/L}$), acidosis, rising urea levels or pulmonary oedema will require dialysis. This can be haemodialysis, haemofiltration or via peritoneal dialysis. However, most patients will not require dialysis and will recover from acute renal failure in 2–3 weeks.

Chronic renal failure is mainly caused by glomerulonephritis, chronic pyelonephritis and diabetes mellitus in Europe, but renal vascular disease, polycystic kidneys, genetic disease, drugs, amyloidosis and systemic lupus erythematosus may also be responsible. The symptoms, signs and investigation are similar to those for acute renal failure, though in chronic renal failure there is polyuria and nocturia as the kidney ceases being able to concentrate urine, and anaemia is a particular feature. Impaired ability of the kidney to produce the active form of vitamin D causes a variety of bone diseases such as osteomalacia and osteoporosis (see Section 2.6.5), and heart disease is common due to hypertension, abnormal lipid metabolism and vascular calcification in the presence of chronic renal failure. As with acute renal failure, the underlying cause should be addressed, and a low protein diet and good blood pressure control instigated. Recombinant human erythropoietin can be given for anaemia, and oral sodium bicarbonate used to treat acidosis. However, most patients with chronic renal failure will continue to lose renal function and ultimately will require dialysis.

2.7.7.1   Dialysis and transplantation for renal failure

The toxins normally excreted by a healthy kidney can be removed from the blood by passing it along semipermeable membrane with dialysis fluid on the other side of the membrane: the principles of osmosis cause urea, creatinine and other waste products to move out of the blood and into the dialysis fluid along the gradient from high to low concentration, thus removing them from blood. Dialysis can be achieved via the process of haemodialysis, in which a patient's blood is extracted from an arm, and dialysis of the blood within a haemodialysis machine, with the dialysed blood returned to the patient's body. This usually takes 4–5 h, three times a week for adult patients, and requires the creation of an arteriovenous fistula (a surgically created connection between an artery and a vein) in the arm. However, chronic ambulatory peritoneal dialysis has now been developed, in which patients infuse dialysis fluid into their own peritoneal cavity — which has a rich, superficial vascular network — and toxins diffuse out of the bloodstream and into the fluid, which is then drained away from the peritoneal cavity. This can often be done overnight as the patient sleeps, but requires the surgical insertion of a permanent catheter into the peritoneal cavity (Figure 2.22). Although peritoneal dialysis allows patients considerably more freedom, potentially fatal infection of the peritoneal cavity is an ever-present danger. Amyloidosis (see Chapter 4) is also a possible complication of renal dialysis, as the amyloid protein is not removed by dialysis membranes.

Renal transplantation offers the only hope of a cure from complete renal failure and survival of a transplanted kidney can be as high as 80% at 10 years posttransplant, though many patients will require further renal transplantation once the initial transplanted kidney fails. Healthy kidneys from organ donors — living relatives in many cases, or organs donated from the victims of serious accidents — are

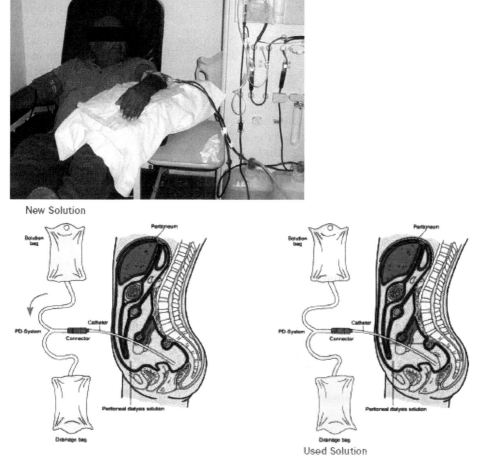

Figure 2.22   Haemodialysis (top) and peritoneal dialysis (bottom) for renal failure

used; the donors must be ABO compatible and a good HLA match with the recipient. Long-term treatment with immunosuppressive drugs such as cyclosporin is used to improve the survival of the transplanted organ, and reduce the chances of rejection.

2.7.8   Tumours of kidney and bladder

2.7.8.1   Renal cell carcinoma

The commonest tumour of the kidney in adults is renal cell carcinoma, which arises from the epithelium of the renal tubule (Figure 2.23). It occurs in males twice as frequently as in females. Most present in patients in their 40s, with haematuria, pain in the loin area, sometimes a palpable mass, and may cause weight loss, fever and polycythaemia. Renal cell carcinoma metastasizes early to the lung, bone and liver. CT, MRI and ultrasound scanning are useful in confirming the diagnosis, and the affected kidney is removed surgically (nephrectomy). Prognosis is good if the tumour has not metastasized.

2.7.8.2   Urothelial tumours

Tumours can arise from the urothelium that lines any part of the urinary tract, including the renal pelvis, ureters, bladder and urethra, though the bladder is most commonly affected. Most occur after the fourth decade of life, and males are affected four times as often as females. Industrial solvents, cigarette smok-

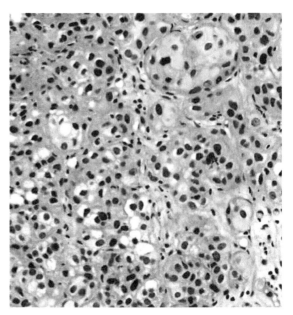

Figure 2.24   Urothelial bladder tumour histology

ing, chronic inflammation from infection and certain drugs (e.g. phenacetin) are known to increase the risk of urothelial tumours, which cause painless haematuria in most patients. Urine cytology is used to demonstrate the presence of malignant cells, and cystoscopy of the bladder will enable examination and biopsy of suspicious lesions in the bladder. Urothelial tumours should be surgically removed and chemotherapy and/or radiotherapy may be recommended (Figure 2.24).

2.7.9   Diseases of the prostate gland

2.7.9.1   Benign prostatic hypertrophy

The prostate gland is found only in males, and benign enlargement of the prostate due to hyperplasia is extremely common in older men. This causes frequent urination with dribbling following urination, and difficulty initiating the flow of urine, with nocturia in many cases. A smooth, enlarged prostate gland can be felt on rectal examination. Serum prostate specific antigen (PSA) levels should be performed to rule out prostate cancer, which has a very similar set of symptoms. Treatment with drugs can be tried initially: α-reductase inhibitors prevent conversion of testosterone to dihydrotestosterone and can decrease the size of

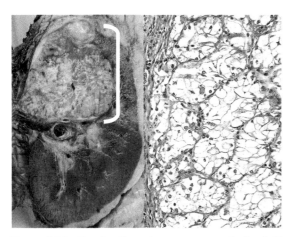

Figure 2.23   Renal cell carcinoma macroscopically (left) and histologically (right)

the prostate gland, but many patients will require surgical resection of the prostate gland.

2.7.9.2 Prostate cancer

The incidence of prostate cancer rises with increasing age in men, and malignant changes are present in 80% of males over 80; however, most of these foci of prostate cancer will remain dormant. Presentation is very similar to benign prostatic hypertrophy in terms of difficulty and frequency of urination, but patients with prostate cancer may present with metastases, particularly to bone. A hard, irregular gland is felt on rectal examination, and measurement of serum PSA will reveal an increased level (and one that is rising, if serial measurements are taken). Small tumours may be managed with watching and waiting, while radical removal of the prostate gland or radiotherapy may be used in other patients. Metastases can be treated with antiandrogen drugs (to remove the stimulus of testosterone to the tumour) such as cyproterone acetate, or synthetic luteinizing hormone analogues to cause negative feedback to the pituitary gland along this pathway.

2.8 Cardiovascular disease

2.8.1 Symptoms of cardiovascular disease

Patients suffering from cardiovascular disease typically present with signs of the struggling system. Chest pain may occur when the heart muscle, the myocardium, is deprived of oxygen and nutrients due to diseased coronary arteries in angina, myocardial infarction, or a lack of blood reaching the coronary arteries in heart valve disease. In some cases chest pain may be experienced due to inflammatory processes of the pericardial membrane (e.g. following some viral infections). Patients may experience palpitations where there are abnormalities in the electrical conduction pathway of the heart or the heart's rhythm, which can in turn lead to fainting and periods of unconsciousness. Some patients experience heart failure, where the heart can no longer fulfil the body's requirements for pumping blood; if the left side of the heart is failing this causes a pooling of blood travelling towards the left side of the heart, causing swelling most noticeable in the ankles. If the right side of the heart is failing, a backlog of blood in the lungs causes pulmonary oedema there, with the excess fluid in the lungs leading to shortness of breath and coughing. Disorders of blood pressure, particularly hypertension (high blood pressure), are common in the general population but are often symptomless. In some cases hypertension causes headaches, and later it may damage other organs (e.g. the kidneys) and generate symptoms subsequent to that.

2.8.2 Investigation of cardiovascular disease

Taking a careful history from the patient with suspected cardiovascular disease will often yield a good idea of the diagnosis, as will a careful examination of the cardiovascular system, including measurements of pulse and blood pressure. A chest X-ray can give important clues about the health of the cardiovascular system, particularly in terms of the size of the heart and in looking for signs of pulmonary oedema that might indicate right-sided heart failure. Echocardiography may be performed, where the heart is visualized in motion using Doppler ultrasound – this can be done from the oesophagus (transoesophageal echocardiography) to get close to the heart for the most detailed information, and this is particularly useful when visualizing heart valves and the septa between the chambers of the heart (Figure 2.25).

Many patients will have an electrocardiogram (known as the ECG in the UK, but EKG in several other countries), which is a recording of the electrical pathway of the heart made using leads across the front of the chest, and from the upper and lower limbs (Figure 2.26). These different electrical 'perspectives' on the heart allow the clinician to pinpoint which part of the heart any such disorder is present in, and how widespread it is.

Some patients will have ECG and blood pressure recordings made while exercising to various degrees on a treadmill ('treadmill test'), particularly during the investigation of angina, and following a myocardial infarction. The heart may also be investigated by visualizing the coronary arteries (Figure 2.27) using a radio-opaque dye injected into the coronary arteries via a catheter threaded through to the heart (usually from a blood vessel in the groin).

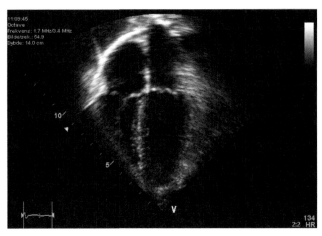

Figure 2.25   Echocardiography

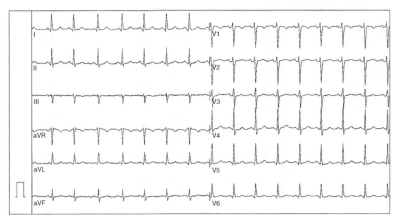

Figure 2.26   A normal 12-lead ECG recording

Blood tests in cardiovascular disease include full blood count and urea and electrolytes as routine (these may show risk factors for cardiac disease, such as polycythaemia), serum cholesterol (since high cholesterol levels are an independent risk factor for coronary artery disease), and measurement of cardiac markers such as troponin, creatine kinase and myoglobin (see Section 2.8.7).

2.8.3   The cardiovascular system before and after birth

Before birth, the cardiovascular system is modified to allow the uninflated lungs to be bypassed, and to enable the acquisition of oxygen and nutrients from the placenta, and the disposal of carbon dioxide and waste products to the maternal circulation via the placenta. The fetal circulation is illustrated in Figure 2.28.

Some blood in the fetal aorta travels into the iliac arteries and then into the two umbilical arteries to the placenta, where oxygen and nutrient supplies are picked up. Blood then passes into the umbilical vein, then into the ductus venosus or via the fetal liver to join the inferior vena cava, and then travels to the right side of the heart.

To enable bypass of the majority of the pulmonary circulation to the uninflated lungs, there is an opening between the right and left atria (the foramen ovale, or 'oval hole'!) which allows the majority of blood entering the right atrium (which is oxygenated and rich in nutrients during fetal life, having recently returned from the placenta) to pass straight into the left atrium and go via the left ventricle back off to the systemic

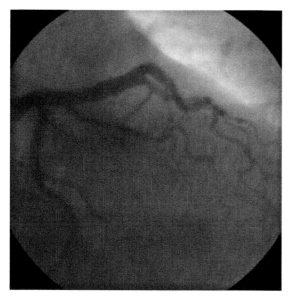

Figure 2.27   Coronary angiography – the coronary arteries are visualized on X-ray after injection of radio-opaque dye

circulation. Blood that does pass into the right ventricle and then the pulmonary artery can also be shunted back into the systemic circulation through the ductus arteriosis, which connects the pulmonary artery to the aorta. In this way, most blood returning to the right side of the heart is redirected into the left side

of the heart, or to the aorta, and is returned to the systemic circulation without entering the pulmonary circulation.

The fetal circulation normally undergoes conversion to the postbirth circulatory system in the minutes, hours and days after birth. Immediately after a baby is born, the umbilical vessels are shut off as the umbilical cord is clamped and cut, and as the baby takes its first breath its lungs are inflated. Inflation of the lungs causes a fall in resistance in the pulmonary circulation, and blood flow into the pulmonary arteries is increased by the 'sucking effect' within the chest from air being drawn into the lungs, with more blood travelling through this pathway to reach the left atrium from the pulmonary veins. The increased pressure in the left atrium and decreased pressure in the right atrium causes the foramen ovale – a structure rather like an open door between the two atria – to flap shut at this point, separating the systemic and pulmonary circulations. The ductus arteriosus between the pulmonary artery and aorta also closes off within 1–2 days of birth after the rise in pO_2 at birth causes the muscle layer in its wall to contract. The umbilical vessels are obviously no longer functional once the umbilical cord is cut, and these degenerate.

2.8.4   Common development defects of CV system

Congenital heart disease is relatively common, with around one in 200 babies born with some form of developmental defect in their cardiovascular system. Congenital heart disease is known to be commoner in babies born to women with diabetes, in babies with Down syndrome, where the pregnancy was affected by rubella infection during early gestation, and with maternal alcohol abuse during pregnancy.

2.8.4.1   Ventricular septal defect

The most common congenital defect is ventricular septal defect (VSD) where the septum between the left and right ventricles is not successfully formed between the fourth and eighth weeks of gestation, causing mixing of the pulmonary and systemic circulations after birth (Figure 2.29). VSD may be asymptomatic if the hole is small, but larger defects put extra strain on the heart as it needs to pump additional blood around the body to compensate

Fetal circulation

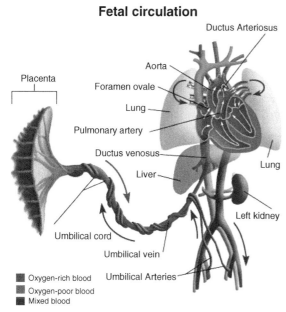

Figure 2.28   The fetal circulation

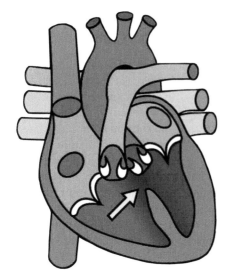

Figure 2.29   Ventricular septal defect

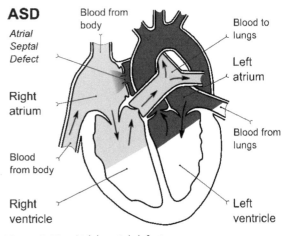

Figure 2.30   Atrial septal defect

for the lowered oxygen levels in the mixed blood, and an increased respiratory effort in oxygenating a greater volume of blood. VSD may be one of the features of other forms of congenital heart disease such as the Tetralogy of Fallot.

A heart murmur will usually be heard with VSD, and a chest X-ray will often show an enlarged heart, while echocardiography will identify the VSD and define its extent. If a VSD is small no treatment may be required, but larger defects may require drugs to lower blood volume and pressure, or surgical repair for the largest defects. In some cases this can now be done without open heart surgery using minimally invasive techniques. Patients with VSD are at risk of bacterial infection of the heart defect (subacute bacterial endocarditis – see Section 2.8.8) and require antibiotic cover for a range of medical procedures.

2.8.4.2   Atrial septal defect

Atrial septal defect (ASD) is also a common congenital heart defect caused by failure of the closure of the foramen ovale between the right and left atria of the heart (Figure 2.30). Like VSD, this results in the inefficient mixing of oxygenated and deoxygenated blood from the pulmonary and systemic circulations, with symptoms dependant on the size of the defect. It can be undiagnosed throughout a person's life if small enough. Larger ASDs cause shortness of breath on exertion, fatigue and recurrent respiratory infections, and in later life may lead to arrhythmias and hypertension in the pulmonary circulation.

Surgical closure of ASD is generally recommended in patients with clinical symptoms and signs, and can often be performed by minimally invasive techniques in which a catheter is threaded through to the heart from a large vein, and a patch that unfolds from the catheter like an umbrella used to seal the ASD. Like VSD, ASD is a risk factor for subacute bacterial endocarditis (see Section 2.8.8).

2.8.4.3   Patent ductus arteriosus

The ductus arteriosus is the normal fetal connection between the pulmonary artery and the aorta, but when it fails to close after birth it is known as a patent (open) ductus arteriosus (PDA). After birth, the systemic circulation is under higher pressure than the pulmonary circulation, and blood will pass from the aorta into the pulmonary artery along a PDA, causing the hypertension in the pulmonary circulation (which is only designed to withstand low pressure blood flow). PDA is commoner in premature babies, and in babies born at high altitudes. It may be symptomless, or cause feeding problems, failure to thrive, and respiratory distress, and even heart failure in some cases. A rapid heart rate and respiratory rate will be seen on examination, and a flow murmur heard when listening to the heart. A chest X-ray may show enlarged pulmonary arteries and heart, while echocardiography will confirm the diagnosis. Initial treatment is often to give antiprostaglandin drugs such as indometacin, as it known that the prostaglandin E2 keeps the ductus arteriosus open during fetal life and this may close a PDA. However, many cases will require surgical closure of the PDA, often by implanting a coil in the vessel via a cardiac catheter. PDA, like ASD and VSD, is also a risk factor for subacute bacterial endocarditis.

2.8.5 Arrhythmias

Any abnormality in the rhythm of the heart is known as an arrhythmia. Arrhythmias can be simply divided into bradycardia and tachycardia.

2.8.5.1 Bradycardia

Bradycardia, is an abnormality in which the heart rate is slow: <60 beats/min. Sinus bradycardia occurs normally during sleep and in athletes, but can also be caused by hypothyroidism, hypothermia, beta blocker drugs and increased pressure inside the skull. Bradycardia may also be seen where the sinoatrial node becomes fibrosed and cannot conduct electrical impulses effectively, or where there is damage to the atrioventricular pathway of electrical conductance (usually caused by ischaemic heart disease). Bradycardia may require a pacemaker to be installed in the chest — this is a small device, about the size of a pack of cards, which has electrodes that will detect pauses in electrical signals in the heart's conductance pathway and deliver their own jolt of electricity to stimulate the ventricles to contract (Figure 2.31).

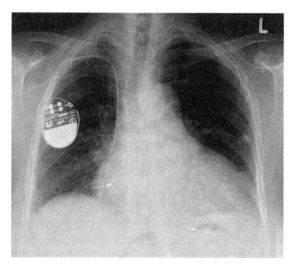

Figure 2.31   A cardiac pacemaker as seen on a chest X-ray

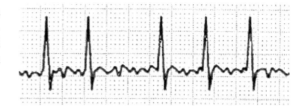

Figure 2.32   The ECG in atrial fibrillation — no clear P waves are seen, just oscillations in the baseline of the trace

2.8.5.2 Tachycardia

Tachycardia, is an abnormality in which the heart rate is fast: >100 beats/min. Tachycardia can be subdivided into supraventricular tachycardia (arising in the atria, or at the atrioventricular junction) and ventricular tachycardia (arising within the ventricles). Tachycardias are usually caused by the presence of damaged cells in the atria and ventricles which trigger extra heart beats, or by abnormal electrical pathways in the heart. Atrial fibrillation is the most frequent form of supraventricular tachycardia (Figure 2.32), occurring in around 10% of people over the age of 65, and causes chaotic, unsynchronized and ineffective atrial contractions (rather like a 'bag of worms'!). This can cause blood to pool in the atria and leads to a risk of a thrombus — a pathological clot — forming, with consequent risk of a section of this (an embolus) breaking off and travelling to the brain to trigger a stroke. People with known atrial fibrillation are thus usually on the anticoagulant drug warfarin.

Arrhythmias may cause dizziness, the sensation of palpitations in the chest, no symptoms at all, or rarely, sudden death. An ECG may fail to show an intermittent arrhythmia, and many patients will have a 24 h ambulatory ECG recording made (in which they spend a day with ECG leads attached, and wear a monitor in a bag to collect the data) and analysed to look for arrhythmias over that period of time. Others may be issued with event recorders, a portable device which the patient can activate when they feel an arrhythmia occurring.

2.8.6 Heart failure

Heart failure is said to occur when the heart can no longer meet the body's requirements for sufficient cardiac output. It affects around one in 100 people over the age of 65. Causes of heart failure may lie within the heart itself (e.g. congenital malformations, ischaemic heart disease, heart valve disease, arrhythmias) or may be due to conditions such as severe anaemia or thyrotoxicosis. The failing heart will initially generate compensatory mechanisms: the sympathetic nervous system will increase heart rate and blood pressure, and the renin–angiotensin system will also increase fluid retention. However, progression of heart failure will eventually overwhelm these systems. Heart failure is conventionally described as predominantly 'right-sided' or 'left-sided', but in practice it is uncommon for just one side of the heart to fail. Heart failure usually develops slowly, but in some cases can develop suddenly over minutes or hours.

2.8.6.1   Right-sided heart failure

Right-sided heart failure usually follows on from left-sided heart failure. The right side of the heart is unable to pump blood effectively, and thus blood travelling to the right side of the heart begins to pool in the extremities (i.e. the ankles in people who are mobile, or around the sacrum in the immobile), with an enlarged liver in many.

2.8.6.2   Left-sided heart failure

Most left-sided heart disease is due to ischaemic heart disease caused by atherosclerosis of the coronary arteries. The left side of the heart is unable to push blood through effectively, causing pulmonary congestion with resultant fatigue, shortness of breath, and an increased heart rate. A chest X-ray will show enlargement of the heart and signs of oedema in the lungs, and an ECG may show signs of ischaemia and arrhythmias. An echocardiogram will enable assessment of how effective ventricular function is, and show any problems with heart valves. A full blood count, urea and electrolytes and thyroid function tests are usually done and other tests such as coronary angiography or radionuclide scans of the ventricles may be used.

Heart failure is treated initially by using diuretic drugs (e.g. furosemide) to lose some of the blood volume and thus relieve pressure on the ventricles and decrease pulmonary and systemic congestion. Vasodilator drugs (e.g. ACE inhibitors such as enalopril) can also reduce constriction of blood vessels and help redistribute fluid. If patients have underlying problems, such as arrhythmias, these should also be treated, for example using digoxin in atrial fibrillation, or surgery on diseased heart valves.

2.8.7   Ischaemic heart disease

Ischaemic heart disease is the result of atherosclerosis of the coronary arteries, which reduces the amount of oxygen and nutrients reaching the highly metabolically active myocardial cells. It affects more males than females, and may have a strong family history. Risk factors include hyperlipidaemia, smoking, obesity, hypertension and diabetes mellitus.

2.8.7.1   Angina

Angina is the term used for chest pain that is caused by myocardial ischaemia — the myocardial cells are still alive and there is thus no infarction. It is described as a 'crushing' pain in the centre of the chest, made worse by exercise and relieved by rest. It often radiates to the left arm and the neck. An ECG usually shows depression of the ST segment, though it is normal in most people inbetween episodes of angina. An exercise ECG on a treadmill may be useful to confirm angina if the diagnosis is in doubt, since symptoms are easily confused with those of gastric problems such as dyspepsia and reflux. Coronary angiography may also be useful. Measurement of cardiac markers for MI should be negative.

Angina is treated by treating underlying causes (e.g. improving the management of diabetes, stopping smoking, losing weight etc.) initially. Angina attacks are treated using sprays or tablets of the powerful vasodilator glyceryl trinitrate (GTN). In frequent attacks, longer acting nitrate drugs such as isosorbide mononitrate may be added, possibly with a beta-blocker (e.g. atenolol) and a calcium channel antagonist such as nifedipine. Aspirin is usually given daily to reduce the risk of myocardial infarction, with a statin drug to lower blood cholesterol levels.

2.8.7.2   Myocardial infarction

Myocardial infarction (MI) — or a heart attack - is the commonest cause of death in the western world, and is almost invariably due to a thrombus forming around a burst atherosclerotic plaque in a coronary artery, totally blocking the vessel and cutting off the supply of oxygen and nutrients to the cardiac muscle that it supplies. It results in severe crushing central chest pain (often radiating to the left arm and neck, though in women symptoms are often less 'classical' of a heart attack) that lasts for hours, and the patient is often grey,

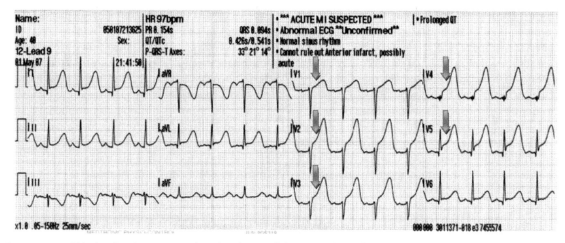

Figure 2.33   ECG showing ST segment elevation (arrowed) in myocardial infarction

sweating and nauseous with it. Some patients have no pain ('silent' MI). The diagnosis is made on the basis of the description of symptoms, and an ECG will show elevation of the ST segment in the leads from the area of the heart where infarction has occurred (Figure 2.33).

Various cardiac markers can be measured to show the leakage of cellular contents from dying myocardial cells, and these can help to distinguish between angina, MI and noncardiac chest pain, particularly when measured serially over hours and days (Figure 2.34). Full blood count, urea and electrolytes, serum cholesterol and blood glucose are also measured to look for reversible risk factors for myocardial infarction.

Cardiac markers for myocardial infarction

Troponin

The amino acid sequences of Troponin I and Troponin T are unique in cardiac muscle (Troponin C is identical in cardiac and skeletal muscle), so these are assayed

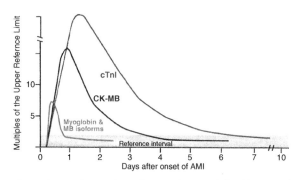

Figure 2.34   Comparison of troponin, creatinine kinase and myoglobin levels following myocardial infarction

when damage to the myocardium is suspected; the death of myocardial cells will allow some of their cellular contents to leak into serum, where they can be measured. In the hours after myocardial infarction, levels of troponin rise well about the 99th percentile and peak around 2 days postinfarction before slowly falling.

Creatine kinase

Creatine kinase (CK) consists of two subunits, M and B, in various combinations in different tissues: CK—BB is found in nonmuscle tissue, CK—MM predominantly in skeletal muscle, while CK—MB is specific to cardiac muscle, so this may also be assayed where cardiac damage is suspected. Levels of CK—MB begin to rise 3—6 h after myocardial infarction and peak around 12—24 h. Peak levels of CK—MB roughly correlate with the size of the infarction; however, small infarcts can be missed.

Myoglobin

Myoglobin levels rise rapidly after myocardial infarction, and are a sensitive indicator of MI. However, myoglobin is found in many tissues of the body, so measurement of myoglobin is a nonspecific indicator of muscle damage — there is no form unique to cardiac muscle.

Treatment of myocardial infarction

Pain relief should be given immediately (e.g. with the opiate analgesic drug diamorphine, and an antiemetic drug to treat nausea such as metaclopramide), along with a GTN spray to dilate coronary blood vessels and aspirin for its antiplatelet action. Within the hour the patient should ideally undergo a reperfusion of the blocked coronary artery (e.g. by using coronary angiography to identify the blocked coronary artery, then inserting a small balloon to push the vessel open, or placing a 'stent' to hold it open), or be treated with a thrombolytic drug such as streptokinase or recombinant tissue plasminogen activator (rTPA) to break down the thrombus in the coronary vessel. The patient should be treated on a coronary care unit (CCU) and kept on a cardiac monitor for several days — 75% of deaths from myocardial infarction occur within the first 24 hours. During rehabilitation on the ward, the patient should be encouraged to mobilize, and started on long-term aspirin (which helps prevent the formation of thrombi), a beta blocker, an ACE inhibitor and a statin drug to lower serum cholesterol levels, as large clinical trials have shown that these measures reduce the chances of subsequent myocardial infarction. An exercise ECG may be done before the patient goes home, or as part of follow-up care following hospital discharge, as well as a coronary angiography if not done as part of the initial treatment.

2.8.8   Valvular heart disease

Heart valves can be ineffective, allowing blood to flow backwards ('regurgitant'), or narrowed and difficult to open ('stenosed'), and both will give a heart murmur that can be heard through a stethoscope. Many people have innocent heart murmurs, particularly the young, and pregnant women. If a heart murmur is heard and a pathological heart valve problem suspected, echocardiography of the heart will identify which valve is affected, and how severely. Defective heart valves may require repair or replacement if causing serious symp-

toms. Heart valves for replacement are other artificial (prosthetic) mechanical valves, or obtained from the hearts of pigs (Figure 2.35). Damaged heart valves and mechanical heart valves are at risk of bacterial infection, and people with mechanical heart valves need to be on anticoagulant drugs (e.g. warfarin) to prevent a pathological thrombus forming around them.

The most clinically important forms of valvular heart disease are stenosis and regurgitation affecting the mitral and aortic valves, that is the valves on the high-pressure left side of the heart.

Mitral regurgitation usually occurs in people who have had rheumatic fever (a now rare inflammatory disease following infection with Group A Streptococci — this triggers an autoimmune reaction in some individuals, which attacks the heart), and results in a mitral valve that cannot keep blood in the left ventricle for ejection via the aorta during systole. A proportion of blood is pushed back into the left atrium via the regurgitant mitral valve; causing increased pressure back along the blood vessels towards the lung with resultant pulmonary oedema, shortness of breath on exertion and tiredness. A heart murmur is heard between the first and second heart sounds (i.e. you would hear 'lub-swoosh-dub' rather than the usual 'lub-dub' heart sounds). A chest X-ray may show pulmonary oedema and an enlarged left atrium, and echocardiography will confirm the diagnosis. Mild cases may be treated with diuretics or ACE inhibitors to reduce the workload of the heart, but more serious cases will require surgical repair or replacement of the valve.

Mitral stenosis also usually occurs after rheumatic fever, but results in a valve that will not open properly to allow blood into the left ventricle, with a reduced cardiac output. The left atrium becomes very dilated trying to push blood through the valve, and this can cause atrial fibrillation (sometimes with a thrombus forming in the atrium in addition, with the resultant

Figure 2.35   Mechanical heart valve (left); pig heart valve (right)

risk of stroke from emboli breaking off it and travelling to the brain). People with mitral stenosis can become very short of breath on exertion, often with a cough from the backlog of blood in the lungs, and a blue-tinged colour (cyanosis) in the face. A heart murmur can be heard. Heart failure may result in severe mitral stenosis. A chest X-ray will show a large heart and signs of pulmonary oedema, and an echocardiogram will identify the stenosis of the valve. An ECG may show atrial fibrillation. Treatment is initially with diuretics for heart failure, and digoxin and an anticoagulant for any atrial fibrillation. The valve may require surgical repair or replacement.

2.8.8.1 Aortic regurgitation

Aortic regurgitation is also generally caused by rheumatic fever, or by infective endocarditis (see Section 2.8.8). It results in blood falling back into the left ventricle from the aorta, with ventricular enlargement and ultimately left-sided heart failure. Patients will have shortness of breath on exertion and fatigue, and a murmur can be heard early in diastole. A chest X-ray will show a large heart, and echocardiography with Doppler (which shows movement of blood) will show the regurgitant aortic valve. Mild cases may be treated with diuretic drugs, but in most cases, surgical repair or replacement of the valve will be required.

2.8.8.2 Aortic stenosis

Aortic stenosis can be due to rheumatic fever, but also from calcification of the valve (which may be congenitally abnormal). It causes obstruction to the left ventricle emptying into the aorta, which causes some ischaemia of the myocardium as the coronary arteries come off the first part of the aorta to supply the heart muscle. Angina, fainting on exertion and shortness of breath on exertion result, and a ventricular arrhythmia may develop and lead to sudden death. There is a murmur heard in systole. Echocardiography should be diagnostic. The treatment is with replacement of the aortic valve.

2.8.8.3 Infective endocarditis

Infective endocarditis is an infection of the endocardium (lining cell layer) of the heart, and is usually a chronic process often also known as subacute bacterial endocarditis (SBE). It can occur on normal valves, but congenitally abnormal and prosthetic valves have a greater chance of developing infective endocarditis. Most cases are caused by infection with *Streptococcus viridans*, though other bacteria and fungi may be responsible. A vegetative mass of bacteria, platelets and fibrin forms along the edge of the valve and destroy it, producing regurgitation and heart failure. The patient with SBE will have a fever, changing heart murmur and may have abscesses elsewhere in the body where the vegetations embolize. Immune complexes form and can cause joint pain and inflammation of kidneys and blood vessels. A raised ESR will be seen on blood tests. Multiple sets of blood cultures must be taken (e.g. six sets over 24 h) to identify the infecting organism, and echocardiography will show the vegetation on a valve. Aggressive treatment with intravenous antibiotics is given initially, with a long course of oral antibiotics thereafter. If there is severe heart failure, the valve may need to be replaced.

Patients with congenital valve disease and prosthetic valves should always be given antibiotic prophylaxis when undergoing a procedure that could introduce bacteria into the bloodstream to protect them from infective endocarditis.

2.8.9 Hypertension

Hypertension is the condition of having chronically elevated blood pressure in the arteries of the systemic circulation, though levels of acceptable blood pressure do vary according to age, gender and ethnicity. Blood pressure readings of >140 mmHg systolic or >90 mmHg diastolic may conventionally be regarded as hypertensive if present on repeated readings. Hypertension is a risk factor for stroke, heart failure, myocardial infarction, kidney failure and aneurysms of arteries, and is thus highly undesirable to have and requires intervention. Around 95% of cases of hypertension are primary or essential hypertension in which no underlying cause can be identified for the rise in blood pressure, though an association with obesity and a sedentary lifestyle is well established. Some 5% of cases of hypertension are secondary to other conditions, such as hyperthyroidism, Cushing's syndrome, kidney disease, pregnancy and tumours of the adrenal glands.

Hypertension is usually symptomless, though very rapid increases in blood pressure can trigger visual problems, vomiting, fits, headaches and even heart failure. On examining most people with hypertension, only increased blood pressure is evident, though some may show changes in the retinal blood vessels on visualizing the back of the eye. A chest X-ray may show signs on heart failure, and measurement of urea and electrolytes in blood may show signs of renal impairment. An ECG may show signs of enlargement of the left side of the heart as it copes with pumping against high pressure. Urinalysis should also be performed to look for signs of renal disease; that is, proteinuria and haematuria. A range of investigations should be performed to look for possible causes of secondary hypertension (e.g. thyroid function tests), though the great majority of cases will be primary (essential) hypertension.

Hypertension is treated according to its severity. Lifestyle measures such as weight loss, reducing alcohol consumption, increasing exercise and cutting down the salt content of food can cure hypertension in some patients, while others will require treatment with antihypertensive drugs. The main groups used are as follows

Diuretics Diuretic drugs increase the loss of water and decrease its reabsorption in the renal tubules, thus lowering blood volume and decreasing blood pressure. A common side effect is hypokalaemia (low K^+ levels). Examples include furosemide and bendrofluazide.

Beta blockers This group of drugs blocks adrenergic beta receptors of the sympathetic system in the heart, causing the heart to slow and stop pumping as hard. However, they will also block beta receptors in the lungs to an extent, which can cause bronchoconstriction – beta blockers must thus not be used in patients with asthma. Common side effects include fatigue, weakness and a slow heart rate. Examples of beta blockers include propranolol, atenolol and metaprolol.

Angiotensin converting enzyme (ACE) inhibitors This group of drugs blocks the renin–angiotensin pathway by preventing conversion of angiotensin I into angiotensin II by inhibiting angiotensin converting enzyme (ACE), thus preventing the formation of the powerful vasoconstrictor angiotensin II and preventing a rise in blood pressure (see also Section 2.3.3 for a revision of the renin–angiotensin system and blood pressure). Common side effects of ACE inhibitors are a dry cough, rashes and hypotension after the first dose. Examples of ACE inhibitors include captopril, enalopril and lisinopril.

Calcium channel blockers This group of drugs acts by blocking calcium channels in the muscular wall of peripheral arterioles, leading to their dilatation and thus lowering blood pressure. Common side effects include headaches and a slow heart rate. Examples of calcium channel blockers are verapamil and diltiazem.

2.8.10 Arterial and venous disease

2.8.10.1 Aneurysms

Aneurysms are abnormal dilatations of blood vessels, and are often formed in the area of atherosclerotic plaques where the vessel wall has been damaged. The aorta is the commonest site of aneurysms, and these are most commonly seen in the abdominal aorta and the thoracic aorta (Figure 2.36).

Aneurysms are usually asymptomatic unless they rupture or 'dissect' (where a tear occurs in the inner lining of the vessel, and blood tracks down the media layer under high pressure to create a false lumen). Dissection or rupture of an aortic aneurysm typically results in severe chest or back pain, and a CT scan will identify the aneurysm. Patients need to undergo emergency surgery to place a graft into the ruptured vessel.

Berry aneurysms

Berry aneurysms are small congenital defects in vessels of the brain, which can rupture and cause a stroke or subarachnoid haemorrhage, with resulting severe headache. A CT scan or angiography can be used to diagnose this, and treatment involves clipping off the vessel or occluding it with a small coil.

2.8.10.2 Deep vein thrombosis

The deep veins of the pelvis and leg are the most common sites of thrombus (pathological clot) formation, though it can occur in any vein. Risk factors for deep vein thrombosis (DVT) include immobility, surgery, pregnancy, the oral contraceptive pill, obesity and cancer. People with a DVT often have no symptoms, but may notice a swollen calf with redness and warmth – measurements are taken around both calves with a tape measure, and compared (Figure 2.37).

The diagnosis is made by scanning the veins of the leg with a Doppler ultrasound scan, which looks for movement of blood in the vessel. Patients with diagnosed or suspected DVT are started on anticoagulants as there is a risk of pulmonary embolus (see below);

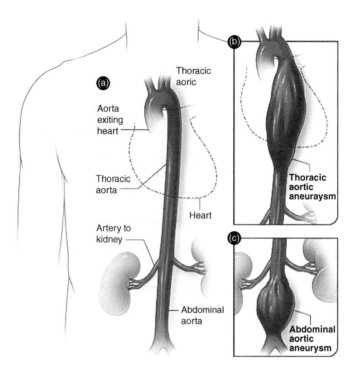

Figure 2.36   Site of aortic aneurysms

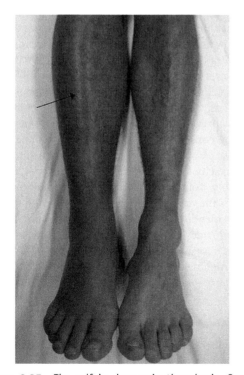

Figure 2.37   The calf in deep vein thrombosis. Source: wikipedia.org/wiki/File:DVT2010.JPG © James Hellman, MD, 2010

heparin and warfarin are started together, and heparin starts to work immediately while warfarin takes several days to achieve its anticoagulation effect. Blood clotting will be measured on a daily basis to tailor doses of heparin and warfarin, and once warfarin has achieved sufficient anticoagulation (i.e. INR in the range 2.0–3.0), heparin is stopped. In cases of massive DVT, drugs to dissolve the thrombus may be required (e.g. streptokinase).

2.8.10.3   Pulmonary embolism

Pulmonary embolism (PE) is usually seen following the formation of a deep vein thrombosis where a part breaks off and travels back to the heart and then onto the lungs, though it can sometimes occur where a thrombus has formed in the right atrium from atrial fibrillation. The presence of an embolus in the lung means that a section is being ventilated but not perfused with blood, which has an adverse effect on gas exchange. A small PE causes shortness of breath, and may also cause chest pain; a very large PE produces severe chest pain, a rapid respiratory and heart rate and shock. A chest X-ray is often normal, and an ECG usually only shows a rapid heart rate. Arterial blood gases may show lowered oxygen levels and CO_2 levels,

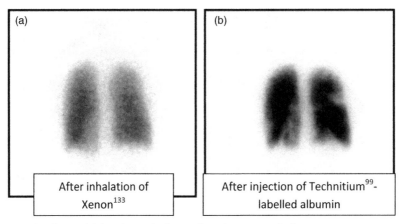

(a) After inhalation of Xenon^{133}

(b) After injection of Technitium^{99}-labelled albumin

Figure 2.38 V/Q scan showing multiple mismatched areas in pulmonary embolism

or may be normal. Measurement of D-dimer levels (which show the presence of fibrin, i.e. a thrombus) are generally elevated. A specialized radionuclide scan known as a ventilation-perfusion scan (V/Q scan) is performed to diagnose PE: a radionuclide-labelled gas is inhaled by the patient, and a gamma camera used to produce an image of all the areas of the lung being ventilated ('V' part of the scan). The patient is also injected with an intravenous infusion of albumin labelled with radioactive Technitium^{99} and the gamma camera used to produce an image of all areas of the lungs being perfused by blood ('Q' part of the scan). The two images are then compared and an area of the lung that is seen to be ventilated but not perfused gives a high probability of PE (Figure 2.38). PE is treated with anticoagulants in the same manner as for DVT. In cases of massive PE, drugs to break down the embolus such as streptokinase may be used.

2.9   Respiratory disease

2.9.1   Symptoms and investigation of respiratory disease

Most patients with lung disease will present with some combination of a cough, sputum production, wheezing, shortness of breath (dyspnoea) or coughing up blood (haemoptysis). Coughing is an extremely common symptom of respiratory tract infection, but also occurs in a range of other conditions such as asthma — in which the cough is generally worse at night — and in chronic smokers, as well as in many people with lung cancer. Dyspnoea is a sensation of shortness of breath, which may be worse upon lying flat (orthopnoea). Wheezing occurs when the flow of air into and out of the trachea is constricted, and is typical of asthma and some allergic reactions. Haemoptysis is a symptom to be taken seriously and investigated thoroughly, since it can signify serious pathology such as lung cancer, tuberculosis and pulmonary infarction.

Investigation of respiratory symptoms, after a full history has been taken from the patient and clinical examination undertaken, often uses techniques such as peak flow measurement and spirometry, as well as chest X-ray. Peak flow meters record the maximum expiratory flow rate during a forced expiration in L/min, which is compared to the expected range for the patient's age, height and gender, and to an individual patient's previous best reading; it is particularly useful in assessing constriction of the airways in asthma (Figure 2.39). A spirometer can be used to measure total lung capacity (TLC), respiratory volume (RV), forced expiratory volume (FEV) and forced vital capacity (FVC) — the volume expired in the first second of FVC measurement is known as FVC_1, and the ratio of FEV: FVC1 provides useful information indicating whether patients have airflow limitation (FEV_1:FVC < 75%) or restrictive lung disease (FEV_1:FVC > 75%).

2.9.2   Asthma

The number of people being diagnosed with asthma has risen in recent years for reasons that are poorly understood. It is a common disorder in many areas of the world, and it is characterized by airway inflammation,

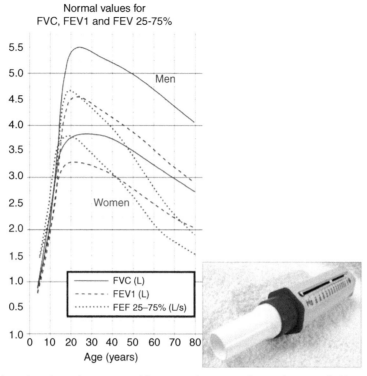

Normal values for
FVC, FEV1 and FEV 25-75%

Men

Women

FVC (L)
FEV1 (L)
FEF 25–75% (L/s)

Age (years)

Figure 2.39   Normal lung function values expected for men and women using a spirometer (left); a peak flow meter (right)

hyperresponsivity of the airways to a number of stimuli, with reversible constriction of the airways. Asthma is known to be related to the degree of 'atopy' of an individual, which is how readily people develop IgE antibodies against common allergens in the environment, such as the house dust mite and pollen. People with asthma often have attacks triggered by dust mites and pollen, but also by exercise, cold air, respiratory tract infections, animal fur and drugs such as β-block-ers or nonsteroidal antiinflammatory drugs. Patients having an asthma attack experience wheezing, cough-ing and shortness of breath, with reduced chest expansion and audible expiratory wheeze, which are thought to be caused by the release of inflammatory mediators from T-cells, mast cells and lymphocytes following antigen binding to IgE on the surface of mast cells. Peak flow measurements will be reduced during an asthma attack and a measurement of <50% of a patient's best peak flow measurement implies a serious attack. Skin prick allergy testing can be used to identify the allergens for a particular patient, which should be avoided where possible, as should cigarette smoke in all cases. Patients are issued with a peak flow meter to monitor their own condition, and managed in a step-

wise approach. Those with occasional symptoms can be prescribed an inhaled β agonist drug such as salbutamol to use as required to induce airway dilatation. For daily symptoms, an inhaled corticosteroid such as flixotide can be added to help prevent airway constriction, and the dose of this can be increased as needed. If symptoms are still not controlled, a long-acting β agonist drug such as salmeterol may be used, with or without the addition of oral steroids (Figure 2.40). In an acute severe asthma attack, patients needing hospital admission will be treated with nebulized salbutamol given along with high-dose O_2 and hydrocortisone, and a mast cell stabilizing drug such as ipratropium bromide may be given in addition. Arterial blood gases will be monitored and patients who are worsening despite all measures may require admission to an intensive care unit for ventilation.

2.9.3 Bronchitis and emphysema

Bronchitis is an inflammation of the airways that causes coughing, retrosternal discomfort and some wheezing. Acute bronchitis is short-lived and usually

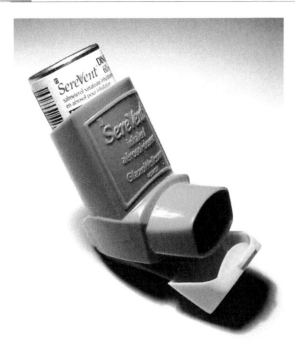

Figure 2.40   A salmeterol inhaler, used for patients with asthma

due to a viral infection, while chronic bronchitis is a cough productive of sputum on most days over at least three months of the year, for more than one year, and is usually associated with cigarette smoking. Chronic bronchitis is in some ways a distinct entity from emphysema; emphysema causes dilatation and destruction of lung tissue in the alveoli leading to a loss of surface area for effective gas exchange. In practice, most patients with chronic bronchitis also have emphysema and together the two conditions form the spectrum of chronic obstructive pulmonary disease (COPD). Cigarette smoking is the major cause, with pollution playing a small role in some patients. Patients with COPD will have a chronic cough productive of sputum, with wheezing and shortness of breath. Frequent chest infections with organisms such as *Haemophilus influenzae* and *Streptococcus pneumoniae* regularly exacerbate their condition. Lung function testing will show a reduced FEV_1: FVC ratio, and a chest X-ray may be normal or show hyperinflation. A full blood count may reveal a secondary polycythaemia (increased haemoglobin level to compensate for hypoxia) and an arterial blood gas measurement may be normal, or reveal hypoxia and hypercapnia (a raised pCO_2). Management comprises advising the patient to

stop smoking, and uses similar drugs to asthma, that is β agonist bronchodilators such as salbutamol, inhaled or oral steroid drugs and continuous oxygen therapy for the worst affected.

2.9.4 Cystic fibrosis

Cystic fibrosis (CF) is a genetic disorder inherited in an autosomal recessive fashion, so sufferers need to inherit two copies of the defective gene, usually from two healthy carrier parents. The gene for CF codes for the cystic fibrosis transmembrane conduction regulator (CFTR) protein, leading to decreased excretion of chloride into the lumen of the airways, along with increased sodium and water reabsorption from the airways; this leads to very thick secretions in the airways, in turn causing dilatation of the bronchi and impaired gas exchange. One in 16 people are carriers in the Caucasian populations, and the high frequency of CF carrier status is believed to be because carrier status conferred a degree of protection against infection with cholera. The gene for CF is very large and particularly prone to new mutations.

CF causes recurrent chest infections, and is often identified in infants with 'failure to thrive' who do not gain weight and feed well, or in infants with small intestinal obstruction due to the presence of very thick meconium ('meconium ileus'). The presence of thick secretions also blocks ducts in the pancreas, and inadequate secretion of pancreatic digestive enzymes causes malnutrition, and diabetes mellitus may result from pancreatic damage. Males with CF are infertile as the vas deferens fails to develop properly.

Treatment of CF is with daily, vigorous lung physiotherapy to enable thick mucus to be expelled, inhaled bronchodilator drugs and corticosteroids, dietary supplementation of pancreatic enzymes and additional nutrients, and with antibiotics for chest infections. Some patients with CF receive heart and lung transplants. Ultimately it is hoped that gene therapy for CF will be possible, with the gene for the normal CFTR protein delivered directly to the airways using a viral vector, and initial research has shown some promise.

2.9.5 Pneumonia

Pneumonia is usually caused by bacterial infection triggering an inflammatory process in the lung tissue;

common organisms responsible for pneumonia acquired in the community include *Streptococcus pneumoniae* and *Mycoplasma* species. In people with preexisting lung disease, *Haemophilus influenzae* is a common cause of pneumonia and *Chlamydia psittaci* can cause pneumonia in people who keep birds. Pneumonia contracted in hospital is often due to infection with *Klebsiella* and *Pseudomonas* species. In immunosuppressed individuals (e.g. HIV positive patients), infection with *Pneumocystis carinii*, Cytomegalovirus and *Aspergillus* species can cause pneumonia.

Pneumonia usually causes a cough, fever and pleuritic chest pain. A full blood count will often show a raised white cell count, and sputum culture should be performed to identify the infecting organism. A chest X-ray will show consolidation, which is areas of a lung that appear denser than normal (due to accumulation of inflammatory exudates in pneumonia). Treatment is with appropriate antibiotics for the specific patient—depending on how sick they are and the setting in which they acquired pneumonia. Most people with pneumonia can be managed at home with oral antibiotics, but some will require hospital admission and intravenous antibiotics.

2.9.6 Tuberculosis

Tuberculosis (TB) is caused by primary infection with the bacterium *Mycobacterium tuberculosis*, which usually occurs in the lung following droplet transmission from another actively infected patient. The lung infection most often occurs in the mid and upper zones of the lung, with an inflammatory exudates of neutrophils appearing; the characteristic caseating granulomas would be seen on histology of a lung biopsy (see also Chapter 4), and in the regional lymph nodes. However, most people recover completely, with just calcified lymph nodes remaining to indicate the TB infection, and this primary infection is often symptomless. Reactivation of TB can occur after several weeks, with symptoms of fatigue, weight loss, fever and coughing. The cough may be productive of sputum, which can be blood-stained; staining of the sputum with the Ziehl—Neelsen stain will reveal mycobacteria, but a bronchoscopy can be performed to acquire washings of affected lobes for staining if no sputum is otherwise available. A chest X-ray will often show round shadows in the upper lobes of the lungs. Sometimes it may be necessary to biopsy lung tissue

and/or lymph nodes to make a definitive diagnosis of TB.

TB is treated with a regime of several antibiotics: isoniazid and rifampicin for 6 months, with pyrizinamide and ethambutol too for the first 2 months of treatment. All can cause a range of side effects, of which hepatitis is common to rifampicin, isoniazid and pyrazinamide. Rifampicin is a liver enzyme inducer that accelerates metabolism of several other drugs (including warfarin and the oral contraceptive pill), and stains tears and urine pink. Isoniazid can cause a polyneuropathy and allergic reactions, while pyrazinamide can trigger gout and arthralgia. Ethambutol can trigger inflammation of the optic nerve and visual problems.

Patients who are found to have multidrugresistant TB (which is occurring with increasing frequency worldwide) will need specialist tailored regimes to eradicate their TB infection and are cared for in a negative-pressure room to minimize the potential for infecting others with a more dangerous form of TB. Close contacts of an individual found to be infected with TB should be screened for infection too using the Mantoux skin test, and treated if found to be infected. A degree of protection from infection with TB is provided by the Bacille Calmette-Guérin (BCG) vaccination, which is currently on offer to groups at risk of TB infection in the UK.

2.9.7 Pneumoconiosis

Pneumoconiosis is a term given to lung disease caused by the inhalation of fine inorganic dust particles that trigger inflammation and fibrosis. Coal dust, asbestos, silica and beryllium are the more common triggers of pneumoconioses, and most forms of this condition are industrial diseases resulting from exposure to these materials in the course of employment. Exposure to coal dust can lead to coal workers' pneumoconiosis, which initially produces small nodules in the lungs (simple pneumoconiosis). However, this can lead on to progressive massive fibrosis of the lungs with further exposure to coal dust, causing large fibrotic masses with shortness of breath, coughing and the potential for complete respiratory failure for which there is currently no treatment.

Asbestos exposure can cause pneumoconiosis after a long latent period (often of several decades), with a spectrum of disease from asymptomatic pleural plaques through to lung cancers. Mesothelioma is the

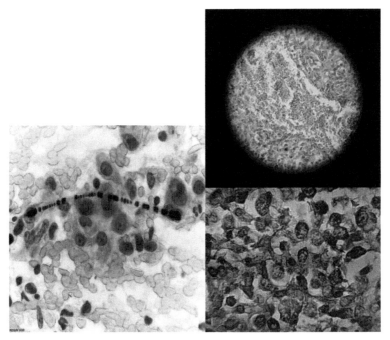

Figure 2.41   Asbestos bodies seen on lung biopsy (left) and mesothelioma (right)

tumour which is characteristic of asbestos exposure, but adenocarcinoma of the lung can also be seen (Figure 2.41).

2.9.8 Lung cancer

Lung cancer is the commonest malignant tumour in Europe, and a leading cause of death after heart disease and pneumonia. It occurs more frequently in males, but the incidence in females continues to rise, reflected in the fact that smoking is the most important factor in the aetiology of lung cancer, and the proportion of female smokers is growing. Passive smoking is also associated with lung cancer, as is asbestos exposure. Most lung cancer is squamous cell carcinoma, arising from the squamous epithelial cells lining the bronchi, followed by small cell carcinoma (arising from endocrine cells in the lung), then large cell carcinoma, adenocarcinoma and alveolar cell carcinoma.

Patients with lung cancer usually develop a cough, chest pain and shortness of breath, and may cough up blood. Local spread of the tumour to the pleura and ribs can cause further pain and rib fractures, while metastasis can produce bone pain from bony metastases, or seizures from brain metastases. A chest X-ray is a useful initial investigation, usually showing a round shadow with fluffy edges, and a CT scan may give further definition and look for evidence of metastatic tumours. A bronchoscopy can be used to obtain biopsies of suspicious lesions, or sputum examined cytologically to look for malignant cells. Liver function testing can be used to indicate the presence of liver metastases, with liver ultrasonography if indicated. Isotope scans of bone may be used to look for evidence of bony metastases. Treatment is with surgical resection of a tumour if possible, and high-dose radiotherapy where this is not possible in most cases. Small cell lung carcinoma can be treated with chemotherapy.

2.10   Endocrine disease

The endocrine system comprises a number of organs which communicate mainly via hormones, chemical messengers which generally act by binding to receptors at a site distant from their origin and triggering specific intracellular reactions. Some hormones act on many — if not most — tissues of the body (e.g. thyroid hormones), while others only target one tissue (e.g. thyroid-stimulating hormone acts only on the thyroid gland).

2.10.1   Symptoms and investigation of endocrine disease

Given the complexity of the endocrine system, the symptoms and investigation of endocrine disease will be considered for the individual organs as we move through this section for the sake of clarity.

2.10.1.1   The hypothalamus and pituitary disease – tumours, hypopituitarism

The hypothalamus produces several 'releasing factor' hormones, which travel to the pituitary gland via the portal circulation in the connection between the hypothalamus and pituitary gland. The posterior pituitary gland stores antidiuretic hormone (ADH) and oxytocin, which are made in the hypothalamus and passed to the posterior pituitary via an axon in the pituitary stalk. As discussed in Chapter 7, most hormones are regulated by some form of feedback system, in which the final hormone product of a pathway has a negative feedback affect on further secretion of the releasing factors and stimulating hormones. Patients who have hormone-secreting tumours will not show this negative feedback, leading to a number of endocrinological diseases.

2.10.2   Pituitary tumours

Most pituitary tumours are benign, but cause symptoms resulting from excess production of a specific pituitary hormone, underproduction of pituitary hormones or from local pressure effects of the tumour. The pituitary gland is situated in a small, saddle-shaped compartment in the skull known as the sella turcica (Latin for 'Turkish saddle'!) in which it has little room to expand. The chiasm of the optic nerve sits just about the pituitary gland, and if it becomes enlarged by a tumour, pressure on the optic nerve at this crossover point produces a specific pattern of visual field loss known as bitemporal hemianopia, in which the outside half of each visual field is lost. Headaches may also occur, or hydrocephalus from interruption of the flow of cerebrospinal fluid.

2.10.3   Hypopituitarism

A lack of pituitary hormones is known as hypopituitarism, and may affect a single hormone secreted by the pituitary, or can affect several hormones together. It may be caused by primary tumours of the pituitary gland or by secondary metastatic tumours from elsewhere in the body, it may result from infection (meningitis, encephalitis or syphilis), problems with blood supply to the pituitary gland, antibodies against pituitary tissues, head injuries or brain surgery, damage from radiotherapy or chemotherapy, infiltration in sarcoidosis, or due to the congenital disorder Kallmann's syndrome (which causes selective deficiency of gonadotrophin hormones). Where several hormones are affected, LH and FSH production is usually decreased first, with TSH and ACTH last.

The symptoms and signs of hypopituitarism depend on which hormones' secretions are affected. Growth hormone deficiency can cause few symptoms in adults, though in children a lack of adequate growth will become apparent. TSH or ACTH deficiency will cause hypothyroidism or adrenal failure, with subsequent tiredness, hypotension and associated symptoms. Gonadotrophin deficiency – lack of FSH and LH – will lead to impotence or amenorrhoea with loss of libido. Investigations will need to target each of the pituitary pathways, with measurement of basal hormone levels and stimulatory tests of the pituitary gland (detailed in later sections). Treatment of hypopituitarism is with replacement of missing hormones as required (either by replacing the pituitary hormones themselves, e.g. with FSH or LH analogues if fertility is required, or replacement of their target organ hormone, e.g. thyroxine is replaced rather than TSH for hypothyroidism).

2.10.4   Male hypogonadism

Male hypogonadism refers to the syndrome resulting from androgen deficiency, meaning a lack of testosterone, which is normally produced in the testicles in response to LH from the pituitary gland. The symptoms and signs depend on when this occurs: if prior to puberty, it causes increased height and a lack of secondary sexual characteristics, with small penis and testicles and decreased muscle bulk. If it occurs after puberty, it triggers decreased libido, impotence and a lack of growth of facial and body hair, with a decrease in the size of the prostate gland. It is most commonly caused by the chromosomal disorder Klinefelter's Syndrome (XXY chromosomes are present rather than the normal male XY), but may also be due to isolated

GnRH deficiency in the hypothalamus, generalized hypopituitarism (as detailed above in Section 2.10.3), from testicular failure following trauma, torsion or radio/chemotherapy to the testicles, from androgen receptor deficiency, or in renal or hepatic failure.

Investigation of hypogonadism includes measurement of testosterone, LH and FSH levels – this enables distinction between primary testicular failure (in which LH and FSH will be high, but testosterone low) and secondary pituitary disease (in which LH, FSH and testosterone levels will all be low). Other investigations such as CT scanning of the pituitary or chromosomal analysis can be undertaken as required. Treatment is by reversing the cause where possible and replacing testosterone, or using LH, FSH or pulses of GnRH analogues if fertility is required and possible.

2.10.5 Gynaecomastia

Gynaecomastia is the development of breast tissue in the male, and is caused by an increase in serum oestrogen levels relative to those of testosterone. It is very common in the neonatal period in baby boys and during puberty, and usually requires no intervention before it spontaneously resolves. However, in older males it requires investigation as it may be a sign of hypogonadism, oestrogen-producing tumours of testis or adrenal gland, or hCG-producing tumours in the lung. Some drugs may trigger gynaecomastia, notably digoxin, cannabis, heroin, cimetidine or spironolactone. It may also be seen in hyperthyroidism and hepatic disease such as liver cirrhosis. However, in many cases no underlying cause will be identified. If possible, drugs that may be responsible should be stopped, or any identified cause treated. Some males with gynaecomastia will require cosmetic surgical excision of breast tissue.

2.10.6 Menopause

Females reach the natural menopause around the age of 45–55, and at this age periods become irregular and finally cease, following a gradual rise in FSH and LH levels over several years. Oestrogen levels then begin to fall, and the menstrual cycle is disturbed. A premature menopause results if the ovaries are removed or damaged (e.g by chemotherapy or ovarian disease) prior to this. Menopause results in the characteristic 'hot flushes', with breast atrophy and loss of bone density, and some women experience weight gain, a loss of libido, and possibly even depression. After the menopause, the oestrogenic protection against ischaemic heart disease is lost, and hormone replacement therapy (HRT) is often given to postmenopausal women partially to protect them from this and the loss of bone density, as well as to relieve other symptoms. HRT always comprises a combination of oestrogens and progestogens, since oestrogens given alone cause an increased risk of endometrial cancer.

2.10.7 Female hypogonadism and amenorrhoea

In females, hypogonadism almost always presents as amenorrhoea (absence of periods), or oligomenorrhoea (abnormally long menstrual cycles). Amenorrhoea may be primary amenorrhoea where a woman has never had periods, or secondary amenorrhoea, in which menstrual cycles previously occurred. The commonest cause of amenorrhoea is weight loss causing abnormal GnRH secretion from the hypothalamus, though isolated GnRH deficiency may occur. Other causes include generalized hypopituitarism, hyperprolactinaemia, premature ovarian failure, defective ovarian development, ovarian tumours, thyroid disease, adrenal tumours and polycystic ovary syndrome (PCOS). Patients with PCOS will have an irregular menstrual cycle with acne, hirsuitism (increased hair, e.g. on the face) and small cysts are seen on both ovaries on ultrasound scan. PCOS is associated with decreased oestrogen production and excess androgens, and is a common cause of subfertility in females.

Investigation of amenorrhoea comprises a careful history and examination, with measurement of basal levels of FSH, LH, oestrogen, progesterone, testosterone and prolactin. Ultrasounds or CT scanning of the ovaries and pituitary gland or ovarian biopsy are performed if indicated. Treatment is of any underlying cause (e.g. surgical excision of an adrenal or pituitary tumour, weight gain, treating hypothyroidism), or providing oestrogen replacement where this is not possible, or FSH/LH for GnRH deficiency or hypopituitarism. Patients with PCOS are treated with the drug clomifene, which stimulates ovarian function, and are advised to lose weight where necessary. They may additionally receive the hypoglycaemic drug metformin, which is known to reduce testosterone production in the ovaries, and injected gonadotrophins if pregnancy is desired and ovulation is not occurring with clomifene.

2.10.8   Hyperprolactinaemia

High prolactin levels are usually caused by a prolactin-secreting pituitary adenoma ('prolactinoma'), but can also be seen in hypothyroidism, and due to drugs such as oestrogens, cimetidine (an antacid drug) and metoclopramide (an antiemetic drug). High prolactin levels cause subfertility with impotence in males (due to negative feedback of prolactin on GnRH production in the hypothalamus), with galactorrhoea (milk production), amenorrhoea or oligomenorrhoea in females. Investigation is with measurement of serum prolactin levels, CT or MRI scanning of the pituitary gland and thyroid function tests. Treatment is by stopping any drugs that might be the cause and using the drug bromocriptine to shrink any prolactinoma. Some prolactinomas will require surgical excision.

2.10.9   Parathyroid hormone disorders

Parathyroid hormone (PTH) is involved in calcium homeostasis and is produced by the parathyroid gland. It causes increased serum calcium levels by increasing calcium absorption from the intestine, increasing calcium reabsorption in the renal tubules and increasing calcium release from bones. Its counterpart in calcium homeostasis is calcitonin, which is produced by the parafollicular cells of the thyroid glands and has the opposite effects to parathyroid hormone.

2.10.9.1   Hypoparathyroidism

A lack of PTH will cause hypocalcaemia, with symptoms of numbness around the mouth and in the extremities of the body, then cramps, twitching, convulsions, and even death if untreated. Hypoparathyroidism may be due to the chromosomal disorder DiGeorge syndrome, in which the parathyroid glands are defective, or due to autoimmune destruction of the parathyroid glands, following removal of the thyroid glands or parathyroid glands, or even due to a severe lack of magnesium, which inhibits the release of parathyroid hormone. In some patients, PTH levels are normal but the tissues are resistant to it ('pseudohypoparathyroidism'). Investigations will reveal a low serum calcium and low PTH levels (or normal levels of PTH in pseudohypoparathyroidism). Treatment is with calcium gluconate infusion initially to correct serum calcium levels if there is severe hypocalcaemia, and then treatment with vitamin D.

2.10.9.2   Hyperparathyroidism

Hypercalcaemia is most commonly caused by hyperparathyroidism, or elevated levels of PTH, though some cases may be due to tumours that secrete a peptide with PTH-like activity. Hypercalcaemia causes a general feeling of being unwell, depression, bone pain and constipation; renal stones may develop.

Hyperparathyroidism may be primary, secondary or tertiary. Primary hyperparathyroidism is caused by overproduction of PTH in the parathyroid glands, often due to an adenoma or hyperplasia of the parathyroid glands (rarely, it is due to adenocarcinoma of a parathyroid gland). Secondary hyperparathyroidism is an elevated level of PTH occurring secondary to hypocalcaemia (e.g. in renal disease), while tertiary hyperparathyroidism occurs where there is hyperplasia of the parathyroid glands following chronic secondary hyperparathyroidism. Investigations for hyperparathyroidism show high serum PTH levels in the presence of hypercalcaemia in primary hyperparathyroidism, and treatment comprises lowering serum calcium levels by rehydration with intravenous fluids in the sick patient with bisphosphonate drugs to increase calcium deposition into bones. Parathyroid tumours should be excised and all four glands removed in parathyroid hyperplasia.

2.10.10   Disorders of growth hormone

Growth hormone is produced by the anterior pituitary gland, and interacts with insulin-like growth factor (IGF-1) at the tissues to produces its effects. Growth hormone deficiency is often asymptomatic in adults, though a cause of short stature in children. If growth hormone is overproduced in children it can lead to gigantism and in adults it will give rise to acromegaly.

2.10.10.1 Acromegaly

Most cases of acromegaly are due to a benign pituitary adenoma that produces excessive growth hormone (GH), though occasionally an excess of GH releasing hormone (GHRH) from the hypothalamus is responsible. Excess growth hormone in adults who have fused epiphyseal plates in long bones (and thus no potential for further increase in height) causes excessive growth of soft tissues, signified by a large tongue, separation of teeth, increase in the size of the hands, feet and nose, possibly with heart failure, hypertension and the symptoms of a pituitary tumour (headaches, visual field defects). Investigation is with a glucose tolerance test: usually, a glucose load will produce a fall in serum GH, but this is absent in acromegaly. Serum GH levels are usually increased, and a skull X-ray will show enlargement of the pituitary fossa in 90% of patients with acromegaly — MRI scanning will show the pituitary adenoma well. Treatment is almost always with surgical removal of the pituitary tumour and followed up with radiotherapy, though some patients may be treated with radiotherapy to the pituitary gland alone.

2.10.11   Thyroid disease

The hypothalamus produces thyrotropin-releasing hormone (TRH), which triggers release of thyroid-stimulating hormone (TSH) by the pituitary gland. The thyroid gland responds to TSH by producing mainly thyroxine (T_4) and some tri-iodothyronine (T_3). T_3 is the biologically active form, and most serum T_3 is produced by conversion from T_4. Serum T_3 and T_4 feedback to the hypothalamus and pituitary gland to decrease production of TRH and TSH in a classical negative feedback loop. The vast majority of thyroid hormones circulate bound to thyroxine-binding globulin (TBG).

2.10.11.1 Hypothyroidism

Underproduction of thyroid hormones is usually due to problems with the thyroid gland itself, rather than the hypothalamus or pituitary gland. The commonest cause is autoimmune disease, in which antibodies are produced against thyroid tissue, causing infiltration of the thyroid gland with inflammatory cells, followed by fibrosis and atrophy of the thyroid. As with all autoimmune diseases, this occurs far more frequently in

females than males (six to one ratio). However, other causes include iodine deficiency (particularly in mountainous regions of the world) or following previous surgery or radiation treatment to the thyroid gland, and rarely, due to the inheritance of genes for defective thyroid hormone synthesis.

Hypothyroidism causes a slow heart rate, mental slowness, dry thin hair, weight gain, and may lead to pericardial effusion, oedema, deafness, anaemia and loss of the outer part of the eyebrows. Some patients have a marked goitre, or enlarged thyroid gland, particularly where the cause of hypothyroidism is iodine deficiency. Measurement of serum TSH will show elevated levels, with low levels of T_4, and thyroid antibodies may be detected in autoimmune causes of hypothyroidism. Treatment is by replacing thyroxine for life, with the response to treatment assessed by regular measurement of thyroid hormones and TSH, particularly initially as it stabilizes.

2.10.11.2 Hyperthyroidism

Hyperthyroidism is a common problem in women in the third and fourth decades of life, affecting an estimated 5% of this group. The symptoms and signs include weight loss, increased appetite, restlessness, tremor, tachycardia (a rapid heart rate), bulging eyes (exopthalmus), a goitre and hypertension (Figure 2.42).

The commonest cause of hyperthyroidism is Grave's disease, in which IgG antibodies bind to TSH receptors in the thyroid gland and stimulate thyroid hormone production. However, a solitary toxic nodule or a multinodular goitre may also be responsible, and

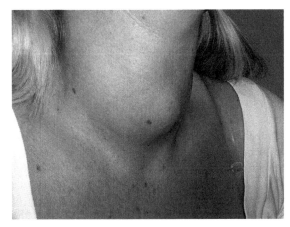

Figure 2.42   Large goitre in hyperthyroidism. Source: en. wikipedia.org/wiki/File:Struma_001.jpg

in some cases hyperthyroidism occurs when acute inflammation of the thyroid gland is seen following a viral infection. Investigation of hyperthyroidism will show a low TSH level (suppressed by high thyroid hormone levels) and elevated T_3 and T_4 levels in serum. If Grave's disease is the cause of hyperthyroidism, serum microsomal and thyroglobulin antibodies will be detected.

Treatment of hyperthyroidism is with antithyroid drugs such as carbimazole, which inhibits the formation of thyroid hormones. A β-blocker drug may be used in the first weeks after diagnosis to control symptoms such as a rapid heart rate and increased blood pressure. In some patients in whom drug treatment is ineffective or unsuitable, radioactive iodine (^{131}I) is used; this accumulates in the thyroid gland and destroys it via local irradiation. Alternatively, most of the thyroid gland can be removed surgically, particularly in patients with a large goitre. However, risks of surgery include damage to a nerve that is involved in speech, hypocalcaemia from damage to parathyroid glands, and hypothyroidism.

2.10.12 Disorders of the glucocorticoid hormones

The adrenal cortex produces the steroid hormones cortisol, aldosterone and androgens. Cortisol is produced pulses in response to adrenocorticotrophic hormone (ACTH) from the pituitary gland, which is itself released in response to corticotrophin releasing factor (CRF) produced in pulses by the hypothalamus; cortisol feeds back on the hypothalamus and pituitary gland to decrease further CRF and ACTH release. Cortisol triggers increased fat and glycogen deposition and sodium retention. Aldosterone secretion by the adrenal cortex is controlled by the renin−angiotensin system (see Chapters 1 and 7 too).

2.10.12.1 Addison's disease

Addison's disease — also known as primary hypoadrenalism — is rare, and is a fall in cortisol and aldosterone levels caused by destruction of the adrenal cortex, usually by autoantibodies. However, it is also seen in tuberculosis of the adrenal gland, infiltration by malignant cells, and acute haemorrhage into the adrenal gland (often after meningococcal septicaemia). It causes gradual weight loss, loss of appetite, depression and tiredness, but can present acutely with vomiting, shock and abdominal pain. Low blood pressure is an important feature due to loss of sodium and water, and there is hyperpigmentation since ACTH levels are elevated, and ACTH stimulates melanocytes. Measurement of urea and electrolytes will show a decreased serum serum with increased serum potassium, low serum glucose and raised urea. Full blood count often shows an increased white cell count, and adrenal autoantibodies can be detected in most cases. The diagnosis can be confirmed with the Synacthen test (also known as the 'short tetracosactrin' or 'synthetic ACTH' test), in which blood is taken for measurement of serum cortisol levels, then a dose of synthetic ACTH is given, with further serum cortisol levels taken at 30 and 60 min after the ACTH dose. In adrenal failure, plasma cortisol levels are low, and fail to rise in response to the ACTH dose. (Note that serum ACTH levels cannot be used in diagnosis in their own right as ACTH production is pulsatile and highly variable.) Treatment of Addison's disease is with lifelong steroid replacement treatment, for example with oral hydrocortisone, and the dose adjusted according to measurements of serum cortisol. The dose may need to be increased when patients are in inherently stressful situations, for example infection, surgery or trauma.

2.10.12.2 Secondary hypoadrenalism

Secondary hypoadrenalism may occur due to suppression of hypothalamic/pituitary CRF/ACTH production by long-term steroid use, or from disorders of the hypothalamus and pituitary gland. It causes the same clinical problems as Addison's disease except without the skin pigmentation, since ACTH levels are low in secondary hypoadrenalism. A longer version of the Synacthen test is used to investigate and will show a delayed but normal response to injected synthetic ACTH.

2.10.12.3 Cushing's syndrome

In Cushing's syndrome, there is an excess of glucocorticoid levels, and most cases are caused by patients taking steroids for chronic conditions such as inflammatory bowel disease. Other causes include excessive ACTH production by a pituitary adenoma (this is specifically called Cushing's disease), other

ACTH-producing tumours such as small cell lung tumours or carcinoid tumours, and from excessive glucocorticoid production by adrenal tumours (adrenal carcinoma and adrenal adenomas). An excess of glucocorticoids causes obesity centred around the abdomen and trunk with a 'moon face', thin skin, hypertension, stretch marks, and pigmentation in cases due to excess ACTH. Investigation is with a dexamethasone suppression test; dexamethasone is a synthetic glucocorticoid, and in normal people its administration causes suppression of cortisol levels. In Cushing's syndrome, suppression of serum cortisol by dexamethasone is seen if the cause is ACTH production by the pituitary gland, but none is seen if ACTH is being produced by tumours elsewhere in the body as the usual negative feedback loop will not work. Injected CRF will show an exaggerated response in pituitary causes of Cushing's syndrome. In nonACTH dependent causes of Cushing's syndrome, very low levels of plasma ACTH will be recorded. A 24 h urine collection will also show raised cortisol levels in urine.

Treatment of Cushing's syndrome involves surgical excision of pituitary or adrenal tumours (or tumours elsewhere that are producing ACTH). Drugs to inhibit cortisol synthesis may be useful if this is not possible (e. g. metyrapone). If Cushing's syndrome is due to steroid treatment, a gradual reduction in steroid dose should be instigated where possible.

2.10.13 Diabetes insipidus

Diabetes insipidus is a condition in which ADH secretion by the pituitary gland is impaired (cranial diabetes insipidus), or ADH secretion is normal but the kidneys no longer respond to it by increasing resorption of water in the tubules (nephrogenic diabetes insipidus). Cranial diabetes insipidus may follow head injury or brain surgery, or be caused by tumours in the hypothalamus or pituitary, infections such as meningitis, or vascular damage to the pituitary gland or hypothalamus, though some cases are idiopathic (no clear cause). Nephrogenic diabetes insipidus can occur in hypercalcaemia or hypokalaemia, or after treatment with drugs such as lithium (a drug used for bipolar disorder) or glibenclamide (a hypoglycaemic drug used in Type II diabetes mellitus).

Diabetes insipidus leads to polyuria (large volumes of urine are produced frequently) and thirst. Urine volume is measured, and urea and electrolytes will often show a high serum sodium level and osmolality due to dehydration, while urine osmolality is low. A water deprivation test with synthetic ADH administration is used to confirm the diagnosis: water intake is restricted for 8 hours and blood and urine osmolality measured hourly. A continuing low urine osmolality is seen in diabetes insipidus after water deprivation, and following a dose of synthetic ADH, this normalizes in cranial diabetes insipidus, but remains low in nephrogenic diabetes insipidus. Treatment comprises treating any underlying condition (e.g. excision of a pituitary tumour) and giving a synthetic ADH (e.g. desmopressin).

2.10.14 Diabetes mellitus

Insulin is a peptide hormone produced by the β cells of the islets of Langerhans in the pancreas in response to a rise in plasma glucose levels. It allows glucose to be taken up by cells, and glucose to be stored as glycogen in the liver and muscles, and triglycerides to be stored as fat. In diabetes mellitus, there is insulin deficiency or a lack of tissue responsiveness to insulin (sometimes both), with hyperglycaemia resulting. Diabetes is divided into Type I diabetes mellitus, in which patients are dependent on insulin injections, and Type II diabetes mellitus, which can often be managed with dietary changes or drugs alone. Type I diabetes occurs when the islets of Langerhans are destroyed by an autoimmune reaction, rendering the patient reliant on injected insulin to survive thereafter. It is typically diagnosed in young patients (with a peak incidence around the age of 12) who are not overweight, and is associated with HLA-DR3 or HLA-DR4 genes. In contrast, Type II diabetes often affects patients older than 40 years old who are usually overweight, and there is no evidence of an autoimmune basis for Type II diabetes. Most cases of Type II diabetes can be managed with diet or oral medication, though some patients will require insulin.

In diabetes, patients develop chronic hyperglycaemia as a result of lack of insulin, or tissue resistance to insulin — glucose is effectively 'locked out' of cells. A major symptom is polyuria as the renal threshold for glucose reabsorption is exceeded with rising plasma glucose levels; glucose in urine drags water with it by osmotic diuresis, triggering frequent, high volume urination, and accompanying dehydration and thirst. Since glucose cannot be used by cells in the absence of glucose, fat and muscle are broken down as alternative fuel sources, with weight loss. In Type I diabetes, ketoacidosis may result if the diagnosis is not made early.

2.10.14.1   Ketoacidosis

Investigation of a patient for diabetes often begins with urine dipstick testing for the presence of glucose in urine, with a fingerprick 'BM' test for blood glucose levels, or with a blood sample submitted for a random blood glucose level. A fasting blood glucose sample allows more accurate assessment, and is obtained in the early morning after a patient has fasted overnight. A random blood glucose ≥ 10 mmol/L, or a fasting blood glucose ≥ 6.7 mmol/L, both allow a presumptive diagnosis of diabetes mellitus to be made. In borderline cases, a glucose tolerance test can be undertaken, in which a patient fasts overnight, a baseline blood glucose sample is taken, and they are then given 75 g of glucose in 300 mL water. A further blood glucose sample is taken at 2 h, and the results interpreted as shown below in Table 2.6.

Treatment of diabetes aims to keep blood glucose levels as normal as possible at all times, and thus avoid the potential complications of diabetes mellitus, which include renal failure, macrovascular disease (strokes, heart disease and gangrene in feet), microvascular disease (which can cause blindness from retinal disease, renal failure when it affects the renal glomerulus, and neuropathy) and infections. All patients with diabetes should be advised on a healthy, balanced diet to include complex carbohydrates rather than simple sugars, enabling the avoidance of rapid fluctuations of blood sugar levels. Type II diabetics will sometimes also require oral hypoglycaemic drugs such as glibenclamide (which increases insulin secretion) or metformin (which reduces glucose absorp-

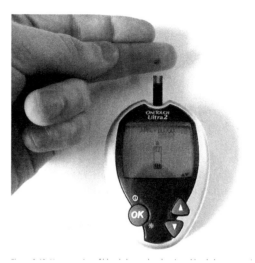

Figure 2.43   Home testing of blood glucose levels using a blood glucose monitor

tion from the gut) if diet alone is failing to control their blood glucose levels. Type I diabetics will require injected insulin — usually recombinant human insulin — in a tailored regime of rapid-acting and long-acting formulations, typically with two to four injections of insulin each day and the dose adjusted according to self-monitoring of blood glucose levels using fingerprick testing and a blood glucose monitor (Figure 2.43).

Since patients can continue to feel well in spite of considerable degrees of hyperglycaemia, home monitoring of blood glucose levels to ensure good control is extremely important. Patients should record their home blood glucose measurements in a log, and take this to hospital or GP surgery clinics for advice from a diabetic specialist. In hospital, the control of blood glucose levels can be assessed over the longer term by measuring haemoglobin A_{1C} levels — HbA_{1C} is produced by glycosylation of HbA by glucose in the bloodstream, and occurs in proportion to average blood glucose levels over the lifespan of normal red blood cells, that is 120 days. The normal range is 4–6.5% (i.e. 4–6.5% of Hb is HbA_{1C}) in nondiabetics, and for diabetic patients the target range is 6.5–7.5% — readings above this range indicate imperfect diabetic control and a need to adjust control and monitoring of diabetes to avoid complications. In patients with abnormal forms of haemoglobin, or abnormal rates of production of haemoglobin, such as sickle cell anaemia or thalassaemia, other glycosylated plasma proteins can be assessed instead.

Table 2.6   Interpreting the results of a glucose tolerance test

	Normal	Diabetes	Impaired glucose tolerance
Baseline blood glucose	<6.7 mmol/L	≥6.7 mmol/L	<6.7 mmol/L
2 h blood glucose	<6.7 mmol/L	≥10 mmol/L	6.7–10 mmol/L

2.10.14.2 Diabetic ketoacidosis

If Type I diabetes is diagnosed late, or in known Type I diabetics who are otherwise ill or have stopped taking insulin, diabetic ketoacidosis can ensue. This occurs where hyperglycaemia has led to osmotic diuresis and dehydration, and lack of insulin has led to fat being used as an alternative fuel source; with lipolysis underway, free fatty acids are released and their liver conversion to acidic ketone bodies leads to a fall in blood pH. The diagnosis is made in patients who have hyperglycaemia (usually blood glucose >20 mmol/L) along with a lowered blood pH — measured on an arterial blood gas sample — and the presence of ketones in plasma. A full blood count may show an increased white cell count when infection is also present, and serum urea and electrolytes will show an increased urea and creatinine (dehydration) with a raised serum potassium, since potassium is also locked out of cells in the absence of insulin.

Treatment of ketoacidosis comprises replacing lost fluids and insulin intravenously, and also giving potassium supplements as potassium will flood back into cells once insulin is given (and rehydration is occurring). Patients with diabetic ketoacidosis are often extremely sick, and should be intensely monitored with very frequent measurement of blood glucose levels, serum urea and electrolytes, pH and bicarbonate levels, and adjustment of fluids, potassium and insulin accordingly; some will require admission to an intensive care unit. Once the patient is well enough to eat and drink, they are converted to subcutaneous insulin injections.

2.10.14.3 Hyperosmotic hyperglycaemic nonketotic coma (HONK)

In Type II diabetics, severe hyperglycaemia can develop without significant ketosis, and results in serious dehydration from osmotic diuresis and elevated plasma osmolality, ultimately leading to a decreased level of consciousness. It is often set off by intercurrent illness in a Type II diabetic patient. Management is as for diabetic ketoacidosis, with rehydration and insulin, but also with anticoagulation with heparin as the risk of thrombosis is high in HONK. Once the patient is well again, their usual diabetic regime can begin again.

Bibliography

This chapter represents a 'whistle-stop' tour of a vast tract of clinical medicine, epidemiology and pharmacology, and students are advised to consult other textbooks for subjects they wish to research in more depth. Some of the author's own favourites are as follows.

Bonita, R., Beaglehole, R. and Kjellström, T. (2007) *Basic Epidemiology*, 2nd edn, World Health Organization.

Kumar, P. and Clark, M. L. (eds) (2009) *Clinical Medicine*, 7th edn, Elsevier.

Longmore, M. *et al.* (eds) (2010) *Oxford Handbook of Clinical Medicine*, 8th edn, Oxford University Press.

Neal, M.J. (2009) *Medical Pharmacology at a Glance*, 6th edn, Wiley Blackwell.

Rang, H.P., Dale, M.M., Ritter, J.M. and Flower, R.J. (eds) (2007) *Rang and Dale's Pharmacology*, 6th edn, Churchill Livingstone.

Chapter 3
Clinical cell biology and genetics

Professor Ray K. Iles, B.Sc., M.Sc., Ph.D., CBiol, FSB, FRSC
Dr Stephen A. Butler, B.Sc., Ph.D.

At the most basic level, biomedical science is an understanding of how cells work and genes are expressed or silenced. At a clinical level this is perhaps not nearly as important as recognizing a disease and applying the correct treatment. Clinical biomedical science attempts to bridge these two extremes. Thus, an understanding of basic cell biology and genetics is essential, but its relevance to human diseases and laboratory medicine must not be lost. This chapter aims to highlight and explain relevant aspects of our growing understanding of basic genetics and cell biology for the clinical biomedical scientist.

3.1   The cell

One of the fundamentals of biology is an understanding of the structures and functions of the cell. In biomedical science it is an appreciation that organelles vary in prominence and indeed may be absent (e.g. no nuclei in red cells and platelets) depending on the function of the cell and tissue architecture in which they are embedded. Cellular organizations are highly regulated specific to the cell type and not a matter of enzymes, proteins or structures floating freely in a cytoplasmic soup (see Figure 3.1).

3.1.1   Cell structures, organelles and function

3.1.1.1   Cell membrane

The plasma or cell membrane is a bilayer consisting of amphipathic phospholipids, that is consisting of a polar hydrophilic head (phosphatidyl choline) and a lipid hydrophobic tail (commonly two long chain fatty acids). The phospholipids spontaneously form bilayers that, as complete oval structures, form an effective barrier which is impermeable to most water soluble molecules. This barrier defines the interior environment of the cell and the exchanges across the plasma membrane are regulated by various proteins which are embedded in the lipid bilayer. The lipids' hydrophobic nature means that the plasma membrane is held together by relatively weak bonds. However, these collective interactions strongly oppose transverse movement by hydrophilic molecules whilst allowing considerable freedom for lateral 'fluid' movement by the hydrophobic molecules, such as membrane proteins, which are embedded within the plasma membrane. This makes the plasma membrane a very dynamic structure (also see Chapter 4, Section 4.1.1).

Biomedical Sciences: Essential Laboratory Medicine, First Edition. Edited by Ray K. Iles and Suzanne M. Docherty.
© 2012 John Wiley & Sons, Ltd. Published 2012 by John Wiley & Sons, Ltd.

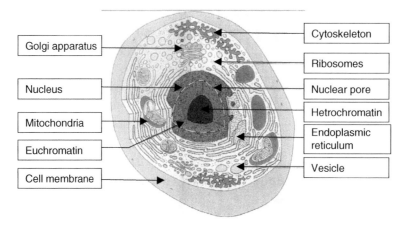

Figure 3.1 Diagrammatic representation of the cell illustrating the major organelles. From Kumar and Clark, *Clinical Medicine,* 6ed 2005 © Elsevier

3.1.1.2 Cytoplasm

This defines the interior of a cell membrane in which the organelles are found; the cytoplasm. Although a fluid compartment, the organelles are held within a scaffolding or cytoskeleton.

3.1.1.3 Cytoskeleton

The cytoskeleton is in fact a complex network of structural proteins, it regulates not only the shape of the cell but its ability to traffic internal cell organelles and move in response to external stimuli. This network even regulates the passage and direction in which the interior solutes and storage granules flow. Although the principal solute is water it contains amino acids, sugars and whole protein complexes which traffic between, and even within, the organelles. The major components are microtubules, intermediate filaments and microfilaments.

3.1.1.4 Nucleus

An organelle containing the human genome and bound by two bilayer lipid membranes constitutes the nucleus of a cell. The outer of the two is continuous with the endoplasmic reticulum. Nuclear pores are present in the membranes allowing the passage of nuleotides and DNA interacting proteins in and mRNA out.

3.1.1.5 Nucleoli

Dense areas within the nucleus, rich in proteins and RNA, are the nucleoli. These regions are chiefly con-cerned with the synthesis of ribosomal RNA (rRNA) and ribosomes.

3.1.1.6 Endoplasmic reticulum (ER)

Interconnecting tubules or flattened sacs (cisternae) of lipid membrane bilayer which are continuous with the nuclear double membrane constitute the endoplasmic reticulum. Areas of the ER may contain ribosomes on the surface (termed rough endoplasmic reticulum (RER) when present, or smooth endoplasmic reticulum (SER) when absent). The ribosomes translate mRNA into a primary sequence of amino acids of a protein peptide chain. This chain is synthesized into the ER luminal or intermembrane where it is first folded and modified into mature peptides. Often tethered to the plasma membrane they are packed into blunt ends of the ER folds that bud of from the ER as vesicles. These transport vesicles often move to the Golgi.

3.1.1.7 Golgi apparatus

A double membrane series of folded flattened cisternae (like a stack of towels) they are similar to, but distinct from, the ER. They are characterized as a stack of cisternae from which many vesicles bud off from the thickened ends. The primary processed peptides of the ER are exported to the Golgi for maturation into functional proteins (e.g. glycosylation of proteins which are to be excreted occurs here) before packaging into secretory granules and cellular vesicles which bud off the end (see Figure 3.2).

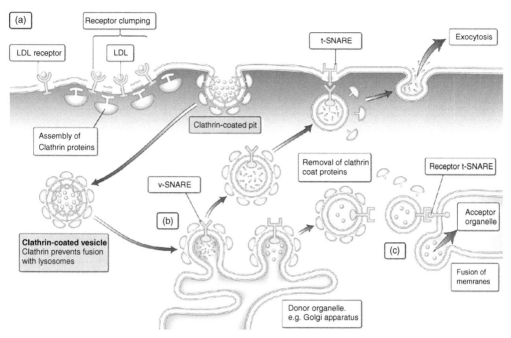

Figure 3.2   Intracellular transport. (a) Receptor-mediated pinocytosis. (b) Trafficking of vesicles containing synthesized proteins to the cell surface (e.g. hormones). (c) Traffic between organelles is also mediated by v- and t-SNARE-containing organelles. v-SNARE, vesicle-specific SNARE; t-SNARE, target-specific SNARE. From Kumar and Clark, *Clinical Medicine,* 6ed 2005 © Elsevier

3.1.1.8   Lysosomes

Dense cellular vesicles containing acidic digestive enzymes — lysozymes — fuse with phagocytotic vesicles from the outer cell membrane, digesting the contents into small biomolecules which can cross the lysosomal lipid bilayer into the cell cytoplasm. Furthermore, lysosomal enzymes can be released outside the cell by fusion of the lysozyme with the plasma membrane. Lysosomal action is crucial to the function of macrophages and polymorphs in killing and digesting infective agents, tissue remodelling during development and osteoclast remodelling of bone. Not surprisingly many metabolic disorders result from impaired lysosomal function.

3.1.1.9   Peroxisomes

Dense cellular vesicles — so named because they contain enzymes that catalyse the breakdown of hydogen peroxide — peroximes are involved in the metabolism of bile and fatty acids; they are primarily concerned with detoxification. For example D-amino acid oxidase and H_2O_2 catalase are key enzymes in this regard.

The inability of the peroxisomes to function correctly can lead to rare metabolic disorders such as Zellweger's syndrome and rhizomelic dwarfism.

3.1.1.10   Mitochondria

This organelle is the power house of the cell — mitochondria contains two lipid bilayer membranes and a central matrix. It also possesses several copies of its own DNA in a circular genome reflecting their evolutionary autonomous origins before symbiotic incorporation with our eukaryotic cell.

The outer mitochondrial membrane contains many gated receptors responsible for the import of raw materials like pyruvate and ADP and the export of products such as oxaloacetate (precursor of amino acids and sugars) and ATP. An interesting caveat to our symbiotic relationship is that proteins of the Bcl-2-Bax family are incorporated in this outer membrane and can release cytochrome C which triggers apoptosis. The inner mitochondrial membrane is often highly infolded to form cristae to increase its effective surface area. It contains transmembrane, enzyme complexes of the electron transport chain, which generate an

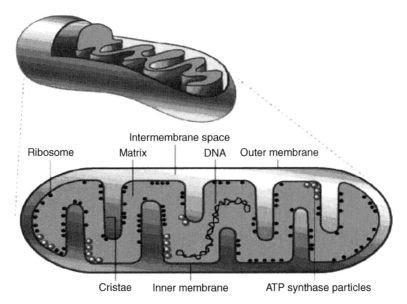

Intermembrane space

Ribosome Matrix DNA Outer membrane

Cristae Inner membrane ATP synthase particles

Figure 3.3 Structure of the mitochondria illustrating the double membrane and presence of its own circular chromosome for autologous synthesis of proteins within the organelle

H^+ ion gradient. This gradient then drives the adjacent transmembrane ATPase complex to form ATP from ADP and Pi. The mitochondrial inner matrix contains the enzymes of the Kreb cycle which generate the substrates of both the electron transport chain ($FADH^{2+}$ and NADH) and central metabolism (e.g. succinyl CoA, α-oxoglutarate, oxaloacetate, see Figure 3.3).

3.1.2 The dynamic cell

Like all systems, the component proteins and even organelles of the cell are continually being formed and degraded. Most of the degradation steps involve ATP dependent multienzyme complexes. Old cellular proteins are mopped up by a small cofactor molecule called 'ubiquitin', which interacts with these worn proteins via their exposed (hydrophobic) leusine residues (Figure 3.4). The ubiquitin acts as a signal for destruction, and a complex containing more than five ubiquitin molecules is rapidly degraded by a large proteolytic multienzyme array termed '26S proteasome'. The failure to remove worn protein can result in the development of chronic debilitating disorders. Alzheimer and Pick's dementia are associated with the accumulation of proteins (prion-like proteins) which are resistant to ubiquitin-mediated proteolysis. Similar proteolytic resistant ubiquinated proteins give rise to inclusion bodies found in myositis

and myopathies. This resistance can be due to point mutation in the target protein itself (e.g. mutant p53 in cancer) or as a result of an external factor altering the conformation of the normal protein to create a proteolytic resistant shape, as in new variant Creutzfeldt–Jakob disease (CJD).

3.1.2.1 The cytoskeleton

The major components are microtubules, intermediate filaments and microfilaments (see Figure 3.5).

Microtubules. These are made up of two protein subunits α and β tubulin (50 kDa) and are continuously changing length. They form a 'highway' transporting organelles through the cystoplasm. There are two motor microtubule-associated proteins (MAP) — dynein and kinesin — allowing antegrade and retrograde movement. Dynein is also responsible for the beating of cilia. During interphase the microtubules are rearranged by the microtubule-organizing centre (MTOC), which consists of centrosomes containing tubulin and provides a structure on which the daughter chromosomes can separate. Another protein involved in the binding of organelles to microtubules is the cytoplasmic linker protein (CLIP). Drugs that disrupt the microtubule assembly (e.g. colchicine and vinblastine) affect the positioning and morphology

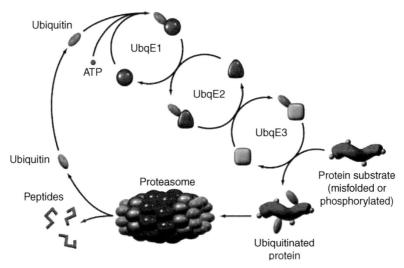

Figure 3.4   26S Proteosome — the reprocessing complex of the cell to which worn proteins are reduced and recycled. Ubiquination of proteins by UBqE3 signals that the target protein is damaged and if it is polyUbiquinated it is transported to the 26S proteosome complex for degradation. The ubiquitin protein is recovered and recycled to interact with other warn and damaged proteins

of the organelles. The anticancer drug paclitaxel causes cell death by binding to microtubules and stabilizing them so much that organelles cannot move, and thus mitotic spindles cannot form.

Intermediate filaments. These form a network around the nucleus and extend to the periphery of the cell. They make cell-to-cell contacts with the adjacent cells via desmosomes and basement matrix via hemidesmosomes. Their function appears to be in structural integrity, being prominent in cellular tissues under stress. The term cytoskeleton was first used to describe the crossing web of these intermediate (thickness) filaments. The intermediate filament fibre proteins are specific to the embryonic lineage of the cell concerned, for example keratin are intermediate fibres only found in epithelial cells whilst vimentin is only found in mesothelial (fibroblastic) cells.

Microfilaments. Muscle cells contain a highly-ordered structure of actin (a globular protein, 42—44 kDa) and myosin filaments, which form the contractile system. These filaments are also present throughout the nonmuscle cells as truncated myosins (e.g. myosin 1), in the cytosol (forming a contractile actomyosin gel) and beneath the plasma membrane. Cell movement is mediated by the anchorage of actin filaments to the plasma membrane at adherent junctions between cells. This allows a nonstressed coordination of contraction between adjacent cells

of a tissue. Similarly vertical contraction of tissues is anchored across the cell membrane to the basement matrix at focal adhesion where actin fibres converge. Actin-binding proteins (e.g. fimbria) modulate the behaviour of microfilaments and their effects are often calcium-dependent. The actin associated proteins can be tissue-type specific, for example actin binding troponin is a complex of three subunits and two of these have isomers which are only found in cardiac muscle. Cardiac troponin I is released into the blood circulation 6 h after the onset of a heart attack. This marker has proved more clinically relevant than creatinine kinase MB. Furthermore, since it is not found in the blood of those with undamaged heart tissue, elevated levels in patients with angina correlates with the likelihood of a heart attack and months of survival. Similarly cardiac troponin T has been shown to identify patients suffering a myocardial infarct and, again, the higher the serum levels the worse the prognosis. More tissue specific actin associated proteins are being identified and the number of serum cardiac status markers is continually increasing.

Alterations in the cell's actin architecture are also controlled by the activation of small ras-like GTP-binding proteins rho and rac. These are important in rearrangement of the cell during division and thus dysfunctions of these proteins are associated with malignancy.

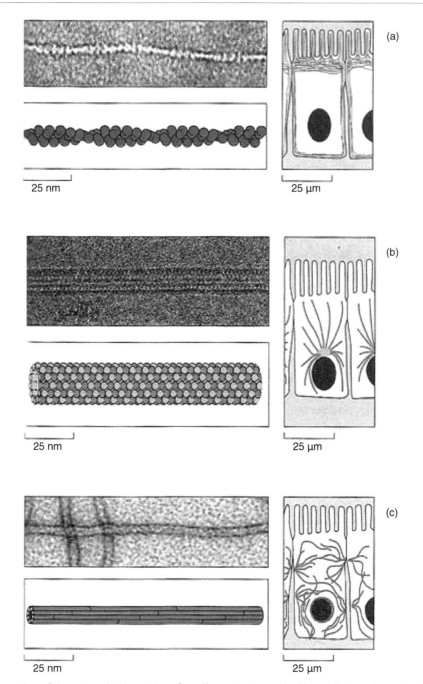

Figure 3.5   Illustration of the cytoskeletal proteins of a cell, contractile actin fibres (a), tram line microtubules (b) and connection scaffolding of microfilaments (c)

3.1.3   Intercellular connections

The cytoskeleton and plasma membrane interconnect and extracellular domains form junctions between cells to form tissues. There are three types of junction between cells: tight, adherent and gap junctions (see Figure 3.6).

Tight junctions (zonula occludens) hold cells together. They are at the ends of margins adjacent to epithelial

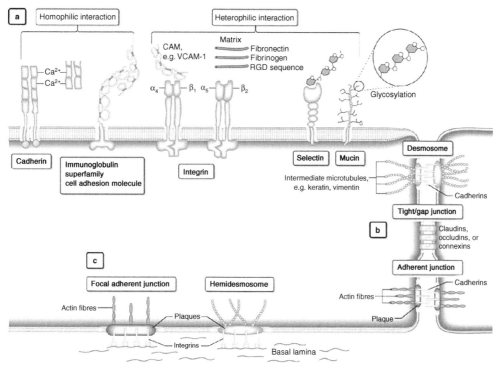

Figure 3.6   Cellular connection and cell adhesion molecules. (a) Five main groups of adhesion molecules. Selectins have a weak binding affinity for specific sugar molecules found on mucins. Mucins are long tandem-repeat peptides which protrude from the cell membrane and are covered in glycosylation moieties. (b) Adjacent cells form focal adhesion junctions. **Desmosomes** are where the membrane forms a proteinaceous plaque containing molecules like desmoglein, from which cadherins protrude and bind cadherins of the adjacent cell. Intracellularly the plaque binds loops of cytoskeleton to intermediate filaments, e.g. keratin in epithelial cells and vimentin in fibroblasts. **Tight junctions** are mediated by integral membrane proteins, claudins and diclaudins, which associate to form subunits bridging the intercellular gap. **Gap junctions** consist of connexin subunits that form a regulated hollow tube. **Adherent junctions** are similar to desmosomes in that cell-to-cell adhesion is mediated by cadherins but the membrane proteinaceous plaque components are different and bind contractile cytoskeletal fibres like actin at the terminus. (c) Basement membrane adhesion. This is similar to desmosomes and adherent junctions in that membrane plaques link intercellular intermediate filaments (e.g. keratin or vimentin) in **hemidesmosomes** and contractile cytoskeleton actin in **focal adherent junctions** to the basement matrix. However, integrins replace cadherins as the surface adhesion molecules. From Kumar and Clark, *Clinical Medicine,* 6ed 2005 © Elsevier

cells (e.g. intestinal and renal cells) and form a barrier to the movement of ions and solutes across the epithelium, although they can be variably 'leaky' to certain solutes. The proteins responsible for the intercellular tight junction closure are called claudins. They show selective expression within tissue and regulate which small ions may pass through the gaps between cells. For example, the kidney displays a differential expression of these claudin proteins. Mutations of claudin-16 (which is only expressed in the thick ascending limb of the loop of Henle where magnesium is reabsorbed) is responsible for some forms of Gitelman's syndrome — a rare inherited

hypomagnesaemia characterized by massive urinary magnesium loss, hypercalciuria and seizures at an early age. Since magnesium reabsorbtion is para-cellular, tight junctions, which contain claudin-16, presumably prevent these divalent ions rapidly diffusing back between the cells into renal tubules.

Adherent junctions (zonula adherens) are continuous on the basal side of cells. These transmembrane junctions contain cadherins and are the major site of the attachment of intracellular microfilaments. Intermediate filaments attach to desmosomes, which are apposed areas of thickened membranes of two adjacent cells. Hemidesmosomes attach cells to the basal lamina and are also connected to inter-mediate filaments. Transmembrane integrins link the extracellular matrix to microfilaments at focal areas where cells also attach to their basal laminae. In blistering dermatological disorders autoantibodies cause damage by attacking tight junction desmo-somal proteins such as desmoglein 3 in pemphigus vulgaris and desmoglein-1 in pemphigus foliaceus.

Gap junctions allow substances to pass directly between cells without entering the extracellular fluids. Pro-tein channels (**connexons**) are lined up between two adjacent cells and allow the passage of solutes up to molecular weight 1000 kDa (e.g. amino acids and sugars), as well as ions, chemical messengers and other factors. The diameter of these channels is regulated by intracellular Ca^{2+}, pH and voltage. Connexons are made up of six subunits surrounding a channel and their isoforms in tissues are encoded by different genes. Mutant connexons can cause disorders, such as the X-linked form of Charcot−Marie−Tooth disease.

3.1.4   Cell adhesion molecules

Adhesion molecules and adhesion receptors are essen-tial for tissue structural organization. Differential ex-pression of such molecules is implicit in the processes of cell growth and differentiation, such as wound repair and embryogenesis. There are four major families of cell adhesion molecules: cadherins, integrins, immuno-globulin superfamily and the selectins (see Figure 3.6).

3.1.4.1   Cadherins

The cadherins establish molecular links between adja-cent cells. They form zipper-like structures at 'adherens junctions', areas of the plasma membrane where cells make contact with other cells. Through these junc-tions, bundles of actin filaments run from cell to cell. Related molecules such as desmogleins form the main constituents of desmosomes, the intercellular contacts found abundantly between epithelial cells. Desmo-somes serve as anchoring sites for intermediate fila-ments of the cytoskeleton. When dissociated embry-onic cells are grown in a dish, they tend to cluster according to their tissue of origin. The homophilic (like with like) interaction of cadherins is the basis of this separation and has a key role in segregating embryonic tissues. The expression of specific adhesion molecules in the embryo is crucial for the migration of cells and the differentiation of tissues. For example, when neural crest cells stop producing N-CAM and N-cadherin and start to display integrin receptors they can separate and begin to migrate on the extracellular matrix. Changes in cadherin expression are often associated with tumour metastatic potential.

3.1.4.2   Integrins

Integrins are membrane glycoproteins consisting of alpha and beta subunits and they exist in active and inactive forms. The integrins principally bind to extracellular matrix components such fibrinogen, elas-tase and laminin. The amino acid sequence arginine−glycine−aspartic acid (RGD) is a potent recognition sequence for integrin binding and integrins replace cadherins in the focal membrane anchorage of hemi-desmosomes and focal adhesion junctions. An impor-tant feature of integrins is that the active form can come about as a result of a cytoplasmic signal that causes a conformational change in the extracellular domain in-creasing affinity for its ligand. This 'inside out' signalling occurs when leucocytes are stimulated by bacterial pep-tides, rapidly increasing leukocyte integrin affinity for immunoglobulin superfamilies structures such as the Fc portion of immunoglobulin. The 'outside in' signalling follows the binding of the ligand to the integrin and stimulates secondary signals resulting in diverse events such as endocytosis, proliferation and apoptosis. Defec-tive integrins are associated with many immunological and clotting disorders such as Bernard−Soulier syn-drome and Glanzmann's thrombasthenia.

3.1.4.3   Immunoglobulin superfamily cell adhesion molecules (iCAM's)

These molecules contain domain sequences, which are immunoglobulin-like structures. The neural-cell ad-hesion molecule (N-CAM) is found predominantly in the nervous system. It mediates a homophilic (like with like) adhesion. When bound to an identical molecule on another cell, N-CAM can also associate laterally with a fibroblast growth factor receptor and stimulate the tyrosine kinase activity of that receptor to induce the growth of neurites. Thus adhesion molecules can trig-ger cellular responses by indirect activation of other types of receptors. The placenta and gastrointestinal tract also express immunoglobulin superfamily mem-bers, but their function is not completely understood.

3.1.4.4   Selectins

Unlike most adhesion molecules — which bind to other proteins — the selectins interact with carbohydrate-ligands or mucin complexes on leucocytes and endothelial cells (vascular and haematologic systems). Selectins were named after the tissues in which they were first identified. L-selectin is found on leucocytes and mediates the homing of lymphocytes to lymph nodes. E-selectin appears on endothelial cells after they have been activated by inflammatory cytokines; the small basal amount of E-selectin in many vascular beds appears to be important for the migration of leucocytes.

P-selectin is stored in the alpha granules of platelets and the Weibel–Palade bodies of endothelial cells, but it moves rapidly to the plasma membrane upon stimulation of these cells. All three selectins play a part in leucocyte rolling.

3.1.5   Receptors

There are many intracellular receptors that bind lipid soluble ligands such as steroid hormones (e.g. progesterone, cortisol, T3 and T4). These cytoplasmic receptors often change shape in response to binding their ligands, form dimers, enter the nucleus and interact directly with specific DNA sequences The nonlipid soluble ligands such as growth hormone (GH), insulin, insulin-like growth factor (IGF) and luteinizing hormone (LH) have membrane surface receptors which pass their extracellular signal across the plasma membrane to cytoplasmic 2° signalling molecules. These membrane bound receptors can be subclassified according to the mechanism by which they activate signalling molecules:

- ion channel linked;
- G-protein linked;
- enzyme linked.

Structurally these plasma membrane receptors can be (see Figure 3.7):

- serpentine (seven transmembrane domains; e.g. the LH receptor);
- transmembrane with large extra- and intracellular domains (e.g. the epidermal growth factor (EGF) receptor);
- transmembrane with large extracellular domain only;
- entirely linked onto the outer membrane leaflet by a lipid moiety known as a GPI (glycanphosphatidylinositol) anchor (e.g. T-cell receptor).

The function of these membrane receptors is to initiate a secondary message which ultimately results in activation of a DNA binding protein. This translocates to the nucleus and initiates transcription of a specific set of genes.

3.1.5.1   Ion channel receptors

Commonly found, but not exclusive on the cell membranes of neurological tissue where a sudden influx of ions causes a depolarization wave or nerve impulse,

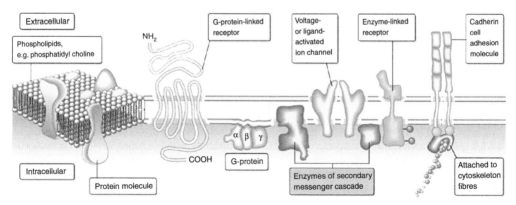

Figure 3.7   Diagrammatic representations of the plasma membrane and integral protein receptors and secondary messenger proteins. From Kumar and Clark, *Clinical Medicine*, 6ed 2005 © Elsevier

ion channel receptors are found as abundant intracellular components and will trigger a release of, for example, Ca^{2+} ions from vesicles/organelles in which they are stored as part of a secondary messaging pathway.

3.1.5.2 G-Protein linked receptors

The G-protein linked receptor, once activated by a ligand, binds a trimeric complex (α, β, γ) which is anchored to the inner surface of the plasma membrane (see Figure 3.7). This complex is a GTP-binding protein, or G-protein. The G-protein binds GTP rather than GDP (see Figure 3.8), and then interacts with enzyme complexes anchored into the inner leaflet of the membrane. These complexes in turn activate one or all three of the secondary messengers:

- cyclic AMP (CAMP).
- Ca^{2+} ions.
- Inositol 1,4,5-triphosphate/diacylglycerol (IP$_3$/DAG).

3.1.5.3 Enzyme-linked surface receptors

These receptors usually have a single transmembrane-spanning region and a cytoplasmic domain that has intrinsic enzyme activity, or will bind and activate other membrane-bound or cytoplasmic enzyme complexes. Four classes of enzymes have been designated as follows.

Guanylyl cyclase-linked receptors, for example the atrial natriuretic peptides, which produce cyclic GMP. This in turn activates a cGMP-dependent kinase (GPCRK), which binds to and phosphorylates serine and threonine residues of specific secondary messengers.

Tyrosine kinases receptors, for example the platelet derived growth factor (PDGF) receptors, which either specifically phosphorylate kinases on a small set of intracellular signalling proteins, or associate with proteins that have tyrosine kinase activity.

Tyrosine phosphatase receptors, for example CD45, which remove phosphates from tyrosine residues of specific intracellular signalling proteins.

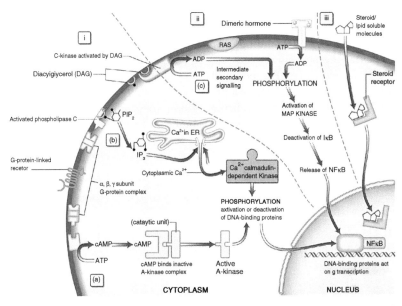

Figure 3.8   Receptor and secondary messengers. (i) G-protein receptor binds ligand (e.g. hormone) and activates G-protein complex. The G-protein complex can activate three different secondary messengers: (a) cAMP generation; (b) inositol 1,4,5-trisphosphate (IP$_3$) and release of Ca^{2+}; (c) diacylglycerol (DAG) activation of C-kinase and subsequent protein phosphorylation. (ii) Dimeric hormone binds receptor subunits bringing them into close association. Intracellular domains cross-phosphorylate and link to the phosphorylation cascades via molecules such as RAS. (iii) Lipid-soluble molecules, e.g. steroids, pass through the cell membrane and bind to cytoplasmic receptors, which enter the nucleus and bind directly to DNA. ER, endoplasmic reticulum; IκB, inhibitory factor kappa B; NFκB, nuclear factor kappa B. From Kumar and Clark, *Clinical Medicine,* 6ed 2005 © Elsevier

Serine/threonine kinase receptors, for example transforming growth factor beta (TGFβ) receptors, which phosphorylate specific serine and threonine residues of intracellular signalling proteins.

3.1.5.4   Cytoplasmic receptor

These are predominantly steroid binding proteins which are free in the cell cytoplasm. On binding the lipid soluble steroid molecule, which simply dissolves across the cell membrane (no need for a membrane bound receptor), they undergo a conformational change, enter the nucleus and bind directly with DNA initiating gene transcription or silencing. Thus, there is no need for a secondary messenger cascade for steroids but all other cell signalling mediated via a cell membrane bound receptor requires an intracellular secondary messenger cascade (see Figure 3.8).

3.1.6   Secondary messengers

Secondary messengers are molecules which transduce a signal from a. bound receptor to its site of action (e.g. the nucleus). There are essentially four mechanisms by which secondary messengers act but they crosstalk and are rarely activated independent of each other (see Figure 3.8). These mechanisms are cyclic AMP/cAMP, IP3/DAG, Ca^{2+} ions changes and protein phosphorylation.

3.1.6.1   G-protein activation, inactivation

Guanine nucleotide binding proteins are a ubiquitous cellular mechanism for coupling an extracellular signal via a serpentine transmembrane receptor to a second messenger such as cAMP. G-proteins have three noncovalently associated subunits: α, β, γ. In the inactive state GDP is bound to the α subunit of the G-protein. When the receptor is activated by ligand binding, the G-protein is activated by the exchange of GDP for GTP. In this active state the α subunit dissociates from the Ry subunit complex. Either of these two complexes can then interact with enzymes producing second messengers such as adenyl cyclase forming cAMP. The α subunit is rapidly inactivated by hydrolysis of GTP to GDP (this is an intrinsic property of a subunit) and then reassociates

with the β and γ subunits resetting the whole system to the inactive state (see figure 3.9). The dissociation of the complex away from the receptor on activation allows multiple G-protein complexes to be activated by a single receptor so amplifying the signal (see Figure 3.8).

3.1.6.2   Cyclic AMP, IP$_3$/DAG and CA^{2+} ions

The generation of cAMP by G-protein-linked receptors results in an increase in cellular cAMP, which binds and activates specific cAMP-binding proteins. These dimerize and enter the cell nucleus to interact with set DNA sequences (the cAMP response elements or CREs found within the promoter region of genes). In addition, cofactors in the cAMP-binding proteins are coactivated and interact with the phosphorylation pathway.

Other G-protein complexes activate inner membrane-bound phospholipase complexes. These in turn cleave membrane phospholipid — polyphosphoinositide (PIP$_2$) — into two components. The first is the water-soluble molecule inositoltriphosphate, IP3. This floats off into the cytoplasm and interacts with gated ion channels in the endoplasmic reticulum (or sarcoplasmic reticulum in muscle cells), causing a rapid release of Ca^{2+}. The lipid-soluble component diacylglycerol (DAG) remains at the membrane, but activates a serine/threonine kinase, protein kinase C (see phosphorylation section below).

Although the cellular calbindin and ion pumps rapidly remove Ca^{2+} from the cytoplasm back into a storage compartments (such as the endoplasmic reticulum), free Ca^{2+} interacts with target proteins in the cytoplasm, inducing a phosphorylation/dephosphorylation cascade, resulting in activated DNA binding proteins entering the nucleus.

3.1.6.3   Protein phosphorylation

Although phosphorylation of the cytoplasmic secondary messengers is often a consequence of secondary activation of cAMP, Ca^{2+} and DAG, the principal route for the protein phosphorylation cascades is from the dimerization of surface protein kinase receptors, which have bound their ligands. The tyrosine kinase receptors phosphorylate each other when ligand binding brings the intracellular receptor components into close proximity. The

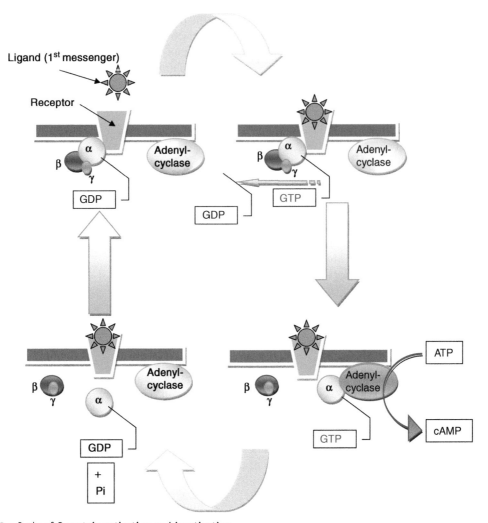

Figure 3.9   Cycle of G-protein activation and inactivation

inner membrane and cytoplasmic targets of these activated receptor complexes are Ras, protein kinase C and ultimately the MAP (microtabule associated protein) kinase, JAK—STAT pathways or phosphorylation of I-KB causing it to release its DNA-binding protein nuclear factor KB (NFKB). These intracellular signalling proteins usually contain conserved noncatalytic regions called SH2 and SH3 (serc homology regions 2 and 3). The SH2 region binds to phosphorylated tyrosine. The SH3 domain has been implicated in the recruitment of intermediates which activate Ras proteins. Like G-proteins, Ras (and its homologous family members

Rho and Rac) switches between an inactive GDP-binding state and an active GTP-binding state. This starts a phosphorylation cascade of the MAP kinase, JAK—STAT protein pathways, which ultimately activate a DNA-binding protein. This undergoes a conformational change, enters the nucleus and initiates transcription of specific genes (see Figure 3.8).

Lipid soluble ligands, for example steroids, do not need secondary messengers and their cytoplasmic receptors, once activated, enter the nucleus as DNA binding proteins and alter gene expression directly (see Figure 3.8)

3.1.6.4 Receptors and disease

Receptors —and indeed the secondary messenger systems they activate — are often the target of a drug therapy. However, illness — be it a genetic malfunction of a component or that an infectious agent targets a receptor/secondary molecular protein — is most often the molecular basis of pathology. There is a vast array of receptor and secondary messenger systems (see Table 3.1) and G-protein deregulation has been studied intensively as the molecular pathology behind not only inborn errors but how/why certain infections are so harmful: Albright's hereditary osteodystrophy or pseudohypoparathyroidism, is where there is generalized resistance to a variety of hormones and dysmorphic features. The molecular cause has been found to be due to a 50% reduction in the activity of the G-protein subunit which activates adenylate cyclase in response to hormones such as PTH, T3 and Gonadotrophins. At the genetic level this has been found to be due to a germ line mutation affecting the G-protein alpha subunit — Gsα. Similarly for pituitary adenoma, somatic mutations activating the Gsα subunit of G-proteins occur in about 40% of patients with acromegaly. This leads to autonomous secretion of GH. In cholera infections it was found that *Vibrio cholerae* secretes an exotoxin which catalyses ADP-ribosylation of an arginine residue on Gsα. This makes the subunit resistant to hydrolysis and the second messenger (in this case cAMP) remains activated and this ultimately leads to the fluid and electrolyte loss characteristic of the disease. In Pertussis the bacterial toxin blocks G-protein function.

Table 3.1   Examples of the variety of receptor/secondary messenger systems and their classification

Receptor type	Examples
Gated ion channel	Acetyl choline receptor (Nicotinic). GABA and glycine receptors
G-protein receptors	The Gonadotropins, ACTH, GHRH, TSH, α1 adrenergic, histamine-1, vasopressin −1, TRH, substance P, endothelins, IL-8, Rhodopsin
Phosphorylation/ kinase receptors	Insulin, Platelet Derived Growth Factor, IGF-1, Macrophage-Colony Stimulating Factor, Nerve Growth Factor, TNF, fas, IFNs, GH, prolactin, IL2-7

3.1.7 Cell cycle and apoptosis

Regulation of the cell cycle is complex. Cells in the quiescent G0 phase (G, gap) of the cycle are stimulated by the receptor-mediated actions of growth factors (e.g. EGF, epithelial growth factor; PDGF, platelet-derived growth factor; IGF, insulin-like growth factor) via intracellular second messengers. Stimuli are transmitted to the nucleus (see below) where they activate transcription factors and lead to the initiation of DNA synthesis, followed by mitosis and cell division. Cell cycling is modified by the cyclin family of proteins that activate or deactivate proteins involved in DNA replication by phosphorylation (via kinases and phosphatase domains).

Thus from G0 the cell moves on to G1 (gap 1) when the chromosomes are prepared for replication. This is followed by the synthetic (S) phase, when the 46 chromosomes are duplicated into chromatids, followed by another gap phase (G2), which eventually leads to mitosis (M) (see Figures 3.10 and 3.11).

3.1.8 Programmed cell death (see Figure 3.12)

Necrotic cell death is where some external factor (e.g. hypoxia, chemical toxins) damages the cell's physiology and results in the disintegration of the cell. Characteristically there is an influx of water and ions, after

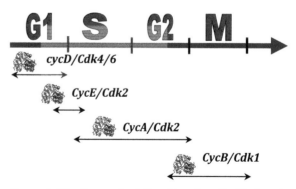

Figure 3.10 Temporal relationship of cyclin (Cyc) and cyclin dependant kinase (Cdk) complex expression during the cell cycle. The thick read arrow represents a linear progression through the cell cycle and the double-headed black arrows represent the period at which a given Cyc/Cdk complex will be expressed. The CycB-Cdk1 complex is also known as the mitosis-promoting factor (MPF). The repeated 3D molecular model is of CycA/Cdk2

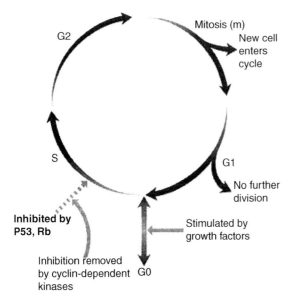

Figure 3.11 Diagrammatic representation of the cell cycle. From Kumar and Clark, *Clinical Medicine*, 6ed 2005 © Elsevier

Apoptotic cell death has characteristic morphological features:

- chromatin aggregation, with nuclear and cytoplasmic condensation into distinct membrane-bound vesicles which are termed apoptotic bodies
- organelles remain intact;
- cell 'blebs' (which are intact membrane vesicles);
- there is no inflammatory response;
- cellular 'blebs' and remains are phagocytosed by adjacent cells and macrophages.

This process requires energy (ATP), and several Ca^{2+}- and Mg^{2+}-dependent nuclease systems are activated which specifically cleave nuclear DNA at the interhistone residues. An endonuclease destroys DNA following apoptosis. This involves the enzyme CASPASE (cysteine-containing aspartase-specific protease) which activates the CAD (caspase-activated DNase)/ICAD (inhibitor of CAD) system which can destroy DNA. It is now recognized that regulated apoptosis is essential for many life processes from tissue structure formation in embryogenesis and wound healing to normal metabolic processes such as autodestruction of the thickened endometrium to cause menstruation in a nonconception cycle. In oncology it has become clear that chemotherapy and radiotherapy regimes only work if they can trigger the

which cellular organelles swell and rupture. Cell lysis induces acute inflammatory responses *in vivo* owing to the release of lysosomal enzymes into the extracellular environment. In apoptosis, physiological cell death occurs through the deliberate activation of constituent genes whose function is to cause their own demise.

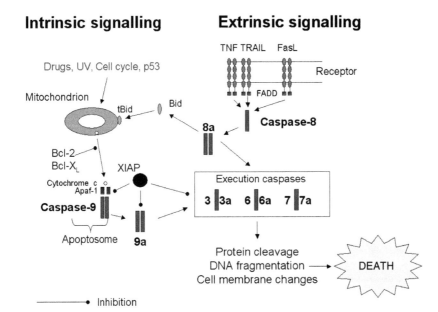

Figure 3.12 Diagrammatic representation of the intrinsic and extrinsic apoptosis pathways and the links between them

tumour cells' own apoptotic pathways. Failure to do so in resistant tumours can result in the accumulation of further genetic damage to the surviving cells. Several factors initiate apoptosis but in general there are two signalling pathways: the extrinsic apoptotic pathway triggered by death receptors on the cell surface and the intrinsic pathway initiated at the mitochondrial level. Death receptors are all members of the TNF receptor superfamily and include CD95 (APO-1/Fas), TRAIL (TNF-related apoptosis ligand)-R1, TRAIL-R2, TNF-R1, DR3 and DR6.

The extrinsic pathway is important in processes such as tissue remodelling and induction of immune selftolerance. Activated receptors with internal death domain complex multiple pro-caspase 8 molecules whose autocatalytic activity results in released of the initiator caspase 8. In turn caspase 8 cleaves pro-caspases 3 and caspase 3, in combination with the other effector caspases, activates DNA cleavage, cell condensation and fragmentation. The intrinsic pathway centres on the release of cytochrome C from the mithochondria. Cellular stress, such as growth factor withdrawal and p53 cell cycle arrest induces the expression of pro-apoptotic Bcl-2 family of proteins, Bax and Bak. These form tetrameric complexes, which imbed into the outer mitochondrial membrane forming permissive pores. Cytochrome C, released from the mitochondria, binds Apaf1, forms a complex known as the apoptosome, which then activates an initiator caspase, in this case, caspase 9. Caspase 9 then activates the effector caspase, caspase 3. Other proteins released from damaged mitochondria, Smac/DIABLO and Omi/HtrA2, counteract the effect of IAPs (inhibitor of apoptosis proteins), which normally bind and prevent activation of procaspase 3. Antiapoptotic Bcl-2 protein, when incorporated as a member of the Bak/Bax pore complex, renders the mitochondrial pore nonpermissive to the release of cytochrome C and the anti IAPs.

There is an amplification link between the extrinsic and intrinsic apoptotic pathways in that caspase 8 cleaves a Bcl-2 family member, tBid, which then aids formation of the Bcl-2/Bax/Bak pore complexes. If this complex is predominately formed from proapoptotic members of the Bcl-2 family of proteins then apoptosome/caspase 9, along with mitochondrial anti-IAPs, amplifies the apoptotic activation of effector caspases 3. Conversely, over expression of antiapoptotic Bcl-2 will not only inhibit intrinsic, but also dampen down extrinsic apoptotic signalling.

Most cells rely on a constant supply of survival signals without which they will undergo apoptosis. Neighbouring cells and the extracellular matrix provide these signals. Cancer, autoimmunity and some viral illnesses are associated with the inhibition of apoptosis and increased cell survival. Metastatic tumour cells have circumvented the normal environmental cues for survival and can survive in foreign environments.

3.2 Genetics

3.2.1 Human chromosomes

The nucleus of each diploid cell contains 6×10^9 bp of DNA in long molecules called chromosomes. Chromosomes are massive structures containing one linear molecule of DNA that is wound around histone proteins into small units called nucleosomes, and these are further wound to make up the structure of the chromosome itself (see Figure 3.13). Diploid human cells have 46 chromosomes, 23 inherited from each parent; thus there are 23 'homologous' pairs of chromosomes (22 pairs of 'autosomes' and two 'sex chromosomes'). The sex chromosomes, called X and Y, are not homologous but are different in size and shape. Males have an X and a Y chromosome; females have two X chromosomes (see Figure 3.14). (Primary male sexual characteristics are determined by the SRY gene — sex determining region, Y chromosome.)

The chromosomes can be classified according to their size and shape, the largest being chromosome 1. The constriction in the chromosome is the centromere, which can be in the middle of the chromosome (metacentric) or at one extreme end (acrocentric). The centromere divides the chromosome into a short arm and a long arm, which are referred to as the p arm and the q arm respectively (see Figure 3.15). In addition, chromosomes can be stained when they are in the metaphase stage of the cell cycle and are very condensed. The stain gives a different pattern of light and dark bands that is diagnostic for each chromosome. Each band is given a number, and gene mapping techniques allow genes to be positioned within a band within an arm of a chromosome. For example, the CFTR gene (in which a defect gives rise to cystic fibrosis) maps to 7q21; that is, on chromosome 7 in the long arm in band 21.

During cell division (mitosis), each chromosome divides into two so that each daughter nucleus has the

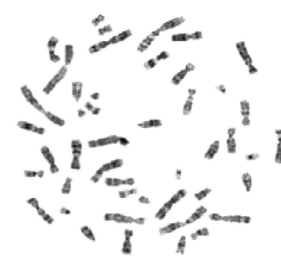

Figure 3.13   G banded chromosome spread

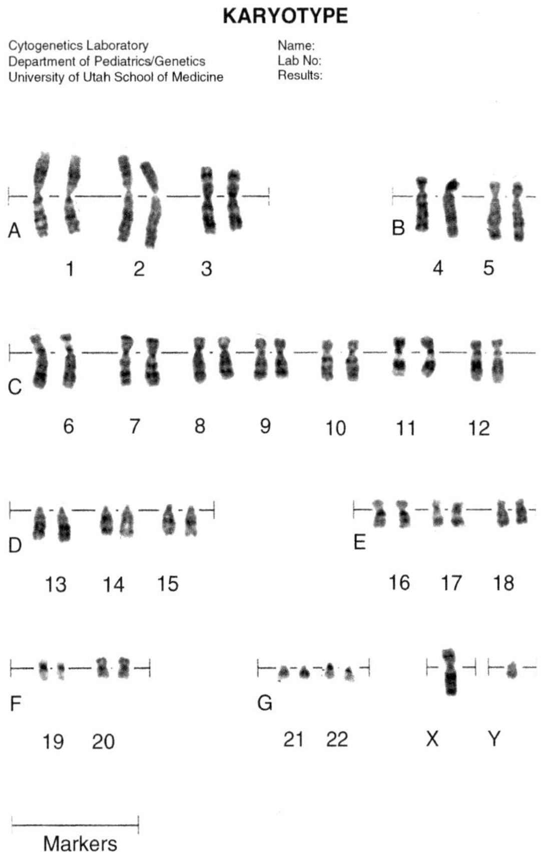

KARYOTYPE

Cytogenetics Laboratory   Name:
Department of Pediatrics/Genetics Lab No:
University of Utah School of Medicine Results:

A
  1   2   3   B   4   5

C
  6   7   8   9   10   11   12

D
  13   14   15   E   16   17   18

F   G
  19   20   21   22   X   Y

Markers

Figure 3.14   46 chromosomes, 44 in 22 pairs (autosomes) and two sex chromosomes

same number of chromosomes as its parent cell. During gametogenesis, however, the number of chromosomes is halved by meiosis, so that after conception the number of chromosomes remains the same and is not doubled. In the female, each ovum contains one or other X chromosome but, in the male, the sperm bears either an X or a Y chromosome.

Chromosomes can only be seen easily in actively dividing cells. Typically, lymphocytes from the peripheral blood are stimulated to divide and are processed to allow the chromosomes to be examined. Cells from other tissues can also be used — for example fetal epithelia in amniotic fluid, placental cells from chorionic villus sampling, bone marrow and skin.

3.2.1.1   The X chromosome and inactivation

Although female chromosomes are XX, females do not have two doses of X-linked genes (compared with just one dose for a male XY), because of the phenomenon of X inactivation or Lyonization (after its discoverer, Dr Mary Lyon). In this process, one of the two X chromosomes in the cells of females becomes transcriptionally inactive, so the cell has only one dose of the X-linked genes. Inactivation is random and can affect either X chromosome.

3.2.1.2   Telomeres and immortality

The ends of chromosomes, telomeres (see Figure 3.15), do not contain genes but many repeats of a hexameric sequence TTAGGG. Replication of linear chromosomes starts at coding sites (origins of replication) within the main body of chromosomes and not at the two extreme ends. The extreme ends are therefore susceptible to single-stranded DNA degradation back to double-stranded DNA. Thus cellular ageing can be measured as a genetic consequence of multiple rounds of replication with consequential telomere shortening. This leads to chromosome instability and cell death.

Stem cells have longer telomeres than their terminally differentiated daughters. However, germ cells replicate without shortening of their telomeres. This is because they express an enzyme called telomerase, which protects against telomere shortening by acting as a template primer at the extreme ends of the chromosomes. Most somatic cells (unlike germ and embryonic cells) switch off the activity of telomerase after birth and die as a result of apoptosis. Many cancer cells, however, reactivate telomerase, contributing to their

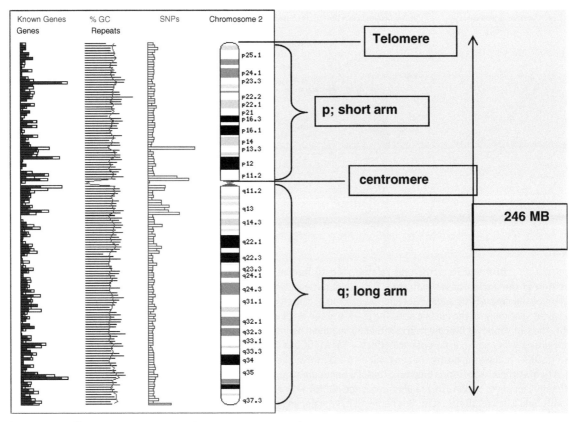

Figure 3.15　Chromosome structure and gene sequence mapping

immortality. Conversely, cells from patients with pro-geria (premature ageing syndrome) have extremely short telomeres. Recent research has shown transient expression of telomerase in various stem and daughter cells as part of their normal biology. The inability to activate telomerase in cells such as those of the immune system can produce disease pathologies, in addition to over expression.

3.2.1.3　The mitochondrial chromosome

In addition to the 23 pairs of chromosomes in the nucleus of every diploid cell, the mitochondria in the cytoplasm of the cell also have their own chromo-somes. The mitochondrial chromosome is a circular DNA (mtDNA) molecule of approximately 16 500 bp, and every basepair makes up part of the coding se-quence. These genes principally encode proteins or RNA molecules involved in mitochondrial function. These proteins are components of the mitochondrial respiratory chain involved in oxidative phosphoryla-tion (OXPHOS) producing ATP. They also have a critical role in apoptotic cell death. Every cell contains several hundred mitochondria, and therefore several hundred mitochondrial chromosomes. Virtually all mitochondria are inherited from the mother as the sperm head contain no (or very few) mitochondria. Disorders are described later.

3.2.2　Nucleic acids

Nucleosides are ribose sugarbases without the phos-phate group (Adenosine, Guanosine, Cytidine, Thymi-dine, Uridine). Nucleotides (nucleosides with phos-phate groups) form dinucleotides by the formation of a phosphodiester bond between the OH group of the sugar on one nucleotide and the phosphate group of another nucleotide; this reaction is catalysed by DNA polymerase. These dinucleotides, in turn, link to other nucleotides with many phosphodiester linkages to form oligonucleotides and then a polynucleotide or nucleic acid. Nucleic acids are either ribonucleic acids (RNA, where the sugar backbone is ribose) or deoxyribonucleic

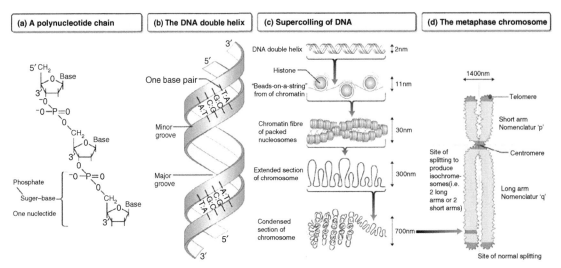

Figure 3.16 **DNA and its structural relationship to human chromosomes.** (a) **A polynucleotide strand with the position of the nucleic bases indicated.** Individual nucleotides form a polymer linked via the deoxyribose sugars. The 5' carbon of the heterocyclic sugar structure links to the 3' carbon of the next via a phosphate molecule forming the sugar—phosphate backbone of the nucleic acid. The 5'—3' linkage gives an orientation to a sequence of DNA. (b) **Double-stranded DNA.** The two strands of DNA are held together by hydrogen bonds between the bases. As T always pairs with A, and G with C, there are only four possible pairs of nucleotides — TA, AT, GC and CG. The orientation of the complementary single strands of DNA (ssDNA) is always opposite; i.e. one will be 5'—3' whilst the partner will be 3'—5'. CG base-pairs form three hydrogen bonds whilst the AT bonds form only two. Thus, CG bonds are stronger than AT bonds, which affects the biophysical nature of different sequences of DNA. In a random sequence of DNA with equal proportions of CG and AT base-pairs, complementary strands form a helical 3D structure. This helix will have major and minor grooves and a complete turn of the helix will contain 12 basepairs. These grooves are structurally important, as DNA-binding proteins predominantly interact with the major grooves. DNA sequences rich in repetitive CG basepairs distort the helical shape whereby the minor grooves become more equal in size to the major grooves, giving a Z-like structure. These CG-rich regions are sites were DNA-binding proteins are likely to bind. (c) **Supercoiling of DNA.** In humans, and other higher organisms, the large stretches of helical DNA are coiled to form nucleosomes and further condensed into the chromosomes that can be seen at metaphase. DNA is first packaged by winding around nuclear proteins — histones — every 180 bp. This can then be coiled and supercoiled to compact nucleosomes and eventually visible chromosomes. (d) **At the end of the metaphase DNA replication will result in a twin chromosome joined at the centromere.** This picture shows the chromosome, its relationship to supercoiling, and the positions of structural regions: centromeres, telomeres and sites where the double chromosome can split. From Kumar and Clark, *Clinical Medicine*, 6ed 2005 © Elsevier

acids (DNA, where the sugar backbone is deoxyribose). RNA also differs from DNA in that Uracil replaces Thymine and generally RNA is single-stranded where DNA is double-stranded (see Figure 3.16(a)).

Watson and Crick first described the double-stranded helix (*Nature*, 1953) based on their earlier work, the work of Erwin Chargaff (complementary base paring) and an X-ray diffraction image produced by Maurice Wilkins and Rosalind Franklin. The DNA model has antiparallel strands made of a sugar phosphate backbone with a nitrogenous base core of both purines and pyrimidines — which pair in a complementary fashion A with T, C with G, that is purines always pair with pyrimidines. DNA has a uniform

diameter (20 Å) with a double right-handed twist where each full twist is 34 Å comprising 10 base pairs (3.4 Å between bases).

DNA is packaged and stored as chromatin in the nucleus of eukaryotic cells. Chromosomes are only visible during cell division when the chromatin condenses during metaphase. Chromatin is therefore very tightly packaged DNA, coiled, wrapped and looped, then coiled again into chromosomes (see Figure 3.16). We therefore consider DNA packaging in three levels: DNA sequence structure, nucleosome structure and chromatin structure.

Each nucleosome is formed from a histone protein octamer wrapped twice by a 140 bp section of DNA and 'locked' in place by another histone protein H1. Each nucleosome is connected to the next by a short section of Linker DNA. Six nuclesomes spiral together to form one cross section of the nucleosome solenoid, where each set of six nucleosomes stack against the next in a spiral. Approximately half the mass of nuclear chromatin is comprised of histones. These proteins are not just structural but are also involved in transcription. The histone octamer is comprised of two of each histones H2A, H2B, H3 and H4.

Collectively chromatin **is** contained in the nucleus, comprises DNA and proteins (histones and topoisomerases). The nucleosome solenoid loops back on itself and is stabilized by the topoisomerases which also play a role in DNA replication and transcription. This whole structure then twists to form the spiral radial loop model which again twists to form the 'arm' of the chromosome (see Figure 3.16(c)).

3.2.2.1   The central dogma

Genetic information is passed only one way from DNA, to RNA to protein, that is DNA is transcribed to RNA, which in turn is translated to protein. The discovery of retroviruses which made DNA from an RNA template with reverse transcriptase challenged the first part of this one way genetic dogma. The next challenge is to see if protein (perhaps prions) challenges the final argument of this classic dogma.

3.2.3   DNA replication

When a cell undergoes cell division the cellular material effectively divides in two, but the nuclear complement of the daughter cells remains diploid. In order to achieve this the cell must generate an additional set of chromosomes and effectively double its DNA. This happens during the S phase of the cell cycle. The human genome has over 3 000 000 000 000 basepairs (bp) incorporating 20—25 000 genes, so DNA replication is a huge and sensitive task but it is essentially the same in prokaryotes and eukaryotes.

One of the last lines in the famous Watson and Crick paper hypothesized that the nature of the structure of DNA, with its complementary basepairing and antiparallel strands, suggested a model for replication. The DNA unwinds and the two strands separate, each strand functioning as a template for the synthesis of a complementary copy strand. What results, therefore,

are two daughter double-stranded DNA helices each with one old strand and one new synthesized strand. This model of DNA replication is therefore called semiconservative (see Figure 3.17(a)).

3.2.3.1   DNA polymerases

The DNA replication process is catalysed by enzymes; these are DNA polymerase I and DNA polymerase III. As always the situation is much more complex and other DNA polymerases exist and are associated with DNA repair and other more specific functions.

DNA polymerases require magnesium ion cofactors and dNTPs (the four deoxyribonucleotide triphosphates — dATP, dCTP, dGTP and dTTP) to make the new strand. The polymerases also need a template DNA strand to copy and a primer oligonucleotide which binds DNA at the start of the replication site and provides polymerase with a recognizable place to start adding new nucleotides.

DNA polymerase I and III both have a $5'{\rightarrow}3'$ polymerase activity meaning they add new dNTPs one by one onto a free $3'$ end starting with that of the primer. Deoxyribonucleotides are added at several hundred per second so at rates such as these mistakes can occur (about 1 in 100 000 000 nucleotides added) and both enzymes have the ability to 'proof read' what they have done and go back to correct the mistake, this is called $3'—5'$ exonuclease activity. DNA polymerase I can also do this in the $5'—3'$ direction.

3.2.3.2   Enzymes of replication and the replication fork

Other enzymes are also involved in the replication process: DNA Polymerases -I, II, III, IV and V (they essentially catalyse phosphodiester bond formation), Helicase, Topoisomerase (I, II — Girase), Primase, Ligase and telomerase (see Figure 3.17(b)).

Polymerase I is slow DNA polymerase but can remove nucleotides as well as synthesize (exonuclease activity). It is not essential for DNA synthesis but essential for faithful replication (involved more in repair and bacteria deficient in Pol I are grossly mutated). DNA polymerase II (along with IV and V) are responsible for repair of DNA damage caused by external influences. DNA polymerase III despite being found last (in the original three enzymes discovered) is the largest. It is a very complex holoenzyme which catalyses DNA synthesis in a dimeric form.

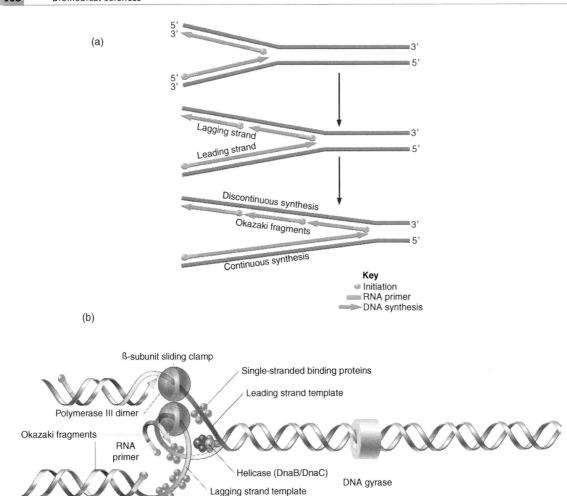

Figure 3.17   DNA synthesis and the interplay of enzymes/protein necessary for replication DNA has antiparalell strands each arranged in an opposite direction (one 5'-3' and the other strand 3'—5'). DNA polymerase III operates in a 5'—3' direction only and can therefore only continuously synthesize a new 3'—5' strand, this is called the leading strand. A problem occurs where the template runs 5'—3' and DNA synthesis has to be discontinuous inserting short pieces of new DNA into what is called the lagging strand. These short strands of new DNA are called Okazaki fragments (b). This 'Y' shaped region between leading and lagging strands is called the replication fork

There are over 20 helicase enzymes and DnaA, DnaB and DnaC are essential. DnaB in association with primase is particularly important for DNA replication and is involved in helical unwinding of double-stranded (ds)DNA into single-standed (ss) DNA. Primases termed ssBPs stabilize the replication bubble.

Topoisomerases are another complex group of enzymes which counteract the effect of unwinding DNA which causes kinks; this is called supercoiling in circular DNA. Topoisomerase II (girase) breaks phosphodiester bonds in either one or both strands of the helix to relax the 'tension' and permit linear strands of DNA to be exposed.

Primases insert the primer RIBOnucleotides and are therefore essentially an RNA polymerase but unlike DNA polymerase it does not require a 3' end to start adding its nucleotides. It adds a short RNA sequence ($\sim$ 10NTPs) which is called a primer and is inserted as DnaB (helicase) unwinds DNA at the replication fork.

Telomerase work in the 3'→5' and therefore can build nucleotides into the exposed ends of replication, for example at the ends of Eukaryotic chromosomes which are linear and not circular packages of DNA as seen in bacteria and viruses.

3.2.4 Transcription

Transcription produces an RNA copy of the template DNA, this copy is said to be complementary. Transcription occurs in three stages: initiation, elongation and termination. The purpose of transcription is to carry the coded message contained in the DNA to the site of translation so that a protein can be synthesized. In prokaryotes this process occurs in the cytoplasm but in compartmentalized eukaryotes transcription occurs in the nucleus and the transcribed RNA is transported to the cytoplasm for translation. In eukaryotes the process has much more complex regulation and control and there are three RNA polymerases, each one responsible for transcribing different RNA molecules:

- RNA Polymerase I is found in the nucleoli, transcribes rRNAs (ribosomal RNA) complex termed 18S, 5.8S and 28S (reflecting their respective molecular masses).
- RNA Polymerase II is found in the nucleoplasm, transcribes hnRNAs (heterogeneous nuclear RNAs — mRNA precursors) and snRNAs (small nuclear RNAs).
- RNA Polymerase III is also found in the nucleoplasm and transcribes 5S rRNA, tRNA (transfer RNA), viral RNAs and possibly other small cellular RNAs.

In prokaryotes RNA polymerase is composed of five subunits (four subunits in the core, plus σ). However, in eukaryotes the arrangement is much more complex involving 12 subunits (in RNA Polymerase II) each with specific functions, some of which are unique to the specific polymerase and some which are common to all three.

Where elongation and termination are essentially similar between prokaryotes and eukaryotes, initiation is much more complex in eukaryotes occurring through interaction and binding of an entire RNA polymerase holoenzyme complex with additional bound transcription factors and is influenced by enhancers and/or silencers. The promoter sequences are specific to each polymerase where: Class I promoters are specific to RNA Polymerase I, Class II promoters are specific to RNA Polymerase II, Class III promoters are specific to RNA Polymerase III.

In the Class II Preinitiation complexes there are the 12 subunit RNA Polymerase II, 6 General Transcription factors (TFIIA, TFIIB, TFIID, TFIIE, TFIIF, TFIIH) where TFIID contains TATA box Binding Protein (TBP) and is common to all classes and at least eight TBP Associated Factors TAFIIs which are specific to Class II preinitiation complexes.

Transcription can only occur when promoters are bound by transcription factors and RNA polymerase subunits. However, the rate of transcription can be affected further by enhancers and silencers. Enhancers are regions involved in stimulation of transcription whereas silencers are regions involved in reducing the rate of transcription. They are common in Class II genes and are not strictly part of the promoter. Enhancers and silencers are generally tissue specific relying on tissue specific DNA binding proteins and can significantly affect transcription. In each case a further DNA binding protein (activator protein) is required to recognize the enhancer/silencer to elicit its effect on the RNA polymerase transcription rate (see Figure 3.18).

3.2.4.1 Insulators

A problem of transcription is how do you stop the RNA polymerase running on through the chromosome activating the promoter of some other gene in the same region of the chromosome, or indeed DNA binding proteins like enhancers turning on the promoters of genes located thousands of basepairs away? One answer is insulator DNA binding proteins/complexes binding to insulator sequences spaced between genes in the genome.

Insulator regions are located between the enhancer(s) and promoter or silencer(s) and promoter of adjacent genes or clusters of adjacent genes and comprise moderately long stretches of DNA (as few as 42 base pairs may do the trick).

Their function is to prevent a gene from being influenced by the activation (or repression) of its neighbours. For example, the enhancer for the promoter of the gene for the delta chain of the gamma/delta T-cell receptor for antigen (TCR) is located close to the promoter for the alpha chain of the alpha/beta TCR (on chromosome 14 in humans). A T-cell must choose between one or the other (see Chapter 7, Section 7.15). There is an insulator between the alpha gene promoter and the delta gene promoter that ensures that activation of one does not spread over to the

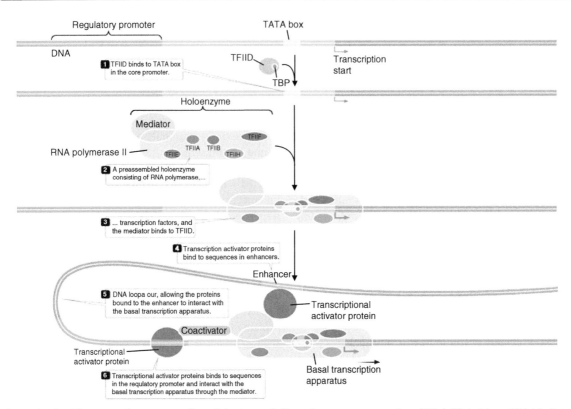

Figure 3.18   Diagrammatic representation of the transcription of a gene sequence to mRNA initiated by a DNA binding protein facilitating the assembly of RNA Polymerase II complex at a close by TATA recognition sequence to which the RNA polymerase complex transcription factor TFIID binds

other (see Figure 3.19). All insulators discovered so far in vertebrates work only when bound by a protein designated CTCF ('CCCTC binding factor', named for a nucleotide sequence found in all the insulators identified).

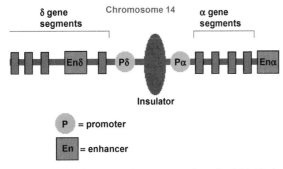

Figure 3.19   Diagrammatic representation of a DNA binding protein recognition sequence for an **insulator** (marked in red), positioned between adjacent promoter regions for the transcription of T-cell receptor subunit genes

3.2.5   Posttranscriptional modifications

During transcription modifications are made almost simultaneously to the mRNA and include: capping, polyadenylation and splicing. These changes result in mRNA and in most cases give rise to termination of transcription and result in transportation of mRNA to the cytoplasm for translation (see Figure 3.20).

Capping (or G capping) occurs at the 5′ end of the RNA and essentially involves the addition of a modified GTP. RNA triphosphatase removes terminal phosphate and guanylyl transferase adds capping GMP (from a dephosphorylated GTP molecule). Another enzyme, methyl transferase, methylates the guanosine base and also the base of the penultimate nucleotide in the sequence. All of this occurs before the transcript is 30 bases long. Capping essentially has four functions:

- ensures proper splicing;
- facilitates transport;

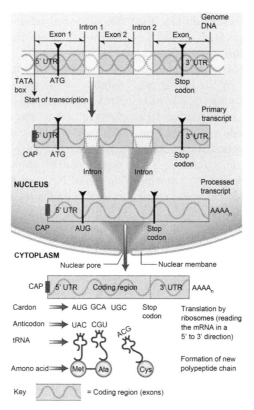

Figure 3.20   Transcription and translation (DNA to RNA to protein). RNA polymerase creates an RNA copy of the DNA gene sequence. This primary transcript is processed: capping of the 5′ free end of the mRNA precursor involves the addition of an inverted guanine residue to the 5′ terminal which is subsequently methylated, forming a 7-methylguanosine residue. (The corresponding position on the gene is thus called the CAP site.)

- protects from ribonuclease degradation;
- enhances translation through facilitation of ribosomal binding.

Polyadenylation (Poly A tail) is literally the addition of a long chain of adenosine nucleotides at the 3′ end of the transcribed RNA. The polyadenylation site is marked by a AAUAAA sequence in the RNA transcript which is closely followed by subsequent GU then U rich regions. Cleavage factors and Poly A polymerase are involved in the addition of 100–300 adenosine nucleotides and marks the end of transcription. Polyadenylation essentially has four functions:

- transcription termination;
- ensures proper splicing;
- provides RNA stability;
- involved in transport to cytoplasm.

Splicing-DNA includes coding (exons) and non-coding regions (introns). Noncoding regions can be repetitive nonsense regions between and within genes as well as regions involved in transcription but not required for translation. These introns are therefore removed in the process of splicing. Small nuclear ribonuclear proteins (snRNPs, or Snurps) bind consensus sequences (specific sequences) in and around the boundary between introns and exons and through the formation of a splicosome 'cuts' out the intron and rejoins the two exons together. The excised intron is degraded and recycled and the snRNPs release and return to splice another intron.

The 3′ end of an mRNA defined by the sequence AAUAAA acts as a cleavage signal for an endonuclease, which cleaves the growing transcript about 20 bp downstream from the signal. The 3′ end is further processed by a Poly A polymerase which adds about 250 adenosine residues to the 3′ end, forming a Poly A tail (polyadenylation). Without these additions, the mRNA sequence will be rapidly degraded $5'-3'$ but the inverted cap nucleotide prevents nuclease attachment.

The activity of specific $5'$ mRNA nucleases to remove the cap is further regulated by the Poly A tail which must first be removed by other degradation enzymes. Splicing out of the introns then produces the mature mRNA (prokaryote genes do not contain introns). This then moves out of the nucleus via nuclear pores and aligns on endoplasmic reticulum.

Ribosomal subunits assemble on the mRNA moving along $5'-3'$. With the transport of amino acids to their active sites by specific tRNAs, the complex translates the code, producing the peptide sequence. Once formed, the peptide is released into the cytoplasmic reticulum for correct folding (from Chapter 3, Iles, R. K. in Kumar Clark *Clinical medicine*, 6th edn, Figure 3.9)

3.2.6   Translation

Translation is mRNA directed polypeptide (protein) synthesis. Once complete, the transcribed mRNA exits the nucleus through pores in the nuclear membrane and is directed towards the site of translation, the endoplasmic reticulum and recruits ribosomes.

Ribosomes are comprised of proteins and RNA (ribosomal RNA, rRNA) arranged in two subunits (large and small). In prokaryotes these are 50s and 30s ribosomal subunits ($=$ 70s Ribosome) and in eukaryotes these are the 60s and 40s ribosomal subunits ($=$ 80s Ribosome).

The mRNA is bound between the large and small subunit of the ribosome and translated by transfer RNA (tRNA). Each ribosome has four binding sites (in addition to the mRNA binding site) EPAT (or TAPE in reverse): transfer site (T) (tRNA lands here), amino site (A) (tRNA anticodon binds codon), polypeptide site (P) (amino acid tags onto peptide) and exit site (E) (tRNA releases amino acid, exits and returns to cytosol) (see Figure 3.21).

Each of the three letters in the mRNA sequence is called a codon and is recognized and bound by the anticodon on the anticodon loop of tRNA (see Figure 3.21). Each tRNA is specific for each amino acid, and through codon/anticodon recognition brings the amino acids to the ER/ribosome complex and presents them to the previous amino acid where they form a peptide bond and the next unit in the polypeptide chain. Thus, the polypeptide is a direct 'translation' of the sequence in the mRNA facilitated by the unique properties of tRNA. This process can be divided into three stages: initiation, elongation and termination.

Initiation: Small ribosomal subunit binds an upstream recognition sequence on the mRNA. A methionine charged tRNA binds AUG start codon forming the

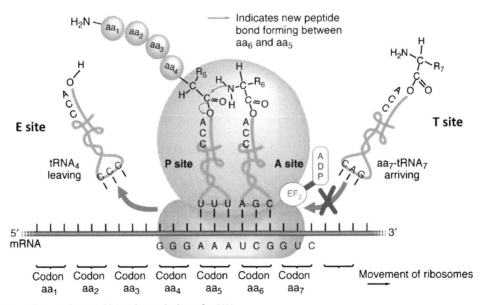

Figure 3.21   Ribosomal assembly and translation of mRNA

initiation complex. The large subunit joins with the first tRNA in the P site.

Elongation: Another tRNA binds the codon in the next position situated in the A site of the ribosome. A peptide bond forms between the original methionine and the new amino acid as it moves to the P site. The free tRNA is released from the P site to the E site as the ribosome moves along mRNA.

Termination: Release Factor (a complex mRNA binding protein) binds the stop codon and frees tRNA from the P site to the E site. The release factor therefore disconnects the peptide from the ribosome and then itself dissociates from the mRNA.

Once the polypeptide is translated it can undergo any one of a number of posttranslational modifications. It is not free circulating and translation is accompanied by the passage of the new peptide chain in to the luminal space of the ER (see Figure 3.22). In this preprotein stage the peptides often have signal recognition particles (SRPs) associated with them that act as escorts to direct the peptide to organelles for further processing via organelle docking channels or SNARES (signal and recognition elements).

3.2.6.1   Protein modifications

There are many types of protein modification and after folding into the correct shape posttranslational modification occurs first in the ER and then the Golgi. The three most common are:

- Proteolysis or protein cleavage which forms individual smaller proteins or subunits of the same protein, for example insulin.
- Glycosylation or addition of sugar residues which forms glycoproteins such as membrane receptors or hormones, for example LH.
- Phosphorylation or addition of phosphate groups by the action of kinases. Amino acids that can be phosphorylated include Serine, Threonine and Tyrosine. Many enzymes and growth factors can be activated in this way.

Other posttranslational modifications include: Methylation, Sulphation, Prenylation, Acetylation and vitamin dependent Hydroxylation/Carboxylation.

3.2.7   Chromatin and regulation of gene expression

Chromatin structure is known to have a major impact on levels of transcription. In eukaryotes, DNA typically exists as a repeating array of nucleosomes in which 146 bp of DNA are wound around a histone octamer. Studies have demonstrated that nucleosomal DNA is generally repressive to transcription in this way, nucleosome structure and DNA-histone interactions usually make the regulatory regions of genes unavailable for the binding of the transcriptional DNA binding proteins and other factors.

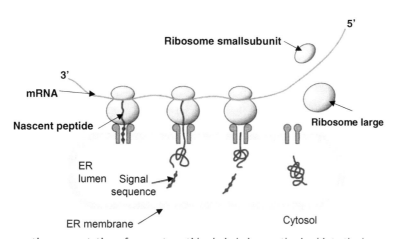

Figure 3.22   Diagrammatic representation of nascent peptide chain being synthesized into the lumen of the endoplasmic reticulum

Several enzymes and protein complexes are now known to bring about changes in the chemical interactions of chromatin, with resultant effects on gene expression. One class of complexes alter the DNA packaging (remodel chromatin) in an ATP-dependent manner; these include the Swi–Snf complex. Another class of chromatin-altering factors acts by covalently modifying histone proteins. These modifications can include phosphorylation, ubiquitination, ADP-ribosylation and methylation, but the best-characterized mechanism is acetylation, catalysed by histone acetyltransferase (HAT) enzymes.

HATs function enzymatically by transferring an acetyl group from acetyl-coenzyme A (acetyl-CoA) to the ε-amino group of exposed lysine side chains within a histone's basic N-terminal tail region. Within a histone octamer, these regions extend out from the associated globular domains, and in the context of a nucleosome, they are believed to bind the DNA through charge interactions (positively charged histone tails associated with negatively charged DNA) or mediate interactions between nucleosomes (Figure 3.23). Lysine acetylation, which neutralizes part of a tail region's positive charge, is postulated to weaken histone-DNA or nucleosome–nucleosome interactions and/or signal a conformational change, thereby destabilizing nucleosome structure or arrangement and giving nuclear transcription factors access to a genetic locus. Thus, acetylated chromatin has long been associated with states of transcriptional activation. Interestingly, certain HATs have also been shown to acetylate lysine residues specifically within transcription-related proteins not just histones.

Histone acetylation is a reversible process, and deacetylase tends to correlate with repression of gene expression. Two opposing regulatory processes work to bring about appropriate levels of transcription. Indeed, some of the histone deacetylases (HDACs) and the proteins with which they associate were previously known as DNA-binding repressors or corepressors.

HAT activities have been isolated and partially characterized (see Table 3.2). Each of these enzymes generally belongs to one of two categories: type A, located in the nucleus, or type B, located in the cytoplasm. Recent studies indicate that some HAT proteins may function in multiple complexes or locations and thus not precisely fit these historical classifications. Nevertheless, B-type HATs are believed to have somewhat of a housekeeping role in the cell, acetylating newly synthesized free histones in the cytoplasm for transport into the nucleus, where they may be deacetylated and incorporated into chromatin. The A-type HATs, on the other hand, acetylate nucleosomal histones within nuclear chromatin.

3.2.7.1 Euchromatin versus heterochromatin

The density of the chromatin that makes up each chromosome (i.e. how tightly it is packed) varies along the length of the chromosome. Dense regions are called heterochromatin, less dense regions are called euchromatin. Heterochromatin is found in parts of the chromosome where there are few or no genes, such as centromeres and telomeres. These areas are greatly enriched with transposons and other 'junk' DNA replicated late in the S phase of the cell cycle and have reduced crossing over in meiosis.

Genes present in heterochromatin are generally inactive; that is, not transcribed and show increased methylation of the cytosines in CpG islands of the DNA, decreased acetylation of histones and increased methylation of lysine-9 in histone H3, which now provides a binding site for heterochromatin protein 1 (HP1), which blocks access by transcription factors.

Euchromatin is found in parts of the chromosome that contain many genes and is loosely-packed in loops of 30 nm fibres. These are separated from adjacent heterochromatin by insulators. The loops are often found near the nuclear pore complexes

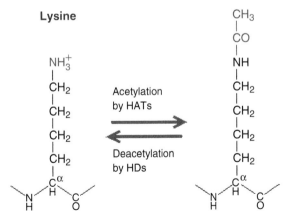

Figure 3.23 Histone acetylation and deacetylation of the lysine residue. Acetylation of the lysine residues at the N terminus of histone proteins removes positive charges, thereby reducing the affinity between histones and DNA

Table 3.2 Summary of known and putative HATs. Histones that are the primary *in vitro* substrates for a given HAT are bold; other histones listed are acetylated weakly or in a secondary manner

HAT	Organisms known to contain the HAT	Known transcription-related functions/effects	HAT activity demonstrated *in vitro*	Histone specificity of recombinant enzyme *in vitro*	Known native HAT complexes and nucleosomal histone specificties *in vitro*
GNAT superfamily					
Hat1	Various (yeast to humans)	None (histone deposition-related B-type HAT)	Yes	**H4**	Yeast HAT-B, HAT-A3 (no nucleosome acetylation)
Gcn5	Various (yeast to humans)	Coactivator (adaptor)	Yes	**H3**/H4	Yeast ADA, SAGA(**H3**/H2B); human GCN5 complex, STAGA, TFTC (**H3**)
PCAF	Humans, mice	Coactivator	Yes	**H3**/H4	Human PCAF complex (**H3**/weak H4)
Elp3	Yeast	Transcript elongation	Yes	ND*	Elongator, polymerase II holoenzyme (**H3**/weak H4)
Hpa2	Yeast	Unknown	Yes	**H3**/H4	
MYST family					
Sas2	Yeast	Silencing	ND		
Sas3	Yeast	Silencing	Yes	**H3**/**H4**/H2A	NuA3^c (**H3**)
Esa1	Yeast	Cell cycle progression	Yes	**H4**/H3/H2A	NuA4 (**H4**/H2A)
MOF	*Drosophila*	Dosage compensation	Yes	**H4**/H3/H2A	MSL complex (**H4**)
Tip60	Humans	HIV Tat interaction	Yes	**H4**/H3/H2A	Tip60 complex
MOZ	Humans	Leukemogenesis, upon chromosomal translocation	ND		
MORF	Humans	Unknown (strong homology to MOZ)	Yes	**H4**/H3/H2A	
HBO1	Humans	ORC interaction	Yes*	ND*	HBO1 complex
p300/CBP	Various multicellular	Global coactivator	Yes	**H2A**/**H2B**/**H3**/**H4**	
Nuclear receptor coactivators					
SRC-1	Humans, mice	Nuclear receptor coactivators (transcriptional response to hormone signals)	Yes	**H3**/H4	
ACTR	Humans, mice		Yes	**H3**/**H4**	
TIF2	Humans, mice		ND		
TAF$_{II}$250	Various (yeast to humans)	TBP-associated factor	Yes	**H3**/H4	TFIID
TFIIIC	Various (yeast to humans)	RNA polymerase III transcription initiation	Yes	**H3**/H4	TFIIIC (**H2A**/**H3**/**H4**)
TFIIIC220	Humans		Yes*	ND	
TFIIIC110	Humans		Yes	ND	
TFIIIC90	Humans		Yes	**H3**	

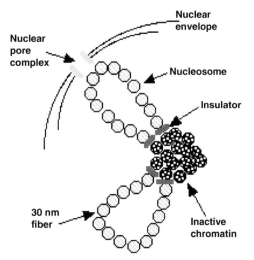

Figure 3.24 Diagrammatic representation of the arrangement of euchromatin and heterochromatin may be organized during interphase in a vertebrate cell

(see Figure 3.24). This would seem to make sense making it easier for the gene transcripts to get to the cytosol, but there is evidence that as gene transcription proceeds, the active DNA actually moves into the interior of the nucleus. The genes in euchromatin are active and thus show decreased methylation of the cytosines in CpG islands of the DNA, increased acetylation of histones and decreased methylation of lysine-9 in histone H3.

3.3 Human genetic disorders

The spectrum of inherited or congenital genetic disorders can be classified as the chromosomal disorders, including mitochondrial chromosome disorders, the Mendelian and sex-linked single-gene disorders, a variety of nonMendelian disorders and the multifactorial and polygenic disorders (Table 3.3). All are a

Table 3.3 Although individual genetic diseases are rare regional variations are enormous — the incidence of Down syndrome varying from 1/1000 to 1/100 worldwide. Collectively they comprise over 15 500 recognized genetic disorders The global prevalence of all single gene diseases at birth is approximately 10/1000

Genetic disease	Congenital malformation
0.5% of all newborns have a chromosomal abnormality	3—5% of all births result in congenital malformations
7% of all stillborns have a chromosomal abnormality	
20—30% of all infant deaths are due to genetic disorders	30—50% of postneonatal deaths are due to congenital malformations
11% of paediatric hospital admissions are for children with genetic disorders	18% of paediatric hospital admissions are for children with congenital malformations
12% of adult hospital admissions are for genetic causes	
15% of all cancers have an inherited susceptibility	
10% of the adult population chronic diseases (heart, diabetes, arthritis) have a significant genetic component	

European incidences per 1000 births				
Single gene disorders:	10	Congenital malformations:	31
Dominant	7.0	Genetically determined	0.6
Recessive	1.66	Multifactorial	30
X-linked	1.33	Nongenetic	~0.4
Chromosomal disorders:	3.5		
Autosomes	1.69		
Sex chromosomes	1.80		

[a]http://www.emro.who.int/publications/TechnicalPublications/Genetics/Part1-Ch1-Frequency.htm

result of a mutation in the genetic code. This may be a change of a single basepair of a gene, resulting in functional change in the product protein (e.g. thalassaemia) or gross rearrangement of the gene within a genome (e.g. chronic myeloid leukemia). These mutations can be congenital (inherited at birth) or somatic (arising during a person's life). The latter are responsible for the collective disease known as cancer, and the principles underlying Mendelian inheritance act in a similar manner to dominant and recessive traits. Both gross chromosomal and point mutations occur in somatic genetic disease.

3.3.1 Chromosomal disorders

Chromosomal abnormalities are much more common than is generally appreciated. Over half of spontaneous abortions have chromosomal abnormalities, compared with only four to six abnormalities per 1000 live births. Specific chromosomal abnormalities can lead to well-recognized and severe clinical syndromes, although autosomal aneuploidy (a differing from the normal diploid number) is usually more severe than the sex-chromosome aneuploidies. Abnormalities may occur in either the number or the structure of the chromosomes.

3.3.1.1 Abnormal chromosome numbers

If a chromosome or chromatids fail to separate ('nondisjunction') either in meiosis or mitosis, one daughter cell will receive two copies of that chromosome and one daughter cell will receive no copies of the chromosome. If this nondisjunction occurs during meiosis it can lead to an ovum or sperm having eitheran extra chromosome, so resulting in a fetus that is 'trisomic' and has three instead of two copies of the chromosome; or no chromosome, so the fetus is 'monosomic' and has one instead of two copies of the chromosome. Nondisjunction can occur with autosomes or sex chromosomes. However, only individuals with trisomy 13, 18 and 21 survive to birth, and most children with trisomy 13 and trisomy 18 die in early childhood. Trisomy 21 (Down syndrome) is observed with a frequency of 1 in 700 live births regardless of geography or ethnic background. This should be reduced with widespread screening. Full autosomal monosomies are extremely rare and very deleterious.

Sex-chromosome trisomies (e.g. Klinefelter's syndrome, XXY) are relatively common. The sex-chromosome monosomy in which the individual has an X chromosome only and no second X or Y chromosome is known as Turner's syndrome and is estimated to occur in 1 in 2500 live-born girls (see Table 3.4).

Occasionally, nondisjunction can occur during mitosis shortly after two gametes have fused. It will then result in the formation of two cell lines, each with a different chromosome complement. This occurs more often with the sex chromosome, and results in a 'mosaic' individual. Very rarely the entire chromosome set will be present in more than two copies, so the individual may be triploid rather than diploid and have a chromosome number of 69. Triploidy and tetraploidy (four sets) result in spontaneous abortion.

3.3.1.2 Abnormal chromosome structures

As well as abnormal numbers of chromosomes, chromosomes can have abnormal structures and the disruption to the DNA and gene sequences may give rise to a genetic disease.

- **Deletions** Deletions of a portion of a chromosome may give rise to a disease syndrome if two copies of the genes in the deleted region are necessary, and the individual will not be normal with just the one copy remaining on the nondeleted homologous chromosome. Many deletion syndromes have been well described. For example, Prader—Willi syndrome is the result of cytogenetic events resulting in deletion of part of the long arm of chromosome 15, Wilms' tumour is characterized by deletion of part of the short arm of chromosome 11, and microdeletions in the long arm of chromosome 22 give rise to the DiGeorge syndrome.
- **Duplications** Duplications occur when a portion of the chromosome is present on the chromosome in two copies, so the genes in that chromosome portion are present in an extra dose. A form of the neuropathy, Charcot—Marie—Tooth disease is due to a small duplication of a region of chromosome 17.
- **Inversion** Inversions involve an end-to-end reversal of a segment within a chromosome, for example abcdefgh becomes abcfedgh.
- **Translocations** Translocations occur when two chromosome regions join together, when they would

Table 3.4   Some of the syndromes resulting from chromosomal abnormalities

Syndrome	Chromosome karyotype	Incidence and risks	Clinical features	Mortality
Autosomal abnormalities				
Trisomy 21 (Down syndrome)	47, +21 (95%) Mosaicism Translocation 5%	1:650 (risk with 20- to 29-year-old mother 1:1000; >45-year-old mother 1:30)	Flat face, slanting eyes, epicanthic folds, small ears, simian crease, short stubby fingers, hypotonia, variable learning difficulties, congenital heart disease (up to 50%)	High in first year, but many now survive to adulthood
Trisomy 13 (Patau syndrome)	47, +13	1:5000	Low-set ears, cleft lip and palate, polydactyly, microophthalmia, learning difficulties	Rarely survive for more than a few weeks
Trisomy 18 (Edwards syndrome)	47, +18	1:3000	Low-set ears, micrognathia, rocker-bottom feet, mental retardation	Rarely survive for more than a few weeks
Sex-chromosome abnormalities				
Fragile X syndrome	46, XX, fra (X) 46, XY, fra (X)	1:2000	Most common inherited cause of learning difficulties predominantly in males Macroorchidism	
Female Turner's syndrome	45, XO	1:2500	Infantilism, primary amenorrhoea, short stature, webbed neck, cubitis valgus, normal IQ	
Triple X syndrome	47, XXX	1:1000	No distinctive somatic features, learning difficulties	
Others	48, XXXX 49,XXXXX	Rare	Amenorrhoea, infertility, learning difficulties	
Male Klinefelter's syndrome	47, XXY (or XXYY)	1:1000 (more in sons of older mothers)	Decreased crown—pubis: pubis—heel ratio, eunuchoid, testicular atrophy, infertility, gynaecomastia, learning difficulties (20%; related to number of X chromosomes)	
Double Y syndrome	47, XYY	1:800	Tall, fertile, minor mental and psychiatric illness, high incidence in tall criminals	
Others	48, XXXY 49, XXXXY		Learning difficulties, testicular atrophy	

not normally. Chromosome translocations in somatic cells may be associated with tumourigenesis.

Translocations can be very complex, involving more than two chromosomes, but most are simple and fall into one of two categories. Reciprocal translocations occur when any two nonhomologous chromosomes break simultaneously and rejoin, swapping ends. In this case the cell still has 46 chromo-somes but two of them are rearranged. Someone with a balanced translocation is likely to be normal (unless a translocation breakpoint interrupts a gene); but at meiosis, when the chromosomes separate into different daughter cells, the translocated chromosomes will enter the gametes and any resulting fetus may inherit one abnormal chromosome and have an unbalanced translocation, with physical manifestations.

Robertsonian translocations occur when two acrocentric chromosomes join and the short arm is lost, leaving only 45 chromosomes. This translocation is balanced as no genetic material is lost and the individual is healthy. However, any offspring have a risk of inheriting an unbalanced arrangement. This risk depends on which acrocentric chromosome is involved. Clinically important is the 14/21 Robertsonian translocation. A woman with this karyotype has a one in eight risk of delivering a baby with Down syndrome (a male carrier has a 1 in 50 risk). However, they have a 50% risk of producing a carrier like themselves, hence the importance of genetic family studies. Relatives should be alerted about the increased risk of Down syndrome in their offspring, and should have their chromosomes checked.

3.3.1.3   Mitochondrial chromosome disorders

The mitochondrial chromosome carries its genetic information in a very compact form; for example there are no introns in the genes. Therefore any mutation has a high chance of having an effect. However, as every cell contains hundreds of mitochondria, a single altered mitochondrial genome will not be noticed. As mitochondria divide there is a statistical likelihood that there will be more mutated mitochondria, and at some point this will give rise to a mitochondrial disease.

Most mitochondrial diseases are myopathies and neuropathies with a maternal pattern of inheritance. Other abnormalities include retinal degeneration, diabetes mellitus and hearing loss. Many syndromes have been described. Myopathies include chronic progressive external ophthalmoplegia (CPEO); encephalomyopathies include myoclonic epilepsy with ragged red fibres (MERRF) and mitochondrial encephalomyopathy, lactic acidosis and stroke-like episodes (MELAS). Kearns—Sayre syndrome includes ophthalmoplegia, heart block, cerebellar ataxia, deafness and mental deficiency due to long deletions and rearrangements. Leber's hereditary optic neuropathy (LHON) is the commonest cause of blindness in young men, with bilateral loss of central vision and cardiac arrhythmias, and is an example of a mitochondrial disease caused by a point mutation in one gene. Multisystem disorders include Pearson's syndrome (sideroblastic anaemia, pancytopenia, exocrine pancreatic failure, subtotal villous atrophy, diabetes mellitus and renal tubular dysfunction). In some families, hearing loss is the only

symptom and one of the mitochondrial genes implicated may predispose parents to aminoglyceride ototoxicity (see Table 3.5).

3.3.1.4   Analysis of chromosome disorders

The analysis of gross chromosomal disorders has traditionally involved the culture of isolated cells in the presence of toxins such as colchicine. These toxins arrest the cell cycle at mitosis and, following staining, the chromosomes with their characteristic banding can be seen. A highly trained cytogeneticist can then identify each chromosome pair and any abnormalities (see Figure 3.12).

New molecular biology techniques have made things simpler: yeast artificial chromosome (YAC)-cloned probes are available and cover large genetic regions of individual human chromosomes. These probes can be labelled with fluorescently tagged nucleotides and used in *in situ* hybridization of the nucleus of isolated tissue from patients. These tagged probes allow rapid and relatively unskilled identification of metaphase chromosomes, and allow the identification of chromosomes dispersed within the nucleus. Furthermore, tagging two chromosome regions with different fluorescent tags allows easy identification of chromosomal translocations (see Figure 3.25).

3.3.2   Gene defects

Mendelian and sex-linked single-gene disorders are the result of mutations in coding sequences and their control elements. These mutations can have various effects on the expression of the gene, as explained below, but all cause a dysfunction of the protein product. Following translation of mRNA a peptide or polypeptide is produced which then undergoes a series of modifications as described earlier but at the same time the protein is being folded and twisted into a shape which is both stable and functional.

The primary structure of proteins is simply the amino acid sequence translated from the mRNA sequence. Each amino acid in the sequence has a similar basic structure but the functional group of each amino acid unit confers charge and polarity and a chain of amino acids is oriented according to the constraints and interactions each amino acid side chain has on its neighbour.

Table 3.5   Examples of mitochondrial genetic disorders

Mitochondrial DNA (mtDNA) deletional syndromes:

Kearns—Sayre syndrome (KSS) — large deletion in mitochondrial DNA (1.3—10kB) from all tissue but particularly muscle

Pearson syndrome — deletions in mitochondrial DNA found particularly in blood cells progressive external ophthalmoplegia (PEO) — deletions in mitochondrial DNA in skeletal muscle only

MELAS (Myopathy, Mitochondrial-Encephalopathy-Lactic Acidosis-Stroke)

The most common mutation in MELAS, present in over 80% of patients, is an A-to-G transition at nucleotide 3243 in the mitochondrial gene *MTTL1* (Mitochondrial tRNA leucine 1)

MERRF (Myoclonic Epilepsy Associated with Ragged Red Fibres)

The most common mutation, present in over 80% of patients, is an A-to-G transition at nucleotide-8344 in the mitochondrial gene *MTTK* (Mitochondrial tRNA lysine)

mtDNA-associated and NARP (Neurogenic muscle weakness, Ataxia, and Retinitis Pigmentosa),
Mutation in the mitochondrial gene *MTATP6* (ATP synthase A chain) are associated with the neurodegenerative energy associated disease NARP

mtDNA-associated Leigh syndrome

Mutations in the mitochondrial genes:

MTATP6 (ATP synthase A chain),

MTTL1 (Mitochondrial tRNA leucine 1),

MTTK (Mitochondrial tRNA lysine),

MTND1 (NADH-ubiquinone oxidoreductase chain 1),

MTND3 (NADH-ubiquinone oxidoreductase chain 3),

MTND4 (NADH-ubiquinone oxidoreductase chain 4),

MTND5 (NADH-ubiquinone oxidoreductase chain 5),

MTND6 (NADH-ubiquinone oxidoreductase chain 6),

MTCO3 (Cytochrome c oxidase polypeptide III),

MTTW (Mitochondrial tRNA tryptophan), and

MTTV (Mitochondrial tRNA valine) are all associated with neurodegenerative Leigh syndrome

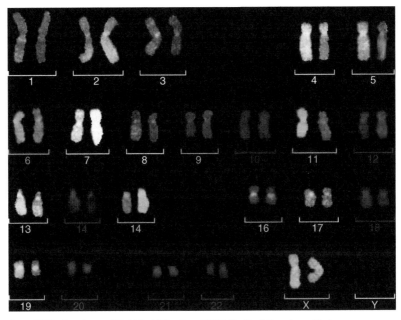

Figure 3.25   24 colour FISH. Courtesy of Debbie Lillington CRUK Medical Oncology Barts

A simple native protein's shape is therefore affected by the amino acids from which it is comprised but is also affected by the environment in which it is formed. The cytosol of the cell is aqueous and as a result a native protein will generally fold with hydrophilic amino acids oriented to external surface of the protein and the hydrophobic amino acids will bury themselves within the protein. The secondary structures, including beta pleated sheets and alpha helices, are formed as a result of these folds and weak interactions such as hydrogen bonding. Some amino acids, namely cysteine and proline have a more pronounced affect on the structure forming disulphide bonds (a strong covalent link) or L-shaped kinks in the polypeptide chain respectively. The eventual three-dimensional shape is critical to the function of the mature protein and this is dependent on the constituent amino acids. If any amino acid is removed or changed the shape of the protein is altered, if the shape is altered the function may be compromised or the protein may have no function at all. These changes come about by changes to the gene sequences or DNA which ultimately codes for proteins. Therefore a change in the DNA sequence in the wrong place can prevent a protein from folding and functioning correctly; if that protein is an enzyme, hormone or receptor, for example, the consequences can be far reaching. Changes to the DNA sequence of an organism are called mutations.

Although DNA replication is a very accurate process with inbuilt repair mechanisms, occasionally mistakes occur to produce changes or mutations in the sequence. These changes can also occur owing to other factors such as radiation, ultraviolet light or chemicals. Many different types of mutation can occur.

3.3.2.1   Point mutation

This is the simplest type of change and involves the substitution of one nucleotide for another, so changing the codon in a coding sequence. For example, the triplet AAA, which codes for lysine, may be mutated to AGA, which codes for arginine. Whether a substitution produces a clinical disorder depends on whether it changes a critical part of the protein molecule produced. Fortunately, many substitutions have no effect on the function or stability of the proteins produced as several codons code for the same amino acid. However, some mutations may have a severe effect; for example, in sickle cell disease a mutation within the globin gene changes one codon from GAG

to GTG, so that instead of glutamic acid, valine is incorporated into the polypeptide chain, which radically alters its properties.

3.3.2.2   Insertion or deletion

Insertion or deletion of one or more bases is a more serious change, as it results in the alteration of the rest of the following sequence to give a frame-shift mutation. For example, if the original code was:

TAA GGA GAG TTT

and an extra nucleotide (A) is inserted, the sequence becomes:

TAA AGG AGA GTT T

Alternatively, if the third nucleotide (A) is deleted, the sequence becomes:

TAG GAG AGT TT

In both cases, different amino acids are incorporated into the polypeptide chain. This type of change is responsible for some forms of thalassaemia. Insertions and deletions can involve many hundreds of basepairs of DNA. For example, some large deletions in the dystrophin gene remove coding sequences and this results in Duchenne muscular dystrophy. Insertion/deletion (ID) polymorphism in the angiotensin-converting enzyme (ACE) gene has been shown to result in the genotypes II, ID and DD. The deletion is of a 287 bp repeat sequence and DD is associated with higher concentrations of circulating ACE and possibly cardiac disease.

3.3.2.3   Splicing mutations

If the DNA sequences which direct the splicing of introns from mRNA are mutated, then abnormal splicing may occur. In this case the processed mRNA which is translated into protein by the ribosomes may carry intron sequences, so altering which amino acids are incorporated into the polypeptide chain.

3.3.2.4   Termination mutations

Normal polypeptide chain termination occurs when the ribosomes processing the mRNA reach one of the chain termination or 'stop' codons (see above). Mutations involving these codons will result in either late or premature termination. For example, Haemoglobin Constant Spring is a haemoglobin variant where instead of the 'stop' sequence a single base change allows the insertion of an extra amino acid.

3.3.3   Single-gene disease

Monogenetic disorders involving single genes can be inherited as dominant, recessive or sex-linked characteristics. Although classically divided into autosomal dominant, recessive or X-linked disorders, many syndromes show multiple forms of inheritance pattern. This is predominately because multiple defects can occur within a given disease associated gene or in separate genes which all contribute to a particular molecular/cellular pathway and thus give rise to the same phenotype. For example in Ehlers−Danlos syndrome we find autosomal dominant, recessive and X-linked inheritance. In addition there is a spectrum between autosomal recessive and autosomal dominance in that having just one defective allele gives a mild form of the disease whilst having both alleles with the mutation results in more severe form of the syndrome. In some cases, such as factor V Leiden disorder, the boundary between dominance and recessive forms is very blurred. Tables 3.6 and 3.7 give examples of autosomal dominant and recessive disorders.

Some diseases show a racial or geographical prevalence. Thalassaemia is seen mainly in Greeks, Southeast Asians and Italians, porphyria variegata occurs more frequently in the South African white population and Tay−Sachs disease particularly occurs in Ashkenazi Jews. The most common recessive disease in the UK is cystic fibrosis. Thus the worldwide prevalence of a disease may be low (as detailed in Tables 3.6−3.9) but much higher in specific populations. Some disorders are rare to the extent that the prevalence is counted in the total number of cases ever reported. Molecular biology has enabled the subclassification of some syndromes according to which gene is giving rise to the disease (e.g. Polycystic kidney disease) and the proportion of cases arising from defects in that particular gene can be estimated from linkage analysis. However, this may not be reflected in the detection rate of a given clinical genetic test, since a single test cannot detect all the possible mutations arising at a particular locus.

3.3.3.1   Autosomal dominant disorders

Each diploid cell contains two copies of all the autosomes. An autosomal dominant disorder occurs when one of the two copies has a mutation and the protein produced by the normal form of the gene cannot compensate. In this case a heterozygous individual who has two different forms (or alleles) of the same gene will manifest the disease. The offspring of heterozygotes have a 50% chance of inheriting the chromosome carrying the disease allele, and therefore also of having the disease. However, estimation of risk to offspring for counselling families can be difficult because of three factors:

- These disorders have a great variability in their manifestation. 'Incomplete penetrance' may occur if patients have a dominant disorder but it does not manifest itself clinically in them. This gives the appearance of the gene having 'skipped' a generation.
- Dominant traits are extremely variable in severity (variable expression) and a mildly affected parent may have a severely affected child.
- New cases in a previously unaffected family may be the result of a new mutation. If it is a mutation, the risk of a further affected child is negligible. Most cases of achondroplasia are due to new mutations.

The overall incidence of autosomal dominant disorders is seven per 1000 live births (see Table 3.6).

3.3.3.2   Autosomal recessive disorders

These disorders manifest themselves only when an individual is homozygous for the disease allele, that is both chromosomes carry the mutated gene. In this case the parents are generally unaffected, healthy carriers (heterozygous for the disease allele). There is usually no family history, although the defective gene is passed from generation to generation. The offspring of an affected person will be healthy heterozygotes unless the other parent is also a carrier. If carriers marry, the offspring have a one in four chance of being homozygous and affected, a one in two chance of being a carrier and a one in four chance of being genetically normal. Consanguinity increases the risk. The clinical features of autosomal recessive disorders are usually severe; patients often present in the first few years of life and have a high mortality. The overall incidence of autosomal recessive disorders is about 2.5 per 1000 live births (see Table 3.7).

3.3.3.3   Sex-linked disorders

Genes carried on the X chromosome are said to be 'X-linked', and can be dominant or recessive in the

Table 3.6 Examples of autosomal dominant disorders with chromosome location and gene defect

Syndrome	Disease Prevalence	Gene-	(Product)	Gene Location	Genetic test detection rate
Achondroplasia	1/15,000 – 1/40,000	FGFR3	(Fibroblast growth factor receptor-3)	4p16.3	99%
Alexander disease	very rare	GFAPP	(Glial fibrillary acidic protein)	17q21	95%
Alagille syndrome (AGS)	1/70,000	JAG1	(Jagged1 protein)	20p12	77%
Alzheimers disease (early onset)	41/10,000 at risk	PSENI	(Presenilin 1)	14q24.3	30–70%
		PSENII	(Presenilin 2)	1q31-q42	<5%
		APP	(Amyloid beta A4 protein)	21q21	10–15%
Aniridia	1-2/100,000	PAX6	(Paired box protein Pax-6)	11p13	90%
Breast cancer—Familial early onset	1/300-1/800	BRAC-1	(Breast cancer type 1 susceptibility protein)	17q21	50–75%
		BRAC-2	(Breast cancer type 2 susceptibility protein)	13q12.3	15–30%
Charcot-Marie-Tooth Disease subtypes:	(30/100,000)				
CMT1	15/100,000	PMP22	(Peripheral myelin protein 2)	17p11.2	~50%
		MPZ	(Myelin P_0 Protein)	1q22	<20%
		LITAF	(Lipopolysaccharide-induced tumor necrosis factor-alpha factor)	16p13.1-12.3	unknown
		EGR2	(Early growth response protein 2)	10q21.1-q22	unknown
CMT2	6-12/100,000	KIF1B	(Kinesin-like protein KIF1B)	1p36.2	unknown
		RAB7	(Ras like protein RAB-7)	3q21	unknown
		GARS	(Glycyl-tRNA synthetase)	7p15	unknown
		NEFL	(Neurofilament triplet L protein)	8p21	unknown
		Unknown		12q23-24	unknown
		Unknown		7q11-21	unknown
Cerebral Cavernous Malformation	Rare				
(Familial) Subtypes CM1	(40% of cases)	CCM1	(Krit-1 protein)	7q21-22	~70%
CM2	(20% of cases)	Unknown		7p15-13	unknown
CM3	(40% of cases)	Unknown		3q25.2-27	unknown
22q11.2 deletion Syndrome (includes DiGeorge syndrome)	1/5,000	**Multiple known & unknown sequences**		22q11.2	100%
Myotonic Dystrophy	1/20,000	DMPK	(Myotonin — protein kinase)	19q13.2—13.3	100%

(Continued)

Table 3.6 (Continued)

Syndrome	Disease Prevalence	Gene-	(Product)	Gene Location	Genetic test detection rate
Ehlers – Danalos Syndrome					
Classic type	1/20,000	COL5A1	(Collagen alpha 1(V) chain)	9q34-34.3	50–75%
		COL5A2	(collagen alpha 2 (V) chain)	2q31	50–75%
Vascular type	1/50,000	COL3A1	(Type III collagen cahins)	2q31	98–99%
Epidermolysis Bullosa Simplex	1/50,000	KRT5	(Keratin 5)	12q13	50–70%
		KRT14	(Keratin 14)	17q12-21	50–70%
Facioscapilohumeral Muscular Dystrophy	6/100,000	D4Z4	(repeat motif domain)	4q35	95%
Familial Adenomatous Polyposis	3/100,000	APC	(adenomatous polyposis coli protein)	5q21-22	95%
Familial Hypercholestrolaemia	1-2/500				
	(95–98% of cases)	LDLR	(low density lipoprotein receptor)	19p13.2	unknown
	(2–5% of cases)	APOB	(apoliprotein B100)	2p23-24	unknown
Familial Periodic Fever/TRAPS	Rare	TNFRSF1A	(serum/soluble TNF receptor)	12p13	unkown
Hemorrhagic telangiectasia (Hereditary)	1-4/50,000	ENG	(endoglin)	9q34.1	50–80%
		ACVRL1	(Activin - Serine/threonine-protein kinase receptor R3)	12q11-14	50–80%
Multiple syndromes and chromosomal deletions give rise to Hirshsprung disease					
Hirschsprung's disease (congenital intestinal aganglionosis)	1/5000	RET	(tyrosine-protein kinase receptor)	10q11.2	10–80%
		GDNF	(Glial cell line-derived neutrotrophic factor)	5p13.1-12	<1%
		NRTN	(Neurturin)	19p13.3	<1%
		EDNRB	(Endothelin B receptor)	13q22	3–5%
		EDN3	(Endothelin-3)	20q13.2-13.3	5%
		ECE1	(Endothelin-converting enzyme)	1p36.1	<1%
Li Fraumeni syndrome	Rare	TP53	(p53)	17p13.1	95%
		CHEK2	(Serine/threonine-protein kinase Chk2)	22q12.1	unknown
Malignant hyperthermia	1-5/50,000	RYR1	(Ryanodine receptor type 1)	19q13.1	70%
		CACNA1S	(Voltage-dependent L-type calcium channel alpha-1S subunit)	1q32	unknown
		Unknown		17q11.2-24	unknown
		Unknown		7q21-q22	unknown

		Unknown		3q13.1	unknown
		Unknown		5p	unknown
Marfans syndrome	1-2/10,000	*FBN1*	(Fibrillin 1)	15q21.1	70–90%
MEN Type 2 (Multiple endocrine neoplasia)	1/30,000	*RET*	(tyrosine-protein kinase receptor)	10q11.2	88–95%
Neurofibromatosis	Type 1 1/3000	*NF1*	(Neurofibromin)	17q11.2	80–95%
Type 2	1/40,000	*NF2*	(Merlin)	22q12.2	~65%
Peutz-Jeghers syndrome	1-10/250,000	*STK11*	(Serine/threonine-protein kinase 11)	19p13.3	65–75%
(AD) Polycystic Kidney disease	1-2/1000 (~85% of cases)	*PKD1*	(Polycystin)	16p13.3-13.12	50–75%
	(~15% of cases)	*PKD2*	(Polycystin2)	4q21-23	75%
(AD) Retinitis pigmantosa	4-7/100,000 (25–30% of cases)	*RHO*	(Rhodopsin)	3q22.1	unknown
	(15–20%)	*PRPF31*	(Pre-mRNA splicing factor 31)	19q3.4	unknown
	(5–10%)	*RP1*	(Oxygen-regulated protein 1)	8q12.1	unknown
	(5–10%)	*RDS*	(Peripherin 2)	6p21.2	unknown
	(3–5%)	*IMPDH1*	(Inosine monophosphate dehydrogenase 1)	7q14.3	unknown
	(3%-Japan)	*FSCN2*	(Retinal fascin homolog 2, actin bundling protein)	17q25	unknown
	(Rare)	*CRX*	(Cone-rod homeobox protein)	19q13.3	unknown
	(Rare)	*PRPF8*	(Pre-mRNA splicing factor C8)	17p13.3	unknown
	(Rare)	*NRL*	(Neural retina-specific leucine zipper)	14q11.2	unknown
	(Rare)	*ROM1*	(Retinal outer segment membrane protein1)	11q12.3	unknown
	(Rare)	*PIMIK*	(Pim-1 kinase)	7p14.3	unknown
Retinoblastoma	~1/20,000	*RB1*	(Retinoblastoma-associated protein)	13q14.1-14.2	~80%
Tuberous sclerosis (TS)	~1/6000	*TSC1*	(Hamartin protein)	9q34	10–30%
		TSC2	(Tuberin)	16p13.3	50–70%
Von Hippel-Lindau syndrome	1/40,000	*VHL*	(von Hippel-Lindau disease tumour suppressor)	3p26-25	~75%
Von Willebrand's disease (Type 1&2)	up to 3/100	*vWF*	(von Willebrand's factor)	12p13.2	unknown
Wilm's Tumour	1/10,000	*WT1*	(Wilm's Tumour protein)	11p13	unknown

Table 3.7 Examples of autosomal recessive disorders with chromosome location and gene defects

Syndrome	Disease Prevalence	Gene- (Product)	Gene Location	Genetic test detection rate
Alpha Mannosidosis	0.5-2/1million	MAN2B1 (Lysosomal alpha-mannosidase)	19cen-q12	100%
Alström syndrome	Rare	ALMS1 (ALMS1 protein)	2p13	unknown
Ataxia-Telangiectasia	1-3/100 000	ATM (Serine-protein kinase ATM)	11q22.3	>95%
Alkaptonuria	1/500 000	HGD (Homogentisic acid dioxgenase)	3q21-q23	90%
Charcot-Marie-Tooth Disease subtype CMT4	<1/100 000	GDAP1 (Ganglioside-induced differentiation-associated protein-1)	8q13-12.1	unknown
		MTMR2 (Myotubularin-related protein 2)	11q22	unknown
		MTMR13 (Myotubularin-related protein 13)	11p15	unknown
		KIAA1985 (unknown)	5q32	unknown
		NDRG1 (NDRG1 protein)	8q24.3	unknown
		EGR2 (Early growth response protein 2)	10q21.1-22	unknown
		PRX (Periaxin)	19q13.1-13	unknown
Cokayne syndrome	(<1/100 000)			
Type A	25% of cases	CKN1 (Cockayne syndrome WD-repeat protein/EECC-6 associated protein)	Ch5	unknown
Type B	75% of cases	ERCC6 (Transcription coupled excision DNA repair protein ERCC-6)	10q11	unknown
Cystic fibrosis	1-10/30 000	CTFR (Cystic fibrosis transmembrane conductance regulator)	7q31.2	50—95%
Cystinosis	1/20 000	CTNS (Cystinosin)	17p13	40—65%
21-hydroxylase deficiency (21-OHD) -Congenital adrenal hyperplasia	1:15,000	CYP21A2 (Cytochrome P450 XXIB)	6p21.3 90—95%	
Congenital Icthyosis	1-2/200 000	TGM1 (Transglutaminase-K)	14q11.2	90%
Familial Mediterranean fever (FMF)	Rare	MEFV (Pyrin)	16p13.3	70—95%
Dysferlinopathy	Rare	DYSF (Dysferlin)	2p13.3-13.1	80%
Ehlers-Danalos-Kyphosoliotic form	1/100 000	PLOD (Procollagen-lysine, 2-oxogluterate 5-dioxgenase 1)	1p36.3-36.2	20%
Fanconi anaemia	1/100 000			

Disorder	Gene (protein)	Frequency	Location	%
	FRANCA (Fanconi Anemia Group A Protein)	(66% of cases)	16q24.3	unknown
	FANCC (Fanconi Anemia Group C Protein)	(10% of cases)	9q22.3	unknown
	FANCE (Fanconi Anemia Protein E)	(10% of cases)	6p22-21	unknown
	FANCG (Fanconi Anemia Group G Protein)	(10% of cases)	9p13	unknown
	FANCD2 (Fanconi Anemia Group D2 Protein)	(Rare)	3p25.3	unknown
	FANCF (Fanconi Anemia Complementation Group F)	(Rare)	11p15	unknown
Free Sialic Acid Storage Disorders	SLC17A5 (Sialin)	Rare	6q14-q15	91%
Galactosaemia	GALT (Galactose-phosphate uridyl transferase)	1/30 0000	9p13	95—100%
Gaucher's disease	GBA (Glucocerebrosidase)	1-2/100 000	1q21	95—100%
Glycine Encephalopathy	GLDC (Glycine dehydrogenase)	1/50 000	9p22	84%
	AMT (Amino methyl transferase)	(80% of cases)	3p21.2-21.1	10—30%
	GCSH (Glycine cleavage system H protein)	(10—15% of cases)	6q24	unknown
Hereditary haemochromatosis	HFE (Hereditary haemochromatosis protein)	1/3000	6p21.3	60—98%
Homocystinuria type I	CBS (Cystathionine β synthetase)	1/200 000	21q22.3	95%
Mucopolysaccharidosis Type I (Hurler's syndrome)	IDUA (α-L-iduronidase)	1-5/500 000	4p16.3	95%
Nieman-Pick disease		1/150 000		
Type 1	NPC1 (Nieman-Pick C1 protein)	(90% of cases)	18q11-12	95%
Type 2	NPC2 (Epididymal secretory protein E1)	(4% of cases)	14q24.3	100%
Oculocutaneous Albanism				
Type 1	TYR (tyrosinase)	1/40 000	11q14-21	71%
Type 2	OCA2 (P protein)	1/40 000	15q11.2-12	80—90%
Phenylalanine hydroxylase deficiency (Phenylketonuria- PKU)	PAH (Phenylalanine hydroxylase)	1-55/150 000	12q23.2	99%
Peroxisome biogenesis disorders (Zellweger's syndrome & Refsum's disease)		1/50 000		
	PEX1 (Peroxisome biogenesis factor 1)	(65% of cases)	7q21-22	80%
	PEX6 (Peroxisome assembly factor-2)	(11% of cases)	6p21.1	unknown
	PEX26 (PEX26 protein)	(7% of cases)	22q11.2	unknown
	PEX10 (Peroxisome assembly protein 10)	(4% of cases)	Ch1	unknown

(Continued)

Table 3.7 (*Continued*)

Syndrome	Disease Prevalence	Gene- (Product)	Gene Location	Genetic test detection rate
	(4% of cases)	PEX12 (Peroxisome assembly protein 12)	17q21.1	unknown
	(1–2% of cases)	PEX3 (Peroxisomal assembly protein PEX3)	6q23-24	unknown
	(1–2% of cases)	PXR1 (Peroxisomal targeting signal 1 receptor)	12p13.3	unknown
	(Rare)	PXMP3 (Peroxisome assembly factor-1)	8q21.1	unknown
	(Rare)	PEX13 (Peroxisomal membrane protein PEX13)	2p15	unknown
	(Rare)	PEX16 (Peroxisomal membrane protein PEX16)	11p12-11.2	unknown
	(Rare)	PXF (Peroxisomal farnesylated protein)	1q22	unknown
(AR) Retinitis pigmantosa	4-5/100 000			
	(10–20% of cases)	Unknown	6q14-15	unknown
	(4–5% of cases)	USH2A (Usherin protein)	1q41	unknown
	(3–4% of cases)	PDE6B (Rod cGMP-specific 3',5' cyclic phosphodiesterase β)	4p16.3	unknown
	(3–4% of cases)	PDE6A (Rod cGMP-specific 3',5' cyclic phosphodiesterase α)	5q33.1	unknown
	(2% of cases)	RPE65 (Retinal pigment epithelium 61kDa protein)	1p31.2	unknown
At least 16 other gene and gene loci have been found associated with autosomal recessive (AR) Retinitis pigmantosa				
Sickle cell disease	Rare (1–2/10 000 in Africa)	HBB (β-globin)	11p15.	100%
β-thalassaemia	Rare (1–2/100 Cyprus)	HBB (β-globin)	11p15	99%
α-thalassaemia	Complex	HBA1 (haemoglobin, alpha 1)	16p13.3	
		HBA2 (haemoglobin, alpha 2)	16p13.3	
Hexoamidase A deficiencies (Tay–Sach's disease)	Rare	HEXA (Beta-hexosaminidase alpha chain)	15q23-24	40–98%
Wilson's disease	1/30 000	ATP7B (copper transporting ATPase 2)	13q14.3-12	unknown
Xeroderma pigmentosum	1-10/million			
	(25% of cases)	XPA (DNA-repair protein complementing XP-A cells)	9q22.3	unknown
	(25% of cases)	XPC (DNA-repair protein complementing XP-C cells)	3p25	unknown
	(21% of cases)	POLH (Error prone DNA photoproduct bypass polymerase)	6p21.1-12	unknown
	(15% of cases)	ERCC2 (TFIIH basal transcription factor complex helicase subunit)	19q13.2-13	unknown
	(6% of cases)	ERCC4 (DNA-repair protein complementing XP-F cells)	16p13.3-13.1	unknown
	(6% of cases)	ERCC5 (DNA-repair protein complementing XP-G cells)	13q33	unknown
	(Rare)	ERCC3 (TFIIH basal transcription factor complex helicase XPB subunit)	2q21	unknown
	(Rare)	DDB2 (DNA damage binding protein 2)	11p12-11	unknown

Table 3.8   Examples of X-linked disorders with chromosome location and gene defects

Syndrome	Disease Prevalence	Gene- (Product)	Gene *Location*	Genetic test detection rate
Recessive				
Alport's syndrome	1/50 000	*COL4A5* (Collagen alpha 5(IV) chain)	Xq22	60%
Adrenal hypoplasia congenital -X	1/13 000	*NROB1* (Orphan nuclear receptor DAX-1)	Xp21.3-21.2	70—100%
Adrenoleukodystrophy- X linked	1-3/50 000	*ABCD1* (Adrenoleukodystrophy protein)	Xq28	98%
Agammaglobulinaemia- X linked	3-6/million	*BTK* (Tyrosine-protein kinase BTK)	Xq21.3-22	98—99%
Androgen insensitivity syndrome	2-5/100 000	*AR* (Androgen receptor)	Xq11-q12	90%
Dystrophinopathies	1-4/20 000 (UK)	*DMD* (Dystrophin)	Xp21.2	65—85%
(Becker's & Duchenne's muscular dystrophies)				
Charcot-Marie-Tooth Disease				
Subtype CMX	3-6/100 000	*GJB1* (Gap junction beta-1 protein)	Xq13.1	90%
Fabry's disease	1-2/100 000	*GLA* (α-galactosidase A)	Xq22	100%
Haemophilia A	1/4000	*F8* (Factor VIII)	Xq28	98%
Haemophilia B	1/20 000	*F9* (Factor IX)	Xq27.1-27.2	99%
Myotubular Myopathy- X linked	1/50 000	*MTM1* (Myotubularin)	Xq28	80—85%
Nephrogenic diabetic insipidus	Rare	*AVPR2* (Vasopressin V2 receptor)	Xq28	97%
α-Thalassaemia, X-linked	Rare	*ATRX* (Transcriptional regulator ATRX)	Xq13	90%
(mental retardation syndrome)				
X-linked severe combined immunodeficiency (X_SCID)	1-2/100 000	*IL2RG* (Cytokine receptor common gamma chain)	Xq13.1	99%
Dominant				
Coffin-Lowry syndrome	1/50 000	*RPS6KA3* (Ribosomal protein S6 kinase alpha 3)	Xp22.2-22.1	35—40%
Incontinentia pigmenti	Rare	*IKBKG* (NF-kappaB essential modulator)	Xq28	80%
Oral-facial-digital				
syndrome type-1	1-5/250 000	*OFD1*(Oral-facial-digital syndrome 1 protein)	Xp22.3-p22.2	unknown
Periventricular heterotopia, X linked	unknown	*FLNA* (filamin 1)	Xq28	83%
Retinitis Pigmentosa- X linked	1-4/100 000			
	(70% of cases)	*RPGR* (Retinitis pigmentosa GTPase regulator)	Xp21.1	unknown
	(8% of cases)	*RP2* (XRP2 protein)	Xp11.2	unknown
Rett Syndrome	1/10 000	*MECP2* (Methyl-CpG-binding protein 2)	Xq28	80%

Table 3.9   Examples of trinucleotide repeat genetic disorders

Syndrome —inheritence pattern	Disease Prevalence	Gene, location and gene-disorder	Genetic test detection rate
Friedreich's ataxia —AR	2-4/100 000	*FRDA,* (Frataxin) 9q13	96%
			-GAA trinucleotide repeat expansion disorder in inton 1 of *FRDA*
Fragile X syndrome- X-linked	16-25/100 000	*FMR1* (Fragile X mental retardation 1 protein) Xq27.3	99%
			-CGG trinucletide repeat expansion and methylation changes in the 5′ untranslated region of *FMR1* exon 1
Huntingdon's disease-AD	3-15/100 000	*HD* (Huntingdon protein) 4p16.3	98%
			-CAG trinucleotide repeat expansion within the translated protein giving rise to long tracts of repeat glutamine residues in HD.
Myotonic dystrophy- AD	1/20 000	*DMPK* (Myotonin-protein kinase) 19q13.2-13.3	100%
			-CTG trinucleotide repeat expansion in the 3′ untranslated region of the *DMPK* gene

same way as autosomal genes. As females have two X chromosomes they will be unaffected carriers of X-linked recessive diseases. However, since males have just one X chromosome, any deleterious mutation in an X-linked gene will manifest itself because no second copy of the gene is present (see Table 3.8).

3.3.3.4   X-Linked Dominant Disorders

These are rare. Vitamin D-resistant rickets is the best-known example. Females who are heterozygous for the mutant gene and males who have one copy of the mutant gene on their single X chromosome will manifest the disease. Half the male or female offspring of an affected mother and all the female offspring of an affected man will have the disease. Affected males tend to have the disease more severely than the heterozygous female.

3.3.3.5   X-Linked Recessive Disorders

These disorders present in males and present only in (usually rare) homozygous females. X-linked recessive diseases are transmitted by healthy female carriers or affected males if they survive to reproduce. An example of an X-linked recessive disorder is haemophilia A, which is caused by a mutation in the X-linked gene for factor VIII. It has recently been shown that in 50% of cases there is an intrachromosomal rearrangement

(inversion) of the tip of the long arm of the X chromosome (one break point being within intron 22 of the factor VIII gene).

Of the offspring from a carrier female and a normal male:

- 50% of the girls will be carriers as they inherit a mutant allele from their mother and the normal allele from their father; the other 50% of the girls inherit two normal alleles and are themselves normal.
- 50% of the boys will have haemophilia as they inherit the mutant allele from their mother (and the Y chromosome from their father); the other 50% of the boys will be normal as they inherit the normal allele from their mother (and the Y chromosome from their father).

The male offspring of a male with haemophilia and a normal female will not have the disease as they do not inherit his X chromosome. However, all the female offspring will be carriers as they all inherit his X chromosome.

3.3.3.6   Y-Linked Genes

Genes carried on the Y chromosome are said to be Y-linked and only males can be affected. However, there are no known examples of Y-linked single-gene disorders which are transmitted.

3.3.3.7   Sex-limited inheritance

Occasionally a gene can be carried on an autosome but manifests itself only in one sex. For example, frontal baldness is an autosomal dominant disorder in males but behaves as a recessive disorder in females.

3.3.3.8   Other single-gene disorders

These are disorders which may be due to mutations in single genes but which do not manifest as simple monogenic disorders. They can arise from a variety of mechanisms, including the following.

3.3.4   Triplet repeat mutations

In the gene responsible for myotonic dystrophy, the mutated allele was found to have an expanded 3′UTR region in which three nucleotides, GCT, were repeated up to about 35 times. In families with myotonic dystrophy, people with the late-onset form of the disease had 20–40 copies of the repeat, but their children and grandchildren who presented with the disease from birth had vast increases in the number of repeats, up to 2000 copies. It is thought that some mechanism during meiosis causes this 'triplet repeat expansion' so that the offspring inherit an increased number of triplets. The number of triplets affects mRNA and protein function giving rise to the phenomenon of 'genetic anticipation' (see Table 3.9).

3.3.5   Mitochondrial disease

As discussed various mitochondrial gene mutations can give rise to complex disease syndromes with incomplete penetrance maternal inheritance.

3.3.6   Imprinting

It is known that normal humans need a diploid number of chromosomes, 46. However, the maternal and paternal contributions are different and, in some way which is not yet clear, the fetus can distinguish between the chromosomes inherited from the mother and the chromosomes inherited from the father, although both give 23 chromosomes. In some way the chromo-somes are 'imprinted' so that the maternal and paternal contributions are different. Imprinting is relevant to human genetic disease because different phenotypes may result depending on whether the mutant chromosome is maternally or paternally inherited. A deletion of part of the long arm of chromosome 15 (15q11–q13) will give rise to the Prader–Willi syndrome (PWS) if it is paternally inherited. A deletion of a similar region of the chromosome gives rise to Angelman's syndrome (AS) if it is maternally inherited. Recently the affected gene has been identified as ubiquitin (*UBE3A*). Significantly maternal chromosome 15 *UBE3A* is expressed in the brain and hypothalamus. Defective maternal ubiquitin in Angelman's syndrome is thus responsible for accumulation of undegraded protein, and hence neuronal damage.

3.3.7   Complex traits: multifactorial and polygenic inheritance

Characteristics resulting from a combination of genetic and environmental factors are said to be multifactorial; those involving multiple genes can also be said to be polygenic (see Table 3.10).

Measurements of most biological traits (e.g. height) show a variation between individuals in a population and a unimodal, symmetrical (Gaussian) frequency distribution curve can be drawn. This variability is due to variation in genetic factors and environmental factors. Environmental factors may play a part in determining some characteristics, such as weight, whilst other characteristics such as height may be largely genetically determined. This genetic component is thought to be due to the additive effects of a number of alleles at a number of loci, many of which can be individually identified using molecular biological techniques, for example studying identical twins in different environments.

A common genetic variation in the 3′untranslated region of the prothrombin gene is associated with elevated prothrombin (up to fourfold), increased risk of venous thrombosis and myocardial infarction. Presumably the 3′mutation induces overexpression of the prothrombin gene by increasing mRNA stability. In some pedigree studies those homozygous for this mutation and heterozygous for factor V Leiden

Table 3.10 Examples of disorders that may have a polygenic inheritance

Disorder	Frequency (%)	Heritability (%)[a]
Hypertension	5	62
Asthma	4	80
Schizophrenia	1	85
Congenital heart disease	0.5	35
Neural tube defects	0.5	60
Pyloric stenosis	0.3	75
Ankylosing spondylitis	0.2	70
Cleft palate	0.1	76

[a]Percentage of the total variation of a trait which can be attributed to genetic factor

mutation have an even greater risk of thrombotic events. Conversely, downregulation of the α1-antitrypsin gene (due to similar mutation in the 3′ and 5′ untranslated regions) is associated with emphysema and cirrhosis. There are sex differences. Congenital pyloric stenosis is most common in boys, but if it occurs in girls the latter have a larger number of affected relatives. This difference suggests that a larger number of the relevant genes are required to produce the disease in girls than in boys. Most of the important human diseases, such as heart disease, diabetes and common mental disorders, are multifactorial traits.

3.3.8 Cell cycle control oncogenes and tumour supressors

Regulation of the cell cycle is complex. Cells in the quiescent G0 phase (G, gap) of the cycle are stimulated by the receptor-mediated actions of growth factors (e.g. EGF, epithelial growth factor; PDGF, platelet-derived growth factor; IGF, insulin-like growth factor) via intracellular second messengers. Stimuli are transmitted to the nucleus where they activate transcription factors and lead to the initiation of DNA synthesis, followed by mitosis and cell division. Cell cycling is modified by the cyclin family of proteins that activate or deactivate proteins involved in DNA replication by phosphorylation (via kinases and phosphatase domains). Thus from G0 the cell moves on to G1 (gap1) when the chromosomes are prepared for replication.

This is followed by the synthetic (S) phase, when the 46 chromosomes are duplicated into chromatids, followed by another gap phase (G2), which eventually leads to mitosis (M).

As shown in Figure 3.26 mutations leading to de regulation of expression or function of any protein in the pathway from growth factor to target replication gene expression can be an oncogene (see Table 3.11).

However, although multiple mutations may arise to give enhanced replication signalling, the cell cycle itself is regulated by two gate keepers which would normally halt the anerrant signal. Tumour suppressor gene products are intimately involved in control of the cell cycle (Table 3.12).

Progression through the cell cycle is controlled by many molecular gateways, which are opened or blocked by the cyclin group of proteins that are specifically expressed at various stages of the cycle. The RB and p53 proteins control the cell cycle and interact specifically within many cyclin proteins. The latter are affected by INK 4A acting on p16 proteins. The general principle is that being held at one of these gateways will ultimately lead to programmed cell death. p53 is a DNA-binding protein which induces the expression of other genes and is a major player in the induction of cell death. Its own expression is induced by broken DNA. The induction of *p53* gene transcription by damage initially causes the expression of DNA repair enzymes. If DNA repair is too slow or cannot be affected, then other proteins that are induced by p53 will affect programmed cell death.

One gateway event that has been largely elucidated is that between the G1 and the S phase of the cell cycle. The transcription factor dimer complex E2F-DP1 causes progression from the G1 to the S phase. However, the RB protein binds to the E2F transcription factor, preventing its induction of DNA synthesis. Other, cyclin D-related molecules inactivate the RB protein thus allowing DNA synthesis to proceed. This period of rapid DNA synthesis is susceptible to mutation events and will propagate a preexisting DNA mistake. Damaged DNA-induced *p53* expression rapidly results in the expression of a variety of closely related (and possibly tissue-specific) proteins WAF-1/p21, p16, p27. These inhibit the inactivation of *RB* by cyclin D-related molecules. As a result *RB*, the normal gate which stops the cell cycle, binds to the E2F-DP1 transcription factor complex, halting S phase DNA synthesis. If the DNA damage is not repaired apoptosis ensues (Figure 3.27).

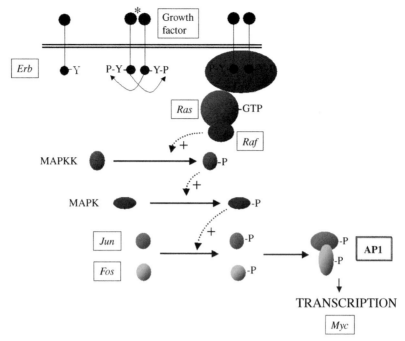

Figure 3.26   Diagrammatic illustrations of protooncogene positions within the growth factor receptor signal transduction pathway

3.3.8.1   Viral inactivation of tumour suppressors

The suppression of normal tumour suppressor gene function can be achieved by disabling the normal protein once it has been transcribed, rather than by mutating the gene. Viruses have developed their own genes which produce proteins to do precisely this. The main targets of these proteins are *RB* and *p53* to which they bind and thus disable. The best understood are the adenovirus E1A and human papillomavirus (HPV) E7 gene products which bind *RB*,

whilst the adenovirus E1B and HPV E6 gene products bind *p53*. The SV40 virus large T antigen binds both *RB* and *p53*.

3.3.8.2   Epigenetics and cancer

The term 'epigenetics' was first coined to explain changes in gene expression that do not involve changes in the underlying DNA sequence. Despite not altering the coding sequence, the effects of epigenetic changes are stable over rounds of cell division and sometimes between generations. There are two principal molecular epigenetic mechanisms implicated in cancer:

- modifications to DNA's surface structure but not to its basepair sequence — DNA methylation resulting in nonrecognition of gene transcription DNA binding domains;
- modification of chromatin proteins (in particular histones) which will not only support DNA but bind it so tight so as to regulate gene expression — at the extreme such binding can permanently prevent the DNA sequences being exposed to, let alone acted on by, gene transcription (DNA binding) proteins.

Table 3.11   Examples of oncogenes

Gene	Function of product
Sis	PDGF growth factor
ErbB/Neu	EGF receptor Truncated
Erb A	Thyroid hormone cytoplasmic receptor
Ras	G protein
Src	Membrane/cytoskeleton- associated tyrosine Kinase
Fes	Cytoplasmic tyrosine kinase
Raf	Serine/threonine protein kinase
Myc Fos Jun	Transcription factor nuclear proteins

Table 3.12 Examples of tumour suppressor genes

Gene	Function of product	Hereditary tumours	Sporadic tumours
RB1	Transcription factor	Retinal and sarcoma	SCLC, breast, prostate, bladder, retinal and sarcoma
P53	Transcription factor	Li-Fraumeni syndrome — breast, osteosarcoma, leukaemia, soft-tissue sarcoma	50% of all cancers
WT1*	Transcription factor	Nephroblastoma	Nephroblastoma
APC	β-catenin function — cell adhesion role	Familial adenomatous polyposis — colorectal cancer	Colorectal cancer, rarely others.
NF1*	GTPase activating protein	Neuroblastoma, phaeochromocytoma	Melanoma, neuroblastoma
NF2	Plasma membrane link to cytoskeleton	Bilateral vestibular schwannomas	Schwannomas, meningiomas, ependymomas, melanomas, breast Ca
BRCA1+2	DNA repair	12% of all women with breast Ca under 40; ovarian	Ovarian ≫ breast
MSH2, MLH1, PMS1+2	DNA mismatch repair	Hereditary nonpolyposis coli - colorectal, endometrial, gastric	Colorectal, endometrial, gastric

Hypermethylation of gene regulatory regions (denoted by CpG islands) whereby cytosine is converted to 5-methyl cytosine has been identified at a number of gene loci and is associated with oncogenesis. The *Hic-1* gene is noticeably silenced by hypermethylation in numerous tumours. Interestingly the enzymes responsible for methylation — DNA methyltransferases (DNMT1 etc) — are attracted to replication foci and preferentially modify hemimethylated DNA CpG regions. The net result is not only 'maintenance' of the methylated status in daughter strands but also its expansion within the loci.

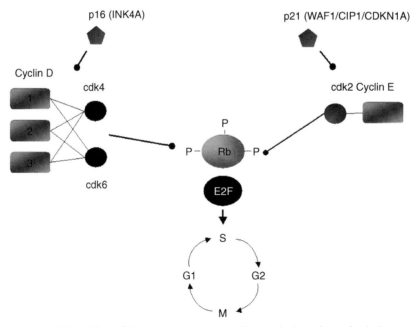

Figure 3.27 Diagrammatic illustration of the tumour suppressor RB protein 'gate keeper' role in preventing cell cycle progression from G1—S phase until progressive phosphorylation by cyclin molecules renders RB unable to bind the cycle specific E2F transcription factor

Methylation is not limited to CpG rich regions of DNA, the associated histones are also modified as follows. Away from the main body of the histone complex the lysine rich tail of histone protein 3 (H3) interacts with DNA by attracting and compensating for the strong polar negative charge — with like negative charge — repulsion of the phosphate backbone of DNA. Effectively it buffers this repulsion allowing DNA to twist and coil tighter. Acetylation of these H3 protein tail lysines removes their buffering positive charge causing the DNA to spring away from the Histone packaging. In this histone influence-free form the DNA helix is able to assume conformational shapes driven by its own internal bond length strains. Thus, where the double helix is under stress from distortion by CpG rich regions this springing away from the nucleosomal histones allows the free assumption of the open Z-form of DNA double helix and hence access to DNA binding proteins that initiate transcription. Deacetylation of the H3 tail lysines result in DNA being pulled onto the histone body and effectively burying the CpG rich transcription recognition regions. However, methylation of the H3 tail lysines results in even tighter coiling and like-with-like attraction of similar modified histones of other nucleosomes. The net result is the tight packing of large regions of DNA in what is termed hetrochromation. Thus DNA methylation can recruit DMNTs to methylate daughter strand loci during replication and their associated histones to padlock a gene region effectively shut. If the region includes tumour repressor gene loci oncogenic progression can proceed without direct mutation of the gene in question.

More epigenetic mechanisms are being discovered that further help explain the phenomena of 'book marking' which has been observed in several biological species. The name refers to the observation that some genes appear to remain active long after the initiating signals for transcription have been removed — much like a book falling open at a much read page once dropped. Recent studies have focused on small RNA species that are the product (or spliced byproduct) of transcriptional activated genes. These RNAs tend to form double-stranded species that keep the gene loci open for continued transcription. This is not an unusual phenomenon in that the protein product of genes such as *Hnf4* and *MyoD* are known to feedback and maintain autotranscription. However, small RNA species can freely diffuse through nuclear pores and into other cells via gap junctions. This unhindered passage makes such RNA species pervasive, up regulating and locking the 'on switch' of oncogenic genes in transformed and neighbouring cells without DNA mutation. In several species the so-called maternal effect phenotype is believed to be due to such small RNA species being laid down in the oocyte by nurse cells and then acting on genes of the resulting offspring as it develops.

3.3.8.3   Hereditary predisposition to cancer

Out of all cancers 1% have a clear hereditary nature though a more subtle genetic predisposition, for example in affecting relative gene product function, will undoubtedly be present in a much higher percentage of patients. The majority are due to inheritance of tumour-suppresser gene mutations. This is not surprising in view of the crucial role in development and differentiation of protooncogenes. Inherited oncogenes would be expected to be lethal though two are recognized — MEN2 and hereditary papillary renal cancer. Only one hit is required to lose the function of the tumour suppressor gene if there is already a hereditary mutation on the other allele. Compared to patients with sporadic tumours, those with inherited predisposition to cancer tend to develop tumours at an earlier age and more commonly bilaterally or multifocal, and often with a more restricted tissue origin.

3.3.8.4   Microsatellite instability

Microsatellites are short (50—300 bp) sequences composed of tandemly repeated segments of DNA two to five nucleotides in length (dinucleotide/trinucleotide/tetranucleotide repeats). Scattered throughout the genome in the noncoding regions between genes or within genes (introns), many of these microsatellites are highly polymorphic. Often used as markers for linkage analysis because of high variability in repeat number between individuals, these regions are inherently unstable and susceptible to mutations. Somatic microsatellite instability (MSI) has been detected in a number of tumours. Detecting MSI involves comparing the length of microsatellite alleles amplified from tumour DNA with the corresponding allele in normal tissue from the same individual. Recent studies indicate that MSI can be detected in approximately 90% of tumours from individuals with hereditary nonpolyposis colorectal cancer (HNPCC). The presence of these additional microsatellite alleles (repeated segments) in tumour cells, results from the inherent susceptibility of these areas to such alterations and from mutations in the DNA mismatch repair mechanism that would normally correct these errors.

3.3.8.5 Tumour angiogenesis

Once a nest of cancer cells reaches 1–2 mm in diameter, it must develop a blood supply in order to survive and grow larger as diffusion is no longer adequate to supply the cells with oxygen and nutrients. As with all tissues solid tumour cancer cells secrete substances that promote the formation of new blood vessels – a process called angiogenesis. Over a dozen substances have been identified that promote angiogenesis. (e.g. Angiopoietin-1, basic fibroblast growth factor [bFGF] and vascular endothelial growth factor [VEGF]). This has lead to the discovery of a number of inhibitors of angiogenesis and some have already advanced to clinical trials as part of a cancer treatment strategy.

Several therapeutic vaccine preparations are under development to produce a range of host immune (humoral, cellular) against proangiogenic factors and their receptors in tumours. One approach has been directed at cell adhesion molecules found in tumour blood vessels. It turns out that the new blood vessels in tumours express a vascular integrin — designated alpha-v/beta-3 — that is not found on the old blood vessels of normal tissues. Vitaxin, a monoclonal antibody directed against the alpha-v/beta-3 vascular integrin, shrinks tumours in mice without harming them. In Phase II clinical trials in humans, Vitaxin has shown some promise in shrinking solid tumours without harmful side effects.

3.4 Important techniques in molecular cell biology

Although routine laboratory biomedical techniques are detailed in the following biomedical discipline specific chapters, an awareness of the techniques peculiar to, or developing in, biomedical research rather than laboratory medicine diagnostic laboratories is essential as some are now entering the clinical laboratory.

3.4.1 Monoclonal antibodies

Myeloma is a malignantly transformed B-cell lineage which secretes a specific antibody. This fact is used to produce specific antibodies directed toward an antigen of choice. A laboratory animal is injected with the antigen of choice against which it mounts an immune response. B-cells are then harvested from the spleen. The cells are fused en masse to a specialized myeloma cell line which no longer produces its own antibody. The resulting fused cells, or hybridomas, are immortal and produce antibodies specified by the lymphocytes of the immunized animal. These cells scan be screened to select for the antibody of interest which can then be produced in limitless amounts. Modification of the mouse antibody is then required for the recognition of the Fc (effector) region of the antibody to initiate human defence mechanisms and to avoid an immune response against the antibody shortening its half-life. Attachment of a human Fc fragment to the mouse Fab fragment to create a chimaeric antibody is called humanization.

3.4.2 Southern hybridization

DNA, which has first been digested with restriction endonucleases (e.g. Eco-R1), can be separated by virtue of the differential mobility of fragments of varying size in an electrical field. This is done in an agarose gel. The gel is then placed on a nylon transfer membrane and the DNA is absorbed onto it by capillary action: this is the process of Southern blotting. (Northern and Western blotting refer to essentially the same process but using mRNA or protein respectively rather than DNA.)

The nylon membrane can then be incubated with a short strand of DNA (the probe) which has been radiolabelled with 32p. If the DNA on the membrane contains sequences homologous to the probe then Watson–Crick basepairing will occur and the probe will stick to the membrane. This can be visualized by exposing the membrane to a standard radiographic film. Thus a probe for a given region of the genome can be used to investigate the DNA of patients to determine the presence or absence of a given mutation if that mutation creates or destroys a restriction enzyme recognition site thereby altering the size of a band on the exposed film.

3.4.3 The polymerase chain reaction

This has led to a revolution in molecular biology. Two unique oligonucleotide sequences on either side of the target sequence, known as primers, are mixed together with a DNA template, a thermostable DNA polymerase (taq polymerase) and purine and pyrimidine bases attached to sugars.

In the initial stage of the reaction the DNA template is heated to 90 °C to make it single-stranded (denature) and then as the reaction cools the primers will anneal to the template if the appropriate, complementary, sequence is present. Then the reaction is heated to 72 °C for the DNA polymerase to synthesize new DNA between the two primer sequences. This process is then repeated on multiple occasions, up to about 30 or so cycles, amplifying the target sequence exponentially. Each cycle takes only a few minutes. The crucial feature of PCR is that to detect a given sequence of DNA it only needs to be present in one copy (i.e. one molecule of DNA); this makes it extremely powerful.

The sensitivity of the technique is dictated by the amount of amplification and thus the number of cycles performed. The specificity relies on the uniqueness of the oligonucleotide sequence of the primers (if the primers bind at multiple sites then multiple DNA sequences will be amplified).

Refinements of PCR include:

- multiplex PCR where multiple pairs of primers are used to amplify several target areas of DNA in parallel;
- nested PCR improves the specificity of the reaction by including a second pair of primers just within the target sequence defined by the first set of primers;
- RtPCR uses reverse transcriptase to form cDNA from mRNA which can then be used for the standard PCR reaction;
- real time RtPCR is a quantitative RtPCR where relative levels of mRNA are determined by monitoring the simultaneous amplification of cDNA of the target gene against that of a housekeeping gene mRNA.

3.4.4 Expression microarrays/gene chips

This is a methodology developed to examine the relative abundance of mRNA for thousands of genes present in cells/tissue of different types or conditions. For example, it can be used to examine the changes in gene expression from normal colonic tissue to that of malignant colonic polyps. The basic technology is the ability to immobilize a sequence of DNA complementary to specific genes or different regions of known genes, onto a solid surface in precise microdot arrays. Total mRNA is extracted from one tissue and labelled with fluorescent tag Cy3-Green and the mRNA from the second tissue with fluorescent tag Cy5-red. The two fluorescent-tagged total mRNA sample are mixed in a 1:1 ratio and washed over the DNA gene chips. The mRNA for specific genes will bind to their complementary microdot and can be detected by laser induced excitation of the fluorescent tag and the position and light wavelength and intensity recorded by a scanning confocal microscope. The relative intensity of Cy5-red: Cy3-green is a reliable measure of the relative abundance of specific mRNAs in each sample. Yellow results from equal binding of both fluorescent-tagged mRNAs. If no hybridization occurs on a dot then the area is black. The power of the system is that many thousands of genes can be screened for not only their expression but relative expression in normal and diseased tissue. A considerable amount of computing power and analysis is required to interpret the thousands of dots on a microarray chip.

3.4.5 Genomics

An international effort to sequence the entire human genome, all 3×10^9 bp on the 24 different chromosomes, was officially launched in 1986 and officially completed on the 14th April, 1983. An initial aim was to construct genetic and physical maps of the chromosomes: 40 three-generation families were studied for microsatellite and restriction enzyme (RFLP) polymorphisms. These landmarks are the first step in positional cloning of a disease gene and are used as genetic markers if they cosegregate with the disease. Now that the entire genome has been sequenced to an accuracy of 99.99% finding genes that are colocated with a genetic disease marker takes a matter of hours instead of years and is in fact a database or adapted 'in silico' cloning exercise.

Only 2% of the genome codes for actual proteins — some 30 000 genes in the human genome of 300 million bases. However, many genetic disorders are the result of deregulation of expression and it is the control elements surrounding the coding that are deregulated. The complete sequence also gives rise to an understanding of how genes are packaged as euchromatin or heterochromatin and why it is organized in the way it is.

3.4.6 Proteomics

A more direct route to understanding genetic and somatic disease is by studying the protein expression characteristics of normal and diseased cells — the proteome. This relies on the separation of proteins expressed by a given tissue by molecular size and charge on a simple two-dimensional display and is achieved by using two-dimensional gel electrophoresis. The pattern of dots corresponds to the different proteins expressed. With the improvement in technology the patterns are reproducible and can be stored as electronic images. Non-, over- and underexpression of a given protein can be detected by a corresponding change on the proteome two-dimensional electrophoresis. Furthermore, posttranslational modifications of the protein show up as a change in either size or charge on the proteome picture. In order to identify the altered protein, and the posttranslational modification it may contain, positively these protein spots are eluted and subjected to modern mass spectrometry techniques such as Matrix assisted laser desorption (MALDI) and Electrospray ionization (ESI) time of flight (TOF), which not only give you the precise mass of the protein, up to ~500 000 Da, but also can sequence its aminoacid, phosphorylation and glycosylation structure. This cannot be detected by genome analysis. Looking for such changes has already lead to the discovery of new protein markers for the diagnosis of Creutzfeldt—Jakob disease, multiple sclerosis, schizophrenia, Parkinson's disease (spinal fluid protein) and Alzheimer's disease (blood and brain proteins). In 2000, the Swiss Institute of Bioinformatics (SIB) and the European Bioinformatics Institute (EBI) announced a major effort to annotate and describe highly accurate information concerning human protein sequences, effectively launching the Human Proteomics Initiative (HPI),

3.4.7 Metabolomics

In the postgenomic era, computing power, statistical software, separation science and modern mass spectrometry has allowed the analysis of complex mixtures as a complete entity and not merely the fluctuation in concentration of one analyte within it. Metabolomics is the study of the repertoire of nonproteinaceous, endogenously-synthesized small molecules present in an organism. Such small molecules include well-known compounds like glucose, cholesterol, ATP and lipid signalling molecules. These molecules are the ultimate product of cellular metabolism and the metabolome refers to the catalogue of those molecules in a specific organism, for example the human metabolome. In terms of clinical biochemistry, the analysis of the pattern of change of such molecules in urine samples of individuals with and without a particular disease and those treated with specific drugs represents a change in the metabolome. It is very likely that, in the future, medicine-regulating authorities will require metabolomic studies on all new drugs.

Bibliography

Alberts, B. *et al.* (2008) *Molecular Biology of the Cell*, 5th edn, Garland Science.

Alberts, B., Bray, D., Hopkin, K. and Johnson, A. (2009) *Essential Cell Biology*, 3rd edn, Garland Publishing.

Baum, H. (1995) *The Biochemists' Songbook*, 2nd edn, CRC Press.

Pritchard, D.J. and Korf, B.R. (2007) *Medical Genetics at a Glance*, 2nd edn, Wiley—Blackwell.

Reed, R., Holmes, D., Weyers, J. and Jones, A. (2007) *Practical Skills in Biomolecular Sciences*, 3rd edn, Benjamin Cummings.

Chapter 4
Cellular pathology

**Dr Christopher M. Stonard, MA, MB, BChir, FRCPath and
Jennifer H. Stonard, B.Sc., MIBMS**

Cellular pathology is the area of biomedical science that combines the disciplines of histopathology and cytopathology. It is an increasingly loose term that refers to the science of the investigation of disease at a cellular level. More specifically, histopathology looks at disease processes at a cellular level in tissues (see below) and its sister discipline of cytopathology looks at individual cells. In the area of healthcare provision, histopathology is chiefly concerned with the diagnosis of disease by the examination of tissue and cytopathology by the examination of a variety of body fluids and aspirates.

In order to understand how the examination of tissues, body fluids and aspirates can indicate diagnostic information it is necessary to review the underlying pathological processes. This is preceded by a brief revision of the nature of cells, tissues and organs. A review of the fundamental principles underlying disease processes forms the first part of this chapter. The second part of the chapter is concerned with laboratory techniques in cellular pathology and how they can be applied to investigating disease.

Part I: Principles of cellular pathology

4.1 Structure and function of normal cells, tissues and organs

Cells are the fundamental building blocks of all tissues. They vary enormously in morphology and function yet all share the same principal constituents (Figure 4.1). These include:

1. *The nucleus.* This contains the DNA code for virtually every protein found in the body. It precisely regulates the expression of each protein and is fundamental to the timing of cell division.
2. *The mitochondria.* These are the energy providers of the cell that convert simple sugars to an energy source that the cell can use at the level of individual

Biomedical Sciences: Essential Laboratory Medicine, First Edition. Edited by Ray K. Iles and Suzanne M. Docherty.
© 2012 John Wiley & Sons, Ltd. Published 2012 by John Wiley & Sons, Ltd.

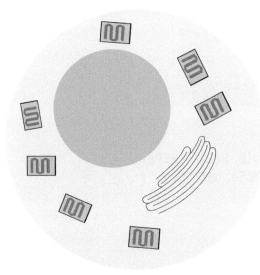

Figure 4.1   A schematic diagram of a cell showing representations of the main organelles including the nucleus, multiple mitochondria and the endoplasmic reticulum

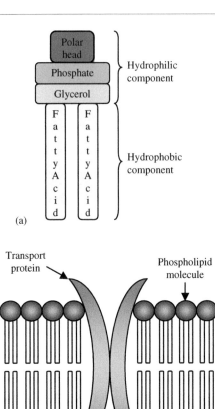

Figure 4.2   The structure of a typical lipid molecule (a) and a membrane (b). The hydrophilic parts of the lipid molecule face the environment and the cytoplasm while the hydrophobic parts face each other to make an impermeable central core

molecular interactions. They also contain a small amount of DNA encoding the few proteins not encoded by the DNA of the cell's nucleus.

3. *The endoplasmic reticulum.* This is effectively the production and assembly line of the cell. The DNA code for cellular proteins is translated and the corresponding proteins assembled here. A large proportion of those proteins destined to become enzymes carry out their actions within the endoplasmic reticulum.

4. *The cytoskeleton.* This is, as the name indicates, the skeleton of the cell that maintains its structural integrity. In mobile cells its ability to be rapidly modified and reshaped gives the cell movement relative to its environment.

5. *Cell membranes.* Cell membranes are composed of a sandwich or bilayer of phospholipid molecules (Figure 4.2) such that the water soluble (hydrophilic) components of each molecule face the watery environment around the membrane and the insoluble (hydrophobic) fatty acid components face each other. The end result is a fragile but impervious barrier to water and water soluble substances.

A continuous membrane surrounds every cell and every discrete functional cellular constituent (or organelle). By inserting pores and transport proteins that allow the passage of very specific substances across the membranes under specific conditions the cell can have

a significantly different internal composition from its environment. In addition it can contain an assortment of intracellular compartments that are effectively sealed off from the bulk of the cytoplasm.

4.2   Tissues and organs

Tissues are populations of cells of the same type that are architecturally organised into a functional structure. Thus aggregates of fibroblasts, with their secreted extracellular proteins, form connective tissue and collections of fatty acid-storing adipocytes compose fatty tissue. When these tissues are organised together to form a discrete structure with one or more defined functions they are termed an organ. In this way an organ such as the small bowel is composed of a variety of tissue including an epithelial layer, connective tissue, smooth muscle, neural tissue and vascular tissue.

4.3 Cellular responses to injury

In this section we will address the nature of damaging stimuli and the corresponding cellular responses.

4.3.1 Damaging stimuli

Given that human cells are protected from their immediate environment only by permeable layers of lipid and protein they are highly vulnerable. Fortunately the body has a wide array of mechanisms for maintaining homeostasis and physical barriers to keep cells in a relatively constant local environment policed by the immune system. A selection of the more commonly encountered means by which injurious stimuli can breach this protection are discussed below.

4.3.1.1 Hypoxia and ischaemia

Hypoxia refers to insufficient oxygen reaching a tissue whereas ischaemia indicates insufficient blood supply. The two usually go hand in hand, although there are circumstances in which hypoxia is the result of inadequately oxygenated blood despite a good blood supply. Cells vary in their response to hypoxia and ischaemia depending on their metabolic demands (e.g. a high demand for heart muscle cells but low for fibroblasts within scar tissue). The mechanisms and consequences of hypoxia and ischaemia are discussed later.

4.3.1.2 Environmental

Damaging environmental factors include an enormous range of toxic chemicals from naturally occurring plant toxins and venoms to common household chemicals and cigarette smoke. Exposure to ionizing radiation is an inevitable fact of life and takes the form of solar ultraviolet light, diagnostic X-rays and unavoidable radiation from the earth around us.

4.3.1.3 Physical

Damaging physical stimuli include heat and cold outside the functional temperature range of cells as well as day-to-day traumatic injuries.

4.3.1.4 Infective and immunological

The body is under constant assault from a variety of infective agents including viruses, bacteria, fungi and parasites. In addition the immune system itself can damage cells, either in the course of mounting an attack against some infective organism or because of some form of dysfunction that results in the body targeting itself (an autoimmune reaction).

4.3.2 Varying vulnerability

The reaction of a cell to an injurious stimulus depends not only on the type, degree and duration of the stimulus but also on the cell type. For example, the more metabolically active the cell the more prone to damage that cell will be by factors that disrupt respiration and blood supply. Thus the highly metabolically active cells of the central nervous system will die within minutes of disruption of their blood supply. Similarly those cells that proliferate rapidly (such as the epithelial cells that line the intestines) are highly susceptible to stimuli that disrupt DNA. Hence they are vulnerable to chemical substances that disrupt DNA replication (such as certain chemotherapeutic agents) and ionizing radiation (which is one reason that diarrhoea is an early sign of radiation sickness).

4.3.3 Reversible versus irreversible injury

If the injurious stimulus is removed the degree to which the cell can recover differs widely between cell types and the nature of the injury. All cells have an essential need to maintain cytoskeletal and membrane integrity, synthesise proteins, keep their own intracellular proteolytic and free radical reactions in check and undergo metabolic respiration in order to survive. If the injurious stimulus allows these essential functions to continue the cell is likely to survive. Disruption of one or more of these functions means that the cell is no longer viable and will die. The two principal methods of cell death (apoptosis and necrosis) are discussed below.

Reversible injury may result in complete resolution of cell function. In cases in which the injurious stimulus is recurrent or remains at a constant non-lethal level the cell and the tissue in which it lies may need to undergo adaptive changes. Tissue adaptive changes are dealt with elsewhere (see Sections 4.7 and 4.13). Such changes include increased production of contractile proteins in muscle undergoing sudden extremes of use or increased production of smooth endoplasmic reticulum in liver cells exposed to a new toxin that they are required to break down.

4.3.4   Necrosis and apoptosis

Apoptosis and necrosis are the two principal mechanisms of cell death. They differ in that apoptosis is an active process mediated by the sequential actions of a set of chemical processes within the cell. Necrosis, on the other hand, is a passive process over which the cell has no control.

4.3.4.1   Necrosis

Necrosis is the manifestation of irreversible cell injury that has progressed to show the effects of enzyme degradation on the cellular constituents. The degrading enzymes may be derived from the cells themselves (termed autolysis) or from acute inflammatory cells recruited to the site of necrosis (termed heterolysis).

Necrosis is often subclassified according to its macroscopic appearance into a variety of groups with colourful descriptive names such as 'caseous' and 'liquefactive' (Figure 4.3). In practical terms by far the most commonly encountered type of necrosis is coagulative necrosis and this is the type typically seen in acute infarction.

Coagulative necrosis manifests on haematoxylin and eosin-stained sections as a gradual loss of nuclear and organelle detail (as basophilia, or blue staining, fades) and increased eosinophilia (or pink staining) of the cytoplasm which subsequently demonstrates a 'spider web' or 'motheaten' pattern of vacuolation. The loss of basophilia is due at least in part to the nonspecific enzymatic degradation of nucleic acids, particularly within the nucleus. The transition of cytoplasm through intense eosinophilia to loss of staining altogether with patchy apparent vacuolation is the result of denaturing of cytoplasmic proteins followed by their degradation into peptides. The terms 'basophilia' and 'eosinophilia' are discussed in more detail in Section 4.20.

Of the other forms of necrosis, the one which tends to imply a specific underlying cause is that of caseous

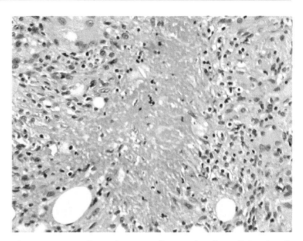

Figure 4.3   An irregular area of necrosis. The cellular detail has been lost and all that remains is an amorphous area of eosinophilic material with some occasional nuclear fragments. A surrounding inflammatory reaction is present and in this case consists of granulomatous inflammation (see Section 4.5). *H&E photographed at ×200*

necrosis. Caseous ('cheese-like') necrosis has a firm texture and cream/yellow colour. Histologically it is surrounded by granulomatous inflammation (see Section 4.5). It is classically caused by infection by *Mycobacterium tuberculosis* and, unless otherwise qualified, its description normally implies tuberculosis.

4.3.4.2   Apoptosis

Unlike necrosis, apoptosis requires the cell to be an active participant in its own destruction through a coordinated process of upregulation of its own enzymes. As well as being a response to injury it is a normal and continuously occurring process in the body [1]. Commonly encountered examples include the deletion of selfreactive T-cells during lymphocyte maturation or the involution of breast tissue after lactation has finished. There is even at least one cell surface receptor (called Fas) that will trigger apoptosis if activated [2]. This is activated for example by T lymphocytes which have detected that the cell is infected with a virus so that the cell can be killed before it spreads its virus to other cells.

Apoptosis may be triggered by the same injurious agents that cause necrosis. This may occur if the amount of damage received by the cell is sufficient to cause irreversible damage to the DNA but the metabolic and synthetic ability of the cell is still sufficiently preserved to undertake the energy-consuming process of apoptosis. If a cell can undergo apoptosis rather than necrosis then this avoids the risk of the leakage of destructive free radicals and enzymes into the extracellular space with consequent damage to surrounding cells. In the case of irreversible cell injury, apoptosis is therefore the lesser of two evils.

Histologically the first features of apoptosis are a shrinking of the cell together with shrinking of the nucleus and intense nuclear basophilia (Figure 4.4). Eventually the nucleus fragments, followed by cytoplasmic fragmentation and uptake by neutrophils and macrophages.

Two important proteins in the expanding list of cellular constituents involved in apoptosis are Bcl2 [3] and P53 [4]. Both appear to be targets for a variety of mechanisms that detect cellular injury and, given sufficient indication of cellular damage, act to initiate a family of proteases called caspases that break down the cell. As discussed in Section 4.14, P53 is heavily involved in the detection of DNA damage and control of cell division. Once the caspases are activated the process of cellular destruction proceeds through a sequence of caspase-mediated nuclear, cytoplasmic, organelle and cytoskeletal degradation. The result is that the cell ends up as a cluster of neatly packaged, harmless degradation products that importantly do not elicit a potentially damaging inflammatory reaction and are ready to be removed, for example, by a macrophage.

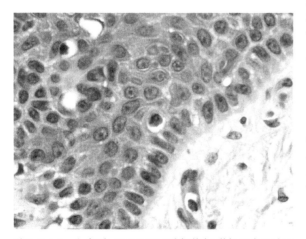

Figure 4.4   A single squamous epithelial cell is undergoing apoptosis (centre). The apoptotic cell has brightly eosinophilic (red/pink) cytoplasm and a nucleus showing condensation and fragmentation. There is no inflammatory reaction. *H&E photographed at ×400*

4.3.5   Summary

The individual cells of the body withstand a constant barrage of damaging stimuli of varying degrees. The individual cellular responses are determined by a wide variety of factors including the type and degree of insult, the metabolic activity of the cell and the line of cellular differentiation. The behaviour of cells in the context of injurious stimuli is linked in a complex fashion to the nature of the tissues of which they are a part. This is explored in the following section.

4.4   Tissue responses to injury: acute inflammation

Acute inflammation marks the first response to any damage to tissues. Although usually followed by a chronic inflammatory response, the two processes overlap and, despite its slightly misleading name, the acute inflammatory process may persist for as long as its chronic inflammatory counterpart (Box 4.1).

Acute inflammation represents a delicate balancing act between the requirements of the immune response to mount a swift and overwhelming attack on any damaging agent (such as an infective organism) and the need to limit the destructive power of the immune system in order to prevent unnecessary damage to the surrounding normal tissue. Each stage of the process must therefore be kept in check by a system of countermeasures and failure at any point can have serious consequences.

The acute inflammatory response can be divided broadly into vascular and cellular components.

4.4.1   Vascular components

The vascular changes in acute inflammation can be viewed as effects on the localised blood flow within the affected tissue and effects on the permeability of the blood vessels within and adjacent to the affected tissue.

The balance between the fluid components of blood and tissues is dependent on pressure and osmotic gradients across the blood vessel wall and the permeability of the vessel wall (Figure 4.5). The osmotic gradient is largely contributed to by large protein molecules within the blood that cannot, under normal circumstances, diffuse freely across the blood vessel wall. The acute

Box 4.1 Acute inflammation

The acute inflammatory response is very frequently a cause of disease but this is usually a manifestation of being the effector arm of another system that is functioning incorrectly. This is illustrated by dysfunction of the immune system that results in inappropriate activation of acute inflammation, for example in chronic diseases such as rheumatoid arthritis or, more dramatically, in the case of anaphylactic shock.

Pure dysfunctions of the acute inflammatory response do occur and often take the form of subtle failures to regulate the degree of response (complete failures to regulate inflammation not being compatible with life).

One example of subtle dysregulation is that of α1-antitrypsin deficiency. α1-antitrypsin is a circulating plasma protein that inhibits the action of neutrophil elastase, one of the neutrophil's hydrolytic enzymes involved in acute inflammation. There are numerous alleles for the gene encoding the protein but one in particular (the Z allele) greatly reduces the amount of functional enzyme inhibitor circulating in the plasma. A person with two Z alleles has only approximately 10–15% of the normal level of circulating enzyme inhibitor.

Most α1-antitrypsin is produced by the liver. Many with the ZZ genotype accumulate the aberrant protein in the liver and this can lead to the development of cirrhosis. A higher proportion however develop emphysema (a disease characterised by destruction of the walls of the smallest airways and airspaces in the lungs that results in enlarged air spaces with loss of functional lung tissue). This is thought to be the result of the gradual destruction of the elastin-rich connective tissue of the lungs by inadequately opposed neutrophil elastin. The process is accelerated by anything that precipitates inflammation within the lungs. Smoking in particular accelerates the develop of emphysema and this may in part be the result of inactivation of the small amount of remaining functional α1-antitrypsin by chemicals within cigarette smoke.

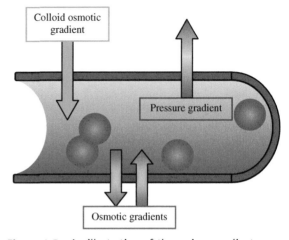

Figure 4.5   An illustration of the various gradients competing to dictate the movement of fluid and ions across the vessel wall. Added to the effect of these gradients is the selective permeability of the vessel wall

inflammatory response results in dilatation of the arterioles supplying blood to the affected area with a net increase in intravascular blood pressure, thus shifting the balance in favour of fluids exiting the vessels and entering the tissues. In addition, the increased permeability of the vessel walls results in the diffusion of some of the blood protein components out into the tissues with an osmotic equivalent amount of fluid. The net result is an increase in blood flow to the affected tissue together with a shift of fluid and certain proteins from blood to tissue. This accounts for the immediate redness (increased blood flow) and swelling (tissue fluid accumulation) seen after an injury.

The blood remaining within the vessels has a lower proportion of fluid and protein together with a higher proportion of cells. It is therefore more viscous than normal and will flow less freely. The stagnation of blood flow subsequently assists the cellular components of acute inflammation.

4.4.2   Cellular components

The cellular hallmark of the acute inflammatory response is the neutrophil (Figure 4.6). The neutrophil has a myriad of actions but its key feature is its ability to engulf and destroy material identified as foreign or unwanted. Its destructive capacity arises from its ability to produce a variety of substances that can break down membranes and constituent proteins. These include a

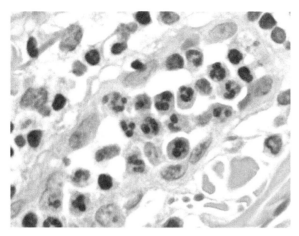

Figure 4.6   Neutrophils within a blood vessel in cross section. Although neutrophils are classically represented as having a trilobed nucleus they rarely appear trilobed in histological section. *H&E photographed at ×400*

range of proteases, such as collagenase and elastase, together with enzymes that can degrade certain plasma proteins to produce more of the chemical mediators of inflammation (see below). In addition the neutrophil possesses enzymes for the rapid formation of a variety of oxygen and chloride free radicals (the so-called oxidative burst) which will react with and disrupt any large molecules in the immediate vicinity.

Most of these destructive substances are stored in precursor form within the neutrophil's cytoplasm in visible bodies called lysosome granules. Upon engulfing some foreign material into a phagocytic vacuole the lysosomal granules are activated and discharged into the vacuole (Figure 4.7). As the destructive capacity of the lysosomal granule components is largely pH dependent the neutrophil can control the activity of these dangerous agents by regulating the acidity within the vacuole.

In addition to destruction of substances within phagocytic vacuoles the neutrophil's proteases and free radicals can be released into the extracellular space. This may happen directly (through neutrophil death and degeneration) or through attempts to engulf material beyond the maximum capacity of the cell (so-called 'frustrated phagocytosis').

The arrival of neutrophils at the site of an injury marks the first characteristic histological feature of acute inflammation. Neutrophils arrive via the circulating blood and their influx peaks approximately 24 h after injury. The arrival of a second population of inflammatory cells (macrophages) over the ensuing

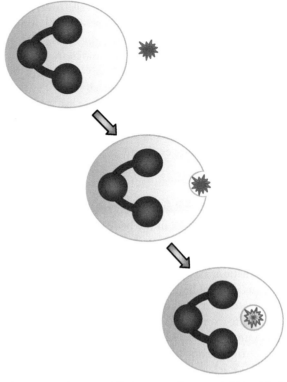

Figure 4.7   A neutrophil phagocytosing material identified as foreign. Destructive enzymes are released into the phagocytic vacuole and the material is degraded

24–48 h heralds the commencement of the overlap between acute and chronic inflammatory phases.

4.4.3   Underlying mechanisms of inflammation

The mechanisms that underlie the initiation and propagation of the acute inflammatory response are enormously complex and detailed discussion is beyond the scope of this review. An examination of the underlying concepts however illustrates not only the complexity but also the intricate ways in which the acute inflammatory response is interlinked with such processes as the clotting cascade, immunity, pain and even behaviour.

Key to the mechanisms of the inflammatory process is the generation of a broad range of chemical mediators. These mediators are chemical signals that diffuse away from the immediate site of injury and act to enhance the various components of the process. They can have effects locally (e.g. on the surrounding cells), over short distances (e.g. to affect the behaviour of neighbouring blood vessels), or even over long distances (e.g. to upregulate the production of inflammatory cells or proteins involved in the inflammatory process by the bone marrow and the liver respectively).

The generation of mediators can arise in a variety of contexts. A good starting point for the exploration of the generation of mediators is to examine the response to injury of a blood vessel wall.

4.4.3.1   The vessel wall and injury

As virtually all tissues in the body contain a complex network of blood vessels an injurious stimulus is likely to involve vessel wall damage. Blood vessels of all sizes and functions have in common a complete and continuous internal lining of endothelial cells. Thus cells and proteins circulating in the blood do not normally come into contact with the connective tissues that make up the surrounding layers of the vessel wall. If an injurious stimulus however disrupts the integrity of the vessel wall then damage to the endothelial cells will result in exposure of the circulating blood to underlying connective tissue components. These connective tissue components are recognised by blood constituents including Factor XII and circulating platelets.

The clotting cascade

Factor XII [5] is a protein synthesised by the liver. It forms a normal circulating blood plasma protein but upon exposure to collagen in connective tissues and basement membranes it is activated to become a serine protease enzyme. This enzyme then acts on a number of circulating proteins to initiate the clotting cascade, the kinin cascade and, indirectly, the complement cascade (Figure 4.8). The end result is that a small amount of activated factor XII results in a large amount of such mediators as bradykinin (which is a local vasodilator, increases local vascular permeability and stimulates local nerve endings to cause pain) and activated components of the complement cascade termed C3a and C5a (which are vasodilators, increase vessel permeability and attract neutrophils). Activation of factor XII also results in the generation of fibrin that is the scaffolding for the formation of a blood clot at the site of injury, thus reducing any bleeding that might be occurring from the damaged vessel, stagnating

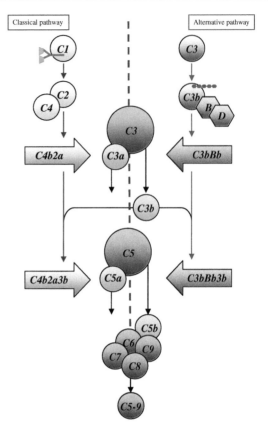

Figure 4.8   The complement cascade is an important initiator and effector of inflammation. Each stage tends to represent an amplification of the inflammatory signal

local blood flow and forming the structure on which repair processes can act (see Section 4.6).

The activation of factor XII is termed the intrinsic clotting pathway. The clotting cascade can also be activated directly by Tissue Factor, a group of substances released by damaged endothelial cells. This latter pathway is termed the extrinsic clotting pathway (see Chapter 8).

Platelets

Platelets are circulating membrane-bound bodies derived from the cytoplasm of large cells within the bone marrow termed megakaryocytes. They possess a cell surface receptor termed GPIa which binds to collagen in exposed vessel wall basement membrane [6]. This stimulates exposure of a second platelet surface receptor complex termed GPIIb/GPIIIa. This complex binds to von Willbrand factor, a multimeric circulating molecule that also binds to exposed basement membranes as well as binding to GPIIb/GPIIIa on other platelets (Figure 4.9). The binding of the GPIIb/GPIIIa to von Willebrand factor activates the platelet to produce a range of mediators via the precursor arachidonic acid (see below). The net result is the production of the mediator thromboxane A2 (which is a vasoconstrictor, activates neighbouring platelets and causes platelets to adhere to each other) and conformational changes in the platelet membrane that promote the clotting cascade. Platelet adhesion and the clotting cascade are interlinked such that the resulting thrombus or blood clot is a complex structure composed of constituents of both pathways.

4.4.3.2 Eicosanoids

Eicosanoids are the metabolites of arachidonic acid which is itself a metabolite of membrane phospholipids. As we have seen in the example of platelet activation, some of the initiators of the acute inflammatory response act by stimulating membrane phospholipases to produce arachidonic acid. The subsequent metabolism of arachidonic acid is via two main pathways

to form two groups of inflammatory mediators, the leukotrienes and the prostaglandins (Figure 4.10).

The leukotrienes collectively cause vasoconstriction, attract inflammatory cells and increase the permeability of vessel walls. The group of prostaglandins (which includes Thromboxane A2) have varied functions depending on subtype. Some cause vasoconstriction whereas others result in vasodilatation and increased vascular permeability. In addition, some are cofactors in stimulating local nerve endings to give the sensation of pain. It is the ability of a nonsteroidal anti-inflammatory drugs (such as Aspirin and Ibuprofen) to block the formation of prostaglandins that results in their properties of reducing pain and swelling.

4.4.4 Chemotaxis

Chemotaxis is the mechanism by which acute inflammatory cells migrate from the circulating blood to the site of injury. We have already seen that an acute

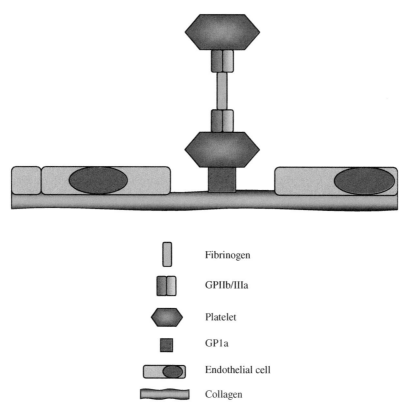

Fibrinogen

GPIIb/IIIa

Platelet

GPIa

Endothelial cell

Collagen

Figure 4.9   The continuous endothelial lining of this blood vessel has exposed collagen. The adhesion of a platelet triggers activation and the adhesion of other platelets via its activated surface receptor

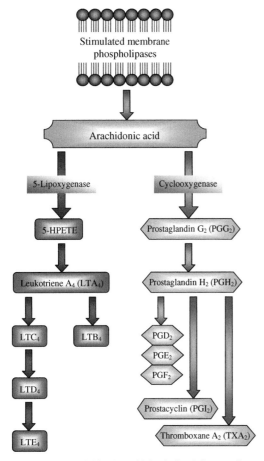

Figure 4.10   Arachidonic acid is derived from cell membrane lipids and forms the basis of a broad range of inflammatory mediators

injurious stimulus results in the production of a range of mediators that alter local blood flow, increase vascular permeability and cause localised clotting. In addition to these actions some of the mediators (e.g. the leukotrienes and activated complement factors C3a and C5a) diffuse away from the site of injury to act on nearby intact vascular endothelium to promote the migration of acute inflammatory cells from blood to the site of damage [7]. Once out of the blood vessels the inflammatory cells then follow the concentration gradient of mediators to reach the site of injury.

The details of the process of migration from blood to inflamed tissue are complex but the underlying process is one of a sequential system of interaction between endothelial and inflammatory cell surface proteins. The histological manifestation is divided into rolling, firm adhesion and transmigration phases (Figures 4.11 and 4.12). The first phase comprises activation of endothelial cells by local diffusion of mediators from a nearby site of injury to produce a set of adhesion molecules that loosely bind surface adhesion molecules on the acute inflammatory cell. This is assisted by sluggish blood flow in the vicinity of inflammation as discussed above. The interaction of adhesion molecules promotes sequential expression of more firmly adhesive surface receptors which in turn promotes conformational changes in both endothelial and inflammatory cells that allow the inflammatory cells to slip between endothelial cells and enter the extracellular space. The acute inflammatory cells are additionally stimulated to produce enzymes that can modify the extracellular matrix and allow migration around the tissues (Box 4.2).

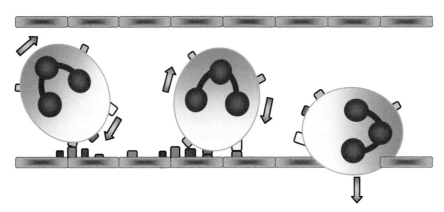

Figure 4.11   Neutrophils adhere to the wall of a blood vessel near the site of injury. They then follow a stepwise series of reactions that result in firmer adhesion and then activation of mechanisms for crossing through the vessel wall

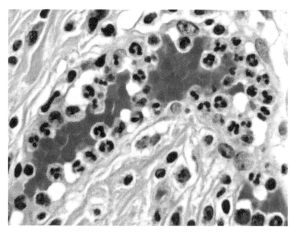

Figure 4.12   Neutrophils in a vessel in cross section. These neutrophils are showing adhesion to the vessel wall and will soon undergo migration into the surrounding tissue. *H&E photographed at ×400*

Box 4.2 Platelets, pain and

The group of medications known as the non-steroidal anti-inflammatory drugs (NSAIDs) share a common action of inhibiting the cyclooxygenase enzyme that metabolises arachidonic acid to a range of inflammatory mediators and cofactors in signalling pain. They are therefore used for their anti-inflammatory and analgesic properties.

Best known of this group of drugs is Aspirin, which has the ability irreversibly to inactivate platelet cyclooxygenase at low dose. Aspirin therefore has a significant effect on reducing the ability of platelets to aggregate.

Myocardial infarction and stroke are two of the western world's biggest killers. Most myocardial infarcts and the majority of strokes are due to the obstruction by thrombus of coronary and cerebral arteries respectively. As thrombus formation involves platelet aggregation, prevention of platelet adhesion would be a logical approach to the prevention of myocardial infarction and stroke. Aspirin has proved extraordinarily successful in this role and remains a first-line treatment in both the acute management of myocardial infarction and in its long-term prevention. More recently developed drugs target the platelet membrane GPIIb/GPIIIa complex in order to prevent platelet activation at the initial receptor binding stage.

4.4.5 *Systemic effects*

In addition to the examples above, the acute inflammatory response results in the secretion by participating cells of a variety of mediators that circulate in the blood and have systemic effects. Chief amongst these are interleukins 1 and 6 (IL-1 and IL-6) and tumour necrosis factor α (TNFα). Each of these has a range of functions including stimulation of the liver to secrete certain proteins involved in the inflammatory response, stimulation of the bone marrow to produce more acute inflammatory cells and a rise in body temperature. The rise in body temperature is presumed to be a useful property in that it may impair temperature-sensitive enzymatic processes in certain infective organisms and hence decrease their ability to replicate.

4.4.6 *Summary*

The acute inflammatory response is initiated by an injurious stimulus which triggers the release of a variety of mediators into the surrounding tissue. These act to alter local blood flow and vascular permeability together with initiating local mechanisms of blood clotting if required. In addition they trigger the mechanisms of attracting acute inflammatory cells (of which the neutrophil is the most important) to exit the circulating blood and attempt to destroy the damaging stimulus and localised dead tissue.

In reality the acute inflammatory response is interlinked with chronic inflammation and the immune response and does not occur in isolation. The fundamental principles however are a common theme in a range of pathological processes and their understanding underlies a great many of the diseases encountered in clinical pathology.

4.5 Tissue responses to injury: chronic inflammation

Chronic inflammation may follow acute inflammation, may run concurrently with the acute inflammatory process or may occur without a preceding acute inflammatory reaction. Like acute inflammation the chronic inflammatory process is intimately linked with immunity. In addition however chronic inflammation demonstrates considerable overlap with the processes of repair and healing.

Chronic inflammation following on from acute inflammation shows a temporal overlap between the dwindling acute process as it fulfils its role and the emergence of the repair processes. When the acute and chronic inflammatory processes run concurrently it is usually because the stimulus that has caused the initial injury has not fully resolved. Given that initiation and coordination of healing and repair form a function of chronic inflammation it is common in this situation to see all three processes occurring simultaneously (Figure 4.13).

If the neutrophil is the characteristic cell of acute inflammation then the macrophage (Figure 4.14) is its chronic inflammatory counterpart.

4.5.1 Macrophages

Bone marrow-derived monocytes circulate in the blood. They are attracted to the site of an inflammatory reaction and transformed to their active form by some of the same mediators as in acute inflammation, together with mediators released from active lymphocytes (particularly interferon γ derived from T lymphocytes). Once extravascular they are termed macrophages or, in histological texts, histiocytes.

Like the neutrophil they have the ability to phagocytose and degrade extracellular material determined

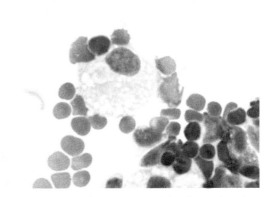

Figure 4.14 A macrophage (centre) with adjacent lymphocytes, red blood cells and poorly preserved cells of uncertain nature. Note the 'foamy' cytoplasm and impression of some intracytoplasmic vacuoles. *Rapid giemsa photographed at ×600*

as foreign or unwanted using a similar array of proteases and free radicals. Macrophages can also produce eicosanoids (see acute inflammation) and secrete a variety of mediators of crucial importance in the orchestration of repair and healing (see Section 4.6). Some of these healing/repair mediators, in particular IL-1 and TNFα, have overlapping functions with acute inflammation (see Section 4.4.5).

A further property of the macrophage is its ability to undergo a functional and morphological change to become an epithelioid macrophage (Figure 4.15). The name is derived from the histological resemblance to an intermediate squamous epithelial cell. An aggregate of epithelioid macrophages is called a granuloma. Classically granulomata form in response to a foreign substance that is too large or too inert to degrade in the usual manner of phagocytosis and release of lysosomal enzymes and free radicals. Such foreign substances might include exogenous material implanted during injury, inert mineral dusts or suture material deliberately implanted during surgery.

Granulomata can also form as a consequence of the secretion of mediators by activated T lymphocytes (see Section 7.1.2), largely in the context of certain infections, the classic example being tuberculosis. In addition granulomata may form for reasons that have not yet been fully established and, by virtue of their space occupying and local destructive effects, cause disease. The best known example of this is the systemic disease sarcoidosis, a disease characterised by the formation of

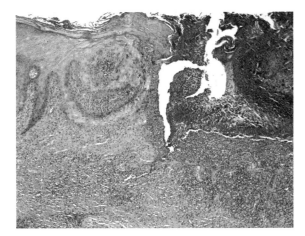

Figure 4.13 An ulcer (right) interrupts the normal continuity of the epidermal surface (left) of this skin biopsy. The base of the ulcer is composed of acutely inflamed granulation tissue with underlying attempts at repair. *H&E photographed at ×40*

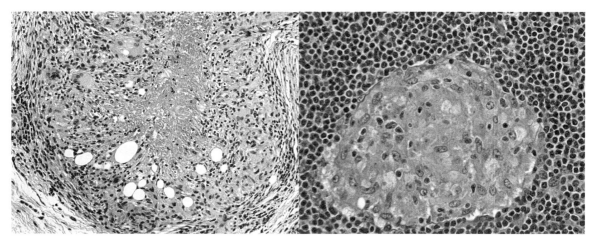

Figure 4.15   The granuloma on the left shows central necrosis with surrounding epithelioid cells, some of which have multiple nuclei. This is typical of tuberculosis. The diagram on the right is composed of epithelioid cells with no necrosis. This is typical of sarcoid. *H&E photographed at ×100*

multiple non-caseating granulomata, particularly within the lungs and lymph nodes. These may be asymptomatic but may affect a broader range of organs and may even be fatal.

The formation of a granuloma effectively encases the provoking agent (whether it is a foreign body or the tuberculous bacillus), thereby preventing it's migration or, in the case of infection, spread of the infective organism. In the case of tuberculosis the infective bacterium may remain safely contained for years, or perhaps even the rest of the patient's life, despite the bacterium remaining potentially infective. Tuberculosis is further discussed in Section 4.11.2.

4.5.2   Summary

Chronic inflammation represents a process that frequently overlaps acute inflammation together with healing and repair. The macrophage is the principal orchestrator and effector cell, showing a remarkable degree of versatility of function from removing the products of the destructive effects of acute inflammation to stimulating the migration of the cells that begin the repair process.

4.6   Healing and repair

Healing and repair after an acute episode of tissue damage typically follow on directly from the acute and chronic inflammatory responses with a large degree of overlap in mechanism and participating cell types. Under normal circumstances the process proceeds through the sequence of angiogenesis, fibroblast migration and proliferation with synthesis of extracellular matrix material and remodelling. This sequence is discussed in more detail below.

4.6.1   Angiogenesis

Angiogenesis is the formation of new blood vessels. It involves a coordinated sequence of events involving the sprouting of cords of endothelial cells from an existing blood vessel, migration through the extracellular space and formation of a central lumen. Like the acute and chronic inflammatory processes, angiogenesis involves the actions of secreted mediators.

The signals to existing blood vessels to sprout, and for endothelial cells to migrate, are transmitted via mediators using their concentration gradient as a stimulus for direction of migration. Several mediators have been shown to be involved, the best characterised of which comprise the vascular endothelial growth factor family (VEGF).

Cells within and surrounding an area of acute tissue damage have a poor blood supply because of local blood vessel disruption. The resulting lack of oxygen may promote these cells to produce a set of hypoxia inducible factors (HIFs) [8]. These are transcription factors that upregulate a variety of processes to mitigate against oxygen deprivation. One of these is the production and secretion of VEGF (Box 4.3).

The result of angiogenesis in an area of acute tissue damage is the formation of a network of rudimentary capillary-sized blood vessels surrounded by the ongoing acute and chronic inflammatory processes. This network of blood vessels is termed granulation tissue (Figure 4.16).

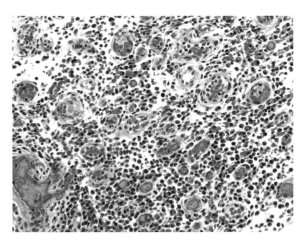

Figure 4.16 Granulation tissue composed of numerous thin-walled blood vessels interspersed with oedematous stroma, neutrophils and chronic inflammatory cells. *H&E photographed at* ×100

Box 4.3   Angiogenesis — friend and ...

Angiogenesis is a vital part of embryonic development and of healing. It is also one of the primary factors in allowing tumours to grow and cancer to kill.

A tumour is as dependent on having an adequate blood supply as any other tissue in the body. Unless a tumour can develop its own blood supply its growth will become limited by the maximum distance over which adequate diffusion of oxygen and other essential cellular requirements can take place to meet the needs of the neoplastic cells. In reality a tumour without its own supply will remain a few millimetres in maximum dimension.

As a tumour cell's demands begin to outstrip the ability of the local blood vessels to provide for their needs they will begin to become functionally hypoxic. This will therefore initiate the same hypoxia inducible factors as in healing and repair. Thus a dedicated vascular supply will develop that is composed of normal, non-neoplastic vascular endothelial cells. Indeed some tumours overexpress angiogenic factors as a result of the cluster of genetic mutations that typically occur in that particular tumour type.

The endothelial cells involved in this process are performing their normal physiological function in response to mediators from an abnormal source. By responding to correct what they interpret as injury they promote tumour growth with potentially disastrous consequences, that is they are helpfully unhelpful. Blocking angiogenesis in this particular situation is an active focus of research and cancer therapy development.

4.6.2   Fibroblasts and extracellular matrix

Fibroblasts are cells that produce the essential structural materials that form the scaffolding of tissues, the chief of which is the family of collagens. Collagens are tough, largely inert polymers that add marked tensile strength to tissues and are found throughout the body, being most densely accumulated in fibrous capsules, tendons, ligaments and in the skin.

In a site of acute tissue injury, fibroblasts migrate in from surrounding tissue following a concentration gradient of mediators. These include such mediators as platelet derived growth factor (PDGF) and transforming growth factor β (TGFβ) [9]. These mediators are secreted by, amongst other cells, macrophages involved in the inflammatory process. The arriving fibroblasts commence secreting extracellular matrix material and the accumulating collagen adds physical strength to the granulation tissue. The transition from granulation tissue to more fibrous tissue marks the beginning of scar formation.

4.6.3   Remodelling

In addition to synthesising extracellular matrix materials such as collagen, fibroblasts are capable of breaking these materials down using a set of enzymes termed matrix metalloproteinases [10]. These enzymes are also produced by other cells involved in the acute and chronic inflammatory processes, most notably macrophages. As the granulation tissue is gradually replaced by fibrous tissue the fibroblasts also break down and reform the new extracellular matrix. This process gradually shapes the fibrous tissue to match the tensile requirements of the tissue and attempts to minimize the amount of fibrous tissue present at the end of the healing and repair process — an important function as fibrous scar tissue is of little use and may even impair the tissue in performing its normal role.

4.6.4   Outcome of healing and repair

The outcome of the healing process is dependent upon a number of factors that influence the degree to which the healing process can restore the tissue to its pre-injury state.

Local factors that might impair the ability to heal include sources of ongoing inflammation or injury such as infection or the presence of foreign material. In addition, a massive area of injury or injury at a site of constant tissue movement such as overlying a joint may be extremely slow to heal.

Systemic factors that might impair healing include general malnutrition or a more specific nutritional deficiency such as vitamin C deficiency (vitamin C being an important factor in the synthesis of collagen [11]) or poor blood supply to the affected area (Box 4.4).

Skin wounds provide an illustration of the importance of local factors on healing. A wound in which the skin edges can be firmly apposed (for example with sutures) and local infection can be prevented may heal with a barely identifiable scar. This healing of a skin wound with apposed edges is termed healing by primary intention. If the wound cannot be closed because of mechanical factors, such as a central area of skin is lost and the edges cannot physically meet, then the central defect is filled by granulation tissue. This gradually turns to scar tissue and the process is termed healing by secondary intention. Although the scar tissue may remodel and slowly draw the wound edges closer together, the result is a large area of scar formation that takes a considerable time to gain tensile strength and may disrupt function, for example if it crosses a joint.

Box 4.4   Collagen and limes

Collagen is a fibrous protein which forms a chief component of connective tissues. It has enormous tensile strength and so is abundant in tendons, ligaments, fibrous capsules and fascia. Part of this tensile strength is derived from extensive crosslinking between its individual component polypeptide strands via proline and lysine residues. Ascorbic acid (vitamin C) is an essential cofactor in the formation of these crosslinks.

The human body has no capacity for synthesising ascorbic acid and so is entirely reliant on dietary intake. Deficiency of ascorbic acid results in failure to form and maintain good quality connective tissue with the result that wounds fail to heal properly, the teeth can become loose and trivial damage to the mucous membranes can result in chronic bleeding and infection. This condition is called scurvy.

Ascorbic acid is fortunately abundant in a wide variety of foods, most notably fresh fruit and vegetables. It is however largely destroyed by intensive cooking and can be degraded during processes designed for long-term food preservation. Because of a reliance on preserved food and a lack of fresh fruit or vegetables, scurvy was for centuries a serious cause of disease and even death on long sea voyages. By the time of the Napoleonic wars however the Royal Navy had begun to implement preventative measures such as the consumption of vitamin C-rich citrus fruits such as lemons and limes. This practice is likely to underlie the slang term 'limey' for British settlers in foreign countries.

4.6.5   Summary

Healing and repair generally follow the sequence of angiogenesis, fibroblast migration and proliferation with synthesis of extracellular matrix material and finally remodelling. Just as the neutrophil is the star of acute inflammation and the macrophage the orchestrator for chronic inflammation, the fibroblast is the principal player in healing and repair. Furthermore, just as mediators and concentration gradients are central to the inflammatory processes, so they are to healing and repair.

The outcome of healing and repair is highly dependent on a variety of factors from the size and position of the focus of destruction to the general levels of nutrition. Under ideal circumstances the final outcome may be virtually undetectable.

4.7   Hyperplasia and hypertrophy

Hyperplasia refers to an increase in the number of cells within a tissue secondary to an increased rate of cell division. This will usually (but not always) result in an increase in the size of the tissue. Hypertrophy refers to an increase in the size of the tissue without an increase in the number of component cells. This is due to an increase in the size of the individual cells.

4.7.1   Hyperplasia

Hyperplasia is frequently a normal event within the body. During pregnancy the smooth muscle cells of the uterus (the myometrium) undergo marked hyperplasia under the hormonal influence of progesterone in order to accommodate the increasing size of the developing fetus, placenta and amniotic fluid. Simultaneous hyperplasia also occurs in the ducts and lobules of the breast in order to prepare for breast feeding.

Hyperplasia also frequently occurs in pathological states. In the case of an acute injury with significant tissue damage, for example, the local endothelial cells undergo hyperplasia in order to form granulation tissue.

On occasion however the hyperplasia may itself be pathological. With increasing age, for example, the glandular and stromal cells of the prostate respond to androgens with nodular hyperplasia (Figure 4.17). Unfortunately, owing to the shape of the median lobe of the prostate the resulting prostatic enlargement can act like a ball valve at the base of the bladder leading to episodes of acute and chronic urinary retention.

A further example is that of the hyperplastic endometrium. Under normal premenopausal circumstances the endometrium demonstrates a burst of proliferation after menstruation which ends at about the point of ovulation. This proliferative phase is stimulated by elevated oestrogen levels during the first part of the menstrual cycle. During the second part of the cycle oestrogen levels subside and progesterone dominates. This change in hormone ratios halts endometrial proliferation and causes endometrium to switch to a secretory function. On occasion however this switch fails to occur and the endometrium proliferates continuously, that is it is hyperplastic. Hyperplastic endometrium is an indicator that there is a significant hormonal abnormality. Although this is usually the result of some sort of imbalance in the complex interconnected pathways or hormonal control it does on rare occasions point to more severe abnormalities such as oestrogen-secreting tumours.

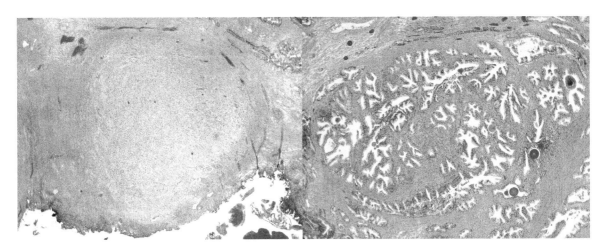

Figure 4.17   Nodules of stromal (left) and glandular (right) hyperplasia in the prostate gland. *H&E photographed at ×40*

4.7.2   Hypertrophy

In tissues in which cellular maturation has resulted in an inability to undergo further rounds of cell division a possible response to functional demand is simply to synthesise more functional components. The result is an increase in the amount of cell cytoplasm with subsequent cellular and hence tissue enlargement. As in hyperplasia this can be a normal process or a response to a pathological process which may itself become intrinsically pathological.

Exercise increases the demands placed on skeletal muscle cells (skeletal myofibres). Skeletal myofibres do not, under normal physiological conditions, undergo cell division once they have reached maturity. The response to cellular metabolic stress and damage caused by exercise is to produce more of the cellular constituents involved in the contraction and respiration processes. Thus the myofibres enlarge and the muscles increase in size.

A similar process occurs in cardiac myofibres placed under increased demand. This can be the result of a wide range of factors that cause an increase in the amount of blood pumped by the heart, an increase in pumping pressure or a decrease in the efficiency of the heart as a pump. Such factors might include profound anaemia, systemic hypertension or valvular disease respectively. The heart myofibres produce more contractile and respiratory elements in order to improve contractility and the heart mass increases. The heart relies on its own blood supply supplied by the coronary arteries and so the greater the size and contractile activity of the heart, the closer that coronary circulation will get to its maximum capacity for supplying blood to the heart muscle. Once this capacity is exceeded the heart muscle becomes starved of oxygen. This results in a rapid deterioration in contractility and, more dangerously, electrical instability that can cause sudden loss of the normal rhythm and death.

4.7.3   Summary

Hyperplasia and hypertrophy represent growth responses in tissues. Whether this response is that of hyperplasia or hypertrophy is dependent on the ability of the cellular components of the tissue to proliferate. In most cases this growth activity is a normal response to a change in the body's requirements of the tissue. In some cases however this response is intrinsically pathological and can therefore result in significant disease.

4.8   Atherosclerosis

Atherosclerosis is the leading cause of death in Europe and North America [12] as the underlying pathological process in ischaemic heart disease, aortic aneurysm and cerebrovascular disease. In its established form an artery affected by atherosclerosis shows irregular thickening of the intima with underlying plaques composed of creamy material with patchy fibrosis and calcification. In small affected arteries the lumen may be narrowed by the presence of plaques almost to the point of complete occlusion.

Atherosclerosis tends to affect the large elastic, non-muscular arteries such as the aorta and common iliac arteries, and the medium to larger calibre muscular arteries such as the carotids, the coronaries and femorals.

4.8.1   Risk factors

A range of risk factors has been established [13], some of which are unavoidable. These include:

- *Gender.* Men are more prone to developing atherosclerosis than women. The difference between men and women for symptomatic atherosclerosis is most marked for premenopausal women. After menopause the difference gradually becomes insignificant. The difference appears to be at least in part related to a protective effect of oestrogen against atherosclerosis
- *Age.* The degree of atherosclerosis varies widely across the population at any given age. The overall trend however is for an increase in amount and degree with age. The earliest features of atherosclerosis (fatty streaks) may become visible in childhood. Symptomatic disease however is rare under the age of 40.
- *Family history.* A predisposition to atherosclerosis appears to run in families. A small number of directly hereditary conditions are strong risk factors for atherosclerosis but in most cases the inherited predisposition is weak and non-Mendelian. This implies a polygenic pattern of inheritance with many potential environmental and behavioural confounding factors.

- *Smoking.* Smoking appears to confer a statistically significant increase in the risk of atherosclerosis. This makes logical sense in view of the proposed underlying pathology (see Section 4.7.3).
- *Hypertension.* As with smoking, hypertension (high blood pressure) appears to confer an increased statistical risk of developing atherosclerosis.
- *Hyperlipidaemia.* While it is true that hyperlipidaemia confers an increased risk of atherosclerosis this is not the whole story. Fatty acids and lipids circulate around the body in a range of complexes with proteins that vary in density. It is the level of low density, cholesterol-rich form of circulating lipid that more specifically correlates with an increase in risk of atherosclerosis. Indeed the high-density forms of lipid actually correlate with a reduced risk. Hereditary forms of hyperlipidaemia, especially hypercholesterolaemia, have been shown to be associated with a greatly increased risk of atherosclerosis.
- *Diabetes mellitus* [14]. Diabetes mellitus is the failure to regulate circulating glucose secondary to an absolute (Type I) or relative (Type II) deficiency in the hormone insulin with the result that circulating glucose levels gradually rise. Diabetes is associated with atherosclerosis and there is substantial evidence that its risk can be significantly reduced by tight control of blood glucose levels by diet, specific drugs or injected insulin.
- *Other risk factors.* A variety of less substantiated risk factors are postulated which range from obesity to personality type. A variety of confounding factors however obscure the picture.

4.8.2   Pathogenesis

Current opinion is that atherosclerosis represents a chronic inflammatory response to low-level continuous damage to the endothelial layer of the arteries [15]. This damage may be the result of shearing forces due to turbulent blood flow, toxins and free radicals directly absorbed via the lungs from cigarette smoke or nonspecific glycosylation secondary to chronic hyperglycaemia. Evidence for the role of shearing forces in turbulent flow is supported by the common distribution of atherosclerotic plaques at branch points in the large arteries.

As we have seen in inflammation, damage to endothelial cells initiates an inflammatory reaction with the consequent recruitment of macrophages to the site of injury. Indeed macrophages are normally found in numbers within areas of atherosclerosis. In an attempt to clean up damaged endothelial cells and blood constituents that have entered the intima and media there will be the inevitable production of some free radicals including reactive oxygen species. These result in the oxidation of low-density lipoproteins (LDLs) from the circulating blood.

Oxidised LDLs are toxic to local endothelial cells and to the smooth muscle cells of the media (Figure 4.18).

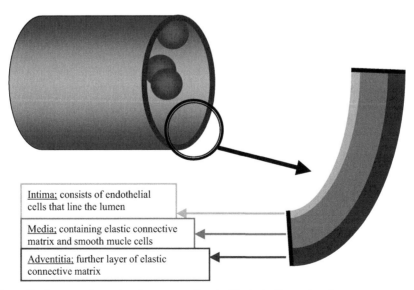

Intima; consists of endothelial cells that line the lumen

Media; containing elastic connective matrix and smooth mucle cells

Adventitia; further layer of elastic connective matrix

Figure 4.18   Schematic diagram of an artery in cross section to illustrate the various layers

The local damage that they cause results in the recruitment of more macrophages and so the cycle of low-grade damage to the blood vessel wall continues. Oxidised free fatty acids will also gradually complex with circulating calcium.

As we have seen in healing and repair, macrophages play a key role in orchestrating the repair process. They can stimulate the proliferation of smooth muscle and the recruitment of fibroblasts to produce collagen and extracellular matrix. In this way the site of chronic arterial intimal injury can evolve into a lesion that consists of oxidised fatty material, macrophages and disordered fibrous and muscular tissue with focal calcification that is an atherosclerotic plaque.

4.8.3   Consequences of atherosclerosis

The principal consequences include the following

- *Thrombosis and embolism.* The intimal surface of an atherosclerotic plaque is uneven, has overlying pockets of turbulent and stagnant blood, and bears damaged endothelial cells. The surface is therefore prone to platelet adhesion which, if unchecked, can result in thrombus formation. The thrombus may eventually block the artery entirely (e.g. the most common cause of an acute myocardial infarct) or may dislodge and block a small calibre artery downstream of the site of atherosclerosis (e.g. carotid artery atherosclerosis can lead to strokes secondary to thromboembolism because a large proportion of the brain's blood supply is derived from the carotid arteries).
- *Plaque rupture.* As plaques enlarge and expand into the media of the artery the overlying layer of intima becomes increasingly at risk of rupturing. The result is a sudden in-rush of blood at pressure into the plaque with rapid expansion which may cause the plaque to occlude the artery. Post-mortem examination demonstrates that this is occasionally the mechanism underlying a myocardial infarct.
- *Aneurysm formation.* As plaques expand and encroach on the media of the artery, the smooth muscle and its surrounding elastic and connective tissue become more and more attenuated, weakening the strength of the artery wall. Under the pressure of the circulating blood the artery may gradually dilate. This abnormally dilated segment of artery is an aneurysm. The principal risk of aneurysm formation is that of rupture. Clearly, as atherosclerosis tends to affect larger calibre arteries, and in particular the aorta, the consequences of aneurysm rupture and haemorrhage are frequently fatal.

4.8.4   Summary

Atherosclerosis has a range of established risk factors and a plausible potential mechanism for its development. The proposed pathogenesis of a response to chronic endothelial damage would appear to be a good example of how the vital processes of acute inflammation, chronic inflammation, healing and repair can occasionally be an inappropriate outcome and cause significant disease. As the pathological mechanism underlying the leading cause of death in the western world and as a widespread cause of morbidity, atherosclerosis is of profound importance. Our understanding of the underlying mechanisms is therefore crucial in determining methods of treatment and prevention.

4.9   Thrombosis and embolism

In order to understand the pathological aspects of thrombosis and embolism it is important to appreciate the underlying principles of haemostasis.

4.9.1   Haemostasis

Any significant injury to a blood vessel will result in localised bleeding which may have fatal consequences if left unchecked. Haemostasis is the process by which the body attempts to curb the bleeding. It has three principal means of action:

- *Local vascular effects.* Damage to an arteriole or small calibre artery results in an immediate constriction of the smooth muscle layer that forms the media (Figure 4.18). This has the obvious benefit of restricting blood flow to the site of injury to reduce any bleeding. The immediate response is due to a combination of reflex nerve stimulation due to injury that signals to the artery to constrict via its own nerve supply and by the action of localised preformed mediators of inflammation.

- *The clotting cascade* — see Chapter 8.
- *Platelets* — see Section 4.4.

The end result of the action of haemostasis is typically a blood clot composed of a meshwork of fibrin and platelets which seals off the site of injury and provides the scaffolding on which repair can take place (see Section 4.6).

4.9.2 Thrombosis

A thrombus is a meshwork of fibrin and adherent platelets. It is formed as a result of the actions of the clotting cascade and platelet adhesion. In essence therefore a thrombus is essentially composed of the same constituents as a blood clot. The principal difference is that a thrombus is formed in moving blood whereas a clot occurs in stagnant blood that has accumulated at the site of a haemorrhage. The result is that thrombi have an ordered structure resulting from successive waves of fibrin deposition and platelet adhesion arranged in an orientation that reflects the direction of blood flow. The structure of a clot however is in contrast apparently disorganised. The precipitants of the formation of a thrombus were described by Virchow [16] and, despite having been added to, still bear his name. Although not strictly accurately named, they are termed Virchow's triad and consist of:

1. endothelial injury;
2. slow blood flow; and
3. hypercoagulability.

The presence of each of these three factors will increase the likelihood of thrombus formation. The role of endothelial injury in the activation of the clotting cascade and platelet aggregation has been discussed in Section 4.4. The role of hypercoagulability is rather more complex. There are inherited deficiencies of some of the factors that inhibit inappropriate clotting that can increase the risk of thrombus formation. There are in addition certain acquired conditions (such as the conversely named lupus anticoagulant syndrome) in which autoantibodies disrupt the clotting cascade and increase the likelihood of inappropriate clotting or thrombosis.

By way of illustration we will examine two examples of pathological thrombus formation as follows.

4.9.2.1 Atherosclerosis

As was discussed in Section 4.8, the intimal surface of the atherosclerotic plaque is a frequent site for thrombus formation. The endothelium overlying the plaque is prone to chronic damage, partially because of the sheer stresses and turbulent flow that were presumed to have initiated the formation of the plaque in the first place and partially because of reactive oxygen species and oxidised LDLs within the underlying plaque. Thus one of Virchow's triad is met and the likelihood of thrombus formation increased.

In addition, the turbulent flow thought to have initiated the plaque, combined with additional turbulence caused by the presence of the plaque, creates eddies and pockets of slow flow over the damaged endothelial surface. Activated clotting factors in fast flowing blood are washed away and immediately diluted but those in slow flowing blood have time to propagate through the cascade. Thus slow blood flow and localised increases in activated clotting factors complete Virchow's triad. It is therefore unsurprising that atherosclerotic plaques are a common nidus for thrombus formation.

4.9.2.2 Deep vein thrombosis

Blood pumped to the lower legs has a considerable distance to travel in order to return to the heart and this is frequently against gravity, even when lying down. In order to assist venous return to the heart the major veins of the lower legs pass through the calf and thigh muscles and have valves to prevent backward flow. Contraction of the calf and thigh muscles therefore acts as a pumping mechanism, effectively functioning as a second heart.

Prolonged immobility (e.g. due to having a leg immobilised after a fracture) reduces the capacity of the leg muscles to pump blood back to the heart. Blood flow through the deep veins therefore slows right down, thereby meeting one of Virchow's triad. Unfortunately the situation is often exacerbated by dehydration (perhaps secondary to blood loss at the time of the injury that resulted in the leg fracture or at the time of surgery to correct the fracture) and hypercoagulability (secondary to a massive acute inflammatory reaction at the site of fracture). Dehydration increases blood viscosity and therefore slows flow even further. Hyper-coagulability increases the chances of inappropriate or disproportionate activation of the clotting cascade. Thus patients with lower limb fractures are at great risk of developing deep venous thrombosis.

4.9.3 Consequences of thrombosis

These can be broadly divided into infarction, organisation and thromboembolism.

4.9.3.1 Infarction

If a thrombus developing in an artery grows beyond a critical size then the tissue supplied with blood by that artery will receive an inadequate blood supply. The consequences depend upon the oxygen demands of the tissue and the anatomy of the local arteries. If the tissue becomes necrotic as a result of vascular occlusion it has undergone infarction (see Section 4.10).

4.9.3.2 Organization

With time endothelial cells, fibroblasts and smooth muscle cells from the vessel wall migrate into the thrombus and proliferate. Branching tracts of proliferating endothelial cells may form an anastomosing network of small blood vessels. Indeed vascular channels may completely traverse the length of the thrombus allowing blood to flow from one side to the other (a process termed recanalisation). Eventually the thrombus is replaced by largely fibrous tissue and increasingly integrated into the wall of the vessel. Organisation may be so successful that old thrombi are only visualised as a smooth-contoured asymmetrical area of thickening in a vessel wall.

4.9.3.3 Thromboembolism

An embolus is material that has dislodged from its site of origin and been carried to a new site via the bloodstream. Although emboli may consist of bone marrow or fat (e.g. after major trauma) or even air, the most common type of embolus is a detached thrombus.

Thrombi that form on atherosclerotic plaque may dislodge, travel down the artery and occlude a smaller branch with consequent infarction. Atherosclerosis of the carotid artery is an occasional source of thromboemboli that lodge in the smaller arteries of the brain and cause stroke.

All peripheral venous blood returns to the right side of the heart and then passes through the pulmonary circulation (the blood supply of the lungs). As the pulmonary arteries transporting blood from the right ventricle of the heart into the lungs branch into smaller arteries they become the site of blockage of thromboemboli arising in the venous circulation. Unfortunately the large calibre deep veins of the legs may produce thromboemboli of considerable size, capable of blocking the main artery to an entire lung, or even the branch point of the pulmonary trunk immediately after it leaves the heart. Pulmonary thromboembolism is therefore a relatively common cause of death.

4.9.4 Summary

Haemostasis is a crucial set of processes that prevent life-threatening haemorrhage from relatively minor traumatic events. Just as in the case of inflammation however there are many occasions on which the normal and vital process activated inappropriately can cause disease and even death. In this way a system that has evolved to prevent major haemorrhage can also result in unwanted thrombus formation that can subsequently lead to thromboembolism. Consideration of Virchow's triad indicates that the circumstances in which thrombus formation would be likely to occur can frequently be predicted. This is of importance because thrombosis of the deep veins of the legs, for example, can lead to pulmonary thromboembolism that is a very common cause of death in the western world.

4.10   Ischaemia and infarction

Ischaemia is inadequate blood supply to a tissue or organ to meet its metabolic needs. If ischaemia persists long enough to cause necrosis in the affected tissue or organ, that tissue or organ has undergone infarction. Ischaemia may be transient and reversible but infarction means permanent damage.

There are essentially two different processes of interruption to the blood supply that can result in infarction as described below.

4.10.1 Arterial occlusion

As discussed above, a thrombus or embolus may occlude an artery. The consequences of occlusion are then highly dependent upon the metabolic demands of the tissue supplied by that artery and on the anatomy of the local blood supply.

Tissues and organs vary in their requirements for continuous blood supply. Fibrous connective tissue has very low metabolic demands and so can tolerate a poor blood supply indefinitely. Organs such as the heart and brain however are massively demanding. Interruption of blood supply will produce dysfunction within seconds and infarction within a matter of minutes. This is the reason that in the treatment of acute myocardial infarction every second counts. When an organ becomes acutely ischaemic the component tissues tend to infarct in order of metabolic demand. Thus in the intestine the highly metabolically active mucosa infarcts long before the less demanding smooth muscle and connective tissue.

Some organs possess an interconnecting arterial supply from several arteries such that blocking one artery simply means that blood has to take a more circuitous route via alternative arterial pathways but will still reach the tissue. In some organs however an area of tissue is supplied by one artery only and if that artery blocks then there is no alternative supply. Unless the blockage is resolved promptly then infarction is inevitable. This is unfortunately true to a great extent in both the heart and brain.

4.10.2 Venous infarction

Venous infarction is much less common that arterial infarction. If a vein is obstructed and no alternative veins are available then blood has no way of leaving the affected tissue. Blood will continue to enter the tissue until the pressure in the small vessels of the tissue equals that of the artery. Once this occurs no more blood can flow into the tissue. The tissue will therefore become ischaemic and eventually infarct. Unlike in arterial infarction, the affected tissue is engorged with blood but the pathology of the underlying necrosis is otherwise the same.

Venous infarction secondary to thrombosis is a rare phenomenon (although can be seen with thrombus formation primarily in the liver and venae cavae). This is largely because atherosclerosis very rarely affects veins but also because the anatomy of the venous system tends much more towards interconnection of vessels than the arterial system. As veins tend to increase in calibre along the direction of blood flow (clearly the opposite of arteries), infarction secondary to embolism is highly unlikely.

The most common scenario for venous infarction is that of torsion. Some organs or parts of an organ are mobile within their body cavity and essentially only anchored in place by their blood vessels and some connective tissue. This is true for example of the ovaries, the testes, the caecum and the sigmoid colon. Under exceptional circumstances it is therefore possible for these organs to twist on their own vascular pedicle. Thin-walled, low pressure veins are vulnerable to compression by the twisting action but the thick-walled muscular arteries at high pressure will continue pumping blood into the twisted organ. A situation may then arise in which the subsequent vascular engorgement swells the affected organ and makes it physically less able to untwist. The result is therefore venous infarction. This is not uncommon in ovaries affected by the growth of an ovarian cyst.

4.10.3 Summary

Ischaemia is the result of insufficient blood supply to a tissue or organ. If the tissue or organ undergoes permanent damage as a result of ischaemia then it is said to have infarcted. Two principal types of infarction have been described, the more common being that of arterial infarction secondary to arterial occlusion. The vulnerability of tissue and organs to infarction is largely dependent on both its metabolic requirements and the anatomy of its arterial supply. Although venous infarction is less common than arterial, mobile organs can be vulnerable to venous infarction under specific circumstances and result is morbidity or even mortality.

4.11 Amyloid and amyloidosis

Amyloid is a heterogeneous group of fibrillary proteinaceous substances, all of which share the same conformation on X-ray crystallography. They also share the same histological appearance and a common pattern of histochemical staining. Amyloidosis is the term for accumulation of amyloid within tissues. Amyloidosis can be localised to the site at which the component proteins are synthesised or can be systemic (i.e. distributed throughout the body).

The three-dimensional structure of amyloid is that of a so-called β-pleated sheet composed of arrays of fibrils consisting of protein chains. The details of the β-pleated sheet conformation are unimportant but the crucial point is that although the component proteins differ between different underlying causes of amyloid formation, the overall structure is always the same. Amyloid is characteristically extracellular and insoluble with an eosinophilic staining pattern on standard haematoxylin and eosin stained histological sections. It is stained a deep red colour by the Congo Red staining technique and subsequently demonstrates a yellow-green dichroism under polarized light.

Although the fibrillary protein strands that make up the β-pleated sheet are the main constituent of amyloid, a range of glycoproteins such as serum amyloid P protein are also minor components.

4.11.1   Types of amyloid

Amyloid is generally classified according to the chemical nature of the fibrillary component proteins as follows.

4.11.1.1   AL amyloid (antibody derived amyloid)

In the case of AL amyloid the component protein is either part of or the complete immunoglobulin light chain. In the vast majority of cases the patient will have an elevated level of circulating monoclonal antibodies or light chains indicative of some sort of underlying B lymphocyte abnormality. In the majority of cases the nature of the B cell abnormality is difficult to ascertain but in a proportion the underlying B cell abnormality will become apparent in the form of a B cell neoplasm (see Section 4.14), the most significant of which is multiple myeloma [17]. This is a malignant neoplasm showing plasma cell differentiation that has a propensity to spread widely through the bones.

4.11.1.2   AA amyloid (serum amyloid-associated protein)

Serum amyloid-associated protein (SAA) is a protein synthesised by the liver that forms a normal constituent of circulating serum proteins. Although normally associated with the transport of lipids the precise role of SAA is still under debate. The production of certain isoforms of SAA is increased in response to mediators of acute inflammation. Thus amyloid derived from SAA tends to be associated with inflammatory conditions that include recurrent or prolonged episodes of inflammation. These include rheumatoid arthritis, osteomyelitis (bone infection) and idiopathic chronic inflammatory bowel disease (Crohn's disease and ulcerative colitis).

4.11.1.3   β$_2$-microglobulin amyloid

β$_2$-microglobulin is a component of the MHC class I molecule (see Chapter 7). Owing to its size and conformation it is not filtered out by conventional dialysis machines and so accumulates in the blood in those on long-term dialysis for chronic renal failure.

4.11.1.4   Localised endocrine amyloid

A variety of endocrine diseases can produce local amyloid deposition. This is best exemplified by medullary thyroid tumours in which the major component of the amyloid is calcitonin. Calcitonin is a hormone produced by parafollicular cells of the thyroid glands and is normally involved in regulating serum calcium levels. Medullary thyroid tumours are composed of neoplastic cells showing parafollicular cell differentiation which therefore produce calcitonin inappropriately and at high levels.

A myriad of other forms of amyloid are recognised but are beyond the scope of this review.

4.11.2   Amyloidosis and disease

Amyloid is often an incidental finding on histological examination of tissue sampled for an unrelated reason. It may however be associated with degeneration of the tissue in which it is found. The exact mechanism for this is uncertain. Given that amyloid is frequently deposited around blood vessels (Figure 4.19) this may be related to localised ischaemia through disruption of transfer across the vessel wall. The amyloid may damage tissue simply because of its physical space-occupying presence. In addition it is possible that the amyloid itself is inert and harmless but it is rather the effect of the presence of high levels of the fibrillary proteins that is

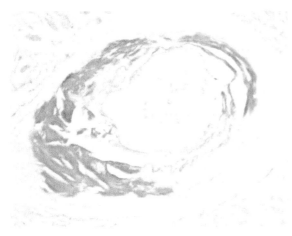

Figure 4.19   Amyloid in the wall of a blood vessel. This stains a deep red colour using Congo Red and demonstrates a yellow-green appearance when viewed under polarized light. *Congo red photographed at ×100*

toxic (and the presence of amyloid is merely an indication that those precursors are present).

4.11.3   Summary

Amyloid is the term given to a diverse array of proteins that share a common three-dimensional conformation, common staining properties on histological examination and a capacity to cause disease. Identification of the actual proteinaceous components often indicates a significant underlying disease process (such as the identification of amyloid composed of immunoglobulins and the presence of multiple myeloma). The precise mechanism by which amyloid results in disease still elicits a degree of debate.

4.12   Infections of histological importance

The laboratory investigation of infective disease is predominantly the territory of the microbiology laboratory. A number of infections however are commonly encountered in the routine workload of the histopathology and cytopathology laboratory. Indeed cervical cytology is almost entirely based around the activity of a single type of viral infection (see Section 4.12.1.1). For the purposes of this review we will focus on a small number of the more commonly encountered infective organisms in the UK.

4.12.1   Viruses

4.12.1.1   Human papilloma virus (HPV)

This extremely commonly encountered double-stranded DNA virus is responsible for common warts and verrucae. There is now overwhelming evidence that a number of serotypes (in particular types 16 and 18) are also involved in the causation of cervical carcinoma. Of particular importance is the production by infected cells of the virally encoded proteins called E6 and E7. Amongst other actions these proteins inactivate p53 and retinoblastoma (Rb) proteins produced by the host cell (see Section 4.14) [18, 19]. The normal action of p53 and Rb proteins is to block replication of the host cell, especially in the context of DNA damage. Removal of this barrier to replication may therefore result in unimpeded proliferation of infected cells irrespective of DNA damage with consequent increased risk of mutations that contribute to the development of malignancy.

The UK cervical screening programme aims to detect through exfoliative cytology the various degrees of abnormality that arise in infected squamous epithelial cells within the cervix so that treatment can be initiated before invasive carcinoma develops (see Section 4.13.2). These changes range from very low risk cells showing features of HPV infection only (Figure 4.20(a)) to grossly abnormal severely dyskaryotic cells (Figure 4.20(b)) at relatively high risk of progressing to malignancy.

Although statistically the most important association, cervical cancer is not the only malignant disease linked with HPV. Anal cancer and penile cancer also demonstrate strong associations with the virus and cutaneous squamous cell carcinoma demonstrates a strong link in the immunocompromised. There is also emerging evidence of subtypes of oropharyngeal carcinoma with a strong association with certain HPV serotypes.

4.12.2   Bacteria

4.12.2.1   Mycobacterium tuberculosis

Tuberculosis is the single most important bacterial disease, with an estimated worldwide prevalence of 1.86 billion in 1997 [20] and worldwide mortality of approximately 5000 people per day. Improvements in sanitation and overcrowded housing and, more recently, effective drug treatments meant that the

(a)

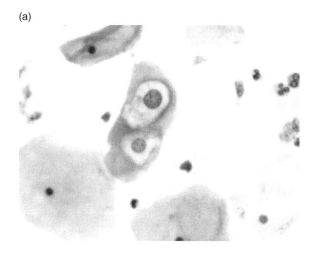

(b)

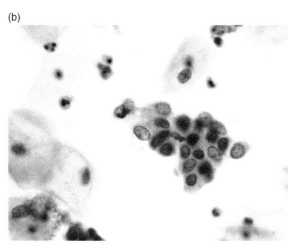

Figure 4.20   (a) Two koilocytes. The characteristic perinuclear halo indicates HPV infection. *Papanicolaou stain
photographed at ×400* (b) A cluster of dyskaryotic cells (right
of centre). The cells are smaller than their neighbouring
counterparts but have larger, more irregular nuclei with
coarse chromatin. *Papanicolaou stain photographed at ×400*

twentieth century saw a great decline in tuberculosis in
the western world. The last 30 years however saw an
end to this decline and a re-emergence of tuberculosis,
largely as a result of the arrival of HIV.

Mycobacterium tuberculosis is an aerobic bacillus
with an ability to survive within macrophages where it
is largely safe from the immune system. Infection is
most commonly by inhalation into the lungs where the
initial inflammatory reaction occurs. *M. tuberculosis*
bacilli elicit a local acute inflammatory response and
are transported to regional lymph nodes at the hilum of
the infected lung within macrophages.

Eventually the immune system becomes activated
and in the immunocompetent a T cell-mediated response kills macrophages infected with the bacilli and
assists uninfected macrophages in forming granulomata and scar tissue to wall off infected foci within the
lungs and associated lymph nodes. Although infection
may be cleared at this point, some bacilli commonly
remain in the walled-off foci. Without oxygen from air
or blood these bacilli are unable to reproduce effectively and so are kept in check by the immune system.

In the frail, debilitated or immunodeficient population or in those infected with a particularly virulent
strain this T-cell mediated response may be inadequate
to prevent systemic infection. This may also occur in
those previously infected with tuberculosis who become relatively immunodeficient, allowing old, dormant bacilli to become active again. The result is
disseminated infection with multiple granulomata
throughout the body and widespread tissue destruction leading ultimately to death.

4.12.2.2   Helicobacter pylori

This gram-negative bacillus is a relatively common
incidental finding in gastric antral biopsies from the
asymptomatic patient. It demonstrates a strong association with chronic gastritis and with duodenal ulceration. In addition there is evidence for an aetiological
link with gastric adenocarcinoma [21] and extranodal
gastric B-cell lymphoma [22].

Helicobacter pylori possesses adhesins to bind to
gastric epithelial cells and secretes a urease enzyme
locally which produces ammonia and therefore protects the bacterium from gastric acids. The curvilinear
shape of the bacterium within the surface mucus of the
gastric epithelium makes *Helicobacter pylori* detectable
on histological examination of gastric biopsies (Figure 4.21). More frequently however *Helicobacter* colonisation is detected on biopsy by simple chemical
testing for urease enzyme activity.

4.12.3   Fungi

4.12.3.1   Candida albicans

This common fungus frequently colonises the mouth
and skin without obviously causing disease. In the
otherwise healthy it may cause a range of relatively
minor diseases secondary to mucous membrane

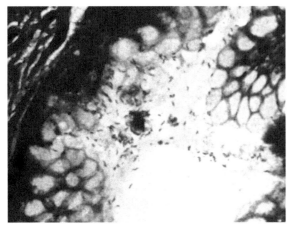

Figure 4.21 Clusters of *Helicobacter pylori* within the mucus of a gastric crypt. *Modified giemsa photographed at ×600*

infection, most commonly manifesting as oral or vaginal white patches that can be removed to reveal inflamed underlying squamous mucosa, better known as 'thrush' (Figure 4.22). *Candida albicans* can also cause an itchy rash in moist skin folds and can result in a chronic, hyperplastic inflammatory process in the oral mucosa.

In the immunocompromised, *Candida albicans* can cause a range of disease from oesophagitis to fatal septicaemia. It may also infect any site in which a foreign body can harbour growth such as intravenous lines, catheters and prosthetic heart valves.

On histological and cytological preparations *Candida albicans* can appear in yeast form, strings of yeast forms arranged in long chains (pseudohyphae) and

true hyphae. Commonly a mixture of the different forms occurs simultaneously.

4.12.4 Protozoa

4.12.4.1 Giardia lamblia

Giardia lamblia is a protozoan parasite that colonises the proximal small bowel including duodenum [23] The effects of infection range from asymptomatic colonisation to severe malabsorption.

The parasite adheres to the intestinal epithelium in its trophozoite form and can be seen in large numbers, individual trophozoites demonstrating a characteristic sickle shape on histological sections (Figure 4.23). Occasional cyst forms develop and pass into the lumen of the small intestine to be shed in the faeces. Outside the gut the cyst form has a lifespan of approximately 2 weeks, during which time it may be ingested in contaminated food or water to infect another small intestine.

4.12.5 Helminths and other larger parasitic organisms

4.12.5.1 Hydatid disease

Hydatid disease is caused by ingestion of the eggs of the dog tapeworm *Echinococcus granulosus.* This also

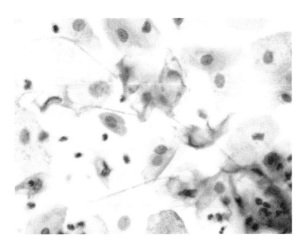

Figure 4.22 The hyphae of *Candida albicans* appear as branching lines on a cervical smear. *Papanicolaou stain photographed at ×400*

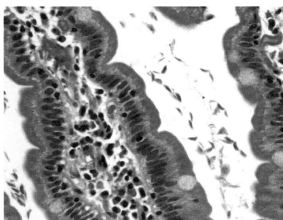

Figure 4.23 *Giardia lamblia* seen colonising the duodenum in this biopsy. The trophozoites have a crescent shape and are slightly detached from the epithelial surfaces in this image. *H&E photographed at ×400*

affects sheep and so is more commonly seen in sheep farming areas such as North Wales and Cumbria in the UK. The ingested eggs hatch in the duodenum and the larvae invade through the intestinal wall to spread widely through the body. The majority of larvae lodge in the liver where those that survive the initial inflammatory response form surrounding cysts that can reach enormous proportions. Occasionally these cysts are biopsied (a risky procedure owing to the potential for anaphylactic shock secondary to a reaction to the leaked contents) because the cause of the cyst is unknown. The cyst wall has a laminated appearance and the refractile scolices are occasionally recognisable.

4.12.5.2   Other larger organisms

It is extremely unusual for larger parasitic organisms to be identified in biopsies in the UK. The main exception is that of the common pinworm *Enterobius vermicularis*, which can occasionally be found in gastrointestinal surgical specimens including the appendix (Figure 4.24). Other rare findings include the scabies mite in diagnostic biopsies of chronic skin rashes and more exotic organisms from foreign travellers. One example of the latter, now rarely found in the UK, is that of the helminth *Strongyloides*, a parasitic worm that is unusual in being able to reinfect its own host repeatedly with the consequence that infection can last decades and may still be found in the dwindling numbers of survivors of the Second World War forced labour camps in the jungles of south-east Asia.

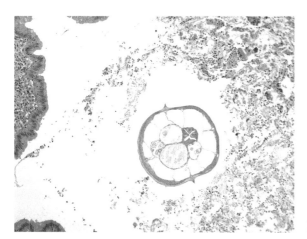

Figure 4.24   Cross section through the body of *Enterobius vermicularis* within the lumen of an appendix. This was an incidental finding within a right hemicolectomy specimen for carcinoma. *H&E photographed at ×40*

4.12.6   Summary

Although primarily the territory of the microbiology department, the identification of infective organisms on cellular pathology preparations can be of diagnostic assistance. Indeed the entire cervical cancer screening programme currently revolves around spotting the cytologic effects of Human Papilloma Virus infection and its subsequent effects on the progression of neoplasia (see Section 4.14). Cellular pathological techniques are normally far less robust than microbiological techniques for identifying organisms but some infective organisms (such as *Mycobacterium tuberculosis*) are difficult (and slow) to culture in a microbiology laboratory and the clues that can be seen on histological techniques allow for sufficient evidence to permit treatment prior to a definitive microbiological result.

4.13   Metaplasia, dysplasia and carcinoma *in situ*

4.13.1   Metaplasia

Metaplasia is the replacement of a mature tissue type usually found at the particular site by another not normally found at this position. The change in tissue type is usually a response to a change in environmental conditions such that the new tissue is better suited than the old. In practice metaplasia is normally observed in epithelial surfaces. The two most commonly encountered examples are present in the oesophagus and the uterine cervix.

4.13.1.1   Oesophageal metaplasia

The oesophagus is normally lined by a layer of squamous epithelium that is continuous with the squamous epithelium of the pharynx and ends abruptly at the gastro-oesophageal junction where it changes to columnar epithelium. Squamous epithelium, best exemplified in the skin, is well suited to chronic, low grade mechanical trauma and assault by mixed bacterial and fungal flora owing to its ability to shed multiple layers of anucleate surface cells without damage to the underlying tissue. Thus the squamous epithelium of the oesophagus can withstand the passage of irregular fragments of food and exposure to the mixed flora of the mouth and sinonasal tract. The glandular surface of the stomach however produces a mucus layer which appears to protect the underlying tissue from acids and digestive enzymes.

Chronic gastro-oesophageal reflux of acid, partially digested food and digestive enzymes is relatively common, increases with age and is exacerbated by such risk factors as obesity and hiatus hernia. With chronic exposure of the oesophagus to stomach contents the squamous epithelium is repeatedly eroded. Eventually the squamous epithelium at the lower end of the oesophagus can become partially or completely replaced by glandular epithelium resembling that lining the stomach and occasionally including some epithelium lining the intestine. This commonly observed condition is termed columnar-lined oesophagus or Barrett's oesophagus (Figure 4.25).

4.13.1.2 Cervical metaplasia

At birth there is a clear histological distinction between the surface of that part of the cervix lining the canal between the vagina and endometrial cavity (the endocervix) and the part of the cervix projecting into the vagina (the ectocervix) (Figure 4.26). The endocervix is lined by mucin-secreting columnar epithelium which is thrown into folds or crypts (usually erroneously referred to as glands). This epithelium produces a

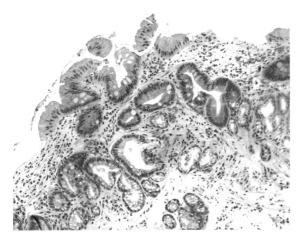

Figure 4.25 This biopsy was taken from the oesophagus. Instead of the normal squamous epithelium the biopsy is lined by columnar epithelium. Most of this is of the type found in the stomach adjacent to the gastro-oesophageal junction. A few scattered cells are also visible that have pale-staining, slightly blue cytoplasm. These are goblet cells of the type seen in the intestine. *H&E photographed at ×100*

mucus plug that forms an effective barrier between the endometrial cavity and the outside world.

The ectocervix is exposed to the varied flora of the vagina and direct contact with the vaginal wall. It is thus more suited to a squamous rather than glandular epithelial layer. During growth and development however the uterus changes conformation with the result that some of the endocervix becomes everted from the canal and effectively part of the ectocervix. The glandular epithelial cell layer is gradually replaced in part or entirely by squamous epithelium to match the rest of the ectocervix. The underlying folds or crypts however remain unchanged and are clues as to the original position of the junction of the ecto- and endocervix. This area of metaplasia is termed the transformation zone. It is the area most prone to undergoing human papilloma virus-associated dysplasia (see Sections 4.12 and 4.13.2) and is therefore the area sampled for cervical screening.

4.13.2 Dysplasia

Dysplasia means abnormal growth and as such is a rather imprecise term that represents different things in different contexts within the field of biomedical science. To the cellular pathologist however it generally refers to an abnormal pattern of cellular and tissue maturation that usually confers an increased risk of malignancy and is the result of underlying genetic damage.

Cells in tissues normally mature from the newly divided state to the state recognised histologically as a mature component of that tissue. This process relies on a stepwise progression of the expression of various cellular proteins as dictated by the cell's genes. If the dividing cell at the beginning of the process has, for whatever reason, lost its ability to prevent genetic mutations then it can be envisaged that the daughter cells will eventually lose the ability to progress through the maturation sequence in an orderly fashion because some of the genes crucial to this process no longer function normally. The cells will effectively become 'stuck' at various stages of immaturity. This process can be illustrated in the uterine cervix, where the process of identifying dysplasia has become the target of a national screening programme.

As we have seen in Section 4.12, infection of squamous cells of the cervix by certain serotypes of the human Papilloma Virus (HPV) leads to the loss of

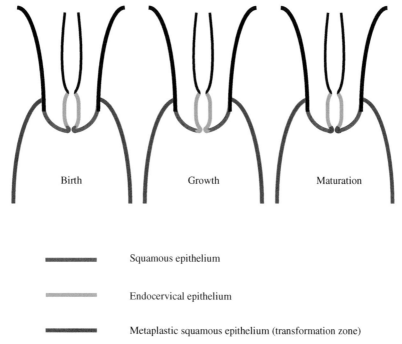

Squamous epithelium

Endocervical epithelium

Metaplastic squamous epithelium (transformation zone)

Figure 4.26   As the uterus matures during childhood and adolescence part of the endocervical canal extends onto the ectocervix. The glandular epithelium is gradually replaced by metaplastic squamous epithelium and forms the transformation zone

cellular checkpoints that stop the cell dividing if there is genetic damage. If a cell therefore acquires a genetic mutation it may well pass this mutation on to its daughter cells before the mutation can be repaired. With the accumulation of genetic abnormalities the progression to cellular maturity becomes disrupted. This becomes apparent histologically (Figure 4.27) in the form of a failure of squamous cells to flatten and their nuclei to shrink with distance from the basal cell layer. The more genetic damage to the component cells, the more likely the cells are to become stalled in a morphologically immature state and the more abnormal (or dysplastic) the growth pattern of the squamous epithelium.

The significance of identifying dysplasia is its association with malignancy (see Section 4.14) and this is both exemplified and utilised in the bowel cancer and cervical screening programmes. It is well established that those with invasive cancer are statistically more likely to have concurrent nearby dysplasia on histological assessment than those without. We also know that those with dysplasia identified are more likely to go on to develop invasive cancer at some point than those without. In addition the more atypical the dysplastic changes, the higher the probability of

developing cancer and the shorter the time interval between the identification of dysplasia and invasive malignancy. Dysplasia would therefore appear to be a risk factor for cancer and it is tempting to infer that dysplasia leads to cancer. Support for this theory comes from finding common patterns of genetic mutation in dysplastic and malignant epithelium in the same patient compared with adjacent morphologically normal epithelium. In reality however a large proportion of identified dysplasia never becomes malignant and some resolves entirely. This knowledge has been used to determine screening age ranges and test intervals in the cervical screening programme (Box 4.5).

4.13.2.1   Carcinoma *in situ*

A final comment should be made regarding the term carcinoma *in situ*. This refers to the most severely atypical and immature end of the spectrum of dysplasia such that the normal architectural pattern of the epithelial surface is lost entirely. Some classification systems of dysplasia make a distinction between severe dysplasia and carcinoma *in situ* as if they are different entities. This would however appear to make little biological sense. Indeed where carcinoma and

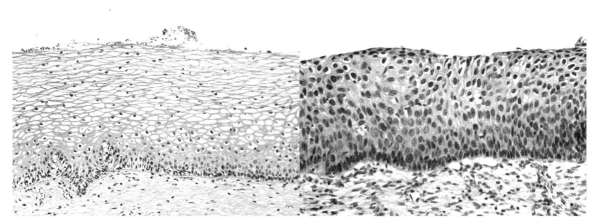

Figure 4.27 Normal ectocervical squamous epithelium (left) shows a transition from polygonal cells with eosinophilic cytoplasm and oval nuclei at the base to flattened cells with clear cytoplasm and tiny, largely inconspicuous nuclei at the surface. Mitotic activity is limited to the basal layer of cells. Dysplasia (right) of the cervical squamous epithelium shows that this orderly transition is partially lost and mitotic activity is no longer limited to the basal layer. Some of the more superficial layers of cells include keratinocytes with hyperchromatic nuclei and a perinuclear zone of cytoplasmic pallor. It is easy to imagine that these would resemble the koilocytes discussed in *Infections of Histopathological Importance*. *H&E photographed at ×100*

'carcinoma *in situ*' coexist, the true carcinoma often appears much less immature and disorganised than the *in situ* component.

4.13.3 Summary

Metaplasia refers to the replacement of one normal mature tissue type (usually epithelial) by another that is, for a variety of reasons, more suitable to that environment. This is frequently because the environment in which that tissue formed during embryogenesis has changed and two good examples are given above.

Dysplasia however refers to disruption in the maturation of a tissue type and, under the circumstances in which it most frequently encountered in cellular pathology, is the product of underlying genetic mutations. Statistical and genetic analyses would suggest that increasing degrees of dysplasia infer an increasing risk of associated malignancy. Consideration of the implications of dysplasia therefore leads into a review of the pathological mechanisms underlying neoplasia.

4.14 Neoplasia

Neoplasia literally refers to new growth. In medical terms however neoplasia is interpreted as new growth that is inappropriate to its context. Unfortunately the common usage of the term has resulted in its meaning becoming synonymous with tumour formation or a clonal proliferation of cells (where 'clonal' refers to all of the cells in question carrying the same set of genetic mutations, presumably derived from a common precursor, and is itself an inaccurate term as we shall see).

4.14.1 Benign versus malignant

Accepting the common use of the term in the context of tumour formation, neoplastic lesions are classically divided into benign and malignant categories. A third category of 'uncertain malignant potential' is employed for extremely rare lesions with an unknown natural history or for more common tumours in which the clinical, radiological and histological features do not give any clue as to how the tumour will behave. In general terms a malignant neoplasm is synonymous with cancer.

The distinction between benign and malignant infers that one type of tumour will do no harm and the other will kill. It is certainly true that malignant tumours will, if left untreated, go on to cause morbidity and perhaps mortality, and that many benign tumours are harmless. There are however numerous exceptions, the most common cancer (basal cell carcinoma

Box 4.5   The UK cervical screening...

The UK cervical screening programme offers cervical smear tests to all women aged between 25 and 64 years. Screening tests are offered every 3 years between the ages of 25 and 49 and every 5 years thereafter.

The cervical smear test consists of examination of the surface epithelium of the cervical transformation zone by means of exfoliative cytology. Although two principal specimen preparation methods are in common use, both essentially consist of the use of a brush to obtain the specimen and then liquid-based cytology to prepare the slides for cytological examination (see Section 4.15).

It is important to understand that although cervical smears can be used to identify cervical cancer this is not the primary purpose of the screening programme. The purpose is to identify the various stages of dysplasia in order to recognise those women most at risk of developing cancer so that treatment can be implemented before cancer ever develops. In this way the programme is estimated to prevent approximately 4500 deaths from cervical cancer per year.[1]

As described in Section 4.13.2, the statistical risk of the development of cervical cancer increases with the degree of dysplasia (the equivalent term being dyskaryosis in cytology) and those with only mild degrees of abnormality are statistically much more likely to revert to normal than develop cancer. The prevalence of mild abnormalities is much higher in younger women, the vast majority of whom will not go on to develop cancer if left untreated. Screening and treating these women will therefore lead to a lot of unnecessary procedures that are of no benefit to the patient, costly to the NHS and not entirely without risk of complication (although this is extremely low). Furthermore, the rate of progression from mild to more severe abnormalities and from severe to cancer is very variable with the potential that cancer could develop between smears if the interval is long enough. Against all of these considerations is the cost of implementing the programme to the NHS. The determination of the age at which to commence screening and the interval between smears is therefore complex and controversial with the inevitable consequence that a small number of women will still develop cancer despite complying with the programme.

[1]NHS Cervical Cancer Screening Programme (2007) *Cervical Cancer: The Facts*, Department of Health Publications (London).

of the skin) being a good example of a malignant tumour that very rarely kills. In addition, certain apparently benign tumours can be fatal. Benign tumours within the brain may kill, simply because they occupy space in a part of the body where there is no room for expansion. The criteria for determining that a tumour is malignant include metastasis and invasion.

4.14.2   Metastasis

Metastasis is the spread of tumour from one site to another with no physical continuity between the two. The spread is presumed to be secondary to detachment of tumour cells from the primary site and successful implantation elsewhere, followed by the ability to proliferate at the new site (Figure 4.28). Spread can be via the circulating blood, via the lymphatics, through the cerebrospinal fluid or across a serous body cavity such as the peritoneal or pleural cavities. With only a few exceptions the ability of a tumour to metastasize defines it as malignant.

4.14.3   Invasion

The majority of tumours of clinical significance occur in organs with an epithelial component. Such organs usually have an ordered structure such that the epithelial component is separated from the nonepithelial component by a histologically identifiable boundary such as a basement membrane, a muscle layer or a layer of myoepithelial cells. A tumour arising in the epithelium remains classified as benign as long as the identifiable boundary remains intact. If that barrier is seen to be breached by the tumour, the tumour is classified as invasive (Figure 4.29). Invasion, like metastasis, is a defining feature of malignancy with only a few exceptions.

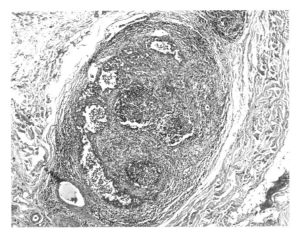

Figure 4.28   Metastatic squamous cell carcinoma that is partially replacing a lymph node. *H&E photographed at ×40*

For some epithelial and all non-epithelial tumours the concept of invasion is more difficult to define because there is no well-defined and histologically identifiable barrier. Some epithelial tumours (such as follicular tumours of the thyroid gland) develop their own capsule of fibrous tissue. Invasion may then be defined by a breach of the fibrous capsule to spread into the surrounding tissue. Invasion may also be defined by the spread into blood or lymphatic vessels outside the bulk of the tumour. Capsular breach and lymphovascular invasion are also defining features of malignancy but, as is so often the case in histopathology, there are exceptions to the rule.

4.14.4   Nomenclature in neoplasia

In general epithelial malignant tumours are termed carcinomas and non-epithelial malignant tumours are called sarcomas. Both benign and malignant tumours are further subdivided into categories that are named after the cell type they most resemble (Figure 4.30). Thus a malignant tumour that most resembles squamous epithelium is called a squamous cell carcinoma. Similarly a malignant tumour that most resembles fatty tissue (also known as lipoid or adipose tissue) is called a liposarcoma.

An important point to make is that tumours are named after the cell type that they most resemble rather than the tissue in which they arise. In most cases a tumour will resemble the tissue in which it arises but this is not always true. On occasion a malignant tumour shows no obvious features to suggest what cell type it is attempting to become and is therefore termed undifferentiated or anaplastic (Figure 4.31).

If a tumour has been classified as malignant an important role of the histopathology department is to help clinicians provide an indicator as to the likely behaviour of that tumour in order to determine what possible treatments can be offered to the patient. The range of possible treatments will, to a large degree, be affected by the site of the tumour, but a number of factors determined at least in part by the cellular pathology laboratory are of great importance. These include differentiation, grade and stage.

4.14.4.1   Differentiation

Determining what cell type the tumour most resembles greatly affects treatment. This is based not only on the histological appearance using conventional H&E staining but also on the use of immunohistochemical staining and several more rarely used ancillary techniques such as electron microscopy and *in situ* hybridisation (see Sections 4.24 and 4.25). A broad variety of treatments are available for cancers including surgery, radiotherapy and chemotherapy. For many tumour

Figure 4.29   A small invasive squamous cell carcinoma in the skin. The normal contour of the skin (seen at the left- and right-hand sides of the fields) with a clear demarcation between the squamous epithelium of the epidermis and the connective tissue of the dermis is interrupted at the centre of the picture by carcinoma. Instead of a smooth junction between dermis and epidermis, the dermis contains irregular, finger-like projections of squamous epithelium. This indicates that the basement membrane of the epidermis has been breached and the lesion is invasive (i.e. malignant). *H&E photographed at ×40*

Non-neoplastic counterpart	Benign	Malignant
Epithelial		
Squamous epithelium		Squamous cell carcinoma
Glandular epithelium		Adenocarcinoma
Non-epithelial		
Lymphoid tissue		Lymphoma
Adipose tissue	Lipoma	Liposarcoma
Smooth muscle	Leiomyoma	Leiomyosarcoma
Skeletal muscle	Rhabdomyoma	Rhabdomyosarcoma
Vascular endothelium	Haemangioma	Angiosarcoma
Cartilage	Chondroma	Chondrosarcoma
Bone	Osteoma	Osteosarcoma

Figure 4.30   Nomenclature in neoplasia

types surgery is the most effective treatment but for some (including many tumours of lymphocytic differentiation) surgery may be ineffective and a combination of radiotherapy and chemotherapy far more appropriate. An increasing number of highly specific treatment strategies are becoming available which are so specific that they target not just a particular tumour cell type but even a subgroup within that subtype. Thus differentiation according to cell type is crucial.

4.14.4.2   Grade

Grading a tumour involves classifying the tumour according to how closely it resembles the mature non-neoplastic tissue type toward which it is differentiating. In other words it is the degree to which the

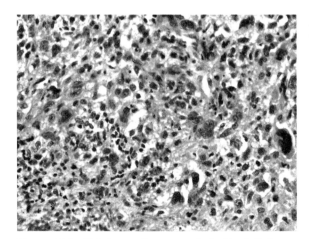

Figure 4.31   This tumour is composed of cells of all shapes and sizes, some of which have enormous and bizarrely shaped nuclei. This tumour shows no morphological evidence of differentiating toward any particular tissue type. The tumour was eventually diagnosed as an embryonal sarcoma of the liver. *H&E photographed at ×400*

tumour is differentiating. By way of example, a low grade squamous cell carcinoma meets the criteria for malignancy but otherwise closely resembles normal squamous epithelium. In contrast, a high grade squamous cell carcinoma may be barely recognisable as squamous epithelium without a prolonged search for histological evidence. The terms 'well differentiated' and 'poorly differentiated' are often used with interchangeable meaning as low grade and high grade respectively.

Apportioning grade to a tumour is largely dependent on the tumour type and there are unfortunately numerous grading systems used in cellular pathology. Some are semiquantitative (e.g. that used for breast carcinoma in the UK Breast Screening Programme) whereas others are entirely subjective. In broad terms however the higher the grade (or poorer the differentiation), the more aggressive a tumour is likely to be if left untreated. Ironically it is the more aggressive tumours that often respond most dramatically to chemotherapy and radiotherapy.

4.14.4.3   Stage

Stage is a description of how extensive a tumour is, grouped into categories according to tumour behaviour, response to treatment or both. The usual parameters considered in attributing stage include tumour size, whether certain physical barriers within the body have been breached and how extensive the metastases. A tumour is usually staged according to the tissue or organ in which it first arose, with every major organ having its own staging system. A carcinoma in the oesophagus will therefore be staged according to a different system from one arising in the stomach, even though the tumour type may be the same and the organs lie next to each other.

As in the case of grading there are many different staging systems in current use. The International Union Against Cancer (UICC) attempted to produce a universal group of site-specific staging systems for carcinoma in order for clinicians all over the world to be able to compare (amongst other things) prognostic statistics and response to therapy. The system for each site was classified according to the properties of the tumour (T stage), the degree of spread to the regional lymph nodes (N stage) and the presence and extent of distant metastases (M stage). This so-called TNM staging system has been widely adopted for carcinomas and, even though other staging systems are often used alongside, it is the one normally quoted by clinicians in the UK when determining treatment options (Box 4.6).

4.14.5 The cellular biology of neoplasia

This is an enormous topic and is beyond the scope of a short review. The broad principles however are fundamental to gaining an understanding of the causes, investigations and treatments of cancer, and underpin much of the activity of the cellular pathology laboratory. As neoplasia is essentially a genetic disease any review needs to address some basic molecular cell biology.

4.14.5.1 The cell cycle (Figure 4.32)

The rate of turnover of cells varies widely between tissue types. Some tissues, for example skeletal muscle and neural tissue of the central nervous system, are composed of cells that have apparently permanently ceased cell division. Others, for example skin and the epithelial lining of the gastrointestinal tract, contain populations of cells that are continuously dividing. Other tissues such as liver can undergo increased cell division in response to injury but normally remain relatively static.

Most cells in the body are in the G_0 or mitotically quiescent state. With appropriate external stimuli some cells can be prompted to undergo division. The process of division is represented by the cell cycle, the significance of which is that once entered the cell is committed to division. There are two important checkpoints in the cycle at which division can be halted. These lie immediately before the cell begins to

---

Box 4.6   The TNM staging system

The TNM stage is composed of three core parameters. These describe the primary tumour (the T stage), the presence or absence of regional lymph node involvement (the N stage) and the presence or absence of distant metastases (the M stage). A variety of prefixes are often included to indicate that the TNM stage has been applied under specific conditions, for example a 'y' prefix indicates that the stage was defined after chemotherapy or radiotherapy and a 'p' prefix indicates that the stage was defined according to the histopathological features.

An example of the TNM stage (as defined by the histopathological features) is that commonly applied to colorectal cancers:

T stage

pT0   no tumour identified
pT1   the tumour invades into the submucosa
pT2   the tumour invades into the muscularis propria
pT3   the tumour breaches the muscularis propria and invades into subserosa or non-peritonealised pericolic or perirectal tissues
pT4   the tumour directly invades other organs and/or involves the visceral peritoneum.

N stage

pN0   no regional lymph node involvement
pN1   one to three regional lymph nodes contain metastases
pN2   more than three regional lymph nodes contain metastases.

M stage

pM0   no distant metastasis
pM1   one or more distant metastases.

Thus a tumour invading beyond the muscularis propria without involving other organs, with two regional node metastases but no distant metastases would be staged as pT3 pN1 pM0.

[1]Sobin, L. and Wittekind, C. (2002) *UICC TNM Classification of Malignant Tumours*, 6th edn, Wiley-Liss (New York).

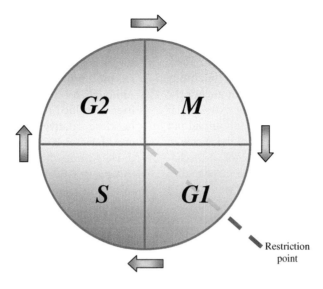

G1	Gap one; preparation for DNA synthesis and contains an important checkpoint for damaged DNA, at this restriction point the DNA is either repaired or the cycle ceases.
S	Synthesis of DNA; the DNA is uncoiled and two daughter chromatids are replicated.
G2	Gap two; the cytoplasm is prepared for mitosis.
M	Mitosis; the nucleus followed by the cell as a whole is divided.

Figure 4.32   Schematic representation of the cell cycle: **G1** Gap one, preparation for DNA; **S** Synthesis of DNA; **G2** Gap two, the cytoplasm is prepared for mitosis; **M** Mitosis

synthesise a duplicate set of chromosomes (G1 phase [24]) and immediately before the nucleus physically divides (immediately prior the M phase diagrammatically) [25].

The first checkpoint is of importance in that it allows the cycle to be interrupted if there is any damage to the DNA. This allows the cell the opportunity to repair the damage before the DNA is copied into the new chromosomes. The second checkpoint allows the cycle to be interrupted if there has been a major error in the DNA copying process. In this way it can be seen that the cell has two important mechanisms during division for detecting damage to its genes in order to avoid passing the damaged genes on to the daughter cells.

Both checkpoints involve the interaction of a set of proteins called cyclins and their corresponding enzymes called cyclin-dependent kinases which trigger the passage of the cell through the checkpoints. The action of the cyclin-dependent kinase enzymes can be blocked by a complicated cascade of proteins involved in the detection of genetic damage. Of note is the involvement of the multifunctional protein p53 [26] which is closely involved in the blocking of the first checkpoint and activation of DNA repair. Also of note is the retinoblastoma protein [27] which is a protein acted on by the active cyclin/cyclin-dependent kinase complex at the first checkpoint with the effect of triggering the transition into the S phase.

Cells with significant genetic damage should not be able to complete the cell cycle and will be halted at one of the two checkpoints. We have seen in Section 4.13.2 however that certain serotypes of HPV are able to disrupt the actions of p53 and retinoblastoma protein and therefore bypass the checkpoints. Similarly mutations in the genes that encode the cyclins, cyclin-dependent kinases, p53, retinoblastoma protein and all of the proteins with which they interact may have the same effect of bypassing the checkpoints. With the checkpoints lost, the dividing cells are free to accumulate genetic mutations without restraint.

4.14.5.2   Stimuli to division

Cells receive a variety of signals from their environment that both promote and inhibit division. These signals range from growth promoting or inhibiting hormones to the membrane proteins that anchor the cell to its neighbours and surrounding extracellular matrix. The range of mechanisms is fairly broad but most of these signalling pathways share the theme of having some sort of extracellular receptor, a mechanism of detecting that the receptor is activated, a mechanism of relaying that message to the nucleus to promote or inhibit entry into the cell cycle and a variety of mechanisms for switching off the whole process when the receptor becomes inactive again.

It is easy to appreciate that any errors in these mechanisms of signalling growth promotion or inhibition can result in uncontrolled or inappropriate growth (i.e. neoplasia). A mutation in a gene encoding a protein in a growth promoting pathway might, for example, lead to spontaneous activation of the receptor or failure to inactivate its signalling pathway after normal activation. Numerous examples have been identified (see Box 4.7) and their identification is gradually assisting in unravelling the normal cellular mechanisms of tissue growth.

4.14.5.3   Tumour growth

A group of cells that has acquired the necessary set of mutations to proliferate without control and to bypass restriction points that halt cell division when genetic mutations arise is an unstable population. The inability to correct mutations before cell division means that when further spontaneous mutations arise then subgroups of cells with the new mutations will appear. If a mutation happens to occur in a gene encoding a protein directly involved in DNA repair then the cell has no chance of repairing mutations and the process of acquiring mutations accelerates.

As spontaneous gene mutations continue to occur, subgroups of cells will arise with new mutations. If these new mutations confer a growth advantage over their neighbours then these cells will begin to dominate the proliferating cell mass. Similarly mutations that produce a subgroup with a growth disadvantage will cause that group to dwindle. As cells within the proliferating mass are usually competing for blood supply, the development of new cell subgroups with varying growth advantages and disadvantages becomes a

> **Box 4.7   Her2 and breast cancer**
>
> Her2 (human epidermal growth factor receptor 2) is a member of a family of cell surface receptor proteins that promote cell division. Although no definite signalling hormone has been identified that specifically activates Her2, its close relative Her1 binds the epidermal growth factor (EGF) hormones. Binding of EGF to Her1 causes the receptor to dimerise and that in turn activates a signalling pathway that promotes entry of the cell into the cell cycle.
>
> In a proportion of several cancers Her2 is expressed at a much higher level than normal. This overexpression greatly increases the probability of spontaneous dimerisation and promotion of cell division.
>
> The most studied cancer in which Her2 is overexpressed is breast cancer and overexpression has been found to be linked to a marked increase in the numbers of copies of the gene in the malignant cells. Her2 overexpression was found to indicate an aggressive form of the cancer with a poor prognosis but a humanised monoclonal antibody (called Trastuzumab) has been developed that specifically blocks its action. For this reason many larger cellular pathology laboratories now routinely look for Her2 overexpression on malignant breast biopsies to determine whether Trastuzumab therapy should be commenced.

process of natural selection. At any one time we can see that the proliferating cell mass will be a composite of numerous different subpopulations of cells with some common mutations and some that differ between subgroups.

Given the correct conditions, some cells within the mass will develop a survival advantage by being able to switch on genes that confer the ability to produce enzymes that break down local cellular and extracellular boundaries, providing more physical space to the neoplastic cells and access to new blood supplies. This might also involve the ability to break down blood vessel walls, to detach from the main tumour mass within blood vessels or even to implant elsewhere and develop new tumours. In such circumstances these cells have crossed the boundary between benign and malignant behaviour.

4.14.6 Genetic mutations and carcinogens

Neoplasia would not occur without genetic mutations and an account of neoplasia should include reference to why they occur. Two principal causes include copy errors and carcinogens.

4.14.6.1 Copy errors

Mitotic cell division requires a complete set of new chromosomes copied from the parent cell's own set. The copying process is extremely accurate considering the amount of DNA material and the number of proliferating cells in the body. Inevitably however mistakes arise and, even though systems to 'proofread' the copied DNA are present and highly effective, some errors will remain undetected. These errors can range from single basepair copying errors to long segments of genes being translocated from one chromosome to another.

4.14.6.2 Carcinogens

We are constantly bombarded by a bewildering array of radiation and chemical substances that can potentially result in DNA damage. Every time we step outside we are exposed to DNA damaging ultraviolet radiation from the sun. The earth around us contains traces of radioactivity and even the air we breathe may contain minute amounts of radioactive gases. An array of chemical agents has been shown under laboratory conditions to cause genetic mutations to a greater or lesser degree and a cocktail of these is found in cigarette smoke. Some albeit weak DNA damaging agents are even commonly found constituents of our food and drink.

Considering the amount of exposure to carcinogens we endure every day and the number of cell divisions that are occurring at any one time it is astonishing that genetic mutations of clinical significance are not more common.

4.14.7 Summary

Neoplasia (according to the most commonly accepted use of the word) is a genetic disease. We have a complex array of cellular mechanisms for the detection and correction of genetic abnormalities and considering the scope for things to go wrong these mechanisms are incredibly effective. When such mechanisms fail however the results can range from a slowly growing benign lump to a rapidly fatal metastatic cancer.

Part II: Clinical application and laboratory techniques

Having reviewed some of the mechanisms of disease we can now discuss the important part that the cellular pathology laboratory plays in diagnosis and treatment. The clinician faced with a patient with a particular disease will formulate a working list of potential diagnoses (a differential diagnosis) and a plan of treatment. In so doing the clinician may take a variety of samples from the patient. The role of the cellular pathology department is to prepare and analyse some those samples to help the clinician narrow the range of potential diagnoses, refine the treatment plan and sometimes to monitor the effects of treatment.

Clinicians in direct contact with patients can choose from a range of techniques or modalities for obtaining samples of tissues or fluids.

4.15 Sampling modalities

A variety of diagnostic sampling modalities are available to the clinician, each of which has its advantages and disadvantages. In general terms the best sampling method is the one that provides the maximum amount of diagnostic information for the minimum amount of discomfort and risk to the patient. This will inevitably vary with the degree of practical difficulty in obtaining the sample (e.g. from a deep-seated organ in the abdomen) and the amount of information needed in order for the clinician to decide the optimum subsequent management of the lesion. The most common sampling modalities are given below.

4.15.1 Analysis of externally secreted bodily fluids

Any body fluid will inevitably contain some cells in varying quantities. Saliva, sputum and semen are cell rich whereas urine is normally relatively cell poor. Cell rich samples can be spread directly onto a slide and stained for visualization under the microscope

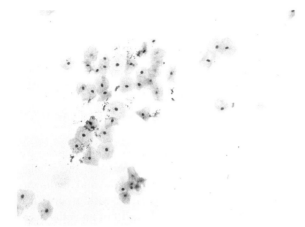

Figure 4.33 The saliva of one of the authors showing the typical content of mucous debris, a few bacterial clusters and numerous superficial squamous epithelial cells. *Rapid giemsa photographed at ×200*

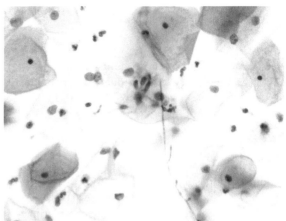

Figure 4.34 An example of exfoliative cytology – this is a cervical smear. *Papanicolaou stain photographed at ×400*

(Figure 4.33). Cell poor specimens can be concentrated by centrifugation or filtration to produce a higher yield of cells.

Body fluids are easily sampled in an outpatient or general practice setting with no obvious risk to the patient. The amount of information that can be obtained may be relatively limited because of degradation of the cells of interest on their way out of the body. In addition many of the lesions of interest that arise in organ systems that produce externally secreted bodily fluids are separated from the excretory tract by some sort of physical barrier such as an epithelial layer. Thus the cells of interest are only secreted if that barrier becomes disrupted.

In the case of sputum the distance travelled from lung to mouth is considerable, the cells of interest become greatly diluted by cells derived from the mouth and upper gastrointestinal tract, and some of the lung lesions of interest simply do not shed their cells into the sputum until a late stage in their natural history. Although sputum is easy to obtain it is therefore of relatively limited diagnostic use.

4.15.2 Exfoliative cytology

This refers to obtaining cells from the surface of the tissue under investigation by some form of scraping, brushing or swabbing. The cells obtained can be spread directly onto slides for viewing or concentrated by centrifugation or filtration as above.

The most familiar form of exfoliative cytology is the cervical smear (Figure 4.34) whereby cells from the cervix are collected on a brush in the general practice or outpatient setting. These cells were formerly collected on a spatula and then spread directly onto a slide but this has now been superseded in the UK by a system of collection on a brush, suspension in fluid and then a combination of centrifugation and filtration. The end result is a thin, concentrated layer of cells from the cervix for analysis of the precursor stages of cervical carcinoma.

Exfoliative cytology has the advantage over other modalities in the ease of the procedure of obtaining the sample and the minimal risk to the patient. In the correct setting (such as the cervical screening programme) the information provided can be extremely useful. The modality does however require that the lesion of interest is either at, or in continuity with, an accessible epithelial surface. This is ideal for the precursor stages of an epithelial surface malignancy but of limited use for deeper lesions or for distinguishing between an *in situ* and an invasive lesion.

Exfoliative cytology specimens can now be combined with endoscopic procedures so that brush samples from bronchus, oesophagus and bile duct are now used in routine diagnosis.

4.15.3 Serous cavity aspiration

Serous cavities are the internal cavities of the body, including pleural, peritoneal and pericardial cavities. In a range of disease processes these cavities, which

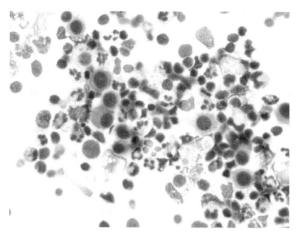

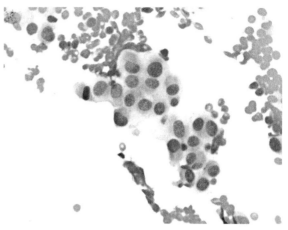

Figure 4.35   An example of a serous fluid aspirate. This is fluid from the pleural cavity in a patient with a pleural effusion. The serous cavities are lined by mesothelial cells that have dense blue cytoplasm on Romanowsky based stains such as this rapid giemsa. Also present are neutrophils, lymphocytes and occasional macrophages. *Rapid giemsa photographed at ×400*

Figure 4.36   An example of a fine needle aspirate. This was a neck lump in a patient with a history of malignant melanoma. The large cells at the centre of the picture are malignant melanocytes. *Rapid giemsa photographed at ×400*

normally contain minimal fluid, can fill with sufficient fluid (termed an effusion) to make direct aspiration with a needle through the skin possible. The fluid is normally relatively cell poor and requires a degree of concentration by centrifugation (Figure 4.35).

When the lining of a serous cavity becomes involved with a metastatic tumour it is common for an effusion to form. The application of serous fluid aspiration is usually limited to a relatively small set of circumstances including the investigation of malignancy in an effusion of unknown cause and to determine if an effusion in a patient known to have cancer is due to involvement of the cavity lining by tumour. In the latter case this may make the difference between a tumour that is potentially curable and one that can be treated only by palliation.

4.15.4   Fine needle aspiration

This modality involves inserting a narrow gauge needle into the lesion of interest and using a few gentle passes of the needle to and fro within the lesion to obtain cells. This can be assisted by gentle suction on the needle or by simply using the capillary action of the needle itself. The cells obtained in this way can be spread directly onto a slide for visualisation or concentrated as above (Figure 4.36).

This procedure can be performed by direct insertion of the needle into a palpable lesion or under image guidance (such as ultrasound) for non-palpable lesions. The needle used is usually sufficiently narrow that the patient doesn't require local anaesthetic and there is little risk of causing significant damage to the surrounding tissues. The principal problems are that the needle may miss the lesion or obtain cells from a small area of the lesion that is not representative of the lesion as a whole. In addition, some lesions simply do not aspirate well or are extremely difficult to classify on cytological appearance alone (e.g. many soft tissue tumours).

4.15.5   Curettage

This is a process of scraping or scooping tissue samples. In contrast to exfoliative scraping, the idea is to obtain tissue fragments rather than individual cells. The resulting preparation is therefore of much larger volume and is histological rather than cytological. If the site of curettage is internal or prone to bleeding the scooping action is accompanied by simultaneous cautery, normally using electrical current (termed diathermy).

This modality lends itself to procedures in which obtaining the tissue for histology is also part of the treatment process, for example in relieving chronic bladder outflow obstruction by removing the central part of the prostate (while simultaneously excluding prostatic carcinoma), or in investigating excessive or prolonged menstrual bleeding by removing a proportion of the endometrium (while simultaneously excluding endometrial polyps, hyperplasia or carcinoma).

4.15.6 Needle core and punch biopsy

This refers to the sampling of a lesion using a wide bore needle or hollow punch, the latter usually being reserved for skin lesions. Deeper lesions can be sampled using a spring-loaded needle that can be triggered to biopsy once in the desired position. The biopsy position can be guided by imaging techniques such as ultrasound and computerised tomography.

The diagnostic uses of this technique overlap with those of fine needle aspiration. Needle biopsy has the advantage of demonstrating the tissue architecture in addition to the cellular morphology. In addition it is a more useful modality for investigating lesions that do not readily aspirate. Its principal disadvantage is that the calibre of the needle is significantly greater and so requires local anaesthetic. In addition there is a greater risk of causing damage to local structures such as large calibre blood vessels and serous cavity walls (e.g. penetrating the pleura during a deep breast or chest wall biopsy). As in the case of fine needle aspiration there is always a risk that the needle will miss the lesion entirely or that the tissue obtained will not be representative of the lesion as a whole.

4.15.7 Excision biopsy

This is the attempted surgical removal of the entire lesion. It is the ultimate form of diagnostic biopsy in that it combines removing the lesion for histological analysis and an attempt at definitive treatment. The entire lesion is available for histological examination, thus avoiding the problem of sampling error. In addition there is usually sufficient lesional tissue for the entire repertoire of histochemical staining, immuno-histochemical staining and cytogenetics if required.

The obvious problem with excision biopsy is the inevitable patient morbidity. This depends on the amount of tissue removed and the site of surgery. A small skin lesion from the back, for example, can be removed in a general practice surgery whereas a lung lesion may require major surgery with a high risk of morbidity and potential mortality.

4.15.8 The role of the laboratory

As indicated above, tissue samples come in all shapes and sizes. In the following sections we will discuss the processes performed in histology (i.e. solid tissue samples) and return to cytology later.

The diagnosis of disease in a tissue sample by a histopathologist may include an important stage of examining the whole intact specimen (see Section 4.17) but is chiefly concerned with the examination of extremely thin slices (or sections) of the tissue under the light microscope. The bulk of the work carried out in a histopathology laboratory is therefore concerned with the complex process of converting a piece of human tissue to a series of thin slices on glass slides for examination under the microscope. The stages of this process include:

1. fixation;
2. specimen dissection (including special techniques to assist dissection such as decalcification);
3. processing;
4. embedding;
5. microtomy;
6. staining and mounting.

In addition, the histopathology laboratory performs a wide variety of ancillary techniques to assist the histopathologist in making a diagnosis based on the appearance of the tissue sections. The core stages of the process will be discussed in turn along with the major ancillary techniques.

4.16 Fixation

4.16.1 The purpose of fixation

Histology seeks to observe tissues in an as near to life-like state as possible. The principal problem with the examination of human tissue however is that of degradation as a result of the cessation of blood and oxygen supply, the release of destructive enzymes within the tissue that are normally safely limited in their actions and attack by enzymes produced by colonising fungi, bacteria and larger organisms. It is therefore necessary to preserve (or fix) the tissue. This is performed using a range of different fixative agents that act at a molecular level to alter the tissue's chemical composition, for example by adding numerous cross-linking bonds between long chain molecules. This has the action of denaturing enzymes that might act to break down the tissues, makes the tissues themselves more resilient to enzymatic breakdown and prevents the colonisation by bacteria and fungi that might degrade the tissue. The earlier the tissue is treated with a fixative agent, the closer it will remain to its original *in vivo* state.

Given that fixation alters the chemical composition of a tissue, we cannot be certain that the tissue viewed after histological preparation is structurally anything like fresh tissue. In other words one cannot be certain how much of what one views under the microscope reflects the appearance of the tissue *in vivo* and how much is artefact introduced during fixation (and other procedures in histological preparation). If we compare several different methods of fixation however those features that are observed on histological examination that are common to each method are logically likely be genuine features of the tissue. In this way we can work out which appearances are likely to be normal for that tissue and which are products of the fixative. We can then select the optimum fixative for that tissue that introduces the fewest artefactual changes (Box 4.8).

In addition to killing off any organisms that might colonise and destroy the tissue, fixation also kills off any organisms already present in the tissue that might infect the laboratory scientist or pathologist. Such infective organisms might include the human immunodeficiency virus (HIV), hepatitis viruses and *Mycobacterium tuberculosis*. All unfixed specimens are therefore potential infection risks and all procedures in which unfixed specimens are handled must be assessed and reassessed regularly to protect staff. During review it may be decided that there is an alternative, safer method that could be employed or the task itself could be deemed unnecessary.

4.16.2 Types of fixative

Precipitating fixatives such as acetone, ethanol and methanol are fixatives that disrupt hydrophobic interactions within the tertiary structure of proteins to reduce their solubility. Cross-linking fixatives are ones that form covalent chemical bonds across long chain proteins, giving them added rigidity. The latter group of fixatives tend to be grouped into the aldehydes (such

---

> **Box 4.8 Fixation**
>
> Fixation is not a new technique that has arisen from the science of histopathology. The preserving properties of such fixatives as acetic acid (vinegar) and alcohol have been exploited since ancient times and the process of 'pickling' foodstuffs has changed little. The preservation of animal tissue has been a reliable tool for naturalists and collectors of animal specimens for centuries.
>
> The application of fixatives to the preservation of human tissue was well developed in ancient China, including the use of mercuric preserving (or embalming) fluids that could be recognised as modern fixatives. The techniques of perfusing the body after death with embalming fluids is still in common use, especially in countries with warm climates or if bodies are to be transported long distances. Famous examples include the preservation of Horatio Nelson in a barrel of brandy for his transfer from Gibraltar to England and the embalming of Mao Zedong, following which he was kept in a mausoleum in Tiananmen Square, China.

---

as formalin) and oxidizing agents such as osmium tetroxide.

Factors which affect optimal fixation include:

1. *pH*. A fixative that is too acidic results in the breakdown of red blood cells with a consequent brown artefact. Fixatives are therefore usually buffered to avoid variations in pH.
2. *Tonicity*. Fixatives must be osmotically similar to the tissues to prevent swelling or shrinking of the cells during fixation.
3. *Thickness of specimen*. Fixatives permeate the specimen by diffusion. The larger the specimen the more time it will take for the fixative to reach the centre of the specimen and therefore the more time the tissue will have at the centre to degrade prior to fixation. Large specimens may therefore be sliced or perfused to allow even fixation.
4. *Volume of fixative*. Given that fixation is a chemical reaction there must be sufficient fixative to complete the process for the whole tissue before it becomes depleted. In the case of the most commonly used (buffered formalin) estimates of an

adequate volume would be at least 10 times that of the specimen.

4.16.3 Drawbacks of fixation

As has been discussed, fixation may introduce artefact into the morphology of the tissue viewed under the microscope. In addition, as fixatives are selected for their ability to react chemically with human tissue, they are all potentially hazardous. Fixative solutions should always be handled with care by trained personnel wearing the appropriate personal protective equipment (PPE). The potential dangers to staff from the most common fixatives include carcinogenic effects, irritant dermatitis and other toxic effects from inhalation, ingestion, eye splash or skin penetration through sharps injuries.

When disposing of fixatives local restrictions must be adhered to and environmental issues such as the effect on wildlife and water treatment facilities must be considered. Some fixatives such as mercuric chloride have added dangers of being corrosive and so familiarity with standard operating procedures (SOPs) and Control of Substances Hazardous to Health (COSHH) guidelines produced by the UK Health and Safety Executive [28] are vital.

Some specimens are required to be in their fresh (i.e. unfixed) state upon arrival at the laboratory, normally because fixation would have a negative impact on specific investigations required of that specimen. Such specific techniques might include enzyme studies (such as testing for acetyl cholinesterase in suspected Hirschsprung's disease) in which fixation would destroy the activity of the enzyme under investigation. Other specimens commonly arriving in the laboratory unfixed include those for direct immunofluorescence (see Section 4.22.5) and those for urgent frozen section examination (see Section 4.21). The majority of unfixed specimens will eventually be fixed and those that are not will either be refrigerated or frozen in order to slow down any degrading enzyme reactions.

Ideally a fixative would penetrate the tissue rapidly and be of relatively low toxicity to staff. Above all else it would cause no artefact in the final tissue histological sections. Unfortunately no one single fixative meets these criteria and a degree of compromise is required. In practice, neutral buffered formalin is the most readily accepted compromise and therefore the most widely used fixative, with the additional benefit of being relatively cheap. For this reason many of the ancillary techniques used in histopathology begin from the starting point of the tissue having been formalin-fixed.

4.17 Specimen dissection

Tissue specimens range enormously in size and shape, from a 1 mm diameter bronchial biopsy to 1 m or more of colon. The various stages of specimen handling in the histology laboratory result in histological slices (or sections) that are slightly less than one cell's width in thickness for examination under the microscope.

A typical section measures approximately 4 μm in thickness and up to approximately 25 by 20 mm in length and width respectively. To examine even a 1 mm biopsy at 4 μm thickness in its entirety on histological sections would take 250 sections. Examination of an entire colon would require hundreds of thousands of sections. It is therefore not realistic to section the entire specimen and we must accept that examination must be limited to a few sections that are interpreted as representative of the tissue and the disease process as a whole. For large specimens that include several different types of tissue, some of which may be entirely normal and some containing the disease process of interest, careful selection of the areas of the specimen to be sectioned is required (Figure 4.37).

The process of specimen dissection (commonly called 'cut up') requires a certain amount of knowledge of:

- what the disease process of interest should look like in this particular specimen;
- what parts of this type of specimen usually contain disease if it is not visible to the naked eye;
- in the context of a particular disease process in that specimen (such as a cancer), what parts of the specimen should be looked at to provide sufficient information to give the clinicians an indication as to treatment and prognosis.

This last point is of great importance in the context of cancer surgery. Often the diagnosis of cancer is already known and histological assessment is more concerned with obtaining additional information such as whether the tumour has been completely excised and whether it has invaded into blood or lymphatic vessels.

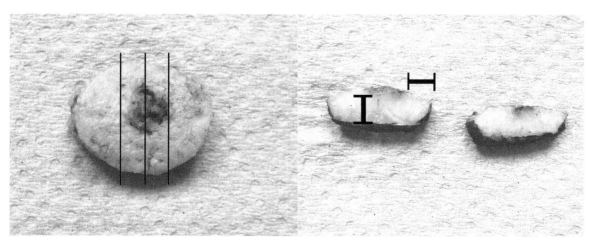

Figure 4.37   A skin ellipse (left) excised from a patient with a suspicious patch of brown pigmentation measuring approximately 5 mm in diameter. Cutting the skin in the planes indicated produces complete transverse slices of the underlying lesion (right). Sections cut from these slices will allow not only diagnosis but measurement of clinically useful parameters such as the distances to the nearest deep and peripheral surgical resection margins (indicated). These parameters are useful for the clinician in determining whether the lesion can be regarded as completely excised

Specimen dissection is normally performed on fixed tissues. It can however be commenced on larger tissues prior to fixation to allow the fixative to permeate the whole tissue evenly.

It should be appreciated that good specimen dissection technique is crucial and that poor selection of the relevant tissue for sectioning can render the rest of the process relatively fruitless.

4.17.1   Decalcification

Calcium salts, predominantly in the form of various calcium phosphates, are commonly found in tissues in both normal and pathological states. They are one of the main structural components of bones and teeth and are also key constituents in stones that form in various organs. Hard, insoluble calcium salts present a significant problem for cutting thin sections and even the hardest and sharpest of blades will struggle. The purpose of decalcification is therefore to remove calcium from tissues prior to processing in order to allow sections to be cut. Calcification is usually recognised at the time of specimen dissection, to which it can be a significant barrier, and steps to permit dissection as well as allow adequate sectioning are usually performed at this stage.

There are two main types of decalcifying solutions available. The purpose of both is essentially to convert calcium ions from an insoluble to a soluble form so

that they are dissolved out of the tissue. The two types are described below.

4.17.1.1   Chelating solutions

Chelating solutions act by a slow process of ion exchange at a pH close enough to that of the fixative such that additional chemical reactions with the tissue are kept to a minimum. They can therefore take a considerable amount of time to act but avoid introducing significant amounts of artefact that might alter morphology, staining pattern or have a detrimental effect on immunohistochemical staining properties. This is important for tissues such as bone marrow in which cytomorphology and immunohistochemical staining are paramount. It is also crucial in cases in which fine disruptions in normal architecture might have very significant consequences (e.g., the examination of paediatric ribs in child abuse cases). As chelating solutions draw out the calcium into the solution itself, the solution must be changed frequently to allow the continuation of the process to completion.

4.17.1.2   Acid solutions

These dissolve the insoluble calcium salts rapidly but result in additional chemical reactions that may alter the staining pattern, introduce artefact and disrupt the potential for ancillary techniques such as immunohistochemistry. They are however useful if a result is required quickly and fine detail is not essential (e.g. to assess whether a tumour that has already been diagnosed in a previous biopsy is involving bone). Different strengths of acid solutions are available from the more mild formic acid up to concentrated nitric acid. In general terms, the stronger the acid, the more rapid the decalcification but the greater the chance of disrupting the soft tissue.

There is a range of solutions available for decalcification and the appropriate one should be selected for each particular tissue type and diagnostic situation. Ethylene-diamine tetraacetic acid (EDTA) is a chelating solution and is useful for bone marrow trephines and paediatric ribs. Trichloroacetic acid (TCA) is an acid decalcification solution and is useful for samples containing bone such as teeth, femoral heads and parts of mandibles.

In a small number of circumstances it is the calcification itself that is of diagnostic significance. An important aim of the breast screening programme is to identify fine areas of calcification within the breast that might signify an accompanying pathological process such as early cancer. In this situation it is necessary to retain the calcium within the tissue sections as much as is technically possible so that its presence can be detected by the pathologist and any associated pathology identified. In such cases decalcification would therefore be inappropriate.

There are several end point tests available that can test for the amount of residual calcium within the tissues either directly or indirectly. Direct testing includes the performance of X-ray, which is not disruptive to the tissue but is expensive and time consuming. Physical testing involves manipulation of the tissue to test its flexibility and ability to be cut with a blade. This is extremely quick and straightforward but can be detrimental to the integrity of the specimen.

Indirect methods are usually less accurate as they are testing the decalcifying solution rather than the tissue itself. An example of an indirect method is the ammonium oxalate test in which ammonium oxalate is added to the decalcifying solution and will react with any calcium present to form a cloudy precipitate of poorly soluble calcium oxalate. If the decalcifying solution is frequently changed then the amount of calcium in the solution after it has been exposed to the tissue will reflect the amount of calcium still left in the tissue. Once calcium stops exiting the tissue into the solution then the tissue has decalcified as much as it ever will.

A sample of the solution will therefore not form a precipitate with ammonium oxalate.

In addition to soaking the specimen in a decalcifying solution prior to processing, decalcification can be performed on the cut surface of tissue blocks embedded in wax at the time of sectioning. The same decalcifying agents will suffice and the procedure is useful either for problematic tissue that did not quite decalcify enough prior to processing or for tissues in which small foci of calcification were not identified at the time of tissue dissection and block selection.

4.17.2 Other obstacles to sectioning

In addition to insoluble calcium salts, other hard substances can be present in tissues that prevent sectioning. These are most commonly artificially implanted items such as sutures and surgical staples that simply require the object to be picked out of the specimen. The principal naturally occurring hindrance to sectioning other than calcium salts is keratin, most commonly in the form of skin horns and nails.

Keratins are structural proteins that form the main constituents of hair and nails in humans as well as claws and feathers in other animals (Box 4.9). Fixation further strengthens the structural properties of keratin but several proprietary agents are available for softening the structural rigidity.

4.18 Processing and embedding

In order to cut very thin sections of tissue for examination under the microscope the tissue must somehow be made sufficiently rigid to maintain structural integrity whilst being sliced with a blade to less than the thickness of one cell, usually of the order of 4 μm. The tissue (which at this thickness is largely colourless and transparent) must also however be made readily available for the application of assorted stains and other ancillary histological techniques. The best means of performing this function is therefore to integrate the tissue into some solid medium that can support the tissue during the sectioning stage but can be removed once the thin tissue section is safely attached to a glass

Box 4.9   Keratin

Keratins are a family of structural proteins that are present in a wide variety of cells but manifest most obviously in forming one of the principal components of skin, hair and nails (as well as scales, claws and shells in other animals).

Occasional surgical specimens include a large amount of tough keratin. Lesions may arise in the vicinity of a nail (e.g. a variant of melanoma has a propensity to arise in the nail bed) and some dysplastic skin lesions result in the formation of a hard mass of overlying keratin. It is therefore sometimes necessary to obtain sections through a tough keratinous structure. This presents the obvious problems of the keratin being difficult to cut and, because whatever tissue the keratin is attached to will be much softer, tearing of the section at the interface between hard and soft. Fixation exacerbates the situation by hardening the keratin and making it less susceptible to softening in water.

A number of softening agents are commercially available for softening formalin-fixed keratin and these are of variable efficacy. A variety of home-made softeners are also used with mixed results. These include alkalinized washing-up liquid and washing powder.

slide so that the tissue is once more accessible for staining.

A wide variety of waxes, resins and plastics can be used for this purpose and each has its advantages and disadvantages. Compromises must be made between the ease of sectioning, rigidity of the medium and the ability to remove the medium as required. In practice the most commonly used medium is paraffin wax which has the principal advantages of being cheap, readily available and easily removed from the tissue using the organic solvent xylene.

The use of paraffin wax presents the principal obstacle of not being water soluble. As we have seen above, fixatives have to be aqueous and be osmotically similar to cells or would otherwise cause catastrophic osmotic damage to fresh tissue that would make histological interpretation impossible. Transferring the tissue straight from its watery fixative into wax however would result in complete failure of the wax to bond to or penetrate the tissue and this in turn

would effectively negate the purpose of wax embedding. The tissue, now fixed and therefore no longer so osmotically sensitive, must therefore be converted from being largely aqueous to being ready to bond to paraffin wax. This sequence of events is called **processing**.

4.18.1   Dehydrating and clearing agents

Routine processing is usually automated and uses machines that are programmable to match the type and size of tissue specimens. After penetration with formalin the tissue must be gradually taken through a series of increasing concentrations of alcohol to reduce the water content. An increasing alcohol series is used because, even in the fixed state, the sudden osmotic changes produced by immersing straight into pure alcohol may introduce artefact.

Absolute alcohol will not mix with paraffin wax and so the tissue is treated with an organic solvent (most commonly xylene, although other agents such as toluene are in common use) that will mix with both alcohol and wax. Xylene and other such organic solvents are also known as clearing agents because they remove lipids and therefore the adipose tissues become optically translucent. The molten paraffin wax can then be introduced to the tissues and will permeate through the xylene. Once fully penetrated by paraffin wax, the tissues can be embedded in a block of wax ready for sectioning.

In the busy histopathology laboratory there is a continuous risk of pieces of tissue from different specimens becoming mixed. If the mixed pieces of tissue happen to be of the same tissue type there is a very real risk of misdiagnosis. For this reason all of the tissue from each specimen is processed in plastic permeable containers (called cassettes), each of which is labelled with a unique number assigned to that particular specimen (Figure 4.38). Thus at any point in the process the tissue in a particular cassette can be matched to the specimen from which it came.

4.18.2   Embedding

Embedding is the process of transferring tissues from each cassette into a mould (Figure 4.39) and setting

Figure 4.38   Two pieces of tissue in a cassette that have undergone processing. They are now ready for embedding

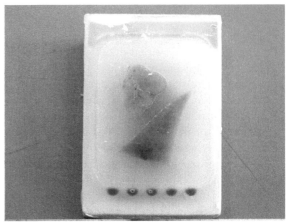

Figure 4.40   Two pieces of tissue embedded in paraffin wax and attached to the original plastic cassette. The tissue is now ready for sectioning on the microtome

them in position within the support medium with which they were finally penetrated during processing. This is then allowed to harden prior to cutting sections. The most common technique in paraffin wax embedding is to place the tissue into a metal mould, fill it with molten wax and then allow it to cool and harden on ice with the plastic cassette placed on top. In this way the block of wax containing tissue remains attached to the cassette in which it was processed so that it is still identifiable by its specimen number label (Figure 4.40).

Where there are multiple pieces of tissue, as much as is practical, the pieces should all be flat, lying within the same plane and with no overlapping or touching of the separate pieces. This enables a clearer examination of the full face of all the pieces within the same section.

Figure 4.39   Two pieces of tissue embedded in paraffin wax within a steel mould

The crucial element of embedding is that it allows an opportunity to orientate the tissue before it is sectioned. Histological sections of skin, for example, normally require all layers of the skin to be apparent for histopathological diagnosis. It is therefore important at the embedding stage to ensure that the skin sample is placed in the hardening wax at the correct angle to produce a complete skin section. Skin samples looking for alopecia may need one piece orientated for the skin layers and another perpendicular to this to observe sections through the hair follicles and this can be positioned in agar by the pathologist prior to processing or by the experienced biomedical scientist at embedding. Tubular structures such as blood vessels, appendix or alimentary tract will also require orientation to allow an examination of the tissue layers through to the lumen.

4.19   Microtomy

Microtomy is a technique whereby thin sections of tissue are sliced from a large block of tissue with great accuracy using a piece of equipment called a microtome. The usual thickness cut is 4 μm which is fractionally less than the diameter of a red blood cell. For delicate structures such as glomeruli in which the fine detail of basement membranes needs to be assessed, thinner sections may be required. Similarly certain investigations require thicker sections for optimal staining.

All microtomes have the same basic components. These consist of a chuck, in which the tissue block can be held securely, a sharp blade and a mechanism for advancing the tissue toward the blade at precise intervals of distance. Although microtomes are always becoming more automated and user-friendly, a certain level of skill and understanding is required to produce a good quality section and to troubleshoot effectively. When troubleshooting, most sectioning problems can be resolved by ensuring the tissue block is cooled well on ice (to keep the wax in a solid and rigid state), that all the components of the microtome are securely attached and that a clean, sharp blade is used.

There are generally two types of microtomes in current use. These are the sledge and the rotary, both of which share the same basic principles of action. The sledge has the blade mounted at the opposite end of the device to the operator. The tissue blocks are presented to the blade by sliding the chuck horizontally towards it. The rotary has a wheel which, when turned, moves the tissue block across the blade in a vertical fashion. Both types of microtome can be used to produce good quality sections, but the sledge microtome is gradually being replaced by the rotary (Figure 4.41) as most people find it easier with which to become familiar.

A complete section of the tissue block consists of an approximately 4 μm thickness slice of tissue and surrounding wax. When multiple consecutive sections are cut then each is usually loosely attached to the previous at one end by wax to produce a ribbon of consecutive sections. Once a ribbon of good quality sections has

Figure 4.42   A ribbon of sections cut from the tissue block photographed earlier floating on a water bath

been cut, they can be transferred onto a warm water bath at around 45 °C, slightly below the melting point of paraffin wax (Figure 4.42). This softens the wax, stretches out any wrinkles within the sections and allows the sections to be carefully manipulated with forceps. The best sections (i.e. ones with no creases, tears, scores, holes and of the correct thickness) are separated with forceps and lifted onto glass microscope slides ready to be stained (Figure 4.43).

Good quality sections are required to give the pathologists the best possible picture of the tissue block so they can make an accurate diagnosis and so any vital information is not obscured by artefacts in the section.

Figure 4.41   An example of the type of rotary microtome commonly in use

Figure 4.43   The same tissue sections have now been floated onto a glass slide ready for staining

4.20 Standard staining methods and procedures

4.20.1 Haematoxylin and eosin

The haematoxylin and eosin (H&E) stain is regarded as the standard stain of clinical histopathology (Figure 4.44). Virtually all solid tissues sent for histological examination will first be stained using H&E before deciding if any ancillary techniques are required.

4.20.1.1 Haematoxylin

Haematin is derived from the *Haematoxylin campechianum* tree. It is the oxidised form of the dye, either oxidised naturally by ultraviolet light or chemically with sodium iodate, that is used in the H&E stain. A mordant such as aluminium potassium sulphate is added that facilitates binding predominantly to nucleic acids such as RNA in the cytoplasm and DNA in the nucleus. The staining property of haematoxylin is commonly described as basic owing to its property of binding to acidic parts of the cell (and hence those components are referred to as 'basophilic').

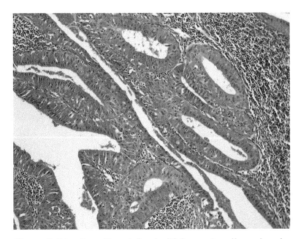

Figure 4.44   A section stained with haematoxylin and eosin showing the contrast between the bright pink cytoplasm of the epithelial cells and the dark blue/purple nuclei of the adjacent lymphocytes. The lesion shown is a Warthin tumour of the salivary gland. This is characterised by strikingly eosinophilic epithelial cells on a background of lymphoid tissue. *H&E photographed at ×100*

Haematoxylin is bright red in colour when mixed with its mordant in a mildly acidic solution at the commencement of the staining process. During staining however the procedure for precipitating the stain onto the tissue in an insoluble form requires neutralising the solution. This has the effect of making the haematoxylin change to the characteristic blue/purple colour that we see on histological sections.

4.20.1.2 Eosin

Eosin is a family of compounds chemically derived from the fluorescent dye fluorescein. It stains a variety of tissue components in varying shades of pink and red. In direct contrast to haematoxylin, eosin has a tendency to stain chemically basic cellular constituents (constituents that are therefore termed 'acidophilic' because of their affinity for acidic stains). Given that many proteins contain basic residues, eosin will stain tissue components that are protein rich such as the contractile elements of muscle and connective tissue components such as collagen.

The quality of the haematoxylin can be checked by ensuring that only the nuclei have stained and not the background, the nuclei are blue and not too red, and that the chromatin detail is visible. The eosin must show the range of shades it can stain and overall not be too light/dark. Overall, there must be a good balance of colour seen in the H&E. Most laboratories have an automated staining machine for this stain because of the number of slides to be stained and for the stain to be standardised so every day a suitable control slide, such as a section of colon, must be stained and checked to ensure the slides for the day are stained accurately.

4.20.2 Major histochemical/ tinctorial stains [29]

Only a few stains are listed here. Most standard histochemistry texts give staining protocols and examples of a wide range of techniques.

4.20.2.1 Periodic acid Schiff (PAS)

The periodic acid Schiff staining technique essentially converts colourless Schiff's reagent to a deep magenta colour when in contact with a variety of carbohydrate compounds (Box 4.10). These include glycogen,

Diastases are a group of enzymes that catalyse the breakdown of complex carbohydrates into simple sugars. In the histopathology laboratory the primary application of commercially available diastase enzymes is the breakdown of glycogen prior to the use of the periodic acid Schiff (PAS) staining procedure. Glycogen is PAS-positive (i.e. a bright magenta colour). Pretreatment of the section with diastase will remove the glycogen. A comparison between sections stained with the PAS technique with and without prior treatment with diastase allows the identification of glycogen and distinction from diastase resistant, PAS-positive carbohydrates such as most neutral mucins.

Human saliva contains the enzyme α-amylase that is a type of diastase enzyme. Prior to the ready commercial availability of diastase enzymes it was common practice for biomedical scientists to spit on sections prior to using the PAS technique. The lack of standardisation between biomedical scientist saliva (as well as a variety of health and safety measures) means that using commercially available diastases is now the preferred option.

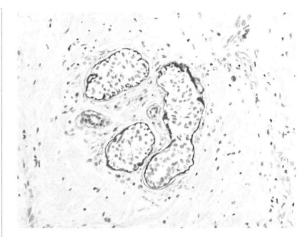

Figure 4.45   A section of skin including part of an eccrine sweat gland. PAS stains the basement membrane of the gland bright magenta. *PAS photographed at ×100*

various mucopolysaccharides and a range of glycoproteins including the constituents of fungal cell walls. Its multiple applications include the identification of glycogen and fungi as well as the examination of basement membranes (Figure 4.45) and neutral mucins.

4.20.2.2   Periodic acid Schiff with diastase (DPAS)

This variant of the PAS method involves pretreating slides with the diastase enzyme before performing the periodic acid Schiff stain. Glycogen is hydrolysed by diastase and will not therefore produce the characteristic magenta colour. A direct comparison between sections of the same tissue stained with the PAS and DPAS techniques can confirm or refute the presence of glycogen (Figure 4.46).

4.20.2.3   Alcian blue PAS (ABPAS)

ABPAS allows the distinction between acid and neutral mucins. Alcian blue will stain acid mucins blue, as it is a basic dye with an affinity for anionic acid mucins. Alcian blue will not however stain nuclei because its molecular conformation results in steric hindrance in staining nucleic acids. As discussed above, the Schiff's reagent will stain neutral mucins magenta and diastase removes any glycogen. The combination is useful in determining whether any tissue mucin identified is acidic or neutral and the relative cellular distributions of the two (Figure 4.47). This has application, for example, in the examination of biopsies from the antrum of the stomach in which the normal gastric epithelium produces neutral mucins. The presence of acid mucin-producing cells of the type normally seen in the small and large intestines (termed goblet cells) indicates a pathological process called metaplasia (see Section 4.13) and is a good indicator of colonisation by the disease-causing bacterium *Helicobacter pylori*.

4.20.2.4   Perls Prussian blue

In the Perls Prussian blue technique hydrochloric acid and potassium ferrocyanide react with ferric ions in the tissue to make the intensely blue pigment ferric ferrocyanide (Prussian blue) (Figure 4.48). This is of particular use in assessing tissues for excess iron deposition as is seen in the disease haemochromatosis. It is also of use in searching for asbestos fibres within lung tissue as these fibres obtain a coating of iron-containing haemosiderin and can be extremely hard to identify on histological sections owing to their relative scarcity.

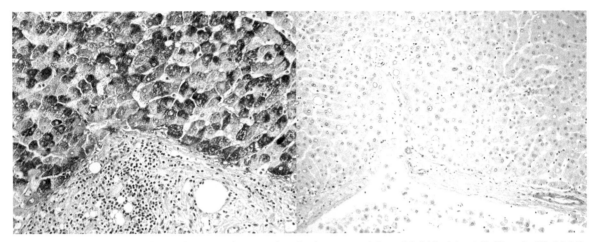

Figure 4.46   Two sections of liver demonstrating cytoplasmic glycogen staining with PAS without (left) and with (right) pretreatment with diastase. *PAS and DPAS photographed at ×100*

4.20.2.5   Trichrome methods

Trichrome (literally three colour) staining methods such as Martius scarlet blue (MSB) and Masson's trichrome are used for demonstrating a range of connective tissue in one stain. Trichrome methods work on the principle that different connective tissues vary in their permeability to pigment molecules of different sizes. The smaller molecular size dyes are applied first and, where their molecular size allows, they are replaced by a larger molecular size dyes that are applied sequentially.

Martius scarlet blue (MSB)

MSB is useful for demonstrating the relative age of the blood clotting protein fibrin. The result of the action of the clotting system together with the adhesion of blood platelets in circulating blood is a solid mass of platelets and fibrin called a thrombus. It is sometimes useful to be able to establish the age of a thrombus. Even after forming a thrombus the constituent fibrin undergoes further chemical modification as the thrombus matures. MSB uses the differences in chemical composition between a newly formed thrombus and a maturing

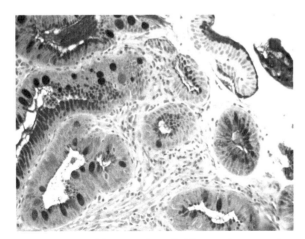

Figure 4.47   This oesophageal biopsy demonstrating Barrett's oesophagus illustrates the PAS-positive neutral mucins in columnar cells resembling gastric epithelium and the Alcian blue-positive acid mucins in columnar cells resembling those found throughout the intestines. *ABPAS photographed at ×100*

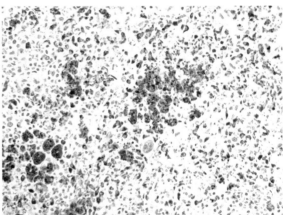

Figure 4.48   This spleen contains an unusual amount of haemosiderin within its resident cells of the monocyte-macrophage lineage. Haemosiderin is a haemoglobin breakdown product that is rich in iron and hence appears bright blue on staining using the Perls technique. *Perls prussian blue photographed at ×100*

thrombus to produce different colours for new and old fibrin. New fibrin stains as yellow strands, the intermediate ages range from orange to red strands and the oldest fibrin stains as blue strands. Collagen stains green/blue, muscle/cytoplasm stains red and the nuclei are dark blue.

4.20.2.6   Silver impregnation techniques

Silver techniques are very delicate and require an experienced eye to determine sufficient impregnation of the tissue and meticulously cleaned glassware as even the smallest contaminant can produce poor results. All silver techniques involve the impregnation of tissues with silver ions which must then be reduced to a visible metallic silver form. The reduction is achieved either through the addition of a reducer (an argyrophil reaction) or through the combination of the tissue components with the silver ions (an argentaffin reaction) such as melanin in the Masson Fontana reaction.

Methenamine silver technique (Grocott's stain)

The methenamine silver technique (MST) can be used to demonstrate the basement membranes within delicate structures such as glomeruli with great effect and is also used to demonstrate the cell walls of the majority of fungi. It is a silver salt-based reaction in which the silver is readily taken up by the basement membranes and fungal cell walls and is directly reduced to precipitate the silver into a visible metallic form in an analogous process to black and white photography. In the case of basement membrane staining the silver may be toned using gold chloride, leaving the collagen golden brown. A counterstain such as a light H&E provides some background for orientation of the tissue (Figure 4.49).

4.20.2.7   Congo red

This staining technique is used for the identification of amyloid (see Section 4.11).

4.20.2.8   Ziehl Neelson

This technique is used for the identification of *Mycobacteria*, the group of bacteria responsible for tuberculosis (the world's single most deadly bacterium) and a host of less common granulomatous diseases. The technique essentially uses carbol fuchsin (bright magenta) to stain the hydrophobic mycolic

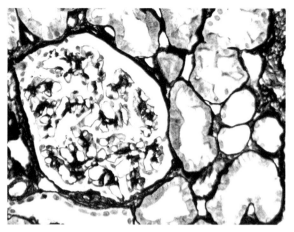

Figure 4.49   An example of MST showing a renal gomerulus and adjacent tubules. The basement membranes stain black. *MST photographed at ×200*

acid constituent of the acid fast mycobacterial cell walls (Figure 4.50). The technique has several variations with more selective staining for various mycobacterial species such as Wade Fite which is modified to stain the acid—alcohol fast bacterium in leprosy.

Once slides have been stained they need to be covered with a glass coverslip to prevent damage and the drying out of the sections which affects its optical properties. The coverslips are held in place by a mounting medium that for the majority of slides will be a resinous medium that is miscible with xylene and has a similar refractive index to that of glass.

Figure 4.50   Ziehl—Neelsen stain showing magenta curvilinear rods in a case of *Myocbacterium tuberculosis* infection. *Ziehl—Neelsen photographed at ×600*

Some staining techniques such as Oil red O and immunofluorescence can be removed or compromised if alcohol or xylene is introduced after staining and so these require an aqueous mounting medium.

4.20.2.9 Control tissues

In performing any histochemical stain in which the presence or absence of a characteristic staining pattern is diagnostic, it is always necessary to stain control tissue under exactly the same conditions as the test tissue. A control is a tissue in which the staining pattern of the particular histochemical stain is already known. A positive control is one in which the stain should be positive if it has worked correctly. Similarly a negative control should show no staining. The simultaneous use of positive and negative controls ensures that the staining pattern of the test slide can be interpreted with a high degree of confidence that the technique has worked correctly. Histopathology laboratories will normally keep a collection of tissues with predictable staining patterns to use as controls.

4.21   Frozen section

Cryotomy (frozen section) is, as the name suggests, the cutting of sections of fresh tissue that has been frozen. For conventional histological examination to be performed the tissue must undergo the time-consuming stages of fixation, processing and embedding. In comparison with these stages the steps of sectioning and staining are relatively quick. Cryotomy bypasses the slow stages by jumping straight from removal from the patient to sectioning. This is extremely useful for swift diagnosis but has the severe drawback that much of the histological detail relied on for reliable diagnosis in conventional histology is not available to the pathologist on frozen section because of distortion during freezing. This technique is therefore most useful when only crude indicators are required rather than specific diagnostic information. Such indications include confirming the nature of the particular tissue that has been removed (useful in distinguishing between parathyroid glands and lymph nodes in the neck that can look very similar to the operating surgeon) and assessing whether a tumour that has already been diagnosed by biopsy and conventional histology has been completely removed at surgery. Frozen sections can be sufficiently rapid to produce for the histological opinion to be given to the operating surgeon within minutes and the surgical procedure can be modified accordingly.

It is essential that tissues are rapidly frozen to reduce the distortion caused by ice-crystal formation and orientated correctly in the embedding medium to ensure the best quality section can be achieved as quickly as possible. Liquid nitrogen is commonly used for rapid freezing but a variety of other options are available including carbon dioxide and various low boiling point organic solvents.

The mechanism of section cutting is similar to that used when using the rotary microtome, but when the sections are mounted onto the glass microscope slides they are done so directly onto the glass without the aid of a water bath. A special low temperature microtome called a cryostat is dedicated to cutting frozen sections (Figure 4.51). To achieve sections without creases, an antiroll bar is used on the cryostat to stop the sections from curling up.

In addition to its use in rapid histological assessment, frozen sections have applications in certain other ancillary techniques. Stains to assess the presence of fat within tissue (such as Oil red O) are not possible on conventional paraffin-embedded tissues because clearing agents such as xylene inevitably dissolve out fatty acids. No clearing agents are required for frozen sections and so fatty acids remain in the tissue and can be stained.

Other ancillary techniques include enzyme histochemistry. As fixation effectively denatures proteins, enzymes are therefore also usually non-functional after fixation and so enzyme histochemistry must be performed on frozen sections. In enzyme histochemistry

Figure 4.51   A typical cryostat

the enzymes present within a cell of interest are used to effect a chemical reaction that results in the appearance of a visible chromagen. The most common application is in the search for ganglion cells in the colon and rectum for the purposes of assessing the evidence for Hirschsprung's disease. This is the failure of normal innervation of a segment of the rectum of variable length. The result is that the affected segment remains severely narrowed and can result in constipation, abdominal distension and even bowel obstruction. Diagnosis is confirmed by a lack of normal ganglion cells within the wall of the affected segment of bowel. This search is greatly assisted by an enzyme histochemical reaction that relies on the presence of the acetyl cholinesterase enzyme in ganglion cells to cause a visible chromagen to form. In the absence of ganglion cells the reaction cannot happen and the diagnosis is confirmed.

It is noteworthy that frozen tissue has not been exposed to the fixing agents that normally eliminate the presence of potentially infective pathogens. Until the tissue is fixed there is therefore a risk of infection to the biomedical scientist.

4.22 Immunohistochemistry

The technique of immunohistochemistry uses antibodies to identify and bind to specific constituents within a cell. The technique then activates a chemical reaction at the site of binding which results in the production of a visible chromagen.

Free antibodies are glycoproteins produced by plasma cells during an immune reaction. The classical example of antibody structure (as depicted by immunoglobulin G but the common theme of all antibody types) approximates to a Y-shape with the principal binding sites on the short arms [30] (Figure 4.52). The areas of the antibody that form the binding site are termed the hypervariable regions owing to the enormous range in structure between antibodies produced by different plasma cells. The remainder of each antibody does not vary significantly between plasma cells.

The structure of the binding site of the antibody is such that it will bind highly specifically with only one particular combination of molecular conformation and component molecular groups, for example a particular short sequence of amino acids with a particular three-dimensional structure. The point at

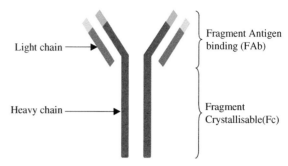

Figure 4.52 The classical Y-shaped depiction of an antibody

which the antibody binds is called the epitope and the molecule that contains the epitope is called the antigen. The binding is via a combination of electrostatic interaction, van der Waals forces and hydrogen bonding. It can be envisaged that the number of structurally different molecules that happen to contain the same epitope is likely to be small and in reality antibodies will bind relatively specifically to just one type of antigen or to a group of antigens that are chemically and structurally closely related.

4.22.1 Clonality

Although individual plasma cells can produce a number of different types of antibody, each type has the same hypervariable region in common. The structure of the hypervariable binding site is encoded by a segment of DNA that has undergone massive recombination and mutation during development as an early B-cell but is fixed in the mature plasma cell form. Thus all of the antibodies produced by a single plasma cell bind to just one specific epitope. These antibodies are described as *monoclonal*.

In the course of an immune reaction to, for example, an infecting bacterium, there may be a great many different plasma cells producing antibodies to a particular bacterial component protein. These plasma cells will all have different genetic sequences for their hypervariable binding sites and will be producing antibodies that bind to different epitopes, but if all of these different epitopes happen to be on the same bacterial protein then they are effectively all binding to the same antigen. Antibodies that recognise different epitopes but all bind to the same antigen are described as *polyclonal*.

4.22.2   Antibodies in histopathology

A wide range of antibodies are commercially available that recognise a variety of cellular constituents. Both polyclonal and monoclonal forms are available for numerous different antigens but the higher specificity, standardisation and consistent reproducibility of staining results of monoclonal antibodies means that the vast majority are monoclonal. Their production involves fusion of reactive B lymphocytes exposed to the antigen in question to a myeloma (a tumour of plasma cells) cell line resulting in a hybrid tumour that secretes a monoclonal antibody [31]. Although there are various types of antibody with different functions and conformations, it is the immunoglobulin G (IgG) antibody isotype that is of use in histopathology because it is secreted as a soluble monomer.

The common uses in histopathology fall into two main groups as follows.

1. *Identification of a particular cell type.* A common problem in histopathology is that many different cell types have a similar morphology. It can be very difficult to distinguish between, for example, a fibroblast and a smooth muscle cell, and even more difficult to distinguish between different types of mature lymphocyte. The problem becomes even more difficult in the case of poorly differentiated malignant tumours. Thus a poorly differentiated carcinoma can look almost identical to a blastic lymphoma but there may be a massive difference between the two in terms of behaviour and treatment.If such a diagnostic dilemma is encountered it is extremely useful to have an antibody to an antigen found in one possible cell type but not the other. Cytokeratins, for example, are filament proteins found almost ubiquitously in epithelial cells but very rarely in significant amounts in lymphocytes. Thus, in the example above of a poorly differentiated tumour that could be a carcinoma or lymphoma based on appearance, an immunohistochemical stain that recognises cytokeratins would stain a carcinoma but not a lymphoma and therefore solve the diagnostic problem.
2. *Prognostication in a known tumour type.* If the tumour type is already know, either because it has a characteristic morphology on histological examination or because we have already used immuno-

histochemical stains to determine the cell type, the behaviour of some types or their response to treatment can be predicted from the expression of certain antigens. One such example is that of the expression of Her2 in breast cancer (see Section 4.14). This cell surface receptor is overexpressed in an aggressive subtype of breast cancer and can be visualised on immunohistochemical staining. In addition to indicating an aggressive form of breast cancer with poor prognosis however the overexpression of Her2 also means that the tumour is likely to respond well to a highly specific form of treatment.

4.22.3   Mechanisms of immunohistochemical staining

Having obtained an antibody to a specific antigen, a mechanism must be found to be able to demonstrate both that the antibody has bound to the histological section and show exactly where the binding has taken place. Several methods are employed including attaching radioisotopes to the antibodies, attaching fluorescent molecules (see Section 4.22.5) or attaching enzymes that can catalyse a colour-changing chemical reaction. This last technique is immunohistochemistry and is the technique most commonly encountered in histopathology.

Antibody techniques can be direct or indirect as follows.

1. *Direct.* In this case the system for identifying the antibody is attached directly to the antibody molecule. This means that each antibody molecule will be attached to one fluorescent molecule or one enzyme molecule. While this technique is quick and gives a semiquantitative indication of the amount of antigen present, a large amount of antigen is required to produce a visible result. The method therefore has relatively low sensitivity.
2. *Indirect.* (Figure 4.53). This makes use of the fact that antibodies differ slightly between mammalian species. It is therefore possible to make antibodies within, for example, a rabbit that react to mouse antibodies. If our antigen of interest has a specific mouse monoclonal antibody bound to it (the primary antibody) then we can use an antimouse IgG antibody conjugated to the colour-generating

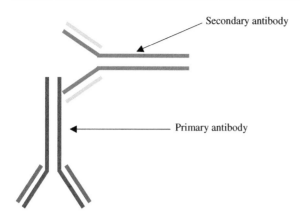

Figure 4.53   Primary and secondary antibodies

enzyme (the secondary antibody) to identify its location. With several secondary antibody molecules bound each primary antibody molecule we have introduced an amplification step (and hence enhanced sensitivity), with the added bonus that only a limited range of conjugated antibodies is required (a costly process) compared with the range of unconjugated primary antibodies.

4.22.3.1   Further amplification steps

In reality the amount of antigen typically present in histological sections is unlikely to be sufficient to make any antibody-conjugated enzyme-mediated colour change reaction easily and reliably identifiable without further amplification. The most commonly employed method of amplification is the avidin-biotin complex method (ABC method) [32].

The ABC method uses the tetrameric protein avidin to bind to the amine biotin (vitamin H). Once bound, the avidin-biotin complex is extremely stable and resistant to separation. Given that avidin is a tetramer and that each monomeric component can bind a biotin molecule, each avidin molecule can theoretically bind up to four biotin molecules.

The ABC method employs a specific primary antibody followed by a secondary antibody conjugated to biotin. Addition of avidin and biotinylated colour-producing enzyme in carefully chosen ratios results in multiple avidin-biotin-enzyme complexes binding to each secondary antibody (Figure 4.54).

The principal problem with the ABC method is that tissues such as liver and kidney contain significant amounts of endogenous biotin and result in extensive nonspecific background staining [33]. This effect can

be reduced through pretreatment methods involving applying avidin followed by biotin and then washing thoroughly but is difficult to eliminate (Box 4.11).

A number of variants on the ABC methods are used but are beyond the scope of this review. Similarly other amplification systems using dextran polymers or even nucleic acids are finding more common application in histopathology but have not yet replaced variants of the ABC technique in common practice.

4.22.3.2   Enzyme and chromagens

The enzyme most commonly used in histopathology is horseradish peroxidase (HRP), a haem-containing enzyme derived from the root of the horseradish. This catalyses the oxidation of the chromagen 3-3′diaminobenzidine (DAB) to form an insoluble dark brown precipitate that is resistant to washing with water or alcohol. Thus most immunohistochemically stained slides appear brown at the sites of antigen (Figure 4.55).

In addition to the HRP−DAB system there are several other enzyme-chromagen systems in use. The most common of these after HRP−DAB is the calf intestine alkaline phosphatase (AP) system that uses the enzyme's alkaline phosphatase activity on naphthol phosphate esters to produce phenols that react with diazonium salts to result in insoluble red azo dyes.

The HRP−DAB system is commercially widely available and entirely satisfactory under most circumstances. If however the antigen in question happens to occur in a cell that naturally produces brown pigment (such as a melanocyte in the skin) then an alternative chromagen technique such as the AP technique is highly useful in reserve.

4.22.3.3   Antigen retrieval

As discussed above, the process of fixation includes the formation of cross-linking bonds within and between molecules within the tissue. Unfortunately these bonds may prevent antibodies from accessing their corresponding epitopes, even though the antigen may be present in abundance. One possible solution is to perform immunohistochemical staining on fresh frozen tissue but this has numerous obvious limitations. The answer is therefore to develop some form of technique of limited effective reversal of the fixation process so that the required epitopes are re-exposed but the tissue does not begin to degenerate. This is termed antigen retrieval.

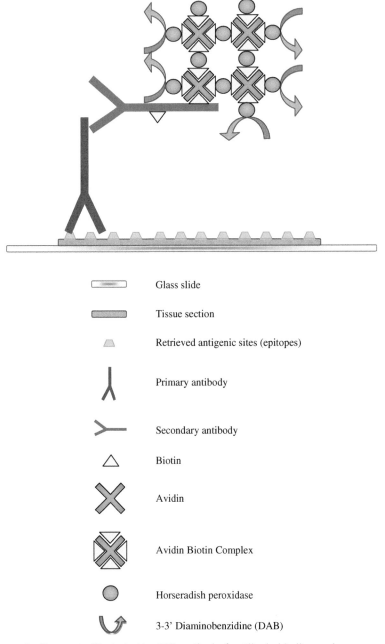

	Glass slide
	Tissue section
	Retrieved antigenic sites (epitopes)
	Primary antibody
	Secondary antibody
	Biotin
	Avidin
	Avidin Biotin Complex
	Horseradish peroxidase
	3-3' Diaminobenzidine (DAB)

Figure 4.54   A schematic diagram to illustrate the ABC method of antibody binding and enzyme-mediated chromagen generation

The principal approaches to antigen retrieval are enzyme digestion and heat:

1. Enzyme digestion is performed using enzymes commonly found in the digestive tract. The most frequently used are the proteolytic enzymes chy- motrypsin and trypsin, normally produced by the pancreas and secreted into the duodenum for the digestion of proteins. These enzymes cleave some of the long proteins and thus expose certain epitopes.

2. Various techniques using a combination of heat and varying pH are employed to break down

Immunohistochemical staining involves the use of relatively expensive reagents such as monoclonal antibodies. Commercially available purified avidin is surprisingly costly and, despite being used in relatively small quantities per test, the expense can mount up in a busy histopathology department.

Egg white is relatively rich in avidin. It transpires that there is sufficient avidin in egg white to allow its use in immunohistochemical staining. Unfortunately the lack of standardisation in egg white avidin concentration indicates that commercial purified avidin is more likely to produce the consistent and reproducible results required in diagnostic immunohistochemistry.

crosslinking bonds. These may be undertaken using pressure cooking or, increasingly commonly, microwaves [34].

The antigen retrieval technique used depends upon the nature of the epitope and antibody in question. In day-to-day practice therefore each of the antibodies used in the laboratory will have its optimised antigen

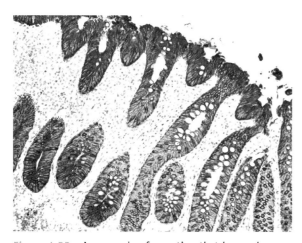

Figure 4.55   An example of a section that has undergone immunohistochemical staining using the ABC method with DAB as a chromogen. This is a section of large intestinal wall stained for a range of cytokeratins (intermediate protein filaments that are abundant in epithelial cells). The brown staining highlights the intestinal crypt epithelium. *Immunohistochemical staining for cytokeratins AE1/AE3 photographed at ×100*

retrieval procedure. It should be remembered however that some antigens do not require any retrieval stage at all.

4.22.4   Control tissues

Just as with histochemical/tinctorial staining procedures it is crucial that immunohistochemical staining is always performed with positive control tissues (ideally on the same slide for a more reliable test) and a negative control tissues is also desirable where time, space and cost allow (see above).

4.22.5   Immunofluorescence

The technique of immunofluorescence is in principle similar to that of immunohistochemistry in that a specific monoclonal or polyclonal antibody is used to identify an antigen on a histological section. Instead of an enzyme-mediated chemical reaction and colour change however immunofluorescence uses a light-emitting marker to indicate the position of the antibody. The most commonly used markers are derivatives of fluoroscein such as fluoroscein isothyocyante (FITC) that emit a yellow/green light when stimulated with light at the ultraviolet end of the spectrum.

As in immunohistochemical staining the marker may be attached to the primary antibody (direct immunofluorescence) or to a secondary antibody (indirect immunofluorescence).

4.22.5.1   Advantages and disadvantages

The principal advantage of direct, and to an extent, indirect immunofluorescence is that the degree of staining is quantitative. This means that tissues can be stained in which the antigen in question is normally present but where an increase in antigen deposition in a particular cellular or tissue location is being sought. On conventional immunohistochemical staining, the amplification processes would probably result in the whole tissue section becoming various shades of brown with little discernable difference in staining intensity in the site of interest. Because eyes are extremely sensitive to differences in intensity of light emission even the most subtle of increased fluorescence in one part of the cell may be visible.

The application of this property of the immuno-fluorescence technique is illustrated by its use in searching for autoantibodies in various tissues, most commonly the kidney. One particular disease of the kidney, for example, is termed IgA nephropathy. In this disease immunoglobulin A (IgA) binds to the basement membrane of the glomerulus (the filtration unit of the kidney) where its presence activates an inflammatory reaction, for example via the complement cascade (see acute inflammation), and this in turn damages the glomerulus. The form of the glomerular damage is not specific to this disease and so it is not possible to distinguish between IgA nephropathy and other glomerular diseases on H&E sections alone. The binding of IgA to the basement membrane however is more specific to this disease and so its presence is frequently diagnostic.

As IgA is always being produced in small amounts somewhere in the body it is effectively a normal low level circulating immunoglobulin and so its presence in the blood at low levels is not abnormal. The kidney is an extremely vascular organ and so some IgA is present throughout the tissue. This means that it is extremely difficult on immunohistochemical staining to demonstrate IgA bound to the glomerular basement membrane without the whole histological section of kidney becoming a mess of brown staining. The eye's extremely sensitive ability to detect contrast in light emission means however that it is often possible to distinguish between a background low level of light emission throughout the tissue and the linear pattern of increased light emission along a basement membrane.

The principal disadvantage of immunofluorescence is that fluoroscein derivatives are relatively unstable and so the immunofluorescent antibody-labelled slide will degrade relatively rapidly after its preparation, unlike the equivalent immunohistochemically stained slide. In addition, viewing immunofluorescent slides is inconvenient in that it requires a mercury lamp to produce the correct wavelength, a microscope fitted with filters to protect the eye from ultraviolet light and a dark room for slide viewing. Furthermore most commercially available antibodies are designed only to work on fresh frozen tissue rather than formalin-fixed, paraffin-embedded tissue. This is in part because formalin has some intrinsic fluorescent properties and clearly introduces the problems of handling of frozen tissue.

4.23   Cytopathology

As indicated earlier, cytopathology is the study of disease by the examination of individual cells rather than tissues. A variety of body fluids and aspirates can be obtained by the clinician in the process of investigating disease. Their subsequent treatment in the cytology laboratory depends upon the nature of the specimen, the site of origin and the disease under investigation (Box 4.12).

Box 4.12   The 'one-stop' clinic

The investigation of the visible or palpable lump frequently involves a three-step process. These steps are:

1. clinical assessment;
2. radiological assessment; and
3. tissue sampling and histological/cytological assessment.

Clinical assessment refers to obtaining information from the patient about the nature of the lump (e.g. how long it has been present, whether it appeared rapidly and whether it is painful) and is commonly performed by a general practitioner followed by a surgeon if required. Radiological assessment refers to any modality of imaging by a radiologist such as X-ray, ultrasound or magnetic resonance imaging. A characteristic clinical or radiological appearance may negate the need for further investigation.

Some medical centres and hospitals have developed a service such that the surgeon, radiologist and pathologist all see the patient during the same attendance. The use of fine needle aspiration, preparation of direct smears of the sample and rapid staining techniques (such as commercially available modifications of the giemsa technique) mean that a cytologist can potentially provide a cytological diagnosis within minutes. Thus the patient can have his or her lump assessed, diagnosed and a treatment plan agreed within the same clinic appointment. The 'one-stop' approach has commonly been applied to the investigation of breast lumps and is increasingly used in the context of head and neck lumps.

4.23.1   Specimen preparation

4.23.1.1   Wet preparation

Certain diseases are characterised by the precipitation of poorly soluble crystals. The most frequently received specimens that require the identification of crystals are joint aspirates. The purpose of these specimens is generally to distinguish between the metabolic disease gout and its various mimics.

Gout is a disease of purine metabolism that is characterised clinically by a destructive and extremely painful arthritis that is often accompanied by the formation of large nodules (called tophi) around the joints and on extensor surfaces of the body. The inflamed joints and tophi contain large quantities of poorly soluble crystals of uric acid that have a characteristic appearance under polarized light.

Wet preparations (as the name suggests) simply involve placing a coverslip over an unstained drop of the joint aspirate on a glass slide. The slide preparation is examined immediately under polarized light and the presence of crystals determined.

4.23.1.2   Direct smears

Specimens that consist of a tiny amount of material, material that is for immediate examination or fluid that is too viscous for realistic further preparation, can be spread directly across a glass slide. A degree of skill is required to ensure that the spread is as thin as possible to avoid the problem of cellular overlap that might obscure cellular detail on microscopic examination. The direct smear is subsequently fixed, stained and mounted for examination.

4.23.1.3   Cytospins (and variants)

Most cytology specimens have a relatively small cellular component compared to the amount of fluid. Direct smears of such samples would produce a lot of slides with sparse numbers of cells that would be enormously time consuming to prepare and examine. In order to increase the cellular component relative to the overall volume these samples are first spun in a centrifuge to separate the cellular component from the fluid, or supernatant. Almost all of the supernatant is discarded and 1 mL of supernatant is retained that is used to resuspend the cellular component. The resuspended cellular component is further concentrated into spots on glass slides that can then be stained. These are called cytospin preparations.

Modified centrifugation processes have been developed that form a cellular monolayer on a filter that is lightly pressed against the slide. These techniques (termed liquid-based cytology) have been automated and rolled out across the UK for use in the cervical screening programme.

4.23.2   Staining

The two most common staining techniques are Papanicolaou's stain (Pap) and various modifications of the Romanowski technique (various type of giemsa) (Figure 4.56).

These two staining techniques differ in that Pap staining is performed on samples that have been fixed when still wet and giemsa techniques are used on samples that have been allowed to dry on the slide prior to fixation. The former approach, in very general terms, preserves excellent nuclear detail whereas the latter allows the cytoplasm to spread out on the slide and so demonstrates cytoplasmic features more clearly. The standard H&E technique is also frequently applied to wet fixed specimens and allows a direct comparison between cytological and histological appearance for those pathologists more accustomed to interpreting histology specimens.

4.23.3   Immunocytochemical staining

Most commercial immunohistochemical staining techniques have been developed and marketed for use on tissue sections. Many of these antibodies can however be used to stain cytological preparation, although a significant amount of experimentation may be required to produce satisfactory results.

The unpredictability of applying immunohistochemical techniques to cytology specimens can be reduced by solidifying part of the cytology specimen (e.g. with agar) and then treating the solid specimen as a tissue specimen.

4.24   Electron microscopy

Light microscopy will normally provide sufficient morphologic detail of tissues and cells to diagnose the

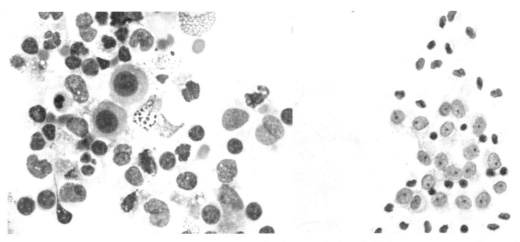

Figure 4.56   Examples of standard cytological stains. The giemsa technique (left) demonstrates cytoplasm and is illustrated by the pair of mesothelial cells to the left of centre showing well-defined blue cytoplasm but indistinct nuclear features. The sheet of mesothelial cells stained with the Papanicolaou technique (right) demonstrated more nuclear detail but the cytoplasm is indistinct. *Left: rapid giemsap Photographed at ×600; right: Papanicolaou photographed at ×400*

vast majority of pathological conditions if supplemented by histochemical, immunohistochemical and immunofluorescence techniques. In terms of morphology however light microscopy has the problem of resolution as its main limitation.

Resolution is the minimum distance between two points that still allows the two points to be recognised as separate from each other when viewed down the microscope. If the points are separate by less than this distance then they will appear to be joined. The greater the resolution power of a microscope, the smaller will be the objects that can be viewed with clarity.

The resolution limit of a microscope depends upon several factors including the quality of the lenses and the set-up of the light source. Ultimately however the very best quality light microscope's resolution is limited by the nature of light itself or, more specifically, the wavelength. The wavelength of light in the visible spectrum is such that resolution much below 1 μm is not technically possible. This means that most cellular organelles are not generally visible in isolation. Resolution can be slightly improved by using a light source at the higher frequency (shorter wavelength) end of the spectrum but this then begins to introduce technical complications to the staining techniques employed.

The practical manifestation of the resolution limit of the light microscope is the maximum magnification that can be obtained. The average histopathology laboratory microscope will achieve magnifications of approximately ×400. This can be improved to approximately ×600 using high quality lenses and oil immersion of both specimen and lens (which reduces light

diffraction). For higher magnifications we must find a source of waves of much shorter wavelength.

A beam of electrons moving at speed through a vacuum has a wavelength several orders of magnitude shorter than visible light. The theoretical resolution of an electron beam is therefore sufficiently small to be able to produce magnifications not only great enough to view cell organelles but even to show the component membranes. Examination of tissues at this magnification would therefore add a new range of histological detail for classifying and investigating pathological processes.

4.24.1   The mechanism of electron microscopy

The electron microscope essentially shares its central design with a cathode ray oscilloscope or cathode ray television (Figure 4.57).

A heated filament (e.g. tungsten wire, similar to a light bulb filament and called the cathode) in a vacuum can be made to emit a beam of electrons towards a cylindrical tube (anode) kept at very high voltage to the cathode. Those electrons passing straight through the anode form a beam which can be deflected or attracted by electrostatic charges and magnetic fields. This beam can be made to pass through a tissue section and the slight deviations in the paths of the component electrons caused by interaction with molecules in the tissue used to form an image on a phosphorescent screen (which produces visible light when struck by electron beams) which can be recorded digitally or on

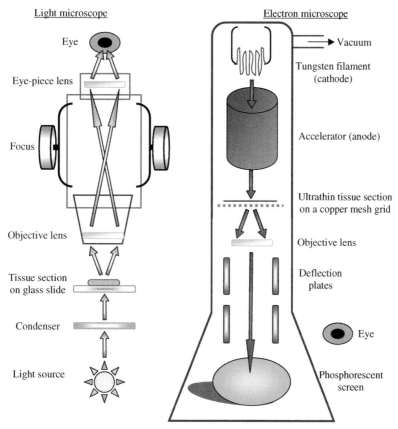

Light microscope

Figure 4.57   The essential components of the electron microscope. Many of the same components are seen in a cathode ray tube television

photographic film. The degree of beam deviation caused by the tissue can be amplified in an analogous fashion to magnification on a light microscope.

The above description is termed transmission electron microscopy. A variant is termed scanning electron microscopy that effectively bounces an angled beam of electrons off the surface of the specimen to form an image. This is seldom used in histological practice.

4.24.1.1   Tissue preparation

It is important to remember that although electron microscopy (EM) involves analogous stages to light microscopy and produces a visible two-dimensional image, the properties of light and an electron beam inside a vacuum are different. The various stages of tissue preparation are therefore different.

Tissue for EM is not generally fixed in formalin because, although entirely adequate for light microscopy, formalin introduces artefacts that become visible at the magnifications used in EM. The most commonly used fixative is glutaraldehyde that introduces fewer artefacts and reacts more quickly with tissues than formalin but is more toxic and makes antigen retrieval far more difficult for general use in histopathology. To assist in rapid fixative penetration and even fixation, tissue for EM is cut into specimens of size of the order of up to about 1 mm in maximum dimension.

As preservation of the fatty acid content of light microscopy specimens is generally unimportant it is of minimal consequence that fat is little affected by fixation and therefore washed out of the specimen during processing. The chief fatty acid component in the cell is generally membranes and, as these cannot be seen on light microscopy, it makes little difference to the appearance of the tissue section if they are washed out. The preservation of fatty acids in such structures as membranes however is important in EM because these structures can be visualised. A stage of fatty acid stabilisation is therefore required and this is most commonly undertaken using osmium tetroxide. The tissue is then dehydrated using an increasing series of alcohol concentrations.

The tissue for EM requires even thinner sections than for light microscopy. Thus the tissue requires an embedding medium that is sufficiently strong to be able to stay relatively rigid in section. Various acrylic or epoxy resins are used and require the tissue to be treated with an appropriate agent (such as propylene oxide) that may dissolve resin in an analogous fashion to treating tissue with xylene prior to embedding in wax.

4.24.1.2   Microtomy and imaging

Cutting sections requires an extremely sharp blade in order to achieve the desired levels of thickness. A 'semithin' section of approximately 0.6 μm in thickness is cut first using a glass knife and stained with Toluidine blue. This is examined under light microscopy for orientation of the electron micrographs. The section for examination is then cut at a thickness of the order of 0.08 μm using a glass or a diamond knife. These 'ultrathin' sections are placed on a copper grid and may be further stained with heavy metal salts (such as uranyl acetate and lead citrate) prior to being examined under the electron microscope. Heavy metals (i.e. metals of high atomic weight) block the path of electrons and so act in a similar fashion to the light-absorbing pigments in light microscopy tinctorial staining. Different cell components will take up the metal salts in varying concentrations and the greater the uptake the more electrons will be absorbed/reflected while those areas that are more lightly stained will allow more electrons to pass through the tissue section (Figure 4.58).

In addition to providing tissue images at extremely high resolution it is possible to perform some ancillary investigations. Immunostaining is possible, for example, by conjugating primary or secondary antibodies to gold microspheres that can be visualised under the electron microscope [35]. The use of antibodies conjugated to gold microspheres of different fixed sizes means that several antibodies can be applied to the same section and their relative staining distributions determined by sphere size.

4.24.2   Applications of electron microscopy

Although electron microscopy has a potentially wide range of applications within the histopathology laboratory that merge into research use, we will consider

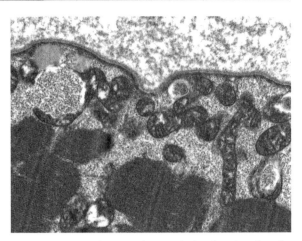

Figure 4.58   An electron micrograph showing the edge of a skeletal muscle fibre. The exatracellular space is visualised at the top of the picture with underlying cell membrane. Multiple mitochondria are seen beneath the membrane, beneath which are the striated contractrile fibres. *photographed at ×4200*

three of the principal diagnostic applications as follows.

4.24.2.1   Glomerular disease

The glomeruli are essentially the filtration units of the kidneys. They are subject to a number of diseases that can manifest as disturbances of renal function and blood pressure and it may become necessary during the process of investigation of renal disease to perform a renal needle core biopsy. In most instances the constellation of clinical features, light microscopic appearance, histochemical/tinctorial staining pattern and immunofluorescence pattern will provide sufficient information to establish a diagnosis. There are occasions however when the light microscopic changes are non-existent, too subtle or the disease too early in its course to allow diagnosis [36]. Fortunately the electron microscopic appearance of the basement membrane of the glomerulus can often indicate the earliest pathological changes and assist in diagnosis.

4.24.2.2   Diseases of skeletal muscle

Certain rare diseases of the skeletal muscles (called myopathies) have characteristic electron microscopic appearances that may not be identifiable on light microscopy [37].

4.24.2.3 The unknown tumour

Although the overwhelming majority of tumours can be characterised using light microscopy with its various ancillary techniques to a sufficient degree to allow treatment, there are occasions on which there are no clues as to what a tumour is differentiating towards. Electron microscopic analysis of the tumour cells may be helpful in that it may be possible to identify ultrastructural clues or characteristic cytoplasmic structures to suggest a direction of differentiation.

It should be recognised that although the range of potential uses of EM is wide, the proportion of cases requiring EM in the general diagnostic histopathology laboratory (the majority of which do not perform renal or muscle analysis) is extremely small. The facilities for EM are therefore generally only available in reference laboratories and the larger tertiary referral centres.

4.25 *In situ* Hybridisation

In situ hybridisation (ISH) is a DNA (or RNA, but for the purposes of this description we will limit the discussion to DNA) identification technique that combines three essential components:

1. identifying which of the cells in the tissue to which the test is applied contain the particular DNA sequence;
2. giving an indication of the number of copies of that particular sequence there are in the DNA of the cell; and
3. indicating whereabouts in the DNA of a particular cell that sequence is located relative to another marked DNA sequence.

Whereas the detection of specific DNA sequences using older techniques had relied on destroying the tissue to extract DNA and then perhaps chopping the DNA into short segments, ISH actually labels the DNA sequence in such a way that not only can one see which cells within a particular tissue contain that DNA sequence but one can gain an indication as to how much of that sequence is found in the DNA of each cell.

4.25.1 Mechanism

We recall that DNA is composed of two strands containing arrays of complementary nucleotide base pairs.

A highly specific way to identify a particular DNA sequence is therefore to attach a distinctive label to a single short strand of DNA with the complementary base sequence to that in question (called a probe), uncoil and separate the DNA being tested, and allow the probe to bind (or hybridise) to the test DNA. If the segment of DNA being sought is present in the test DNA sample then the probe will bind with high affinity and specificity (Figure 4.59(a)).

Although the concept is relatively straightforward, the reality is somewhat complex and problematic. The design of the probe itself is the first major stumbling block. The DNA sequence of interest is usually a particular gene. The probe thus needs to be a complementary sequence to part of that gene. The shorter the probe, the greater the risk that an identical or very similar stretch of DNA will appear in other genes with the result that the probe will bind to the wrong gene. Having a very long probe reduces this risk but introduces problems of expense and the technical difficulties of poor tissue penetration. In practice probes are commonly of the order of 40–50 basepairs in length and are most commonly constructed using RNA rather than DNA.

Labelling the probe for subsequent illustration of the position of its binding produces the same problems and shares some of the same solutions as immunohistochemistry (see Section 4.22). Radioactive isotopes are commonly used in research applications whereas an alternative technique of attaching a specific antigen (commonly digoxigenin) allows the use of the same process of immunohistochemical staining described above. The technique most commonly encountered in the diagnostic histopathology laboratory is to label using a fluorescent marker. This technique is termed fluorescent *in situ* hybridisation (FISH).

The use of ISH suffers some analogous problems to that of immunohistochemistry in that the processes of tissue formalin fixation and paraffin embedding introduce technical drawbacks. The cross-linking of proteins caused by fixation hinders probes in accessing DNA with the result that a number of pretreatment stages are required in an analogous fashion to antigen retrieval.

4.25.2 Applications

Although there are many applications of FISH in the laboratory, the most common applications in the diagnostic histopathology laboratory fall into two main groups as follows.

1. *Detection of increased gene copy numbers.* Some diseases are associated with an abnormally large number of copies of a single gene within the nucleus. In current diagnostic use the detection of gene copy numbers is routinely applied to the investigation of breast cancer [38]. An aggressive variant of breast cancer overexpresses the protein Her2/neu (also known as Erb-B2) and this over-expression is most commonly associated with an

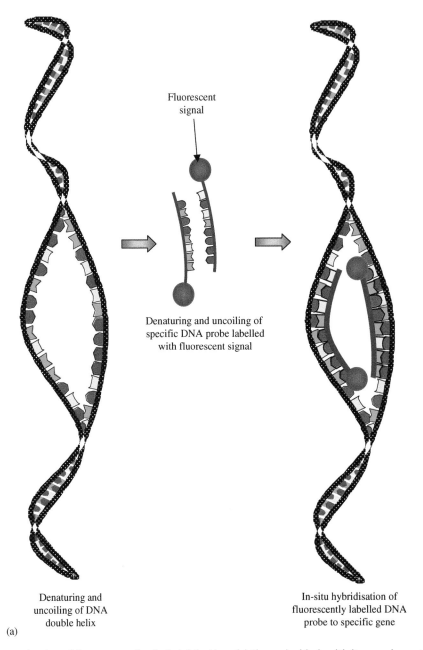

Fluorescent
signal

Denaturing and uncoiling of
specific DNA probe labelled
with fluorescent signal

Denaturing and
uncoiling of DNA
double helix

(a)

In-situ hybridisation of
fluorescently labelled DNA
probe to specific gene

Figure 4.59   The mechanism of fluorescence *in situ* hybridization: (a) the probe binds with its complementary segment of DNA; (b) the translocation between chromosomes 9 and 22 results in the Philadelphia chromosome. The use of red (chromosome 9) and green (chromosome 22) probes in the vicinity of the breakpoint demonstrates colour overlap (resulting in yellow fluorescence) in the Philadelphia chromosome

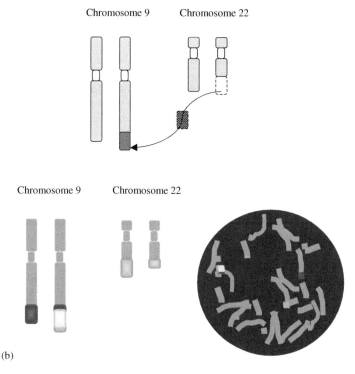

Chromosome 9   Chromosome 22

Chromosome 9   Chromosome 22

(b)

Figure 4.59   (*Continued*)

increase in the number of copies of the corresponding gene (see Box 4.7). These are sometimes present in the form of small chromosome-like segments of DNA termed double minutes. The detection of increased fluorescence and an abnormal number of sites of fluorescence using Her2 probes is a good indicator of an aggressive variant of breast cancer and a potentially good response to the highly specific drug Trastuzumab.

2. *Detection of translocations.* FISH has the advantage that several different probes and colours of fluorescent marker can be used simultaneously. Some diseases are associated with predictable chromosomal translocations, for example the overwhelming majority of cases of chronic myelogenous leukaemia (CML) are associated with a translocation that brings together the bcr gene on chromosome 22 and the abl gene on chromosome 9 to form a fusion gene (and a characteristic new chimeric chromosome called the Philadelphia chromosome). If probes are applied which label the vicinity of the bcr gene one colour and the abl gene another then normal cells will contain distinctly separate coloured

dots within the nucleus and cells that have undergone the translocation will produce dots of different colours adjacent to each other with some colour overlap [39] (Figure 4.59(b) and (c)).

At present the relatively limited range of uses in diagnostic histopathology and the expense of staffing and maintaining this specialised technique mean that it is currently centralised to a relatively small number of reference laboratories. With an increasing caseload requiring FISH (especially in the field of breast pathology) and an increasing number of highly targeted drug therapies there will inevitably be an expansion of the applications of FISH in common diagnostic use.

Bibliography

1. Kerr, J.F., Wyllie, A.H. and Currie, A.R. (1972) Apoptosis: a basic biological phenomenon with wide-ranging implications in tissue kinetics. *Br. J Cancer*, **26**, 239–57.
2. King, L.B. and Ashwell, J.D. (1993) Signaling for death of lymphoid cells. *Curr. Opin. Immunol.*, **5**, 368–73.

3. Vaux, D.L. (1993) Toward an understanding of the molecular mechanisms of physiological cell death. *Proc. Natl Acad. Sci. USA*, **90**, 786–9.

4. Levine, A.J., Momand, J. and Finlay, C.A. (1991) The p53 tumour suppressor gene. *Nature*, **351**, 453–6.

5. Ratnoff, O.D. (1974) Some recent advances in the study of hemostasis. *Circ Res.*, **35**, 1–14.

6. Nieuwenhuis, H.K., Akkerman, J.W., Houdijk, W.P. and Sixma, J.J. (1985) Human blood platelets showing no response to collagen fail to express surface glycoprotein Ia. *Nature*, **318**, 470–2.

7. Ryan, G.B. and Majno, G. (1977) Acute inflammation. A review. *Am. J. Pathol.*, **86**, 183–276.

8. Maxwell, P.H. and Ratcliffe, P.J. (2002) Oxygen sensors and angiogenesis. *Semin. Cell Dev. Biol.*, **13**, 29–37.

9. Raghow, R. (1991) Role of transforming growth factor-beta in repair and fibrosis. *Chest*, **99**, 61S–65S.

10. Woessner, J.F. Jr (1991) Matrix metalloproteinases and their inhibitors in connective tissue remodeling. *FASEB J.*, **5**, 2145–54.

11. Pinnell, S.R. (1985) Regulation of collagen biosynthesis by ascorbic acid: a review. *Yale J. Biol. Med.*, **58**, 553–9.

12. Heron, M.P. *et al.* (2009) Deaths: Final data for 2006. *National vital statistics reports*, Vol. **57**, *No. 14*. National Center for Health Statistics, Hyattsville, Maryland, USA.

13. Solberg, L.A. and Strong, J.P. (1983) Risk factors and atherosclerotic lesions. A review of autopsy studies. *Arteriosclerosis*, **3**, 187–98.

14. Beckman, J.A., Creager, M.A. and Libby, P. (2002) Diabetes and atherosclerosis: epidemiology, pathophysiology, and management. *JAMA*, **287**, 2570–81.

15. Ross, R., Glomset, J. and Harker, L. (1977) Response to injury and atherogenesis. *Am. J. Pathol.*, **86**, 675–84.

16. Virchow, R.L.K. (1856) Gesammelte Abhandlungen zur Wissenschaftlichen Medicin. Meidinger Sohn & Co., Frankfurt, in Virchow, R.L.K. (1998) Thrombosis and emboli (1846–1856) translated by Matzdorff, A.C. and Bell, W.R. Canton, Science History Publications, 5–11, 110.

17. Pasqualetti, P. and Casale, R. (1997) Risk of malignant transformation in patients with monoclonal gammopathy of undetermined significance. *Biomed. Pharmacother.*, **51**, 74–8.

18. Scheffner, M. *et al.* (1990) The E6 oncoprotein encoded by human papillomavirus types 16 and 18 promotes the degradation of p53. *Cell*, **63**, 1129–36.

19. Dyson, N., Howley, P.M., Münger, K. and Harlow, E. (1989) The human papilloma virus-16 E7 oncoprotein is able to bind to the retinoblastoma gene product. *Science*, **243**, 934–7.

20. Dye, C. *et al.* (1999) Global burden of tuberculosis. Estimated incidence, prevalence, and mortality by country. *JAMA*, **282**, 677–86.

21. Parsonnet, J. *et al.* (1991) Helicobacter pylori infection and the risk of gastric carcinoma. *N. Engl. J. Med.*, **325**, 1127–31.

22. Wotherspoon, A.C., Ortiz-Hidalgo, C., Falzon, M.R. and Isaacson, P.G. (1991) Helicobacter pylori-associated gastritis and primary B-cell gastric lymphoma. *Lancet*, **338**, 1175–6.

23. Mims, C. *et al.* (1993) *Medical Microbiology*, 1st edn, pp. 17–18, Mosby, London.

24. Pardee, A.B. (1974) A restriction point for control of normal animal cell proliferation. *Proc. Natl Acad. Sci. USA*, **71**, 1286–90.

25. Elledge, S.J. (1996) Cell cycle checkpoints: preventing an identity crisis. *Science*, **274**, 1664–72.

26. Levine, A.J. *et al.* (1994) The 1993 Walter Hubert Lecture: the role of the p53 tumour-suppressor gene in tumorigenesis. *Br. J. Cancer*, **69**, 409–416.

27. Wiman, K.G. (1993) The retinoblastoma gene: role in cell cycle control and cell differentiation. *FASEB J.*, **7**, 841–5.

28. UK Health and Safety Executive COSHH guidance is available at http://www.coshh-essentials.org.uk/

29. Bancroft, D. and Gamble, M. (2007) *Theory and practice of histological techniques*, 6th edn, Churchill Livingstone, London.

30. Roitt, I., Brostoff, J. and Male, D. (1998) *Immunology*, 4th edn, Mosby, London.

31. Köhler, G. and Milstein, C. (1975) Continuous cultures of fused cells secreting antibody of predefined specificity. *Nature*, **256**, 495–7.

32. Dodson, A. (2002) Modern methods for diagnostic immunocytochemistry. *Curr. Diag. Path.*, **8**, 113–22.

33. Mount, S.L. and Cooper, K. (2001) Beware of biotin: a source of false-positive immunohistochemistry. *Curr. Diag. Path.*, **7**, 161–7.

34. Jasani, B. and Rhodes, A. (2001) The role and mechanism of high-temperature antigen retrieval in diagnostic pathology. *Curr. Diag. Path.*, **7**, 153–60.

35. Wall, J., Langmore, J., Isaacson, M. and Crewe, A.V. (1981) Colloidal gold: a cytochemical marker for light and fluorescent microscopy and for transmission and scanning electron microscopy. *Scan. Electron Microsc.*, Pt 2: 9–31.

36. Shore, I. and Moss, J. (2002) Electronmicroscopy in diagnostic renal pathology. *Curr. Diag. Path.*, **8**, 207–15.

37. Sewry, C. (2002) Electronmicroscopy of human skeletal muscle: role in diagnosis. *Curr. Diag. Path.*, **8**, 225–31.

38. Mitchell, M.S. and Press, M.F. (1999) The role of immunohistochemistry and fluorescence in situ hybridization for HER2/neu in assessing the prognosis of breast cancer. *Semin. Oncol.*, **26**, 108–16.

39. Amiel, A. *et al.* (1993) Clinical detection of BCR-abl fusion by in situ hybridization in chronic myelogenous leukemia. *Cancer Genet. Cytogenet.*, **65**, 32–4.

Chapter 5
Clinical chemistry

**Professor Ray K. Iles, B.Sc., M.Sc., Ph.D., CBiol, FSB, FRSC
and Dr Stephen A. Butler, B.Sc., Ph.D.**

Introduction

In the broadest sense clinical chemistry is the area of pathology that is concerned with analysis of bodily fluids. Since boundaries are always blurred (and blood is by definition a bodily fluid) many hospitals link their clinical chemistry and haematology sections. However, a more differential definition is that clinical chemistry (or clinical biochemistry in some hospitals) is the analysis of the soluble noncellular components of bodily fluid. These components vary from small inorganic compounds – such as salts and ions – to larger organic compounds – such as lipids and steroids and even drugs – to macromolecules such as albumin, enzymes and protein hormones.

Part I: Analytical methods

The biochemical tests used in a clinical chemistry laboratory are designed to accurately quantify these specific components. In order to reduce variation in laboratory measurement to an absolute minimum, standard operating procedures (SOPs) and rigid quality control testing regimes are enforced in all tests conducted.

5.1 Sample collection

The principal biological fluid examined in clinical chemistry is blood serum as distinct from blood plasma which still contains clotting factors including fibrinogen. However, urine is also collected for analysis and is usually a 24 h collection in a plastic container supplied from clinical chemistry and containing an antibacterial agent. Early morning urine (EMU) is often collected as it is considered to be more concentrated whilst midstream urine (MSU) is often collected for microbiological purposes.

5.1.1 Blood

In general blood, if not sampled by medical staff on the ward or the accident and emergency department, will be sampled at a clinical/haematology blood clinic collected by a qualified phlebotomist and many biomedical scientists are also phlebotomy trained.

Samples are collected in tubes appropriate to the test required. For most analytes to be measured by immunoassay serum is the optimal fluid and the blood is simply allowed to clot in a plain tube thereby allowing the cellular components fibrinogen and clotting factors

Biomedical Sciences: Essential Laboratory Medicine, First Edition. Edited by Ray K. Iles and Suzanne M. Docherty.
© 2012 John Wiley & Sons, Ltd. Published 2012 by John Wiley & Sons, Ltd.

to coagulate leaving the clotting factor-free yellow/straw serum on top and a retracting clot of blood cells on the bottom. Centrifugation simply helps maximize the volume of serum harvested. However, vigorous centrifugation is not recommended as this can lyse some of the red cells releasing soluble haemoglobin that will give a red contamination to the serum sample. This can interfere with immunoassay results, as can not allowing the clot to fully form (usually 1 h on the bench at room temperature is sufficient). Harvested before a clot has properly formed results in a serum that is viscous due to fibrinogen partial polymerization and a sample that can artificially precipitate antibodies or labelled tracers in an immunoassay system.

However, plasma is needed when measuring the levels of clotting factors in certain haematological disorders such as haemophilia and von Leiden factor 8 deficiency (see Chapter 8, Section 8.8.5), or indeed if the unclotted cellular components are to be analysed by a Coulter or flowcytometer (see Chapter 8, Section 8.2.1). In order to minimize the total amount of blood collected from a patient the sample will be collected in a tube containing an anticoagulant so that the spin down cellular component can be analysed by haematology and the plasma sent for simultaneous analysing for common clinical chemistry analytes. Clotting can be prevented by the presence of sodium citrate, heparin or EDTA in the blood collecting tube. However, the anticoagulant used will determine which range of simultaneous test can be performed. Heparin irreversibly blocks clotting in such samples and so is no good if a clotting test is needed. EDTA chelates ions and, although reversible by addition of excess magnesium and calcium, will compromise plasma analysis for certain enzymes. Clotting can also effect the measurement of blood glucose and adrenaline/catecholamine so plasma is often collected instead of serum in

certain circumstances for specific soluble analytes (see Table 5.1). In most hospitals blood collection is via Vacutainer tubes and the leading manufacturer is Becton Dickinson. The BD Vacutainer blood collection system is described as a closed evacuated system, which consists of a double-ended needle with a safety valve. Essentially, the BD Vacutainer holder is a plastic sleeve into which a double-ended needle assembly screws. Blood is collected by screwing the sleeve-covered end of the needle into the holder, then puncturing the patients' vein with the other end. After performing venepuncture the tube is then pushed down into the holder, and the premeasured vacuum of the tube allows the required volume of blood to be drawn (see Figure 5.1). The tubes are colour coded according to the appropriate additive and international standards (ISO 6710, also see Table 5.1).

5.1.2 Urine

After blood, urine is the second most commonly analysed biological fluid (with CSF, saliva, ascites, follicular fluid and inflammatory exudates much more rarely sampled). The dynamics of this fluid are considerably different from most blood analysis. In the majority (60—80%) of blood-based tests the analytes are expected to be maintained within a relatively narrow concentration band in healthy individuals — to be kept in homeostasis — but significantly altered by disease processes. Urine, however, is an excretory media in which the concentration of analytes within any given sample varies dramatically depending on water volume intake, food or drug consumption and the rate of metabolism of molecules. All of these confounding factors vary throughout the day and, in the case of metabolism, can be up or down as regulated

Figure 5.1   Example of Vacutainer venepuncture and the various colour coded Vacutainer tubes with appropriate anticoagulants added

Table 5.1   Collection tube and associated anticoagulant (and colour) for clinical laboratory test.

Investigation	Lithium heparin (green)	EDTA (pink)	Fluoride oxalate (grey)	Sodium citrate (blue)	Clotted (red)
Full blood count		✓			
U&Es	✓				
Group and save/crossmatch		✓			
Clotting				✓	
Liver function tests	✓				
Paediatric renal profile	✓				
Bone profile	✓				
Thyroid function tests	✓				
Alkaline phosphatase	✓				
Aldosterone		✓			
17-alpha OHP	✓				
3-Hydroxybutyrate			✓		
Angiotensin converting enzyme	✓				
ACTH	✓				
Alpha fetoprotein	✓				
Albumin	✓				
Alcohol	✓				
Alkaline phosphatase isoenzymes	✓				
Ammonia	✓				
Amylase	✓				
Anti neutrophil cytoplasmic antibody	✓				
Blood group antibody screen		✓			
AST	✓				
Antithrombin III				✓	
Anti Xa assay				✓	
APTT				✓	
Autoantibody screen (ANA, liver AI screen)	✓				
Autoantibody screen (adrenal, islet cell, salivary gland)	✓				
Vitamin B12		✓			
Folate		✓			
Bicarbonate	✓				
Bilirubin (direct and conjugated)	✓				
Biotinidase	✓				
Blood film		✓			
C3/C4	✓				
Caffeine	✓				
Carbamazepine	✓				
Calcium	✓				
Chloride	✓				
Cholesterol	✓				
Cholesterol HDL	✓				

(*Continued*)

Table 5.1 (*Continued*)

Investigation	Lithium heparin (green)	EDTA (pink)	Fluoride oxalate (grey)	Sodium citrate (blue)	Clotted (red)
Cholinesterase	✓				
Creatinine kinase	✓				
Coeliac screen (TTG)	✓	✓			
Cold agglutinin screen		✓			
Copper					✓
Complement	✓				
Cortisol	✓				
C-Peptide	✓				
Creatinine	✓				
CRP	✓				
Cryoglobulins					
Cyclosporin		✓			
D-Dimer	✓			✓	
DHEAS	✓				
Digoxin	✓				
Direct Coombs test (DCT)	✓	✓			
DsDNA antibodies	✓				
ESR		✓			
Ferritin	✓				
Fibrinogen				✓	
Tacrolimus levels		✓			
Free fatty acids			✓		
Free T3	✓				
Free T4	✓				
FSH	✓				
G6PD		✓			
Gal-1-PUT	✓				
Gamma GT	✓				
Gastrin	✓				
Glandular fever screen		✓			
Glomerular basement membrane (GBM) antibody	✓				
Glucose			✓		
Haemoglobinopathy screen		✓			
Haemoglobin S quantitation		✓			
Haemolysin (ABO HDNB) Screen	✓				
Factor v Leiden				✓	
Haptoglobin	✓				
HbA1C		✓			
Human chorionic gonadotrophin (HCG)	✓				
HLA antibody screen	✓				
Homocysteine		✓			

Table 5.1 (*Continued*)

Investigation	Lithium heparin (green)	EDTA (pink)	Fluoride oxalate (grey)	Sodium citrate (blue)	Clotted (red)
Iron binding capacity	✓				
IgA	✓				
IGF-1	✓				
IgG	✓				
IgM	✓				
Immunoglobulins	✓				
Insulin					
Iron	✓				
Lactate			✓		
LDH	✓				
LDH isoenzymes	✓				
LH	✓				
Lipoprotein electrophoresis	✓				
Lithium	✓				
Magnesium	✓				
Malaria parasite screen		✓			
Methotrexate	✓				
Pregnancy test (blood)	✓				
Oestradiol	✓				
Osmolality	✓				
Paracetamol	✓				
Phenyalanine	✓				
Phenytoin	✓				
Phosphate	✓				
Plasma amino acids	✓				
Progesterone	✓				
Prolactin	✓				
Protein (total)	✓				
Protein C level				✓	
Protein S level				✓	
Prothrombin time				✓	
Parathyroid hormone (PTH)		✓			
Renin		✓			
Reticulocyte count	✓				
Rheumatoid factor (RF)	✓				
Salicylate	✓				
Sex hormone binding globulin	✓				
Sickle screen	✓				
Suxamethonium sensitivity	✓				
Testosterone	✓				
Theophylline	✓				

(*Continued*)

Table 5.1 (*Continued*)

Investigation	Lithium heparin (green)	EDTA (pink)	Fluoride oxalate (grey)	Sodium citrate (blue)	Clotted (red)
Thyroid antibodies	✓				
Tissue typing		✓			
Transferrin	✓				
Transfusion reaction					
Triglyceride	✓				
Troponin I	✓				
Tyrosine	✓				
Urate	✓				
Valproate	✓				
Vitamin A	✓				
Vitamin E	✓				
White cell cystine	✓				
White cell enzymes		✓			

by prior exposure. For clinical chemistry purposes the concentration of urine (90% of urinary analytes) has to be corrected against a common factor such as time (e.g. 24 h total excretion/collection) or a known factor such as creatinine (creatinine correction). The only real exception to this rule is human chorionic gonadotropin (hCG), the mere detection of which in a urine sample will diagnose either pregnancy, malignancy or even exogenous doping abuse in athletes.

Creatinine correction is based on an assumption that creatinine excretion is relatively constant by healthy humans and compared to all other urinary excretory product it is. However, there is a significant variation and the normal range for men is 9–18 mmol/24 h and 7–16 mmol/24 h for women.

5.1.3 Labelling and sample management

Perhaps the biggest cause of error in clinical sample analysis is incorrect labelling of the sample. Particular care must be taken in reception and logging samples. Patients' full name and hospital reference numbers must be recorded as the consequence of mixing up the test results of a patient with the same or similar name can be so serious that it leads to inappropriate treatment and patient death. Most clinical samples are now barcoded so that individual patient identification is

never guessed at – if in doubt about similar names the codes have to match that of the patient.

5.2 Analytical methods in clinical chemistry laboratories

Automated analysis is common in many laboratories and will often link to sample-handling systems which will take the blood sample, directly separate serum, plasma and cellular components and then send the separate sample fluids through to the appropriate analysis system. Often these systems will introduce a dilution and aliquoting step in the sample processing allowing multiple analyses to be carried out on a single blood or other biological sample (see Figure 5.2).

The vast majority of chemical analysis systems are optics-based in which a chemical feature or reaction coupled to a chromatophore, is an optical measure proportional to the analytes concentration, that is light emission, scattering, absorption, fluorescence and luminescence. Since these light measures are detected by photomultiplier tubes the signal is rendered electronic and therefore easily processed by computer systems to generate reportable clinical values with inbuilt quality control checks. The variety of optics-based systems used to measure clinical chemistry analytes are detailed below.

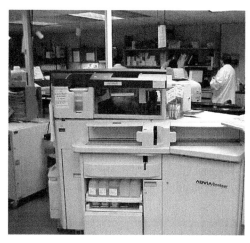

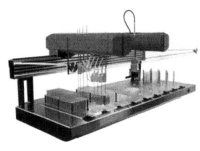

Figure 5.2 Example of an autoanlyser in a clinical chemistry laboratory and a bench top sample dilution and distributor system

5.2.1 Flame photometry

Flame photometry works on the principles of flame atomic emission spectrometry. A spectral scan of the visible portion of the electromagnetic spectrum produced when metals are excited in a flame produces highly intense lines of emission at particular wavelengths (see Figure 5.3). The intensity of emission at these characteristic wavelengths is proportional to the amount of sample present in the fluid. However, there is no need to produce a complex spectral analysis as the combined spectral emission will change the source flame a characteristic colour, for example potassium – violet (lilac); sodium – golden yellow; calcium – brick red.

In its simplest form flame photometry is a quick, economical and simple way of detecting and quantifying traces of metal ions. Simply spray the fluid containing the metal through a flame and measure the light intensity emitted through an appropriate colour filter (see Figure 5.4). The instrument is calibrated by running standards of known concentration and constructing a calibration curve from which the running of a sample of unknown can be read against giving the value of the analyte in the sample (see WHO standard operating procedures flame photometric assay). In clinical chemistry primarily sodium and potassium are measured by flame photometry although lithium, calcium and barium can also be measured in a biological fluid by this method.

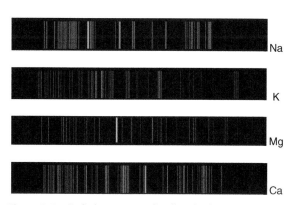

Figure 5.3 Emission spectra of sodium (Na), potassium (K), magnesium (Mg) and calcium (Ca)

Figure 5.4 Example of a simple flame photometer

WHO standard operation procedure for clinical flame photometric assay of Potassium and Sodium

Reagents - All chemicals must be Analar grade.

Sodium chloride (NaCI) and potassium chloride (KCI) should be dried for 2–3 hours at about 100°C before use. Before weighing, the chemicals must be allowed to cool to room temperature either in a desiccators or in a container with a tight-fitting lid with a small air space.

- **Stock Sodium 1000 mmol/L** — Weigh out 29.25 g dried NaCl, dissolve in about 400 ml of distilled water taken in a 500 ml volumetric flask and then make up to 500 ml with distilled water. Store in a pyrex glass bottle at 25–35°C. Stable for one year.
- **Stock Potassium 100 mmol/L** — Weigh out 0.746 g dried KCI, dissolve in about 80 ml of distilled water taken in a 100 ml volumetric flask and then make up to 100 ml with distilled water. Store in a pyrex glass bottle at 25–35°C. Stable for one year.

- **Working Standards** —
  - Low standard for Sodium 100 mmol/L: Dilute 10 ml of stock sodium to 100 ml with distilled water. Stable for 6 months at 25–35°C.
  - Combined standard for sodium and potassium $140Na^+/5K^+$ mmol/L: Dilute 14 ml of stock sodium and 5 ml of stock potassium together to 100 ml with distilled water. Store in a pyrex bottle at 25–35°C. Stable for 6 months.
  - Aspiration standard for sodium - 1.0 mmol/L: Dilute 1.0 ml of working standard to 100 ml with distilled water. Prepare fresh each time.
  - Combined aspiration standard for sodium 1.4 mmol/L and potassium 0.05 mmol/L. Dilute 1.0 ml of working standard (combined standard for Na^+/K^+ 140/5 mmol/L) to 100 ml with distilled water. Prepare fresh each time.

Equipment — Two point calibrated clinical flame photometer with appropriate filter for potassium and sodium colour emission.

Sample dilution
Dilute each serum sample 1:100 with distilled water by mixing 0.1 ml sample with 9.9 ml distilled water.

Procedure for simultaneous measurement of Na^+ & K^+ in the flame photometer (digital flame photometer)

- Switch on the flame photometer. Digital display should turn on.
- Turn the set '(full scale) F.S. coarse and fine controls' into maximum clockwise position.
- Select appropriate filter with the help of filter selector wheel (Na^+ on the left side and K^+ on the right side).
- Switch on the compressor and check the air pressure. Adjust it to read between 0.4 and 0.6 kg/cm^2.
- Open the gas cylinder, remove the trapper at the rear of the flame photometer and ignite the flame.
- Adjust the gas regulator to get a maximum height non-luminous blue flame with 10 distinct cones (5 on each side of the burner head).
- Feed distilled water to the atomizer and wait for at least 30 seconds.
- Adjust the 'Set Ref Coarse' and Fine controls' to zero digital readout for K^+ only.
- Aspirate 1.0 mmol/L Na^+ solution. Wait at least 30 seconds and then adjust the Set Ref Coarse and Fine controls' to a digital read out of I 00 for Na^+ only.
- Aspirate the combined standard solution (1.4/0.05, Na^+/K^+) and wait at least for 30 seconds. Adjust 'F. S control' on Na^+ side for readout 140 and that on K^+ side for a digital readout of 50.

- Repeat steps 9 and 10 once again. The flame photometer now stands calibrated.
- Now feed diluted test sample/QC to the atomizer for at least 30 seconds before recording the readings for Na^+ and K^+.

Calculation

After aspirating the standard solution, the digital reading for Na^+ is adjusted to 140 and that of K^+ to 50. This is done in order to represent Na^+ and K^+ values in undiluted serum. Since the test sample/QC is diluted initially 1:100 and then aspirated, the initial standard values for Na^+ & K^+ (1.4 & 0.05 mmol/L) must be multiplied by 100 to represent 140 mmol/L Na^+ and 5 mmol/L K^+. In the case of K^+, in order to improve the sensitivity of the assay the digital reading for the standard is further multiplied by 10 to show a reading of 50.

Analytical reliabilities

Since Na^+, K^+ are very commonly analysed parameters in a laboratory, it is recommended that internal QC (normal QC pool) be included with every batch of samples analysed in a day, irrespective of the number of samples in a batch. Further, even when a single sample is analysed as an "emergency" sample at any time of the day or night, it is essential to include an internal QC. From the QC results obtained for the day, mean, standard deviation and % CV can be calculated to ensure that **within-day precision** is well within the acceptable limit, i.e. 4%. The mean value of internal QC for the day can be pooled with the preceding 10 or 20 mean values obtained in the previous days and **between-day precision** can be calculated and expressed as % CV. Ensure that this is well within the acceptable limit, i.e, 8%.

"Assayed" QC sera with stated values (ranges) are available from several commercial sources, viz. Boehringer Mannheim, BioRad & Randox.

5.2.2 Colorimetric assay

The classic example of a simple colorimetric assay in clinical chemistry is the measurement of urea in which its reaction with specific compounds (chromogens) yields a colour product. Often called BUN this refers to assay of blood urea − blood urea nitrogen. Urea reacts directly with diacetyl monoxime under strong acidic conditions to give a yellow condensation product. The reaction is intensified by the presence of ferric ions and thiosemicarbazide. The intense red colour formed is measured at 540 nm (yellow green) on a standard colorimeter (see Figure 5.5 and WHO standard operating procedure for urea).

Glucose is also measured by colorimetric assays but unlike urea the colour produced is not a result of the direct interaction of the target analyte with the chromogen. Indeed it is somewhat removed as the colour reaction is a bystander effect of an introduced chromatophore reacting via an enzyme with a metabolic product of glucose being acted on by a specific glucose degrading enzyme. Nevertheless, the intensity of colour or indeed fluorescence generated is still directly proportional to the concentration of glucose present − a coupled reaction

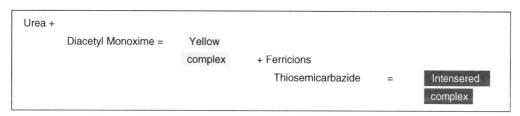

Figure 5.5   Illustration of the chemical colour changes of diacetyl monoxime with urea and ferric ions

WHO standard operation procedure for Urea analysis by colorimetric assay

Diacetyl monoxime method

Principle of the method — Urea reacts directly with diacetyl monoxime under strong acidic conditions to give a yellow condensation product. The reaction is intensified by the presence of ferric ions and thiosemicarbazide. The intense red colour formed is measured at 540 nm/yellow green filter.

Specimen types, collection and storage — Serum is the specimen of choice. Store samples for no longer than 8 hours at room temperature ($25-35\,°C$) and 7 days at $2-8\,°C$. For a longer duration, store in the freezer. If the samples show evidence of bacterial contamination, do not use these for urea estimation. Plasma could also be used for urea estimation.

Reagents — All chemicals must be Analar grade.

- **Stock acid reagent** — Dissolve 1.0 g of ferric chloride hexahydrate in 30 ml of distilled water. Add 20 ml orthophosphoric acid and mix. Store in a brown bottle at room temperature ($25-35\,°C$). Stable for 6 months.
- **Mixed acid reagent** — Add slowly 100 ml of Conc. H_2SO_4 to 400 ml distilled water taken in a 1-litre flat-bottom conical flask kept in an icecold waterbath. Mix well and add 0.3 ml of stock acid reagent. Mix and store in a brown bottle at room temperature ($25-35\,°C$). Stable for 6 months.
- **Stock colour reagent A** — Dissolve 2 g Diacetyl monoxime in distilled water and make the volume up to 100 ml in a volumetric flask. Store in a brown bottle at room temperature ($25-35\,°C$). Stable for 6 months.
- **Stock colour reagent B** — Dissolve 0.5 g thiosemicarbazide in distilled water and make up to 100 ml in a volumetric flask. Store in a brown bottle at room temperature ($25-35\,°C$). Stable for 6 months.
- **Mixed colour reagent** — Mix 35 ml of stock colour reagent A with 35 ml of stock colour reagent B and make up to 500 ml with distilled water. Store in a brown bottle at room temperature ($25-35\,°C$). Stable for 6 months.
- **Stock urea standard** — Weigh 1.0 g of analytical-grade urea and dissolve in 100 ml of benzoic acid (1 g/dl). Use a 100 ml of volumetric flask for preparing this. Store at room temperature ($25-35\,°C$). Stable for 6 months.
- **Working standard 50 mg/dl** — Dilute 5.0 ml of stock urea standard to 100 ml with benzoic acid. Store at room temperature ($25-35\,°C$). Stable for 6 months.

Dilution of Standards (S1-S3), Test & QC
Pipette the following into appropriately labelled 13×100 mm tubes

	S1	S2	S3	Test	QC
Distilled Water (ml)	1.9	1.8	1.7	1.9	1.9
50 mg/dl Urea (ml)	0.1	0.2	0.3	-	-
Test sample/QC (ml)	-	-	-	0.1	0.1

Mix Well

• Colour Development
The colour reagent is prepared fresh at the time of analysis by mixing distilled water, mixed acid reagent and mixed colour reagent in the ratio 1:1:1.

Pipette the following into another set of appropriately labelled 18×150 mm tubes.

	Blank	S1	S2	S3	Test	QC
Colour reagent (ml)	3.1	3.0	3.0	3.0	3.0	3.0
Respective diluted standard (ml)	-	0.1	0.1	0.1	-	-
Diluted test/QC (ml)	-	-	-	-	0.1	0.1

Mix all tubes well. Keep them in a boiling waterbath for 15 minutes. Remove from waterbath and cool the tubes for 5 minutes. Set the spectrophotometer/filter photometer to zero with blank at 540 nm/yellow green filter and measure the absorbance of the other tubes.

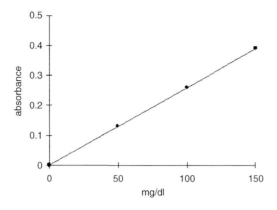

Calculation and calibration graph

Concentration of standards:

$$S1 = 50 \text{ mg/dl}; \quad S2 = 100 \text{ mg/dl};$$
$$S3 = 150 \text{ mg/dl}$$

Plot the absorbance values of standards against their respective concentrations. The measurable range with this graph is from 10 to 150 mg/dl. A calibration graph should be constructed whenever a new set of reagents is prepared. Plot absorbance values of test/QC on the calibration graph and read off the concentrations. Once linearity is proved, it will be enough if S3 is set up every time that patients' samples are analysed and the results calculated using the formula:

$$\text{Urea in test sample} = \frac{\text{Absorbance of test}}{\text{Absorbance of Standard}} \times 150 \text{ mg/dl}$$

Analytical reliabilities

Since urea is one of the most common analytes measured in a laboratory after glucose, it is recommended that an internal QC (normal QC pool) be included with every batch of samples analysed in the day, irrespective of the number of samples in a batch. Further, even when a single sample is analysed as an "emergency" sample at any time of the day or night, it is essential to include an internal QC. From the QC results obtained for the day, mean, standard deviation and %CV can be calculated to ensure that *within-day precision* is well within the acceptable limit, i.e. 4%.

The mean value of internal QC for the day can be pooled with the preceding 10 or 20 mean values obtained in the previous days and between—day precision can be calculated and expressed as % CV. Ensure that this is well within the acceptable limit, i.e. 8%.

"Assayed" QC sera with stated values (ranges) are available from several commercial sources, viz. Boehringer Mannheim, BioRad & Randox. If a laboratory uses QC sera from a commercial source, it is important that the company certifies that their QC materials are traceable to international reference materials.

Hazardous materials

Most of the chemicals used in this method are acids. Care should therefore be taken to avoid mouth pipetting and contact with skin.

WHO standard operation procedure for Glucose analysis by colorimetric assay (Glucose Oxidase Method)

Principle of the method — Glucose present in the plasma is oxidized by the enzyme glucose oxidase (GOD) to gluconic acid resulting in the liberation of hydrogen peroxide, which is converted to water and oxygen by the enzyme peroxidase (POD). 4 aminophenazone, an oxygen acceptor, takes up the oxygen and together with phenol forms a pink coloured chromogen which is measured at 515 nm.

Specimen type, storage and collection — Plasma is the specimen of choice for glucose estimation. Plasma glucose levels have been found to be quite stable for 6 hours at room temperature (25−35 °C) in but it is important that plasma should be separated from the cells soon after collection, preferably within 1 hour. About 2 ml of the patient's blood should be collected by venipuncture into a anticoagulant tube containing a mixture of potassium ethylene diaminetetraacetate (EDTA) sodium fluoride at a ratio 1:2 (W/W). The tube should be gently but thoroughly shaken for complete mixing.

:: **Reagents** — All chemicals must be Analar grade

- **Phosphate Buffer: 100 mmol/L. pH 7.0**
  To 800 ml of distilled water add the following in the order:
  - Disodium hydrogen phosphate dihydrate [Na2HPO4 2H2O] 12.95 g
  - Anhydrous potassium dihydrogen phosphate [KH2PO4] 4.95 g
  - Sodium azide [NaN3] 0.5 g

Add one by one, dissolve and finally make up to 1 litre with distilled water. Stable for 3−4 months, at 2−8 °C. Check that the final pH is 7.0 ± 0.05 with a pH meter.

- **Colour Reagent**
  To 100 ml of phosphate buffer add the following in the order and then mix to dissolve:
  - 4 amino phenazone 16 mg
  - GOD [Sigma G 7016] 1800 units
  - POD [Sigma P 8250] 100 units
  - Phenol 105 mg
  - Tween 20 [Sigma P 1359] 50 ml

Reconstitute the purchased GOD & POD powder with phosphate buffer. Dispense separately into vials so that each vial represents the requisite number of units. Store the vials frozen. Stable for 2 weeks at 2−8 °C. Store in a brown bottle.

- **Benzoic acid 1 g/l.** — Dissolve 1.0 g of benzoic acid in water and make up to 1 litre with water. This solution is stable indefinitely at room temperature.
- **Stock glucose solution, 1 g/l.** — Before weighing, dry the glucose at 60−80 °C for 4 hours. Allow to cool in a dessicator. Dissolve 1 g of glucose in benzoic acid solution and make up to 100 ml in a volumetric flask. Stable for six months at room temperature (25−35 °C). **DO NOT FREEZE THE STANDARD**
- **Working glucose standard 100 mg/dl.** — Dilute 10 ml of stock glucose (use either a volumetric pipette or a burette) to 100 ml with benzoic acid in a 100 ml volumetric flask. Mix well. Stable for 6 months at room temperature (25−35 °C).
- **Dilution of standards (S1−S5), Test & QC**

Pipette the following into appropriately labelled 13 × 100 mm tubes

	S1	S2	S3	S4	S5	Test	QC
Distilled Water (ml)	1.9	1.8	1.7	1.6	1.5	1.9	1.9
100 mg/dl glucose (ml)	0.1	0.2	0.3	0.4	0.5	-	-
Test sample/QC (ml)	-	-	-	-	-	0.1	0.1

Mix well

- **Colour development**

Pipette the following into another set of appropriately labelled tubes.

	Blank	S1	S2	S3	S4	S5	Test	QC
Colour reagent (ml)	1.2	1.2	1.2	1.2	1.2	1.2	1.2	1.2
Distilled water (ml)	0.1	-	-	-	-	-	-	-
Diluted Standards (ml)		0.1	0.1	0.1	0.1	0.1	-	-
Diluted Test Sample/QC (ml)	-	-	-	-	-	-	0.1	0.1

Mix all tubes well. Incubate at 37 °C in a waterbath for 15 minutes.

Remove from water bath and cool to room temperature. Set the spectrophotometer/filter photometer to zero using blank at 510 nm/green filter and measure the absorbance of Standards, Test and QC.

This protocol is designed for spectrophotometers/filter photometer that require a minimum volume of reaction mixture in the cuvette of 1 ml or less. Economical use of reagents is possible with this protocol, thus the cost per test can be kept to the minimum. However, if a laboratory employs a photometer requiring a large volume of the reaction mixture for measurement, viz. 5 ml, it is advisable to increase the volume of all reagents mentioned under Tabulation "(b) Colour development" proportionately.

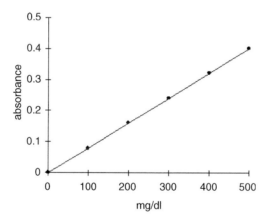

Calculation and calibration graph

Since the protocol for standard tube S1 and test is identical, the standard S1 will represent a concentration of 100 mg/dl. The glucose concentrations represented by other standard tubes are S2 = 200; S3 = 300; S4 = 400 & S5 = 500 (mg/dl).

Plot the absorbance values of standards against their respective concentrations. The measurable range with this graph is from 10 to 500 mg/dl.

Plot absorbance values of Test/QC on the calibration graph and read off the concentrations.

Once linearity is proved, it is not necessary to prepare the standard graph every time that patients' samples are analysed. It will be adequate if standard S2 is taken every time and patients' results are calculated using the formula:

$$\frac{\text{Test absorbance}}{\text{Standard absorbance}} \times 200 \text{ mg/dl}$$

Analytical reliabilities

Since glucose is the most common analyte measured in a laboratory, it is advisable to include an internal QC (normal QC pool) with every batch of samples analysed in the day, irrespective of the number of samples in a batch. Further, even when a single sample is analysed as an "emergency" sample at any time of the day or night, it is essential to include an internal QC. From the QC results obtained for the day, mean, standard deviation and %CV can be calculated to ensure that ***within-day precision*** is well within the acceptable limit, i.e, 5%.

The mean value of internal QC for the day can be pooled with the preceding 10 or 20 mean values obtained in the previous days, and ***between—day precision*** can be calculated and expressed as % CV. Ensure that this is well within the acceptable limit, i.e, 8%.

At least once a day analyse another QC serum from either a low QC or high QC pool.

"Assayed" QC sera with stated values (ranges) are available from several commercial sources, viz. Boehringer Mannheim, BioRad & Randox. *If a laboratory uses QC sera from a commercial source, it is important that the company certifies that their QC materials are traceable to international reference materials.*

:: Hazardous materials – *This procedure uses sodium azide and phenol, which are poisonous and caustic. Do not swallow, and avoid contact with skin and mucous membranes*

(see Figure 5.6 and WHO standard operating procedure for glucose assay).

Enzymes have frequently, in the past, been measured by colorimetric assay in which the catalytic activity of the enzyme itself is used as the definition of quantification: an artificial chromogenic substrate directly or indirectly acted on as a result of the target enzyme is added in access and the amount/concentration of the product (a colour intensity) produced per unit of time. Enzymes are therefore measured in arbitrary 'units' which is the amount of enzyme required to convert a defined amount of substrate to product (e.g. mmol) in a defined time period (i.e. per minute) at a set temperature (i.e. 20 °C). This approach assumes that the enzyme is robust and catalytically stable. This is not always true and now enzymes are being increasingly assayed by immunoassay where

the target is to measure the amount of protein that is the enzyme and not its intrinsic enzymic activity.

5.2.2.1   Other blood analytes commonly measured by colorimetric assay

Calcium

Serum calcium is not usually measured by flame photometry as it lacks sensitivity; however, direct colorimetric assays are sufficiently sensitive. Calcium forms a purple-coloured complex with ortho-cresolphthalein complexone in an alkaline medium. The inclusion of HCl helps to release calcium bound to proteins and 8 hydroxy-quinoline eliminates the interference by magnesium. 2-amino, 2-methyl, 1-propanol (AMP) provides the proper alkaline medium

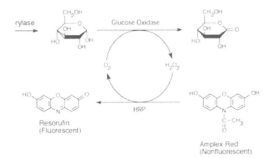

> Glucose present in the sample is oxidized by the enzyme glucose oxidase (GOD) to gluconic acid resulting in the liberation of hydrogen peroxide, which is converted to water and oxygen by a peroxidize (POD) enzyme (here it is horse radish peroxidize-HRP). The POD reaction can only occur if an acceptor substrate is available. Many substrates have been developed that become coloured or, as in this case, fluoresce.

Figure 5.6   Hydrogen peroxide generation enzyme coupled reaction for colorimetric assays

for the colour reaction. The intensity of the colour is measured at 540 nm.

Phosphates (Stannous Chloride reduction colorimetric assay method)

Phosphorus in serum reacts with ammonium molybdate to form phosphomolybdate, which is then reduced by stannous chloride and hydrazine sulphate to molybdenum blue. The intensity of the colour is measured at 640 nm.

Creatinine

Often assayed by the Jaffe method, creatinine present in serum or plasma directly reacts with alkaline picrate resulting in the formation of a red colour, the intensity of which is measured at 505 nm/green filter. Protein interference is eliminated using sodium lauryl sulphate. A second absorbance reading after acidifying with 30% acetic acid corrects for nonspecific chromogens in the samples.

Bilirubin

Often assayed by the Jendrassik and Grof method, this is another direct chromogen reaction colorimetric assay. Serum bilirubin is present in two forms conjugated (liver processing of bilirubin results in mono- and diglucuronide conjugation) and serum protein bound. Conjugated bilirubin in serum can directly reaction couple with diazotized sulphanilic acid to form a red coloured compound. The coupling reaction is rapid and is stopped with ascorbic acid in order to prevent excessive reaction with haemoglobin interfering with the assay. Protein bound bilirubin is released by a caffeine benzoate solution and can then reaction-couple with diazotized sulphanilic acid. In both cases, after reactive coupling with diazotized sulphanilic acid, tartrate buffer alkation converts the red acid—bilirubin complex to a green coloured compound which shows peak absorbance at 607 nm. At this wavelength the absorbance due to haemoglobin or carotene is minimal. Bilirubin measure is reported as conjugated and total bilirubin as the distinction has clinical relevance (see Section XX).

Total proteins (biuret direct colorimetric assays)

Proteins form a purple coloured complex with cupric ions in alkaline solution. The reaction takes its name from the simple compound biuret which reacts in the same way. The intensity of the purple colour is measured at 540 nm/yellow green filter and compared with a standard serum of known protein concentration.

Albumin (BCG dye binding colorimetric method)

Albumin binds quantitatively with bromocresol green at pH 4.15 resulting in the formation of a green colour which can be measured at 630 nm/red filter.

Enzymes

Examples of colorimetric assay for serum enzymes:

- *Alkaline Phosphatase (Total).* Paranitrophenyl phosphate, which is colourless, is hydrolysed by alkaline phosphatase at pH 10.5 and 37 °C to form free paranitrophenol, which is coloured yellow. The addition of NaOH stops the enzyme activity and the final colour shows maximum absorbance at 410 nm.
- *Transaminases.* Transamination is the process in which an amino group is transferred from amino acid to an a-keto acid. The enzymes responsible for transamination are called transaminases. The substrates in the reaction are a-ketoglutaric acid (a KG) plus L-aspartate for AST, and a KG plus L-alanine for ALT. The products formed by enzyme action are glutamate and oxaloacetate for AST and glutamate and pyruvate for ALT. Addition of 2,4, dinitrophenyl hydrazine results in the formation of hydrazone complex with the ketoacids. A red colour is produced on the addition of sodium hydroxide. The intensity of colour is related to enzymic activity.

5.2.3   Immunoassay

Immunoassays are an immunological detection system for quantifying specific analytes in all manner of complex biological fluids. The principle of these assays mimics the body's immune system's ability to recognize foreign antigens when mounting an immune response (see Chapter 7, Section 7.1.2). During an immune response antibodies are recruited to recognize, bind and remove potentially pathogenic antigens — or their source — from the circulation. It is this ability of antibodies to specifically target and bind antigens which has been exploited in immunoassays and lies at the root of the immunoassay detection system.

5.2.3.1 Antibodies and antisera

Antibodies are complex proteins which come in a number of subtypes (see Chapter 7, Section 7.1.4) the immunoassay employs the 'Y' shaped IgG (immunoglobulin G) type in either a polyclonal antiserum or a monoclonal antibody solution.

Polyclonal antiserum, as the name suggests, is made up of antibodies from 'many clones' and indicates that the antibodies within the serum are not all necessarily identical in their specificity. Polyclonal antibodies are generated by a repeat immunization of a host animal (rabbit, sheep or goat are common) which is then bled and the serum separated. This antiserum will have a potent 'memory' for one type of antigen and group of epitopes but may also be nonspecific, that is the serum will also include antibodies which recognize other antigens and epitopes. However, repeat immunization means that the amount, or titres, of antibodies specific for the target antigen are boosted to exceed any other by millions. Usually after such booster immunization the polyclonal antibody response is dominated by one or two highly responsive clones.

In contrast, monclonal antibodies are derived from one B-Cell (essentially an immortalized B-cell clone) and are therefore specific to a single antigen epitope. Immunoassays can employ either polyclonal antisera or monoclonal antibodies or a combination of both in their design.

5.2.3.2 Enzyme immunoassays (EIA)

The majority of immunoassays used today are one form of EIA, more commonly referred to as an enzyme linked immunosorbant assay (ELISA); these tests use the combined specificity of an antibody for antigen with that of enzyme and substrate for detection. Essentially, when an antibody binds antigen the 'linked' enzyme produces a coloured product from a proportional change in substrate, these substrates are referred to as chromogenic because they undergo a colour change.

The sandwich ELISA is by far the most common and involves the use of two distinct antibodies which bind the same antigen but at different epitopes, effectively forming an antibody—antigen—antibody 'sandwich'. The first antibody is often referred to as the capture antibody and is immobilized to a solid matrix such as the bottom of a plastic 96 well plate; the plastic well is then 'blocked' with a protein or detergent solution to

reduce nonspecific binding to the charged plastic surface. Once this solution is removed the sample is added and the immobilized antibody begins to bind the antigen in the sample. The remaining sample is washed away using a mild detergent solution leaving an antibody—antigen complex immobilized to a plastic plate. A second antibody, the detection antibody, with a conjugated enzyme is subsequently added to the well and again any unbound or nonspecific binding is washed away. At this stage there is an invisible molecular sandwich of antibody— antigen —antibody—enzyme all immobilized to the bottom of a plastic plate. The final step is the addition of a chromogenic substrate which becomes coloured in an intensity proportional to the quantity of enzyme present; the colour change is quantified spectrophotometrically. The quantity of enzyme present is proportional to the quantity of antigen captured so by using a set of standards of known concentration, unknown samples can be compared and accurate estimations of antigen concentration can be made.

Indirect ELISAs can also be employed to detect antibody concentration — antibodies are proteins and can therefore be antigens too. An antibody which detects and binds other antibodies is generally specific to different species antibodies (antispecies antibodies — ASA) — human antimouse antibodies (HAMA) are an example of this. Indirect ELISAs employ a captured antigen which is immobilized and an enzyme-linked ASA to detect the captured antibody in an antigen—antibody—antibody—enzyme sandwich. The substrate reaction is the same as the sandwich ELISA and the colour change is proportional to the captured antibody concentration (see Figure 5.7).

ELISAs were developed from earlier radioimmunoassays (RIA) and immunoradiometric assays (IRMA) which were developed in the 1970s (this development resulted in the Nobel Prize for Medicine and Physiology being awarded in 1977 to Roger Guillemin, Andrew V. Schally and Rosalyn Yalow). An IRMA is essentially an ELISA sandwich but rather than using an enzyme-linked antibody to generate a signal a radioactive signal was used and data generated in counts per minute (cpm). The RIA differs from other detection systems in that the radioactive decay signal — cpm — is indirectly proportional to the concentration of antigen in a sample because it has to compete with a defined amount of radioactive antigen (tracer) in the test tube for a single antibody. Defined amounts of radioactive tracer antigens,

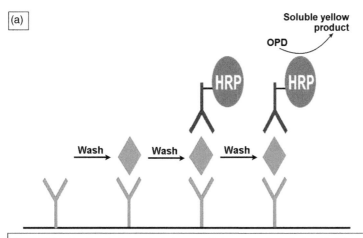

DIRECT ELISA – Specific antibodies are absorbed on to a solid phase such as the plastic of a multiwall plate. Unbound antibody is washed away. Sample is then added and any antigen is bound to the immobilized solid phase antibodies and all other components washed away. A second antibody, conjugated to an enzyme (such as HRP), binding to the same antigen but at a sterically different epitope, is added. This also becomes immobilized to the solid phase via the antigen, any unbound enzyme conjugated secondary antibody will be washed away. The captured enzyme conjugated antibody is detected by adding a coloured reactive substrate (e.g. OPD). The intensity of colour produced is proportional to the enzyme present which in turn is directly proportional to the amount of antigen captured.

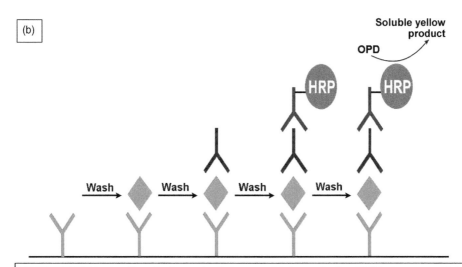

INDIRECT ELISA – Specific antibodies are absorbed on to a solid phase as for direct ELISAs; however, purified and concentrated antigen is added to deliberately saturate all the solid-phase antibody binding sites (excess is washed away). Sample serum is added and any endogenous antibodies in the sample (generated because the patient has mounted an immune response to the antigen) are captured via the antigen to the solid phase. After washing away all unbound molecules an enzyme conjugated (e.g. HRP) secondary antibodies binding to human antibodies (can be specific anti IgM, IgG, IgE, etc.) is added. If human antibodies to the antigen are present, after washing away the unbound, the captured enzyme conjugated antibody is detected by adding a coloured reactive substrate e.g. OPD. The intensity of colour produced is proportional to the enzyme present which in turn is directly proportional to the amount of endogenous human antibody.

Figure 5.7   Direct (a) and indirect enzyme linked immunometric assay - ELISA (b) format with a typical standard curve (c)

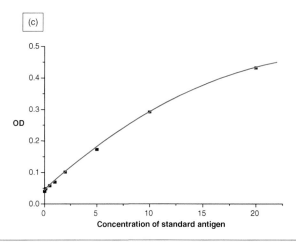

(c)

TYPICAL STANDARD CURVE OF AN ELISA – Both direct and indirect ELISAs produce standard curves in which the optical density of a specific colour of the substrate reaction, plotted again ststandard concentrations of purified antigen reacted in the assay system, are directly proportional.

Figure 5.7   (*Continued*)

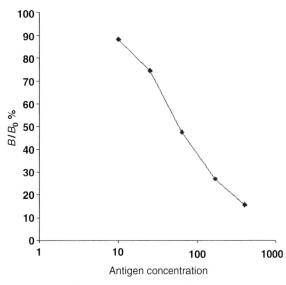

Figure 5.8   Illustration of a typical calibration curve from a competitive immunoassay (RIA or EIA) in which the signal is inversely proportional to the antigen concentration. The antigen concentration range is large and plotted on a logarithmic *x*-axis. The signal can be expressed as simply cpm for RIA of colour intensity in EIA. However, it is best expressed as the percentage of signal found precipitated (B) in the standard/sample tube compared to when no antigen is present to compete with the capture antibody in the 0 standard/control antigen tube $- B_0$. This allows for the decay in signal intensity of the tracer over time and produces more consistent curve plots

antibody and sample are mixed together in a test tube and the antigen−antibody complex that forms is precipitated. Any unbound and therefore unprecipitated, tracer is simply removed by aspirating (sucking off) or decanting the supernatant. If there is no antigen present in a sample then the tracer antigen is bound by the antibody and found in the precipitate − giving a high radioactive count. However, if the sample contains the target antigen it will compete and effectively displace the radioactive tracer, for the limited amount of antibody. The more antigen present in a sample the more effectively the tracer is competed for antibody binding and therefore displaced from the antigen−antibody precipitate (i.e. the lower the cpm the higher the antigen concentration, see Figure 5.8). As for all immunoassays a series of known concentrations are tested in the system to construct a standard. It is also possible to construct competitive EIAs using the same RIA approach.

5.2.4   Electrophoresis

Most undergraduate experience of (gel) electrophoresis would suggest to them that 'electrophoresis' separates molecules (particularly DNA/RNA and proteins) according to molecular size. This is a misconception. The principal determinant of electrophoretic separation is charge and the response of the

The term, electrophoresis, was coined in 1909, by *Michaelis* (*Biochem. Z.* 16, 1909, 81).

It is derived from the Greek elektron meaning amber (i.e. electricity) and phore meaning bearer.

'A mainly analytical method in which separations are based on the differing mobilities (*i.e. rate plus direction of movement*) *of two or more charged analytes* (*ionic compounds*) *or particles held in a conducting medium under the influence of an applied d.c. electric field.'*

Most current methods are derived from the work of *Tiselius* in the 1930s (awarded Nobel Prize in 1948).

Figure 5.9   Brief history and mathematical definitions of electrophoresis

charge on the given molecule (polarity — positive or negative and magnitude, 1+, 2+, etc.) to the applied electrical field. The speed of mobility of a molecule in an electrical field is determined by the mass to charge ratio (m/z) (see Figure 5.9).

The secondary determinant of movement or separation is frictional drag. For small molecules moving in a fluid or even a gel this is negligible (and can be virtually ignored), but as the molecule becomes larger and the matrix/gel in which the molecule (i.e. protein or nucleic acid polymer) is moving increases in solidity, frictional drag becomes a more important determinant of separation (see Figure 5.10).

Various electrophoretic techniques have been developed which exploit these two forces (m/z and frictional drag) and the peculiarities of the molecules being separated with respect to environment buffer altering the molecules' charge properties.

5.2.4.1   Protein electrophoresis

In an electric field, a protein or other charged macromolecule will move with a velocity that depends directly on the charge on the macromolecule and inversely on its size and shape. Gel electrophoresis is carried out in some supporting media, usually polyacrylamide or agarose, with pores big enough to allow the passage of

the macromolecule (see Figure 5.11). Being either acidic or basic is dependent on the pH of the environment the protein is in as this will be very significant in determining the net charge. This is a factor exploited in electrophoretic separation of haemoglobin (Hb) variants. Amino acid substitutions give an overall charge change in the Hb molecule and the magnitude of this charge change can be amplified by altering the surrounding pH of the electrophoresis buffer to effect clear resolution of the various Hb variants.

There are four important variant forms of electrophoresis:

- zonal electrophoresis;
- isoelectric focusing — IEF;
- isotachophoresis; and
- sodium dodecyl sulphate polyacrylamide gel — SDS PAGE.

These in turn have been combined to give two electrophoretic techniques commonly used in proteomics:

- isotachophoresis combined with SDS-PAGE gives rise to discontinuous SDS PAGE; and
- combining SDS PAGE and IEF gives 2D electrophoresis.

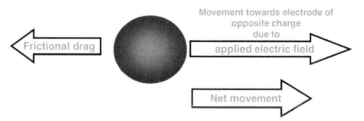

Figure 5.10   The variable importance of frictional drag on electrophoretic mobility

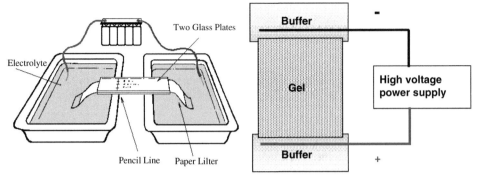

Figure 5.11   Basic organization of the functional components in an electrophoresis tank illustrating the simplicity of connecting electrical charge to the separation platform (gel) via conductive analyte/buffer which establishes the pH environment for the electrophoresis

5.2.4.2   Zonal electrophoresis

In this classic of the clinical chemistry laboratory, proteins are allowed to move in or upon a support medium (acetate sheet or gel) according to their charge with very little consideration of frictional drag. This can easily resolve and indeed defines classes of serum protein (see Figure 5.12). It exploits the fact that an amino acid substitution in a protein may not alter its mass to any significant degree (unless analysed by a mass spectrometer) but the effect on the overall charge of the molecule is often considerable. This is most clearly seen in the detection of haemaglobinopathies (see Figure 5.13).

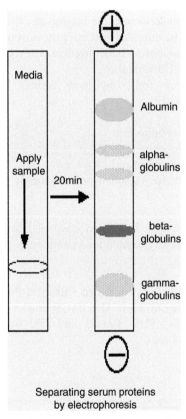

Figure 5.12   Zonal electrophoretic separation of serum proteins migration is according to m/z, thus under these prevailing conditions albumin is strongly negatively charged and migrates to the positive electrode the fastest, gamma globulins are net positively charged and migrate to the negative electrode

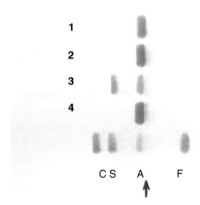

Figure 5.13   Zonal electrophoretic separation of haemo-globin: Hb sickle variants (labelled C and S) migrate separately to both normal haemoglobin A and fetal haemoglobin F; the arrow shows the position line point of application. Patient sample 3 contains both normal haemoglobin A and haemoglobin F and is hence heterozygote (a carrier) of ickle cell disease

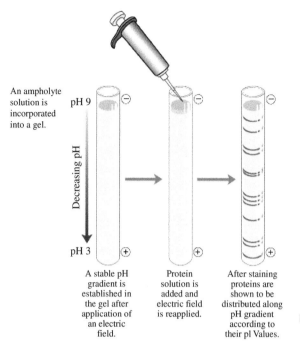

An ampholyte solution is incorporated into a gel.

pH 9

Decreasing pH

pH 3

A stable pH gradient is established in the gel after application of an electric field.

Protein solution is added and electric field is reapplied.

After staining proteins are shown to be distributed along pH gradient according to their pI Values.

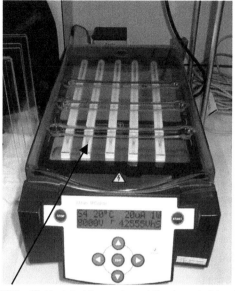

IEF 'coffin' tank holding an immobilized ampholyte strip.

Figure 5.14   Illustration of the original tube IEF gels and the modern strip system of individual tanks nicknamed coffins

5.2.4.3   Isoelectric focusing

The charge of a protein is dependent on its environment. For example a protein that is negatively charged at neutral pH 7.4 can be changed to a positive charge in low pH solution such as pH 4. This is because environments in which there is abundant hydrogen ion concentration carboxyl grouping ionization (negative charge) is suppressed. However, amide residues readily ionize taking on the H^+ ions (giving a positive charge). The opposite occurs in low hydrogen ion concentration environments. At a specific pH the suppression of amide and carboxyl groups within the molecule balance and the molecule reaches its non-charged state or isoelectric point (pI). It is possible to create a gradient of pH due to a group of molecules called ampholyltes.

In the original isoelectric focusing 'electrophoretic' systems a pH gradient was set up first using a purchased mixture of ampholytes (different molecules designed to have a range of pIs) which had to be preelectrophorezed on the gel to form the pH gradient. A mixture of molecules (proteins) was then applied, the electric field turned on, and each protein moves to the position (pH) at which its net charge is zero, that is it is pI. In newer systems the ampholytes have now been immobilized on a solid support, such as a filter strip. These can be placed in an individual sample tank or coffin and the sample applied on to the strip and electrophoresed immediately and easily removed for analysis (see Figure 5.14).

5.2.4.4   Sodium dodecyl sulphate polyacrylamide gel electrophoresis (SDS PAGE)

Polyacrylamide gels are tougher and more resilient gels than agarose. They also have the capacity to form a large range of gel pore sizes through which large molecules, such as proteins and DNA/RNA, can be separated according to their molecular size. This is simply achieved by varying the percentage of acrylamide powder (~5–30%) in the polymerization mix. Thus, in terms of electrophoretic separation the frictional drag component is precisely controlled and defined. However, charge is the dominant variable of a protein determining migration velocity (force) and direction of movement. SDS overcomes this and gives all proteins the same net charge (negative) and mass/charge ratio such that they all move in the same direction (towards the cathode) with the same motive force. Under these circumstances frictional drag is the

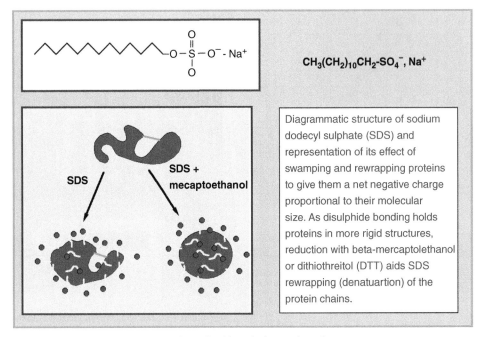

$$CH_3(CH_2)_{10}CH_2\text{-}SO_4^-, Na^+$$

SDS

SDS + mecaptoethanol

Diagrammatic structure of sodium dodecyl sulphate (SDS) and representation of its effect of swamping and rewrapping proteins to give them a net negative charge proportional to their molecular size. As disulphide bonding holds proteins in more rigid structures, reduction with beta-mercaptolethanol or dithiothreitol (DTT) aids SDS rewrapping (denatuartion) of the protein chains.

Figure 5.15   Separation of proteins by SDS polyacrylamide gel electrophoresis

sole determinant of separation (see Figure 5.15). When frictional drag is through a rigid inert polymer such as acrylamide gels then drag is due to the ability of the molecule to penetrate the pores; this is proportional to the molecular size of the molecule. Since SDS has refolded and wrapped proteins into dense spheroids, molecular size closely approximates to molecular mass (see Figure 5.16).

5.2.4.5   Capillary electrophoresis

Used in many automated systems capillary electrophoresis is a liquid-based electrophoresis method that completes run times in a matter of seconds but to achieve this applies very high voltages. Although you can run classic electrophetic systems at very high voltage to make the analytical run time faster, the heat generated from the resistance will melt the gel or electrophoresis paper and set the entire apparatus on fire. The solution was to make small microcapillaries which very effectively dissipate the heat generated from the high voltages (see Figure 5.17).

A surprising finding was that the high voltage induced electro-osmotic flow which was so significant in microcapillaries it overcame the movement of molecules based on charge. Thus eventually all molecules introduced would wash towards the cathode even if negatively charged; thus negatively charged molecules would be the last molecules eluting from the system but hence would still be resolved in the chromatographic separation. Samples are loaded on the capillaries in nano litre quantities but because of the physics of electro-osmotic flow the molecules for analysis separate and concentrate in tightly eluting plugs of intense concentration (see Figure 5.18).

5.2.5   High performance liquid chromatography (HPLC)

HPLC is a chromatographic technique that can separate a mixture of compounds very rapidly (in seconds and minutes) and, for the analytical chemist, give not only characteristic elution times aiding identification, but chromatographic peak heights that are directly proportion to the amount present — *aiding identification and quantification*. This is particularly useful for chemists wishing to quantify small organic compounds such as drugs in biological fluids or indeed small organic biochemicals such as lipid-soluble vitamins, steroids and their various metabolic derivatives.

The heart of all HPLC systems are the columns (see Figure 5.19) and they can be built to exploit all the physical characteristics of a molecule: size, charge, hydrophobicity and reactive group affinity.

Separations of proteins in SDS PAGE are due to molecular size so measuring the relative position of a protein's migration from a set reference point (i.e. choose the top of the gel) can be plotted against the log molecular mass of the calibration proteins to yield a standard curve. The migration (Rm) of unknowns can be used to estimate their molecular mass to within a 10–20% margin of error.

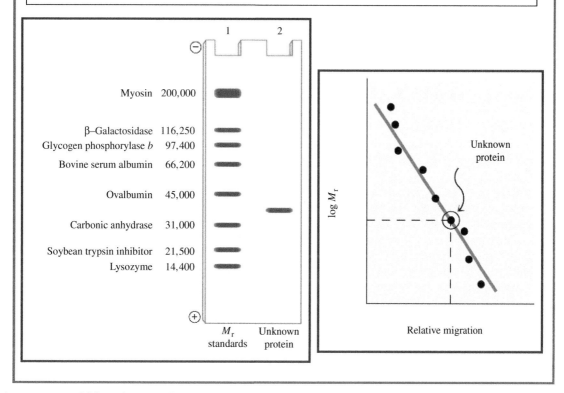

Figure 5.16   Refolding of proteins/denaturation by sodium dodecyl sulphate (SDS)

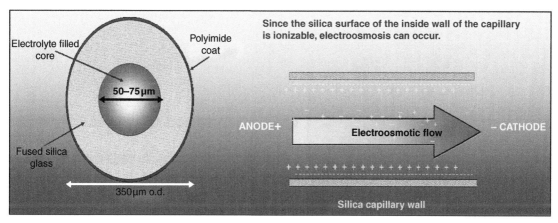

Figure 5.17   Cross section and construction of electrophoretic capillary and the physics of electrosmotic flow

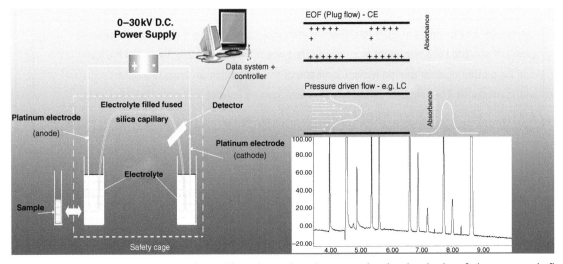

Figure 5.18   Diagrammatic schematics of a capillary electrophoresis set up, the chemicophysics of electro-osmotic flow separation chromatography and an example of the resulting chromatogram

The advantage of HPLC is that separation is rapid (minutes) using small sample sizes (μl) rather than hours; for example a low pressure size exclusion column can take all day and require 1 ml plus of sample. In HPLC the sample to be analysed is introduced in a small volume to the stream of mobile phase. The analyte's motion through the column is slowed by specific chemical or physical interactions with the stationary phase as it traverses the length of the column. How much the individual analyte molecule is slowed depends on the nature of the molecule and on the compositions of the stationary and mobile phases. However, once a method has been established and characterized the retention time is considered a reasonably unique identifying characteristic of a given molecule. The use of smaller particle size column packing (which creates higher backpressure) increases the linear velocity giving the components less time to diffuse within the column, leading to improved resolution in the resulting chromatogram. Common solvents used include any miscible combination of water or various organic liquids (the most common are methanol and acetonitrile). Water may contain buffers or salts to assist in the separation of the analyte components, or compounds such as trifluoroacetic acid which act as an ion-pairing agent.

In clinical chemistry the analytes we tend to measure by HPLC are generally drugs or metabolites as other methods such as immunoassays are fast, inexpensive and easily performed methodologies (a fundamental of clinical testing). HPLC does require both time and a skilled operator if it is to be reliable. Thus for

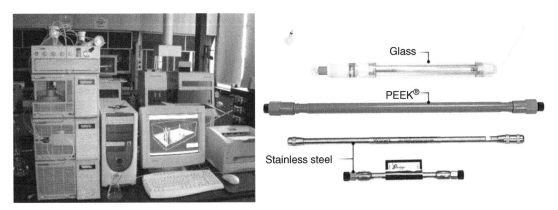

Figure 5.19   Illustration of HPLC equipment and the core component — the HPLC column

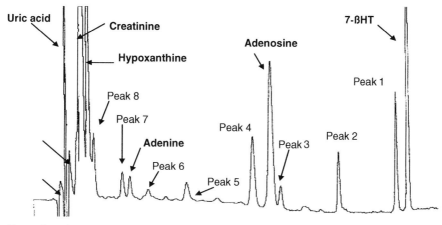

Figure 5.20   Illustration of small bioorganic molecules found in human follicular fluid resolved by RP-HPLC; only a few of the peaks have been identified

detection and measurement of the more common biochemical markers requested by the physician HPLC is not routinely employed even though it can be adapted to measure most of the important common clinical chemistry analytes.

The common columns used exploit small organic molecules' differences in charge (polarity) and hydrophobicity. The first methods developed for HPLC were normal phase or HILIC (hydrophilic interaction chromatography) where a solid phase was very polar (positive, negative or both) and a mobile phase was a water soluble nonpolar solution (e.g. 1–100% methanol). The highly polar molecules which paired with the charge(s) on the solid phase were retarded to varying degrees and neutral molecules rapidly passed through the column eluting first at the detector.

Reverse phase worked on the bases of hydrophobicity of a molecule and a solid phase with nonpolar hydrophobic chemical groupings like phenol chemically bonded to their surface. In very polar starting buffers (i.e. high salt) the hydrophobic elements of any given molecule induce it to associate with the hydrophobic groups coating the solid phase; but as the polarity of the mobile phase decreases, such as lower salt and increasing concentration of nonpolar solvents like acetonitrile, the analyte molecules elute from the column and are registered at the detector.

The detectors are often UV-visible spectrometers that scan at a set or an array of wavelengths. However, sensitivity is dependent on the light absorbing properties of the target analyte so electrochemical and fluorescence detectors are also employed (see Figure 5.20).

5.2.6   Mass spectrometry

Mass spectrometers have developed so rapidly that their application to routine clinical chemistry will be dramatic in the next few years. The principal use has been as the end detector of a HPLC — not only will the mass ion peak give detection and quantification of an eluting fragment but the mass spectrum confirms the analytes molecular Identity. In many cases stripped down instruments are built with the sole aim to detect the characteristic mass ion signature (could be as little as a single parent ion peak) of certain drugs in a complex mixture without the need of resolving the component first by HPLC (see Figure 5.21 for the detection of cocaine).

The science of mass chromatography has now developed such that the process of ionization is no longer destructive and that huge molecules such as proteins of a million Daltons can be given a charge and analysed by a mass analyser. These new instruments give the precise molecular weights of proteins within 0.01% margin of error. Instead of ion bombardment which fragments and therefore ionizes molecules, matrix assisted laser desorption ionization (MALDI, see Figure 5.22) and electrospray Ionization (EI, see Figure 5.23) deposit charges on molecules without destroying their structure.

To match this advance in ionization the old huge sector instruments, which had a strong magnet distorting the flight path of the ions generated, have been replaced in all but the most specialist MS system. Modern mass analysers are smaller and more versatile:

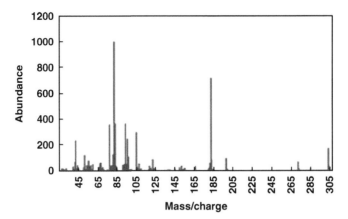

Figure 5.21 Mass chromatograph of cocaine

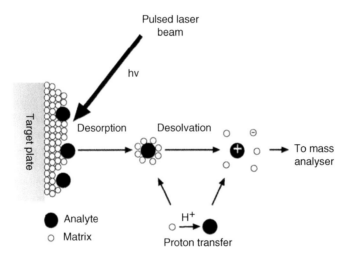

Figure 5.22 Ionization of analytes by MALDI. The cocrystal of matrix and sample is targetted by a laser causing a desorption process followed by desolvation and then introduction into the mass analyser. From Hoffbrand *et al*, *Postgraduate Haematology, Sixth Edition* 2010, reproduced by permission of John Wiley & Sons Ltd

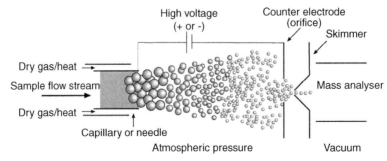

Figure 5.23 Electrospray ionization (EI) process: a fine mist of charged analytes is formed from the sample stream as droplets formed from the capillary needle eject into what is termed the Taylor cone; the charged droplets desolvate by Coulombic explosion and ions free from solvent can then enter the mass spectrometer via an intermediate vacuum region

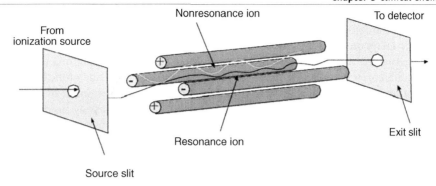

Figure 5.24 Schematic of a quadrupole mass analyser. Ions are fed in from the electrospray and the radiofrequency of the charged rods varied or tuned to particular *m/z*s. Resonant ions will travel down through the poles towards the detector. Nonresonant ions will strike the poles and become neutralized. The quadrupole sweeps though the *m/z* resonance range in seconds scanning the sample

the quadrupole focuses ions of specific mass over charge (*m/z*) on to a single target detector and sweeps through a mass range from 0–2000 Daltons. Fed ions from electrospray ionization the same molecule may have 1 to 100 charges placed on it so the mass data from the quadrupole is mathematically processed – deconvoluted – to give a single mass peak (see Figure 5.24).

Ion traps are similarly fed ions from electrospray but collect all the ions generated in a ring magnet containment chamber and then only eject those of set *m/z* to the detector in a set analysis pattern. The data still has to be deconvoluted but the system is much more sensitive as it captures all ions and loses none when detecting (see Figure 5.25).

The final mass analysis is time of flight (*ToF*) which measures the nanoseconds an ion takes to travel from source down a flight tube (usually 1 m or more) to the

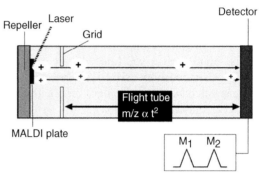

Figure 5.26 The time of flight (*ToF*) mass analyser. Ions will separate according to their velocity along the flight tube from the MALDI plate towards the detector, the larger the molecule (strictly speaking the *m/z*) the longer the flight time

detector. The time it takes is proportional to the ions' *m/z* but since *ToF*s are usually coupled to MALDI the ions generated are mostly only singly or doubly charged (see Figure 5.26). Thus the huge computational deconvolution of the data is not needed. This means that extremely large molecules – proteins of 1 million Daltons – can be analysed.

5.3 Summary: common clinical tests for sample analytes

Although electrophoresis, HPLC and mass spectrometry are more and more used in clinical laboratories,

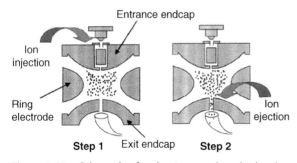

Figure 5.25 Schematic of an ion trap quadrupole showing an overview of MS-in time. Step 1: a trapping rf amplitude is applied for 0–30 ms during which ions are formed from sample molecules and stored. Step 2: an rf amplitude is ramped over the period 30–85 ms during which mass-selective ion ejection and mass analysis occurs

the most performed methodologies are flame photometry and colorimetric assays on autoanalysers. Indeed, although immunoassays are the next highest used methodology, they still represent only a small fraction of the number of performances of a flame photometry or colorimetric assay. This is because the most common requested tests (see Table 5.2) require quick, robust and inexpensive methods for detection and quantification to meet demands.

Part II: Clinical assessments

The molecular structural categorization of bodily fluid components may influence the biochemical tests used in their detection and quantification but clinical chemistry is subdivided more along clinical/biological functions relevant to the practice of medicine: urea and electrolytes, gastro-intestinal markers, renal function markers, bone assessment, heart disease and lipid disorders, liver function markers, endocrine assessment, reproductive endocrinology, therapeutic drug monitoring and cancer biomarkers.

5.4   Urea and electrolytes (U and Es)

The so called U and Es involves the measurement of urea, potassium, sodium and the anion gap, carried out routinely on serum or plasma samples. The test is useful in numerous situations and utilized in some of the most common interpretative diagnostic challenges.

Urea contributes most of the body's nonprotein nitrogen, accounting for about 45% of the total. It is the major end-product of protein catabolism in humans. It is synthesized in the liver, released into blood circulation and excreted by the kidneys. Measurement of urea in blood is a useful indicator of renal and hepatic integrity.

Electrolytes are minerals that are found in body tissues and blood in the form of dissolved salts. As electrically charged particles, electrolytes help move nutrients into and wastes out of the body's cells, maintain a healthy water balance, and help stabilize the body's acid/base (pH) level. Electrolytes are usually measured as part of a renal profile which measures the main electrolytes in the body, sodium (Na^+), potassium (K^+), together with creatinine and/or urea, and

may occasionally include chloride (Cl^-) and/or bicarbonate (HCO_3^-) as well.

Most of the body's sodium is found in the extracellular fluid (ECF), outside of the body's cells, where it helps to regulate the amount of water in your body. Potassium is found mainly inside the body's cells. A small but vital amount of potassium is found in the plasma, the liquid portion of the blood. Monitoring potassium is important as small changes in the K^+ level can affect the heart's rhythm and ability to contract. Chloride travels in and out of the cells to help maintain electrical neutrality, and its level usually mirrors that of sodium. The main job of bicarbonate, which is excreted and reabsorbed by the kidneys, is to help maintain a stable pH level and, secondarily, to help maintain electrical neutrality.

Your diet provides sodium, potassium and chloride; your kidneys excrete them. Your lungs provide oxygen and regulate CO_2. The balance of these chemicals is an indication of the functional well-being of several basic body functions, including those performed by the kidneys and heart.

A related 'test' is the anion gap, which is actually a calculated value. There is more than one formula: one definition is sodium ion concentration minus the sum of chloride plus bicarbonate ion concentration, and the other is sodium plus potassium minus the sum of chloride plus bicarbonate ion concentration. The occurrence of an abnormal anion gap is nonspecific but can suggest certain kinds of metabolic abnormalities, such as starvation or diabetes, or the presence of a toxic substance, such as oxalate, glycolate or aspirin.

5.4.1   Hypo- and hypernatraemia, kalaemia and urea

Hyponatraemia ($Na < 120$ mmol/L) may be due to either loss of sodium or overload of water or a combination of both. The following list (Table 5.3) contains the relatively common and important causes of hyponatraemia. For more complex cases, it is necessary to have simultaneous samples of both blood and urine for estimation of osmolality and electrolytes.

Increased serum potassium level is observed in anoxia, metabolic renal tubular acidosis, shock or circulatory failure. Low serum potassium values are observed due to low intake of dietary potassium over a period of time or increased loss through kidney, vomiting or

Table 5.2 Common clinical tests for sample analytes.

Analyte	Preferred sample	Anticoagulent	Test principle	Reference range	Analytical tips
Sodium (Na)	Serum (then plasma)	N/A (Heparin)	Flame photometry	130–145 mmol/L	Do not refrigerate (potassium leaks from red cells) separate within 3 h and discard haemolysed serum samples
Potassium (K)	Serum (then plasma)	N/A (Heparin)	Flame photometry	3.5–5.0 mmol/L	
Urea	Serum or plasma	Any	Colorimetric assay	15–40 mg/dL	
Calcium	Serum	EDTA, oxalate or citrate must not be used as these preservatives cause removal of calcium by chelation	Colorimetric assay	8.5–10.4 mg/dL	Haemolysed and heparinized samples are unsuitable
Creatinine	Serum or plasma	Any	Colorimetric assay	Male 0.7–1.4 mg/dL Female 0.4–1.2 mg/dL	Avoid using haemolysed or lipaemic samples
Glucose	Plasma	EDTA – Na fluoride	Colorimetric assay	Fasting: 70–110 mg/dL Postprandial: 80–140 mg/dL Random: 60–140 mg/dL	Remove from cellular fraction within 1 h as glucose uptake by blood cells reduces levels
Bilirubin	Serum	N/A	Colorimetric assay	Serum conjugated Bilirubin – up to 0.5 mg/dL Serum total Bilirubin – 0.2–1.0 mg/dL	Bilirubin is unstable and light sensitive assay should be carried out within 2 h of collection or refrigerated
Phosphates	Serum	N/A	Colorimetric assay	3.5–5.0 mg/dL	Nonhaemolysed serum separated from erythrocyte as soon as possible
Total protein	Serum (then plasma)	N/A	Colorimetric assay	6.5–8.5 g/dL	Haemolysed and lipaemic sera interfere strongly with the measurement
Albumin	Serum (then plasma)	N/A	Colorimetric assay	3.5–5.0 g/dL	Haemolysed and lipaemic sera interfere strongly with the measurement
Tranaminases GGT	Plasma	Heparin	Colorimetric (enzyme) assay	AST 5–40 U/L ALT 7–56 U/L GGT 0–51 U/L	Haemolysed and lipaemic sera interfere strongly with the measurement

Table 5.3   Common causes of hyponatraemia.

Hyponatraemia and oedema	CCF
	Liver failure
	Nephrotic syndrome
Hyponatraemia and dehydration	Diarrhoea
	Addison's disease
	Renal salt wasting
Hyponatraemia and normal volume	Diarrhoea and vomiting followed by drinking excess water
	Diuretic therapy may occur within 1/2 weeks of initiation of therapy but may develop after many years of therapy
	ACE inhibitors (see Chapter 2, Section 2.7.9)
Inappropriate ADH (Section 1.7.2) secretion	Drug therapy
	Chest disease
	CNS disease

Table 5.4   Common causes of hypo- and hyperkalaemia and altered urea levels.

Hypokalaemia (K < 3.5 mmol/L)	Diuretics
	Diarrhoea and vomiting
	Ileostomy
	Chronic purgative abuse
	Excessive liquorice
	Elderly patients with poor nutrition (catabolic state)
Hyperkalaemia (K > 5.5 mmol/L)	Haemolysis and delayed separation overnight storage in a fridge
	Renal failure
	Potassium retaining diuretics
	ACE inhibitors
Low urea	Pregnancy
	Starvation
	Chronic liver disease
High urea	Renal failure
	Dehydration
	Transiently after eating high-protein meal
	Following a significant gastrointestinal bleed

diarrhoea. Increased secretion of adrenal steroids or some diuretics may also promote the loss of potassium.

Elevated serum urea levels may be due to prerenal, renal or postrenal disease processes. Prerenal causes could be cardiac related or due to increased protein catabolism, and dehydration. Renal causes include glomerulonephritis, chronic nephritis, nephrotic syndrome and other kidney disease. Postrenal causes include obstruction of the urinary tract. Decreased serum urea levels could be due to pregnancy, intravenous infusion, low antidiuretic hormone (ADH) secretion, low protein intake, severe liver diseases, inborn errors of urea cycle and SIADH (syndrome of inappropriate ADH secretion).

The cause of elevated or lowered potassium (hypo- and hyperkalaemia) and urea are summarized in Table 5.4.

5.4.2   Hypo- and hypercalcaemia

Approximately 99% of total body calcium is deposited in the skeleton. A higher proportion of nonskeletal calcium is present within cells than in extracellular fluids, and most of this intracellular calcium is bound to proteins in the cell membrane. Intracellular ionized calcium is physiologically active and functions as an intracellular messenger by binding to — or being released from — specific intracellular proteins, a process that changes protein conformations and hence its activity or function. Release of calcium into the blood circulation is controlled by parathyroid hormone (PTH) and its fetal analogue PTH-like hormone. Increased serum calcium levels are associated with primary hyperparathyroidism, multiple myeloma, metastatic bone lesions and hypervitaminosis D. Hypocalcaemia is associated with hypoparathyroidism, nephrotic syndrome, rickets and renal failure.

5.5   Metabolism and gastrointestinal markers

There is considerable overlap in the functional association of abnormal analyte-marker levels with particular body systems; but certain common clinical chemistry are more associated with gastrointestinal and metabolism disorders than others and are detailed below.

5.5.1 Creatinine

Creatinine is a waste product formed in muscle from a high-energy storage compound, creatine phosphate. Creatine phosphate can be stored in muscle at approximately four times the concentration of adenosine triphosphate. In muscles it spontaneously undergoes degradation to form a cyclic anhyride-creatinine. The blood concentration of creatinine and its excretion in urine are remarkably constant in healthy individuals. Therefore serum creatinine level is used as an indicator for assessing kidney function.

Serum creatinine concentration is related to muscle mass and the values are lower in children. Increased serum creatinine is associated with decrease in glomerular filtration rate (GFR), whether the cause is prerenal, renal or postrenal. Prerenal factors include conditions such as congestive heart failure, shock, diarrhoea, uncontrolled diabetes mellitus, use of diuretics and so on. Renal factors involve mainly damage to the glomeruli. Postrenal factors may be prostatic hypertrophy, calculi blocking the ureters or neoplasms compressing the ureters. The serum creatinine concentration is monitored closely after a renal transplantation because a rising concentration, even though small, may be an indication of graft rejection.

5.5.2 Glucose

Glucose is a reducing monosaccharide that serves as the principal fuel of all the tissues. It enters the cell through the influence of insulin and undergoes a series of chemical reactions to produce energy.

Lack of insulin or resistance to its action at the cellular level causes diabetes. Therefore, in diabetes mellitus the blood glucose levels are very high. Some patients with very high blood glucose levels may develop metabolic acidosis and ketosis caused by the increased fat metabolism, the alternate source for energy. Hyperglycaemia is also noted in gestational diabetes of pregnancy and may be found in pancreatic disease, pituitary and adrenal disorders. A decreased level of blood glucose, hypoglycaemia is often associated with starvation, hyperinsulinaemia and in those who are taking high insulin dose for therapy and hypoadrenalism.

Thus elevated plasma glucose levels are expected in a variety of clinical conditions, especially diabetes mellitus, Cushing's syndrome and hyperadrenalism. Decreased plasma glucose levels are observed in hyperinsulinism, antidiabetic treatment and anorexia.

5.5.3 Bilirubin

Bilirubin is formed from the haem fragment of haemoglobin released by aged or damaged red blood cells. Liver, spleen and bone marrow are the sites of bilirubin production. Bilirubin formed in spleen and bone marrow is transported to the liver. In the liver it is converted into bilirubin conjugates — bilirubin mono and diglucuronides. Any liver disease affects the above systems, and hence bilirubin accumulates in serum leading to jaundice.

Hyperbilirubinaemia is characteristic of jaundice. An increase in unconjugated bilirubin is observed in haemolytic and neonatal jaundice. In viral and toxic hepatitis there is impaired hepatocellular conjugation and excretion of bilirubin with a major rise in conjugated and a lesser rise in unconjugated bilirubin in serum. In cirrhosis there is overall damage to liver cells and hence the ability of the liver to form conjugated bilirubin, resulting in an increase in unconjugated bilirubin in serum. In obstructive jaundice there is an increase in predominantly conjugated bilirubin in serum.

5.5.4 Phosphates

Eighty per cent of body phosphorus is laid down in bone matrix as insoluble salts. The organic phosphate esters are primarily confined within cells, associated with nucleoproteins, hexoses and purines. Phosphate forms high-energy bonds in ATP, GTP and creatine phosphate. Inorganic phosphate ions are mostly confined to the extracellular fluid where they are part of buffer systems. The plasma phosphate concentration is regulated by parathyroid hormone (PTH) and vitamin D3. PTH stimulates the kidney to excrete phosphate while conserving calcium. In chronic renal disease, phosphate retention occurs because of impaired glomerular filtration.

An increase in serum phosphorous is found in chronic nephritis progressing with increased renal failure. A moderate increase is observed in hypoparathyroidism and vitamin D excess. A decrease in serum phosphorous is observed in rickets or osteomalacia and also in hyperparathyroidism. Hypophosphatemia may result due to disorders of renal tubular reabsorption.

5.5.5 Total protein

Serum total protein represents the sum total of numerous different proteins, many of which vary independently of each other. Proteins are present in all body fluids but the protein concentration is normally high (>3 g/dL) only in plasma, lymphatic fluids and some exudates. Protein concentration in the cerebrospinal fluid of normal subjects is <45 mg/dL, whereas the urine contains only trace amounts. Measurement of serum-total protein is useful in conditions relating to changes in plasma or fluid volumes, such as shock and dehydration. In these conditions concentration of serum-total protein is elevated indicating haemoconcentration. Haemodilution is reflected as relative hypoproteinemia, which occurs with water intoxication or salt retention syndrome, during massive intravenous infusions.

In addition to dehydration and diarrhoea, increased serum-total protein levels are observed in multiple myeloma. Hypoproteinemia is generally seen in conditions associated with hypoalbuminemia.

5.5.6 Albumin

Serum albumin represents approximately 60% of the total protein found in serum. It is synthesized exclusively in the liver and functions as a regulator of blood oncotic pressure, as a carrier for many cations and water insoluble substances, and as a pool of amino acids for caloric or synthetic purposes. Serum levels of albumin are used to assess nutritional status and have important influences on the metabolism of endogenous substances such as calcium, bilirubin, fatty acids and hormones and on the metabolism of drugs. Hyperalbuminemia has little diagnostic significance except in dehydration. Hypoalbuminemia is very common in many illnesses which impair its liver synthesis (also liver disease), increased catabolism, reduced amino acid absorption, protein loss in urine, malnutrition, protein losing enteropathy and in hospital patients with acute illness.

5.5.7 Alkaline phosphatase (total)

Phosphatases are enzymes which catalyse the splitting of a phosphate from monophosphoric esters. Alkaline phosphatase (ALP), a mixture of isoenzymes from liver, bone, intestine and placenta, has maximum enzyme activity at about pH 10.5. Serum ALP measurements are of particular interest in the investigation of hepatobillary and bone diseases. Serum alkaline phosphatase (adults) − 40−125 U/L (levels up to three times this may be normal in children). Liver, bone and placenta contain very high concentrations of ALP. Increases in total ALP activity are usually related to hepatobiliary and bone disorders. Increased ALP levels are observed in liver diseases, osteomalacia, rickets and bone disorders. Moderate elevations are sometimes noted in congestive heart failure, intestinal disease and intra-abdominal bacterial infections.

5.5.8 Transaminases

The two transaminases of diagnostic importance are serum glutamic oxaloacetate transaminase (SGO) also called aspartate amino transferase or AST, and serum glutamic pyruvate transaminase (SGPT) also called alanine amino transferase or ALT. While AST is found in every tissue of the body, including red blood cells, and is particularly high in the cardiac muscle, ALT is present at moderately high concentration in liver but is low in cardiac, skeletal muscle and other tissues. Both AST and ALT measurements are useful in the diagnosis and monitoring of patients with hepatocellular disease.

5.6 Renal function tests

Renal function is usually explored by measuring blood urea nitrogen (BUN), serum creatinine and urinary creatinine. From the creatinine values the glomerular filtration rate (GFR) can be calculated. The creatinine clearance test compares the level of creatinine in urine with the creatinine level in the blood. It should be noted that the creatinine clearance test is actually an *estimate* of the true GFR as a small amount of creatinine is released by the filtering tubes as a result of their own metabolic activity so it is not all a filtered material from blood. Thus creatinine clearance usually overestimates the GFR; this is particularly true in patients with advanced kidney failure.

Although a single blood sample is taken, total 24 h urine has to be collected since the volume as well as collective creatinine concentration is required. The total amount of creatinine in the 24 h sample will be calculated and, given the determined blood concentration of creatinine for the patient, the total amount of blood needed to be filtered by the kidney to give that amount of creatinine in the urine is calculated. Thus, creatinine clearance GFR (Cr-GFR) is reported as millilitres/minute (mL/min). Normal values are:

- Male: 97 to 137 mL/min.
- Female: 88 to 128 ml/min.

Normal value ranges may vary slightly among different laboratories. Also there is a cheat in that not always 24 h collection is taken, the urine excreted in 4 h might be taken and extrapolated to 24 h. Furthermore, multiple correction charts allow for age and formulas to adjust 'normal' for body surface area have been devised.

5.6.1 Implications of abnormal results

Abnormal results (lower-than-normal creatinine clearance) may indicate:

- acute tubular necrosis;
- bladder outlet obstruction;
- congestive heart failure;
- dehydration;
- glomerulonephritis;
- renal ischemia (blood deficiency);
- renal outflow obstruction (usually must affect both kidneys to reduce the creatinine clearance);
- shock;
- acute renal failure;
- chronic renal failure;
- end-stage renal disease.

There are other methods for calculating GFR based on serum and urinary molecules such as sodium and including proteins such as $\beta 2$ microglobulin.

5.7 Liver function tests

Many of the biochemicals and enzymes routinely assayed are markers of liver disease, alanine amino-transferase (ALT) – a marker of hepatitis; aspartate aminotransferase (AST) – a marker of liver damage, as is albumin. Similarly direct bilirubin and total bilirubin are markers for infant jaundice, whilst alkaline phosphatase (ALP) is often elevated when the bile duct is blocked. Another very important enzyme is gamma-glutamyltransferase (GGT). Isolated elevation of GGT level may be induced by alcohol and aromatic medications, usually with no actual liver disease.

In determining the pathology of liver disorders the pattern of change in the various biomarkers indicates the nature of disease.

5.7.1 Hepatocyte necrosis

In acute hepatitis, toxic injury or ischemic injury results in the leakage of enzymes into the circulation. However, in chronic liver diseases such as hepatitis C and cirrhosis, the serum ALT level correlates only moderately well with liver inflammation. In hepatitis C, liver cell death occurs by apoptosis as well as by necrosis. Hepatocytes dying by apoptosis presumably synthesize less AST and ALT and this probably explains why at least one third of patients infected with hepatitis C virus have persistently normal serum ALT levels despite the presence of inflammation on liver biopsy. Patients with cirrhosis often have normal or only slightly elevated serum AST and ALT levels. Thus, AST and ALT lack some sensitivity in detecting chronic liver injury. Of course, AST and ALT levels tend to be higher in cirrhotic patients with continuing inflammation or necrosis than in those without continuing liver injury.

As markers of hepatocellular injury, AST and ALT also lack some specificity because they are found in skeletal muscle. Levels of these aminotransferases can rise to several times normal after severe muscular exertion or other muscle injury. In fact, AST and ALT were once used in the diagnosis of myocardial infarction.

5.7.2 Cholestasis (lack of bile flow)

Reduction in the outflow of bile often results from the blockage of bile ducts but also from a disease that impairs bile formation in the liver itself. AP and gamma-glutamyltransferase (GGT) levels typically rise to several times the normal level after several days of bile duct obstruction or intrahepatic cholestasis. The highest liver AP elevations, often greater than 1000 U/L, are found in diffuse infiltrative diseases of the liver such as infiltrating tumours and fungal infections.

Both AP and GGT levels are elevated in about 90% of patients with cholestasis. The elevation of GGT alone, with no other liver function test (LFT) abnormalities, often results from enzyme induction by alcohol or aromatic medications in the absence of liver disease. The GGT level is often elevated in persons who take three or more alcoholic drinks (45 g of ethanol/three units or more) per day.

5.7.3 General markers of liver health

Since the secretion of conjugated bilirubin into bile is very rapid in comparison with the conjugation step, healthy individuals have almost no detectable conjugated bilirubin in their blood. Liver disease mainly impairs the secretion of conjugated bilirubin into bile. As a result, conjugated bilirubin is rapidly filtered into the urine, where it can be detected by a dipstick test. The finding of bilirubin in urine is a particularly sensitive indicator of the presence of an increased serum conjugated bilirubin level. Unconjugated bilirubin rises rapidly with haemolytic disease such as in new born due to rhesus incompatibility of mother with fetal/infant blood group. Mild hemolysis, such as that caused by hereditary spherocytosis and other disorders, can also result in elevated unconjugated bilirubin values.

Although the serum albumin level can serve as an index of liver synthetic capacity, several factors make albumin concentrations difficult to interpret. The liver can synthesize albumin at twice the healthy basal rate and thus partially compensate for decreased synthetic capacity or increased albumin losses. Albumin has a plasma half-life of 3 weeks; therefore, serum albumin concentrations change slowly in response to alterations in synthesis. Furthermore, because two thirds of the amount of body albumin is located in the extravascular, extracellular space, changes in distribution can alter the serum concentration.

5.7.3.1 Gilbert syndrome

A major confounding factor in interpreting mildly elevated serum unconjugated bilirubin is Gilbert syndrome. In many healthy people unconjugated bilirubin can be mildly elevated to a concentration of $34-51\,\mu mol/L$, especially after a 24 h fast. If this is the only LFT abnormality and the conjugated bilirubin level and complete blood count are normal, the diagnosis is usually assumed to be Gilbert syndrome, and no further evaluation is required. Gilbert syndrome was recently shown to be related to a variety of partial defects in uridine diphosphate-glucuronosyl transferase, the enzyme that conjugates bilirubin.

5.7.3.2 Prothrombin time

The liver synthesizes blood clotting factors II, V, VII, IX and X. The prothrombin time (PT) does not become abnormal until more than 80% of liver synthetic capacity is lost. This makes PT a relatively insensitive marker of liver dysfunction. However, abnormal PT prolongation may be a sign of serious liver dysfunction. Since factor VII has a short half-life — about 6 h — it is sensitive to rapid changes in liver synthetic function. Thus, PT is very useful for following liver function in patients with acute liver failure. An elevated PT can result from a vitamin K deficiency. This deficiency usually occurs in patients with chronic cholestasis or fat malabsorption from disease of the pancreas or small bowel.

5.8 Heart disease and lipid disorder tests

Lipid disorders, high blood cholesterol and triglycerides, increase the risk for atherosclerosis, and thus for heart disease, stroke and hypertension. There are three main serum forms of cholesterol which are measured:

- total cholesterol;
- high-density lipoprotein (HDL) cholesterol; and
- low-density lipoprotein (LDL) cholesterol.

Along with triglycerides and Apolipoproteins (Apo A-1, etc.), cholesterols are usually measured as a panel of tests.

The laboratory test for HDL or LDL measures how much cholesterol is present in high-density or low-density lipoprotein fractions, not the actual amount of HDL or LDL in the blood. Thus, density centrifugation or zonal electrophoresis would separate high- and low-density lipoprotein fractions in a blood sample and cholesterol would be determined. Lipoprotein levels themselves can also be measured by specific immunoassay.

Cholesterol can be measured by linked colorimetric enzymatic reactions. For example cholesterol esters are hydrolysed by cholesterol esterase, to produce cholesterol which is then oxidized by cholesterol oxidase producing hydrogen peroxide (H_2O_2). H_2O_2 is quantified by the Trinder reaction where 4-aminoantipyrine, 4-chlorophenol produces a coloured product catalysed by hydrogen peroxidase (POD) monitored at 546 nm (see colorimetric assays, Section 5.1.2).

Triglycerides are similarly measured by linked colorimetric enzymatic methods based on the hydrolysis of triglyceride by lipase to glycerol and free fatty acids. Glycerol, in a reaction catalysed by glycerol kinase, is converted to glycerol-3-phosphate. In a third reaction, glycerol-3-phosphate is oxidized by glycerol phosphate oxidase to dihydroxyacetone phosphate and hydrogen peroxide. H_2O_2 is then quantified by the **Trinder** reaction.

Several genetic disorders lead to abnormal levels of cholesterol and triglycerides:

- familial combined hyperlipidemia;
- familial dysbetalipoproteinemia;
- familial hypercholesterolemia; and
- familial hypertriglyceridemia.

However, abnormal cholesterol and triglyceride levels are more often caused by:

- being overweight or obese (metabolic syndrome);
- medications, including birth control pills, oestrogen, corticosteroids, diuretics, beta blockers, and antidepressants;
- pre -existing diseases, for example diabetes, hypothyroidism, Cushing's syndrome, polycystic ovary syndrome and kidney disease;
- excessive alcohol consumption;
- diets that are high in saturated fats (red meat, egg yolks and high-fat dairy products) and transfatty acids (found in commercial processed foods);
- lack of exercise and sedentary lifestyle;
- smoking (depresses HDL levels).

5.8.1 HDL and LDL

The main function of HDL is to transport cholesterol from the walls of blood vessels to the liver, where it is broken down and removed in the bile. In general, the risk for heart disease, including a heart attack, increases if HDL cholesterol levels are <40 mg/dL. An HDL >60 mg/dL is associated with reduced risk of heart disease. Women tend to have higher HDL cholesterol than men and a low HDL level may indicate increased risk for atherosclerotic heart disease. A low HDL level may also be associated with familial combined hyperlipidemia; noninsulin-dependent diabetes (NIDD) and certain drugs.

High levels of LDL may be associated with an increased risk of atherosclerotic heart disease and familial hyperlipoproteinemia. However, very low levels of LDL may be caused by malabsorption (inadequate absorption of nutrients from the intestinal tract) or malnutrition. A healthy LDL level is one that falls in the optimal or near-optimal range:

- optimal: less than 100 mg/dL (less than 70 mg/dL for persons with a history of heart disease);
- near optimal: 100−129 mg/dL;
- borderline high: 130−159 mg/dL;
- high: 160−189 mg/dL;
- very high: 190 mg/dL and higher.

5.8.2 Triglycerides

Generally we try to keep fats to a minimum but these obviously fluctuate after a meal so like the lipoproteins fasted triglycerides are measured as a true indicator of underlying health and condition. High triglyceride levels may be due to:

- cirrhosis;
- diet low in protein and high in carbohydrates;
- familial hyperlipoproteinemia (rare);
- hypothyroidism;
- nephrotic syndrome;
- pancreatitis; or
- poorly controlled diabetes.

Low triglyceride levels may be due to a low-fat diet, hyperthyroidism, malabsorption syndrome and malnutrition. A healthy triglyceride level is one that falls in the optimal or near-optimal range:

- normal: less than 150 mg/dL;
- borderline high: 150–199 mg/dL;
- high: 200–499 mg/dL; or
- very high: 500 mg/dL or above.

5.8.3 Apolipoproteins

Specific to lipid metabolism apolipoproteins are found in the HDL and LDL complexes. Apolipoprotein A-I is the major protein component of HDL in plasma. The protein promotes cholesterol efflux from tissues to the liver for excretion. It is a cofactor for lecithin cholesterolacyltransferase (LCAT) which is responsible for the formation of most plasma cholesterol esters. ApoA-I was also isolated as a prostacyclin (PG_{I2}) stabilizing factor, and thus may have an anticlotting effect. Defects in the gene encoding are associated with HDL deficiencies, including Tangier disease, and with systemic non-neuropathic amyloidosis. Apolipoprotein B (ApoB) is associated predominately with LDL and therefore high levels are associated with heart disease. It is reported that brain apolipoprotein E (ApoE) **levels** influence amyloid-β (Aβ) deposition and thus the increased risk for Alzheimer's disease.

5.8.4 Markers of myocardial infarction

For many years the determination of serum creatine kinase (CK) activity was the mainstay of biochemical tests for acute myocardial infarction. Isoelectrophoresis demonstrated that the increase was due to a specific isoform MB – MBCK. With antibody technology specific immunoassays for the MB and even a myocardial MBCK were developed. However, biomarker research eclipsed the decades of evaluation of CK with the discovery that a protein specifically associated with the microfilament and cytoskeleton of myocytes was released into serum as a result of injury to the heart muscle, that is as a result of ischemia. Termed Troponins, immunoassays specific to these cytoskeleton associated proteins – Troponin I and

Troponin T – developed, Troponin T being specific to the cytoskeletal filaments found in heart muscle cells. Troponin I and T are now the main stay markers of myocardial infarction:

- blood creatine kinase (CK or CPK) levels: male, 38–174 units/L; female, 96–140 units/L;
- MBCK level: normally less than 5% of total CK;
- Troponin I and Troponin T: normally undetectable very high in MI, but slight to moderate elevation during any damage episode to heart tissue such as inflammation – myocarditis.

5.9 Pancreatic function tests

The mainstay clinical chemistry tests for assessing the endocrine and exocrine function of the pancreas are amylase c-peptide and insulin assay.

5.9.1 Amylase

In acute pancreatitis, amylase in the blood increases often to four to six times higher than the highest reference value (upper limit of normal). The increase occurs within 12 h of injury to the pancreas and generally remains elevated until the cause is successfully treated. Then the amylase values will return to normal in a few days. In chronic pancreatitis, amylase levels initially will be moderately elevated but often decrease over time with progressive pancreatic damage. A urine amylase test may also be ordered. Typically, its level will mirror blood amylase concentrations, but both the rise and fall occur later. Amylase levels may also be significantly increased in patients with pancreatic duct obstruction, cancer of the pancreas, and gallbladder obstruction or infection. Measured by colorimetric assay the normal range is 23–85 U/L.

5.9.2 Insulin and c-peptide

Both Insulin and c-peptide are measured by immunoassay. Insulin determination is used to diagnose insulinoma (a tumour of the islets of Langerhans) and to evaluate patients with fasting hypoglycaemia. It is often combined with a fasting plasma glucose test to increase its diagnostic value with respect to diabetes.

Proinsulin, is transported to the Golgi apparatus where it is packed into secretory granules. Maturation of the secretory granules is associated with enzymatic cleavage of proinsulin to insulin and C-peptide. Secretion of insulin into the bloodstream is accompanied by the release of small amounts of proinsulin. Although the absence of insulin/proinsulin is diagnostic of insulin dependent diabetes, the fact that many insulin assays use antibodies that cross react significantly with proinsulin is not always helpful. In pathological conditions elevated apparent 'immunoreactive' Insulin may be nonfunctional because what is released is proinsulin which has not been cleaved to active insulin (and c-peptide). However, c-peptide immunoassays are specific for the free c-peptide cleavage fragment which is also released. Thus, an elevated plasma glucose level in the presence of apparently high-insulin levels may be due to incorrect processing and release of proinsulin (some insulinomas). This can be diagnosed by the absence of c-peptide. If both are elevated this may indicate insulin resistance – noninsulin dependent diabetes (see Figure 5.27).

A significant use of c-peptide assay is in forensics: self abuse with injected insulin or indeed injected insulin induced coma death can be determined by the absence of c-peptide in the serum.

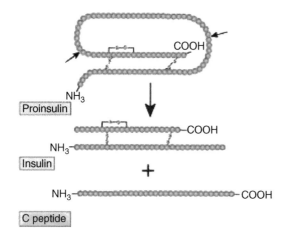

Figure 5.27 Diagrammatic representations of proinsulin and its processing to active insulin and c-peptide

5.10 Bone disease assessment

Classically clinical chemistry assessment of bone disease has been by measuring calcium, phosphate and alkaline phosphatase activity as discussed above. More recently newer markers of bone metabolism have been discovered and are moving from research to routine clinical assessment, primarily in the assessment of primary bone cancers (Ewings sarcoma and oesteosarcoma) and bone metastasis (from primarily breast and prostate cancers).

These new markers are cleavage fragments from the synthesis (oesteoblast activity) or destruction (oesteoclast activity) of collagen complexes that forms the base matrix to which calcium is deposited to form calcified bone. As matrix collagen is synthesized by oesteoclasts it has a preprocessing N-terminal and C-terminal propeptide regions (PINP and P1CP, see Figure 5.28(a)) which are cleaved as the collagen α chains are linked into twisted fibrils and finally together by cross-linking telopeptides which link the N-terminal and C-terminal regions of the fibril α chains complex into large collagen fibres in the extracellular matrix (see Figure 5.28(b)). Cleavage of the C-terminal peptide – PICP – is an early event in collagen synthesis but cleavage and release into circulation of the N-terminal peptide – PINP – appears to be a late process in the formation of new bone. During bone destruction C-Telopeptides (CTX) and N-Telopeptides (NTX) are released into circulation and further metabolized to pyridinoline excreted in the urine (see Figure 5.28(c)).

Immunoassays to CTP, PINP, PICP and NTX are being evaluated as new markers of bone disease.

5.11 Endocrinological assessments

Endocrinological assessments cover a very broad group of medical disciplines and disorders, from neuroendocrinology to reproduction in which markers of the hypothalamic-pituitary adrenal axis, the pituitary thyroid axis and the pituitary gonadol axis are equally important. The endocrine functioning of the other endocrine glands such as the parathyroids are also important to specialists in metabolic disorders.

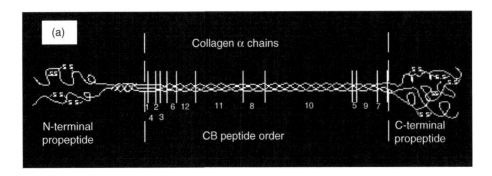

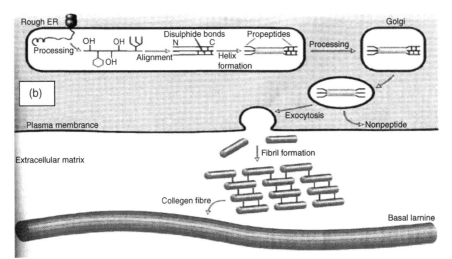

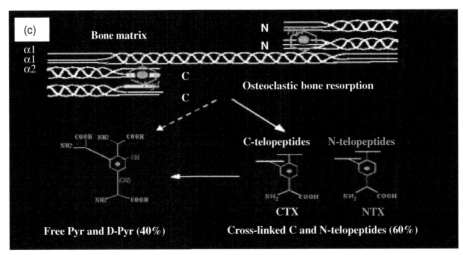

Figure 5.28   Demonstrating the structure of procollagen positions of N-terminal (PINP) and C-terminal (PICP) propedtides (a), their cleavage when laid down as large complex fibril units of extracellular collagen fibres (b) and the structure and position of C and N terminal telopeptides which join the collagen fibres together, but are released during bone metabolism (60% as free telopeptides) and further degraded (40%) to pyridinoline and deoxy- (c) pyridinoline

5.11.1 Thyroid function tests

The hypothalamus senses low circulating levels of thyroid hormones – Triiodothyronine (T3) and Thyroxine (T4) and responds by releasing thyrotropin releasing hormone (TRH). This stimulates the pituitary to produce thyroid stimulating hormone (TSH). The TSH, in turn, stimulates the thyroid gland to produce thyroid hormones T3 and T4 until levels in the blood return to normal. Thyroid hormone exerts negative feedback control over the hypothalamus as well as the anterior pituitary thus controlling the release of both TRH from the hypothalamus and TSH from the anterior pituitary gland (Figure 5.29).

The thyroid hormones T3 and T4 affect cells' metabolism by increasing metabolic rates. In clinical chemistry usually only TSH and T4 levels are measured by specific immunoassay to assess thyroid pathology. However, T3 has to be measured to confirm T3 toxicosis; given the nonpeptide biochemical nature of T3 and T4, HPLC and HPLC-MS methods are sometimes used to measure these molecules:

- TSH normal range: 0.3–3.5 mLU/L;
- free T4 normal range: 10–25 pmol/L;
- free T3 normal range: 3.5–7.5 pmol/L.

5.11.1.1 Disease associated with disturbance of the pituitary thyroid axis

In thyrotoxicosis TSH levels are suppressed (<0.05 mLU/L), T4 and T3 levels are increased, whilst in primary hypothyroidism TSH levels are increased, T4 levels decrease and T3 levels are either normal or low. In T3 toxicosis TSH levels are suppressed but T4 levels are normal and T3 levels are significantly Increased. In patients with TSH deficiency disorders TSH levels are significantly decreased, T4 levels are also decreased and T3 levels are normal or low.

5.11.2 The hypothalamic–pituitary–adrenal (HPA) axis

The hypothalamus contains neuroendocrine neurons that synthesize and secrete vasopressin/ADH and corticotropin-releasing hormone (CRH). These two peptides regulate the function of proopiomelanocortin (POMC) endocrine secretory cells of the anterior lobe of the pituitary gland. In particular, CRH and vasopressin stimulate the secretion of adrenocorticotropic hormone (ACTH). ACTH in turn acts on the adrenal cortices, which produce glucocorticoid hormones – mainly cortisol. The glucocorticoids in turn act back on the hypothalamus and pituitary to suppress CRH and ACTH production in a negative feedback cycle (see Figure 5.30).

Vasopressin is also a powerful antidiuretic hormone and hence is also called ADH. It is released when the body is dehydrated and has potent water-conserving effects on the kidney. It is also a potent vasoconstrictor. Interestingly ACTH is derived from the cleavage of a much larger peptide

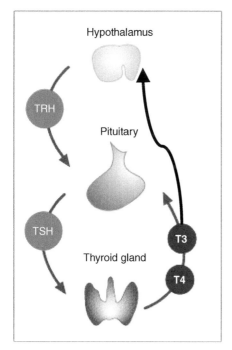

Figure 5.29   The hypothalamic pituitary thyroid axis

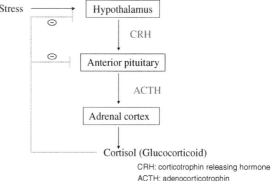

Figure 5.30   The hypothalamic–pituitary–adrenal axis

proopiomelanocortin (POMC) of which the other major products are β-endorphins.

Cortisol has widespread effects on cell metabolism, fatigue and moods; levels dramatically increase after waking slowly decreasing thereafter and is further increased during periods of physical stress. It also dampens down immune reactions.

All hormones of the HPA axis can be measured by specific immunoassay but given the nonpeptide biochemical nature of the glucocorticoids HPLC and HPLC—MS methods are sometimes used to measure these molecules. However, levels fluctuate markedly during the day and between individuals. Thus many of the assessments are measuring the dynamics of the changes in levels after stimulation or specific suppression.

5.11.2.1 Diseases associated with disturbance of the HPA axis

Addison's disease — ACTH stimulation test — in normal individuals 30—60 min after being injected with an ACTH anolog drug plasma cortisol should nearly double from 300—600 nmol/L to 500—1200 nmols/L but in Addison's disease cortisol levels will essentially remain unchanged.

Cushing's syndrome — dexamethasone suppression test — in normal individuals after administration of dexamethasone plasma cortisol will rapidly drop from 300—600 nmols/L to a reference range in the order of undetectable to ~75 nmols/L.

Adrenal insufficiency — ACTH stimulation test — a response intermediate between normal and full-blown Addison's disease.

Diabetes insipidus — major cause is deficiency in vasopressin or target tissue insensitivity to vasopressin; often diagnosis is excessive urination and monitoring a high plasma osmolarity and low urinary osmolarity.

Stress — elevated basal cortisol.

5.11.3 The growth hormone — IGF anabolic endocrine cascade and axis

Also referred to as the somatotrophin—somatomedin system, growth hormone-releasing hormone (GHRH, also known as GRF — growth hormone releasing factor — or sommatocorinin) is a 44 amino acid peptide hormone produced by neuroendocrine hypothalamic cells. GHRH, is secreted in bursts, particularly during sleep, and plays a key role in regulating the body's major anabolic endocrine pathway. GHRH secreted from the hypothalamus is carried to the anterior pituitary gland where it stimulates the production and release of growth hormone/somatotrophin (GH). There are a number of growth hormone isoforms, all of which are secreted in response to GHRH (prolactin is a structural analog of GH but is not controlled by GHRH).

The actions of GHRH are opposed by 'growth-hormone-inhibiting hormone', more commonly termed somatostatin. Somatostatin is released from neurosecretory somatostatin neurons, and is carried to the anterior pituitary where it inhibits GH secretion. Somatostatin and GHRH are secreted in alternation, giving rise to the markedly pulsatile secretion of GH (see Figure 5.31).

Pulses of GH released from the pituitary have numerous growth-enhancing and metabolic effects and also stimulate the secretion of insulin-like growth factor-1/somatomedin (IGF-1) from the liver. IGF-1 has major direct anabolic effects on a range of tissues and also causes the release of numerous other growth factors. IGF-1 and other growth factors exert negative feedback effects which control and limit secretion of GH and GHRH. This mechanism is important in preventing excessive production and release of GH.

Growth hormones and the IGFs are usually measured by immunoassay. Normal ranges: GH - fasting 0—5 μg/L but arginine stimulated it will be greater than 7 μg/L; IGF-1 ranges from 570 to 800 μg/L.

5.11.3.1 Diseases commonly associated with disturbance of the GH—IGF axis

Overproduction of growth hormone causes excessive growth. In children, the condition is called gigantism. In adults, it is called acromegaly. Acromegaly is manifest by growth of soft tissue but in childhood the epiphyseal (growth) plates are still active and thus excessive skeletal growth also occurs. However, it most commonly affects middle-aged adults and can result in serious illness and premature death. Once recognized, acromegaly is treatable in most patients, but because of its slow and often insidious onset, it frequently is not diagnosed correctly.

There are multiple types of growth retardation and dwarfism — some simply due to a lack of GH others are due to a deficiency in the IGF-1 response to GH which are now termed severe primary IGF deficiencies.

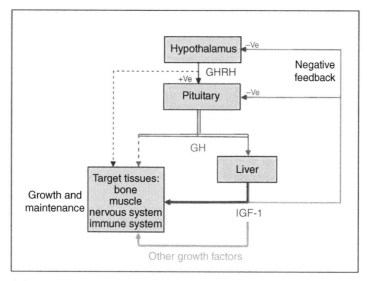

Figure 5.31   The growth hormone — insulin-like growth factor/somatotrophin-somatomedin system

Patients with severe primary IGFD typically present with normal to high GH levels, height below -3 standard deviations (SD), and IGF-1 levels below −3SD. Severe primary IGFD includes patients with mutations in the GH receptor, postreceptor mutations or IGF mutations. In Laron dwarfism there are no GH receptors. As a result, IGFD patients cannot be expected to respond to GH treatment.

5.11.4   Prolactin

Hypothalamic neuronal dopamine release stimulates anterior pituitary prolactin secreting cells to release prolactin. Prolactin stimulates differentiated breast ductal epithelia into lactation (milk production). It also has many other functions, including essential roles in the maintenance of the immune system and disrupting the menstrual cycle. Measured by immunoassay prolactin normal ranges in females are up to 20 µg/L and in males are up to 15 µg/L.

5.11.4.1   Diseases commonly associated with disturbance of prolactin secretion

Hyperprolactinemia, abnormally high prolactin, can delay puberty, in women interfere with ovulation, decrease libido in men and decrease fertility. Elevated prolactin may be due to a benign tumour in the pituitary gland — a prolactinoma.

Hypoprolactinemia, abnormally low prolactin (or prolactin deficiency), can cause menstrual disorders and lead to inadequate lactation. Low prolactin is due to impaired production by prolactin secreting cells of the anterior pituitary gland.

5.11.5   Parathyroid hormone and calcitonin

Parathyroid hormone (PTH), is secreted by the parathyroid glands as a mature polypeptide of 84 amino acids. It acts to increase the concentration of calcium (Ca^{2+}) in the blood, whereas calcitonin — a hormone produced by the parafollicular cells (C cells) of the thyroid gland — acts to decrease serum calcium concentration. PTH targets parathyroid hormone receptors in three parts of the body:

• bone — release of calcium;
• kidney — increased reabsorption of calcium; and
• intestine — increased absorption of calcium.

This latter action is indirect and via another metabolic pathway of the kidneys. Vitamin D activation occurs in the kidney. PTH upregulates 25-hydroxyvitamin D_3 1-alpha-hydroxylase, the enzyme responsible for 1-alpha hydroxylation of 25-hydroxy vitamin D, converting vitamin D to its active form (1,25-dihydroxy vitamin D). This activated form of vitamin D increases the absorption of calcium (as Ca^{2+} ions) by the intestine via calbindin.

A fetal variant (PTH-like hormone) is larger and consists of a propeptide of 139 aminoacids (but splice variants also give propeptides of 173 and 141 amino acids). This is in turn subject to proteolytic processing to give three PTH related peptides (PTHrp) a 1-36 amino acid N terminal and PTH-like fragment, a nuclear localization midregion fragment (AAs 38–94) and a C - terminal oesteostatin peptide (AAs 107–139).

Measured by immunoassay PTH levels are normally 1.1–7.5 pmol/L. Vitamin D is measured by HPLC and HPLC MS and the normal range is 30.0–74.0 μg/L.

5.11.5.1 Diseases commonly associated with disturbance of parathyroid hormone secretion

Hypoparathyroidism, low PTH levels, may be caused by inadvertent removal of parathyroids during routine thyroid surgery, autoimmune disorders and inborn errors of metabolism.

Hyperparathyroidism, high PTH levels can be caused by parathyroid adenoma, parathyroid hypeplasia and parathyroid cancer all termed primary hyperparathyroidism. Chronic renal failure – secondary hyperparathyroidism – is where serum calcium levels are decreased because of failed kidney reabsorption and vitamin D activation, which causes the hypersecretion of PTH from the parathyroid glands.

Apparent primary hyperparathyroidism can occur due to ectopic expression of PTH-like hormone by common cancers and the action of the PTHrps giving rise to parathyroid hormone related peptide is humoral hypercalcemia of malignancy (HHM) by nonparathyroid tumours.

5.11.6   Pituitary gonadol axis and reproductive endocrinology

Hypothalamic pulsitile release of gonadotrophin releasing hormone (GnRH) results in stimulation of anterior pituitary gonadotrophic cells to produce and release luteinizing hormone (LH) and follicle stimulating hormone (FSH). The classical roles of LH and FSH are the control of folliculogenesis and ovulation as well as the regulation of the spermatogensis (see Figure 5.32 and Table 5.5).

Table 5.5   Normal ranges of reproductive hormones.

Hormone		Normal range
Testosterone	Men younger than 50	10—45 nmol/L
	Men older than 50	6.2—26 nmol/L
	Females	0.5—2 nmol/L
Dihydrotestosterone	Adult men	8—35 nmol/L
Estradiol	Follicular phase	400—1500 pmol/L
	Luteal phase	70—600 pmol/L
	Postmenopause	<130 pmol/L
Progesterone	Follicular phase	2.2—9 nmol/L
	Luteal phase	17—92 nmol/L
FSH	Adult men	1—8 IU/L
	Female follicular phase	5—8 IU/L
	Luteal phase	1.5—5 IU/L
	Ovulatory peak	5.5—15 IU/L
	Postmenopause	>35 IU/L
LH	Adult men	2—14 IU/L
	Female follicular phase	3—7 IU/L
	Luteal phase	<1—6 IU/L
	Ovulatory peak	20—75 IU/L
	Postmenopause	>20 IU/L

From menarche to menopause, the menstrual cycle produces a mature oocyte once a month. Primary oocytes arrest during the first meiotic prophase and are held within immature follicles. The human menstrual cycle takes 28–30 days and only a few follicles grow every cycle during the following three phases:

- a follicular phase, consisting of preantral and antral stages first 10–12 days;
- an ovulatory(/preovulatory) stage, next 1–2 days; and
- a luteal phase of 12–15 days.

The initial stages of preantral development occur independently of any extraovarian controls but it is generally believed that cytokines induce graffian follicle development. During the antral stage GnRH production by the hypothalamus results in LH and FSH production by the pituitary. LH induces the synthesis of androgens from the follicular theca interna (an inner glandular, highly vascular layer surrounding the preantral follicle) and FSH stimulates aromatization of androgens to oestrogens by the follicular granulosa

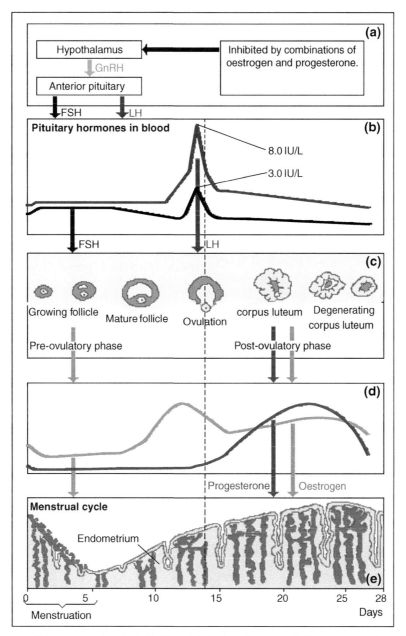

Figure 5.32   Summary of the menstrual cycle indicating: (a) the hypothalamo-pituitary-ovarian interactions, (b) serum LH and FSH levels, (c) the morphology of follicular development, (d) serum oestrogen and progesterone levels, and (e) the sequence of endometrial changes

cells (the cellular population outside the oocyte contacting cells of the zona pellucida). A number of cytokines also function at that point: activin suppresses androgen output and increases the aromatization capabilities of the granulosa cells, while inhibin has the opposite effects of stimulating androgen output and suppressing aromatizing activity. Endometrial changes that eventually lead to endometrial maturation occur as soon as oestrogens are produced and the endometrium prepares for the implantation of any fertilized oocyte.

Midway through a cycle and after the LH surge preovulatory growth commences of a primed immature graffian follicle, as part of the next menstrual cycle. During this stage, the secondary oocyte is formed following the reinitiation of meiosis. During the second half of meiosis, the oocyte is arrested, again at the metaphase point, and this is the maturation state of the oocyte at day 14 ovulation. During preovulatory growth the granulosa cells stop synthesizing oestrogen from androgen and start synthesizing progesterone; LH, therefore, stimulates granulosa cell differentiation and ovulation simultaneously. Postovulation the remaining tissue of the ruptured follicle becomes the corpus luteum; granulosa cells undergo hypertrophy to form large lutein cells. The cells of the theca also form lutein cells, although they are smaller in size. Lutein cells secrete progesterone and oestrogens.

5.11.6.1   Spermatogenesis

Testicular functions in the male are governed by neuroendocrine mechanisms similar to those that regulate ovarian activities. GnRH production by the hypothalamus in males causes the secretion of the gonadotrophins LH and FSH which in turn regulate the endocrine and spermatogenic activities of the testis.

LH stimulation of the testicular Leydig cells initiates the production of testosterone, the key androgen in spermatogenesis, which feeds back down regulating GnRH and LH production by the pituitary. The absence of testosterone leads to hypogonadotropism. FSH acts on specific receptors on the Sertoli cells, leading to Inhibin-B production, which selectively negatively regulates FSH secretion (see Figure 5.33).

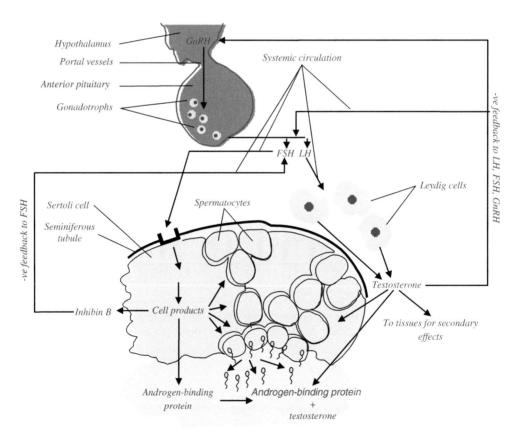

Figure 5.33   In the control of spermatogenesis, GnRH is the key neuropeptide acting on gonadotrophs of the anterior pituitary, which in turn produce LH and FSH. The testicular gonadol steroid testosterone and protein hormone inhibin B products, feed back to negatively downregulate Hypothalamic GnRH and pituitary FSH secretion

Most reproductive hormones are measured by immunoassay although HPLC mass spectrometry is used to measure steroid levels in criminal and athletic doping as the positive identification and quantitative accuracy of this method is deemed more evidential than immunoassays.

5.11.6.2 Diseases commonly associated with disturbance of the pituitary—gonadol axis

Given the nature of the pituitary—gonadol axis any reproductive conditions will alter gonadol/sex steroid levels be it testosterone, progesterone, estradiol and dihydrotestosterone. Often it is more informative to combine steroid measurement with that of gonadotropin levels; for example poor ovarian reserve — also known as premature ovarian aging — is characterized by high FSH levels but not LH high levels. In polycystic ovarian disease LH levels are elevated but FSH levels are lowered. Other conditions associated with altered gonadotropin levels are listed in Table 5.6.

Table 5.6   Diseases associated with high and low LH and FSH levels.

Altered gonadotropin levels	Conditions/pathological cause
Consistently high LH and FSH	• Premature menopause
	• Gonadal dysgenesis, Turner syndrome
	• Castration
	• Swyer syndrome
	• Polycystic ovary syndrome
	• Certain forms of congenital adrenal hyperplasia
	• Testicular failure
Low LH and FSH	• Kallmann syndrome
	• Hypothalamic suppression
	• Hypopituitarism
	• Eating disorder
	• Female athlete triad
	• Hyperprolactinemia
	• Gonadotropin deficiency
	• Gonadal suppression therapy (GnRH antagonist/agonist)

5.12   Pregnancy tests and pregnancy clinical chemistry

Pregnancy tests are based on immunoassay technology (Section) and therefore rely on antibody specificity for test specificity. The antibodies used in pregnancy tests are raised against the hormone human chorionic gonadotrophin (hCG) and can be poly- or monoclonal. hCG is a unique hormone which is produced in only a very few clinical situations namely pregnancy and some cancers. For this reason a qualitative test is often sufficient to determine pregnancy.

hCG is a heterodimeric glycoprotein hormone comprised of an alpha and beta subunit. The alpha subunit is shared with the other three members of the glycoprotein hormones: luteinizing hormone (LH), follicle stimulating hormone (FSH) and thyroid stimulating hormone (TSH). The beta subunits are, however, distinct but display considerable homology in their amino acid sequences and three-dimensional structures. hCG is produced by the trophoblast cells of the early blastocyst and after implantation hCG circulates within the maternal blood to exert its endocrine function on the ovary. hCG binds the LH receptor, which again reflects the homology between the two closely related molecules, and stimulates the expression of 20α-hydroxylase side chain cleavage enzymes and 3β-hydroxsteroid dehydrogenase/isomerase to convert cholesterol to pregnenolone and pregnenolone to progesterone respectively. This increase in the production of progesterone results in the maintenance of a thickened and receptive endometrium and prevents menstruation. In essence hCG (produced by the placenta) takes over from LH in the stimulation of progesterone to allow pregnancy to continue. In the absence of hCG menstrual cycling would continue. hCG is therefore critical to the initiation of pregnancy in humans and, as such, it is logical to use hCG as a single test diagnostic marker for pregnancy.

5.12.1   Pregnancy physiology

hCG rises throughout the first weeks of pregnancy doubling every 2 days until around week 10 where it declines and stabilizes at a much lower concentration until term. Other hormones, progesterone, hPL and estrogens also rise through pregnancy. These significant changes in the endocrinology bring about physiological changes in the maternal cardio-vascular, respiratory, renal, haematological, gastrointestinal and reproductive systems, and also signal changes in the nutritional requirements of the woman (see Table 5.7).

5.12.2 Prenatal diagnosis for fetal aneuploidy

This is offered to all pregnant women in the UK. Practice varies in different maternity units, with some

Table 5.7   Pregnancy changes in major organ system characteristics and associated clinical chemistry.

Body system	Comments	Specific changes
Cardiovascular system	Blood pressure — peripheral vascular resistance falls and there is normally a fall in BP during the second trimester (5–10 mmHg systolic, 10–15 mmHg diastolic), and then returns to normal during the third trimester. Many of the effects of the altered cardiovascular system mimic heart failure.	Cardiac output increases about 30–50% (from 4.5 to 6.0 L/min)
	Cardiac output increases early in gestation and is maintained at an increased level until delivery.	Stroke volume increases about 10–15%
		Pulse increases about 15–20 bpm
		Systolic ejection murmur and S3 gallop is common (about 90% of pregnant women)
Respiratory system	Respiratory rate, vital capacity and inspired reserve volume remain unchanged but other factors decrease.	Functional residual capacity (−20%)
		Expired reserve volume (−20%)
		Residual volume (−20%)
		Total lung capacity (−5%)
Renal system	Anatomical changes see an increase in kidney size and weight, with ureteric dilatation (right > left), the bladder becomes an intraabdominal organ.	GFR increases by 50%
		Renal plasma flow increases by 75%
		Creatinine clearance increases to 150–200 mL/min
Metabolic changes	—	Serum creatinine decreases by about 25%
		Plasma osmolarity decreases about 10 mOsm/kg H_2O
		Increase in tubular reabsorption of sodium
		Marked increase in renin and angiotensin levels, but markedly reduced vascular sensitivity to their hypertensive effects
		Increase in glucose excretion
Haematology	Plasma volume increases by about 50% and RBC volume increases by about 30% This result is 'dilutional anaemia of pregnancy', such that the mean haemoglobin level during pregnancy is about 11.5 g/dL. WBC count increases during pregnancy and platelet count decreases, but stays within normal limits. Large blood loss is well tolerated (up to 1.5 L) thus maternal vital signs cannot be relied on.	Coagulation system — 'hypercoagulable state'
		Increased level of fibrinogen, factors VII–X
		The placenta produces plasminogen activator inhibitor

only offering screening to high-risk mothers. Prediction factors for high-risk include women who are older than 35 years and those with a history or family history of chromosomal abnormalities.

First trimester:
- ultrasound (high-resolution) for nuchal translucency to exclude major chromosomal abnormalities (e.g. trisomies and Turner syndrome);
- maternal serum for pregnancy-associated plasma protein-A (PAPP-A from the syncytial trophoblast) for trisomy 21 — This is more accurate than the triple test at 16 weeks.

Second trimester
- ultrasound for structural abnormalities (e.g. neural tube defects, congenital heart defects) — serum screening has been superseded by high-resolution ultrasound (above) in specialist centres;
- triple test for chromosomal abnormalities: this consists of a serum α-fetoprotein (low), unconjugated oestradiol (low) and human chorionic gonadotrophin (high);
- α-fetoprotein (high for neural tube defects).

All markers are corrected for gestational ages a multiple of the mean (MOM) value for the appropriate week of gestation. If abnormalities are detected, it is necessary to continue investigations with chorionic villus sampling (CVS) at 11—13 weeks, or amniocentesis at 15 weeks, under ultrasound control to sample amniotic fluid and fetal cells necessary for cytogenetic testing. These tests may well be superseded by salvage of fetal cells from the maternal blood sample at 12 weeks.

5.13 Therapeutic drug monitoring and toxicology

Many medications are used without monitoring of blood levels, as their dosage can generally be varied according to the clinical response. In a small group of drugs, this is difficult, as insufficient levels will lead to under treatment or resistance, and excessive levels can lead to toxicity and tissue damage. Therapeutic drug measurement of medication levels in blood is essential under these circumstances. Thus, the main focus of therapeutic drug monitoring is on drugs with a narrow therapeutic range.

5.13.1 Prescribed drugs

Gentamicin is a vestibulotoxin, and can cause permanent loss of the sense of balance, caused by damage to the vestibular apparatus of the inner ear, usually if taken at high doses or for prolonged periods of time. Some individuals have a normally harmless mutation in their mitochondrial RNA that allows the gentamicin to affect their cells. The cells of the ear are particularly sensitive to this, sometimes causing complete hearing loss and in some cases gentamicin completely destroy the vestibular apparatus after 3—5 days of high exposure. Ideally levels are kept at less than 1 mg/L of blood.

Other commonly monitored therapeutic drugs:

- antiepileptic, for example carbamazepine, phenytoin and valproic acid;
- mood stabilizers, for example lithium citrate;
- antipsychotics, for example pimozide and clozapine.

Given the small organic nature of most drugs HPLC—mass spectrometry (quadrupole) is often used as the detection and quantification methodology. Although other colorometric and immunoassay based systems are used depending on the nature of the drug.

Drugs with narrow therapeutic range can often be accidental toxins if taken by children or even deliberately administered in homicides (see Table 5.8).

Aside from drugs that have a very narrow therapeutic range which can easily be exceeded into their toxicity range, other common, over the counter, drugs can be taken at deliberate vast excess and in to the toxic range:

- salicylate — asprin overdoses from 150—300 mg/kg; mild-to-moderate toxicity serum levels being between 300—500 mg/L;
- iron — particularly children taking their mother's iron tablets; children can show signs of toxicity with ingestions of 10—20 mg/kg of elemental iron.

5.13.2 Over the counter drugs—paracetamol poisoning

Paracetamol poisoning is a common toxicological test in clinical laboratories. First used in the late nineteenth century, in the 1960s the first reported liver failure in

Table 5.8 Prescribed drugs — therapeutic drug monitored.

Therapeutic drug	Medical condition	Therapeutic level
Digoxin	Atrial fibrillation	Plasma concentration 1.0—2.6 nmol/L
Lithium	Antipsychotic	Plasma concentrations 0.6—1.2 mmol/L
Theophylline	Mast cell inhibitor used to treat asthma	Serum concentration 10—20 mg/L
Carbamazepine	Antiepileptic is also used as a treatment for patients with manic-depressive illness	Plasma concentration is 4—12 mg/L
Methotrexate	Chemotherapy but used to downregulate immune reaction in arthritis	Therapeutic range 0.01 μmol—10 μmol over 24 h
Phenobarbital	Barbiturate abuse	Toxicity is seen at plasma levels >40 mg/L
Phenytoin	Anticonvulsant	Therapeutic plasma concentration <80 μmol/L
Thyroxine	Treats hypothyroidism	Lowest recorded toxic dose to a child has been 14—20 μg/kg

humans was attributed to paracetamol poisoning. There had been studies on rats and cats but little was known about the side effects but as use of the drug increased, so too did the numbers of overdoses. In 1973 scientists first established the biochemical basis of paracetamol toxicity which led to the development of an antidote: N-acetylcysteine.

The toxicity of a paracetamol overdose is as a result of one of the pathways by which the liver metabolizes this drug (Figure 5.34). In normal doses, paracetamol is conjugated by glutathione to glucuronide or sulphate in the liver. A small amount is also converted to N-acetyl-para-benzoquinoneimine (NABQI) — an alkylating agent that reacts with glutathione This gives an intermediate that is metabolized to nontoxic cysteine and mercapturic acid conjugates. When there is an overdose of paracetamol (>10 g) normal levels of glutathione are depleted and the toxic intermediates metabolite — NABQI binds to important liver cell constituents and inhibits

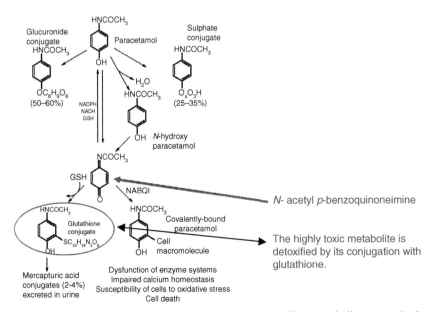

Figure 5.34   Liver metabolism of paracetamol indicating the toxic intermediate metabolite N-acetyl p-benzoquinoneimine (NABQI) and the glutathione dependent final metabolism of NABQI for safe excretion

Table 5.9   Common environmental toxins.

Environmental poison	Comments
Carbon monoxide	Poisoning — carboxyhaemoglobin — CO has a greater affinity for haemoglobin than O_2. Up to 2% COHb can be found in the blood of city dwellers, and up to 10% in smokers
Ethanol	Blood alcohol levels important to determine driving ban
Ethylene glycol (antifreeze)	Neuro and metabolic toxin lowest recorded oral lethal dose is 786 mg/kg
Lead	Common accumulative neurotoxin — nonoccupationally exposed blood levels 1.2 μmol/L; occupationally exposed 2.9 μmol/L
Mercury	Accumulative neurotoxin — normal mercury levels are considered to be <10 μg/L in blood and <20 μg/L in 24 h urine
Methanol	Neuro and metabolic alcohol toxin — binds to cells and has to be displaced by ethanol as this is less toxic
Arsenic	Accumulative metabolic toxin — the reported lethal dose of arsenic ranges from 120 to 200 mg in adults and is 2 mg/kg in children
Paraquat	Weed killer but also a human toxin if a dose of 35 mg/kg is taken

the activity of enzymes. NABQI is also nephrotoxic and will cause renal failure. There is greater toxicity when overdoses are taken by people on low-protein diets or those who consume alcohol regularly — both these states deplete glutathione levels further. *N*-acetylcysteine and oral methionine promotes glutathione synthesis.

5.13.3   Environmental toxins

Toxicology monitoring is not only medicines but environmental chemicals that poison the body. The common environmental toxins are listed in Table 5.9.

5.14   Clinical chemistry at the extremes of age

Clinical chemistry results differ from those of adults in the newborn, infant and child and for quite different reasons in the elderly. The most obvious differences are endocrinological, largely in the reproductive hormones, being low before puberty and high afterwards but declining with age or changing dramatically as in the menopause.

However, there are three other very notable sets of common clinical chemistry markers that change with age: acid—base balance of electrolytes associated with oxygen tension, renal function and metabolism markers.

5.14.1   Blood gases and electrolytes

The trauma of labour and birth results in a significant degree of asphyxia in the infant. Acidosis (blood pH and bicarbonate levels) should return to normal within 1 h. However, premature infants with respiratory distress syndrome usually become acidotic monitored by the increased level of P_aCO_2.

5.14.1.1   Normal newborn arterial blood gas saturation

- $PaCO_2$ (mm Hg)   27—40;
- PaO_2 (mm Hg)   65—80.

Sodium and potassium may be slightly elevated in the newborn and infant and chloride can be 1 or 2 mmols more markedly elevated compared to adults. However, the major differences are in bicarbonate levels which are much lower:

- infant serum bicarbonate   19—24 mmol/L;
- child serum bicarbonate   21—27 mmol/L;
- adult serum bicarbonate   23—32 mmol/L.

5.14.1.2   Old age blood gases and electrolytes

As an adult our arterial blood gases should be 95—99% but as we progress into old age arterial blood PaO_2 decreases 5% every 15 years and $PaCO_2$ increases 2% every 10 years. Sodium and chloride levels remain relatively constant but potassium levels also increase with age.

Table 5.10 Changes in ammonia, urea, creatinine and uric acid with age.

	<1 year	1–5 years	5–19 years	Male adult	Female adult
Ammonia μmol/L	–	10–40	11–35	11–35	11–35
Urea mg/L	60–450	50–170	80–220	100–210	100–210
Creatinine mg/L	2–10	2–10	4–13	5–12	40–90
Uric acid mg/L	10–76	18–50	30–60	40–90	30–60

Table 5.12 Changes in glucose and insulin levels with age.

	Young adult	>65 years
Basal levels blood plasma glucose	8–95 mg/dL	90–110 mg/dL
Peak levels glucose tolerance test plasma glucose	120–140 mg/dL	150–170 mg/dL
Basal levels blood plasma Insulin	5–10 U/L	10–20 U/L
Peak levels glucose tolerance test plasma insulin	50–60 U/L	60–80 U/L

5.14.2 Renal function

The normal neonate has immature renal function, glomerular filtration rate is only 25% of that in the adult and therefore a lower glomerular filtration rate.

In the elderly, major organs begin to lose functional capacity; particularly noticeable is the decline of renal function due to a loss of nephron and metabolic tubular cells. This increases the incidence of pathological processes leading to a decrease in capacity to preserve H_2O so dehydration leads to an increased motality rate. The net result in terms of clinical chemistry is a reduced creatinine clearance and GFR and increased blood urea and creatinine.

Thus standard creatinine clearance changes from 135–145 mL/min/1.73 m^2 in your late 30s to 95–100 in your 80s.

Indeed this is predicted by the following formula:

$$\text{Creatinine clearance(mL/min)} =$$
$$\frac{(140 - \text{age in years}) \times \text{weight in kg}}{72 \times \text{serum creatinine}} \times$$
$$\text{(for women, multiply result by 0.85)}$$

5.14.3 Markers of metabolism

Rapid metabolic changes occur during development not only in maturation but also due to the increase in body mass; this is reflected in nitrogenous metabolites ammonia, urea, creatinine and uric acid (see Table 5.10).

5.14.3.1 Bilirubin and newborn jaundice

Infants have increased bilirubin at birth (see Table 5.11), during the first week there is a further breakdown of haemoglobin, thereby increasing bilirubin levels further. In the newborn the liver is immature and unable to break down bilirubin, so jaundice appears.

5.14.3.2 Glucose and insulin levels

A feature of old age is poor glyceamic control (see Table 5.12), with up to a 25% increase in both basal glucose and glucose response level and, due to reduced sensitivity, an equally high increase to both basal and glucose challenge response levels of insulin.

One consequence of this increase in plasma glucose is that the percentage of glycate haemoglobin increases from 6–9% to 8–12% in the elderly.

5.15 Cancer biomarkers

In clinical chemistry an antigen or protein, which is secreted by the tumour itself or by the surrounding tissues in response to the tumour and can be determined in the serum samples of the patient, is a cancer

Table 5.11 Changes in bilirubin levels with age.

	1 week	Up to 1 year	>1 year old
Total bilirubin mg/L	1–12	2–12	2–14
Conjugated bilirubin mg/L	0–12	0–5	0–2

biomarker (formerly called tumour marker). They are used in screening programs and clinical practice to:

- confirm **diagnosis** of malignancy;
- indicate severity and hence **prognosis**;
- provide posttreatment information to **monitor** therapy.

Several decades ago the criteria of an 'Ideal Tumour Marker' was established. They should be:

- tumour specific and common to tumour type;
- absent from healthy individuals or benign disease;
- easily detected at the early stage of tumour development;
- directly proportional to tumour mass;
- levels parallel response to therapy.

However, in practice **tumour markers** never fulfilled these criteria (and hence the name was changed to **cancer biomarker**) and such markers are:

- Rarely tumour or type specific, even rarer is 100% expression.
- Most cancer associated proteins and molecules are expressed regardless of tumour status. A low level is often found in benign conditions which will be intermediate between that found in healthy controls and that of those with early — late stage malignant disease.
- Most cancers only produce high levels of any given tumour marker in advanced stage and there is considerable variation between patients.
- Although tumour marker levels will parallel response to therapy this is an individual patient issue and there is often a lag response that may be preceded by 'tumour marker flare' whereby large amounts of the marker are released due to cancer cell death.

No tumour/cancer biomarker matches all these criteria but the nearest is hCG when used as a marker of gestational trophoblastic disease.

5.15.1 Classification of cancer biomarkers

Clinical chemistry cancer markers can be defined by their biochemical nature: organic and inorganic compounds and metabolites, peptide hormones, monoclonal defined cancer antigens and oncofetal antigens. There is a tendency for the ectopic expression of proteins to be associated with highly malignant tumours (see Figure 5.35).

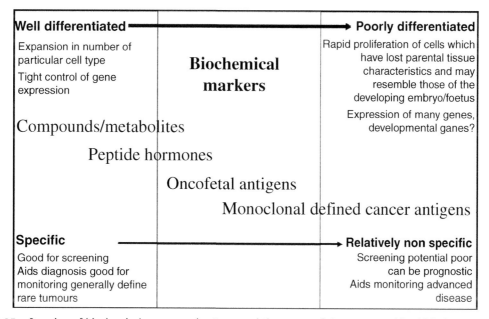

Figure 5.35   Overview of biochemical cancer marker types and the nature of the tumours with which they are associated

5.15.1.1 Organic and inorganic compounds and metabolites

These markers are highly specific, often urinary and define well-differentiated slow growing tumours. In the main they aid diagnosis following clinical observations.

A good example is Phaeochromocytoma, an adrenal gland tumour one in 10 of which are cancerous. Given the nature of the tissue, adenal medullary cells, a clinical consequence of the expanded cell population is elevated production of adrenal hormones (adrenaline/epinephrine). If the tumour arose in the adrenal cortex rather than the medulla then it would be increased cortisol.

Although the adrenalins have potent vasodilatory and 'flight or fight' effects, biochemically they are rapidly degraded. Thus, to detect such increased secretion we measure urinary Catecholamines (breakdown products of the adrenalins) the normal ranges of which are:

- Normetadrenalin — up to 2.8 mmol/24 h;
- Metadrenaline — up to 2.2 mmol/24 h;
- (5HAA) 5-hydroxyindole-3-acetic acid — 8–50 mmol/24 h;
- HMMA (4 Hydroxy, 3 Methoxy Mandelic Acid);
- VMA (Vanilylmandelic acid) — 8–35 mmol/24 h;
- HVA (Homo Vanillic Acid) — up to 40 mmol/24 h.

It should be noted that the relationship to elevated metabolites and steroids are not always direct. For example Carcinoid tumours (neuroendocrine tumours found in the small intestine or the appendix — often benign) release ACTH which induces adrenal corticol cells to produce and elevate serum cortisol levels.

5.15.1.2 Peptide hormones

These markers when elevated are usually specific, serum or urine-based measurement and usually define well-differentiated slow growing tumours. Again primary diagnosis is largely from clinical features. However, nonbioactive forms and fragments of the peptide hormone tend to define aggressive poorly differentiated tumours. For example elevated PTH, PTH-RP, LH, GH and ACTH are diagnostic of parathyroid and pituitary tumours. Hypothalamic tumours require additional clinical investigation, but post chemotherapy or surgery, monitoring of levels is indicative of response to the therapy.

5.15.1.3 Monoclonal defined cancer antigens

They are the result of raising antibodies (monoclonal and polyclonal) against human tumour cells and extracts thereof, immunized in mice, sheep, rabbits, and so on. As such the molecules are defined by the antibody that detects them, for example CEA (Carcinoembryonic Antigen), CA125 (Cancer Antigen 125), CA15.3 (Cancer Antigen 15), PSA (Prostate Specific Antigen), and so on. Characterization has revealed all to be normal proteins inappropriately expressed as a result of oncogenesis. These do tend to be interesting functional molecules almost exclusively large molecular weight glycoproteins (see Table 5.13).

Thus the reality of this category of **tumour markers** is that:

- None are tumour specific but more commonly associated with a particular cancer.
- Low levels are found in normal individuals, higher during pregnancy and occasionally during menses. Often expressed by more than one cancer.
- Found in serum, urine or tissue.
- Can be detected at all stages of tumour development but usually correlate with advancing disease stage.
- They are generally not particularly diagnostic of very early stage disease.
- They are prognostic and used for monitoring disease.

5.15.1.4 Oncofetal antigens

These are peptide hormones and proteins specifically expressed by the placental–fetal unit, for example PLAP (Placental Alkaline Phosphatase), AFP (Alphafetoprotein), hCG (human Chorionic Gonadotropin). They can be measured in serum or urine and when ectopically expressed by cancers tend to define poorly differentiated fast growing/malignant tumours and diagnosis is more reliant on the levels of these markers. Indeed, hCG is diagnostic of gestational trophoblastic disease (GTD), particular forms of testicular germ cell tumours and some nontrophoblastic tumours (see Table 5.14). Additional clinical investigation is required, but monitoring of levels postchemotherapy or surgery is extremely useful in detecting response and

Table 5.13   Examples of major monoclonal antibody defined cancer antigens in clinical use.

Antibody defined Ca marker	Clinical significance/characteristics
CEA	• Define colorectal cancers — but also breast, lung, GI tract and liver • Normal levels <5 mg/L • Elevated in pregnancy, smokers, respiratory disorders • Not reliable for diagnosis but some prognostic value in colon cancers and in monitoring posttreatment levels fall $T^1/_2$ of 2–3 days

CA125	• Ovarian cancer 85–90%
	  — but also adenocarcinomas of breast, lung and GI tract
	• Also elevated in other gynaecological conditions
	• Large screening study completed — not proven effective policy for asymptomatic screening
	• Not reliable for diagnosis but some prognostic value in ovarian cancers and in monitoring
	  — postsurgical levels fall $T^1/_2$ of 5 days

CA15.3	• Breast cancer — up to 35% of early disease
	  — also in lung, GI tract, reproductive system cancers
	• >25 kU/L indicates disease but not really diagnostic
	• Not really prognostic
	  — levels often indicate mass rather than malignancy
	• 5 × cut off indicates metastatic disease
	• Drop in levels posttreatment indicate good response and elevations over 25% indicate recurrent disease

PSA	• Prostate cancer (prostatic carcinoma)
	  — also BPH (benign prostatic hypertrophy) and prostatitis
	• Normal <50 years 2.5 µg/L > 50 < 70 years 4 µg/L
	• No use in screening or diagnosis alone
	• Some use in prognosis
	• Good in monitoring
	  — $T^1/_2$ of 2.2 days to undetectable levels after radical prostatectomy

Table 5.14   Oncofetal antigens in clinical use.

Oncofetal antigen	Clinical significance/characteristics
PLAP	Elevated in trophoblastic and nontrophoblastic dedifferentiated tumours. Particularly associated with seminomatous germ cell tumours of the testis and the female ovarian equivalent — dysgerminomas • seminomas 90% PLAP positive • dysgerminomas 75% PLAP positive

AFP	Elevated in trophoblastic, germ cell and nontrophoblastic dedifferentiated tumours:
	Nontrophoblastic/germ tumours
	  — Hepatocellular carcinoma
	  — Yolk sac tumours
	• Diagnostic
	• Prognostic

(Continued)

Table 5.14 (*Continued*)

Oncofetal antigen	Clinical significance/characteristics

—High levels the more extensive the tumour burden

• Monitor
  — Levels correspond to therapy
  — Persistent or resurgent levels = resistance/recurrence

Testicular germ cell tumours

• Diagnostic of nonseminomatous type that contain yolk sac elements
• Prognostic
  — In combination with hCG
• Monitor
  — Levels correspond with therapy
  — Persistent levels = resistance/recurrence

hCG   Elevated in trophoblastic, germ cell and nontrophoblastic dedifferentiated tumours.

Testicular germ cell tumours

• Diagnostic
  — Seminoma (a radiosensitive tumour) — hCG positive 8% (AFP 0%)
  — Nonseminoma (a radioresistant tumour) — hCG positive 50% (AFP 66%)
• Prognostic
  — Nonseminoma
• hCG > 10 000 mLU/mL = 53% 3 year survival
• hCG <1 000 mLU/mL = 85% 3 year survival
  — Seminoma
• 80—97% overall survival
• Monitor
  — Levels correspond to therapy
  — Persistent or resurgent levels = resistance/recurrence

Gestational trophoblastic disease (GTD)
  — Including hydatidiform mole
  — Partial molar pregnancy
  — Placental site tumour
  — Choriocarcinoma

hCG for this disease is virtually an ideal tumour marker:

• Diagnostic — In young women it is generally diagnostic of GTD if not from any other source (e.g. pregnancy) and all GTD will produce hCG

  A CSF: serum ratio of greater than 1: 60 is indicative of brain metastasis

• Prognostic — the higher the levels the higher the tumour burden

  High pretreatment levels = poor prognosis

• Monitor
  — Levels correspond to therapy
  — Persistent or resurgent levels = resistance/recurrence

recurrent disease prior to other clinical symptoms or imaging measures.

In monitoring GTD response to chemotherapy hCG has proven invaluable, as all GTDs express hCG and at high levels patients are monitored for response to therapy almost daily until in remission defined as hCG not detected for three consecutive weeks. They are then monitored by hCG serum assay every 2 weeks for 6 months then monthly for 5 years. Levels of hCG determine when and how chemotherapy is administered and in the absence of clinically apparent disease, persistently elevated serum hCG reflects the existence of microtumours that if untreated will progress clinically.

Bibliography

Marshall, W.J. and Bangert, S.K. (2008) *Clinical Chemistry*, 6th edn, Mosby.

Gaw, A. *et al.* (2008) *Clinical Biochemistry*, 4th edn, Churchill Livingstone.

Hughes, J. and Jefferson, J.A. (2008) *Clinical Chemistry Made Easy*, 1st edn, Churchill Livingstone.

Chapter 6
Medical microbiology

Dr Sarah J. Furrows, MBBS, M.Sc., MRCP, FRC Path
and Catherine S. Fontinelle, B.Sc. Hons., ARCS, FIBMS, CSci

Introduction

Microbiology, the study of microorganisms, is both a clinical speciality and a study of natural history. Microorganisms existed long before human beings and are an integral part of our everyday existence. If an alien were to land on Earth and study us, it is doubtful whether he would even recognize our bodies as human – our bodies contain 100 000 000 000 000 microbial cells, which is approximately 100 times more than the number of human cells in our bodies. These normal flora, often described as 'friendly bacteria', are widespread in our mouth, intestines and skin and act, as part of the skin, as a barrier to infection,

Microorganisms may be beneficial in other ways. They are essential in industry for fermentation processes used to make alcohol, cheese, yoghurt and vinegar, and are used routinely in food waste and sewage treatment.

Medical microbiology is a rapidly developing field, which incorporates the study of microorganisms along with their impact on the human body and the range of treatments available. In the 300 years since bacteria were first identified under a microscope, hundreds of bacteria of medical importance have been described. In addition there are viruses, fungi, protozoa, parasites and prions, all of which may cause disease.

Formerly a backroom or 'Cinderella' speciality, microbiology has been thrust into the limelight in recent years, largely due to a resurgence of public interest in 'superbugs' such as meticillin-resistant *Staphylococcus aureus* (MRSA) and *Clostridium difficile* (Box 6.1). Add to this the emergence of pathogens from animal reservoirs such as swine influenza and SARS (severe acute respiratory syndrome), travel-associated infections, and the continuing threat from old enemies such as tuberculosis, and it is clear that microbiology will remain a busy speciality for the foreseeable future.

6.1 Overview of microorganisms

Cells can be divided into two kinds: prokaryotic (where the DNA is free within the cell) and eukaryotic (where the DNA is kept within a nucleus). Bacteria are prokaryotes. Eukaryotic pathogens include protozoa, yeasts and multicellular organisms such as tapeworms. Viruses do not have cells but instead enter the cell of the host organism. Table 6.1 gives an overview of commonly encountered cellular microorganisms, classified

Biomedical Sciences: Essential Laboratory Medicine, First Edition. Edited by Ray K. Iles and Suzanne M. Docherty.
© 2012 John Wiley & Sons, Ltd. Published 2012 by John Wiley & Sons, Ltd.

It was once thought that antibiotics could eradicate all infections. In 1967, the Surgeon General of the USA stated that 'we could close the book on infectious diseases' because of the success of antibiotics. We now know that microorganisms and infections are likely to be with us always.

Table 6.1   Overview of medically important cellular microorganisms

Prokaryotes

Bacteria

Gram-positive cocci	*Staphylococcus, Streptococcus, Enterococcus*
Gram-negative cocci	*Neisseria, Moraxella*
Gram-negative rods	Coliforms: *Escherichia coli, Klebsiella, Proteus, Salmonella, Shigella*
	Fastidious organisms: *Haemophilus, Bordetella*
	Pseudomonads: *Pseudomonas*
	Vibrios: *Vibrio*
	Spirilla: *Campylobacter, Helicobacter*
Gram-positive rods	*Corynebacterium, Listeria, Bacillus*
Anaerobes	Spore-forming: *Clostridium.*
	Non-sporing: *Bacteroides, Fusobacterium*
Spirochaetes	*Treponema, Leptospira, Borrelia*
Mycoplasmas	*Mycoplasma, Ureaplasma*
Rickettsiae and chlamydiaeceae	*Rickettsia, Chlamydia, Coxiella*
Mycobacteria	*Mycobacterium*

Eukaryotes

Protozoa

Sporozoa	*Plasmodium, Cryptosporidium, Toxoplasma, Isospora*
Flagellate	*Giardia, Trichomonas, Leishmania*
Amoebae	*Entamoeba, Acanthamoeba, Naegleria*
Other	*Babesia, Balantidium*

Fungi

Filamentous	*Aspergillus, Trichophyton, Epidermophyton, Microsporum*
True yeast	*Cryptococcus*
Yeast-like	*Candida*
Dimorphic	*Histoplasma*

mainly by cell type. An overview of viruses is given in Section 6.6.1.

6.1.1   Basic principles of infection: host—parasite relationship

In the phrase 'host—parasite relationship', the host is the person and the parasite is any microorganism which takes food or shelter from the host. The balance of the relationship may tip one way or the other depending on the pathogenicity of the organism and the immune status of the host.

Everyone is colonized with bacteria but this does not normally cause disease. These colonizing microorganisms are called commensals or normal flora and the host—parasite relationship is **commensalism**, with no harm to either party (Table 6.2). However, it is possible for these organisms to cause disease in certain circumstances, for example. coagulase-negative staphylococci may infect medical devices such as IV lines or prosthetic joints.

An infection occurs when microorganisms invade host tissues or release toxins, causing the host to become unwell. The host—parasite relationship is now **parasitism**, because the microorganism is taking from the host without giving anything back.

Microorganisms which cause infections are called **pathogens**. They may be **obligate pathogens**, which have to cause disease in order to survive and propagate, or **opportunistic or conditional pathogens**, which do not usually cause infections but may do so in certain circumstances, for example if inoculated into a wound or if the host is immunocompromised. Mankind has evolved host defences to fight off infection (see Chapter 7). The natural history of most infections is a short-lived infection followed by recovery and, in many cases, immunity to further infection by that agent. Some agents, such as hepatitis B, may cause a chronic infection causing man and micro organism to coexist for years.

The **infectivity** of a microorganism is its ability to enter, survive and multiply within the host. It can be measured as the number of persons infected, divided by the number who were susceptible and exposed.

Pathogenicity is the ability of an organism to cause disease. Highly pathogenic organisms generally possess one or more structural or biochemical or genetic traits, known as **pathogenicity determinants**, which help them infect the host (Table 6.3).

Table 6.2   Examples of commensals and pathogens

Organism	Location	Infection
Commensals		
Coagulase-negative staphylococci	Skin	Usually colonize. May infect medical devices.
Lactobacilli	Vagina, small intestine	Colonize
Obligate pathogens		
Mycobacterium tuberculosis	Lung and other sites	Tuberculosis
Treponema pallidum	Genital tract and other sites	Syphilis
Opportunistic/conditional pathogens		
Pneumocystis jiroveci (PCP)	Lung	PCP pneumonia in AIDS
Bacteroides fragilis	Intestine (usually a commensal)	Peritonitis and abscesses if it reaches the peritoneal cavity

Virulence is the ability of an organism to cause **serious** disease. This is a subtle distinction from pathogenicity: think of it as referring to the **degree** of pathogenicity of an organism. For example, *Streptococcus pneumoniae* is a conditional pathogen whose natural habitat is man, but which only causes disease in a minority of cases; but some strains of *S. pneumoniae* have a different capsular type, which causes more serious disease and therefore increases its virulence.

Bacteria may produce **toxins**, which damage the host and cause disease (Table 6.4). Toxin production is a pathogenicity determinant. There are two types of toxin: **endotoxin** is the lipopolysaccharide component of the Gram-negative cell wall, and its release causes fever and septic shock; **exotoxins** are proteins secreted by the bacterium into the host, and may cause local or distant damage. Finally, many bacteria secrete enzymes which help them invade the host, such as hyaluronidase, collagenase, DNAase or streptokinase (Box 6.2).

6.1.2   Disease transmission

Transmission of a pathogen, resulting in infection or colonization, requires these six essential links known as the **chain of infection**:

1. an **infectious agent** (bacteria, viruses, fungi, etc.);
2. a **reservoir** for the organism (people, animals, food, water, equipment, etc.);
3. a **portal of exit** from the reservoir (droplets, excreta, secretions, etc.);
4. a **mode of transmission** (such as contact or inhalation — see below);

Table 6.3   Examples of pathogenicity determinants

Pathogenicity determinant	Bacteria with this determinant	How it acts
Fimbriae	*E. coli, Neisseria sp., Pseudomonas sp*, other Gram-negatives	Mediates attachment to host cell surface
Flagella	*Vibrio cholerae, Campylobacter*	Mediates attachment to host cell surface
Capsule	*Streptococcus pneumoniae, Haemophilus influenzae, Neisseria meningitidis, E. coli*	Avoids phagocytosis
M protein	Group A Streptococcus	Prevents complement deposition
Ig A protease	*Neisseria, Helicobacter*	Break down mucosal IgA
Antigenic variation	*Trypanosoma*	Evades immune response
Siderophores	*E. coli, Klebsiella pneumoniae*	Scavenge iron from mammalian iron-binding proteins

Table 6.4 Some important toxins

Organism	Toxin	Effect of toxin
Staphylococcus aureus	Toxic shock syndrome toxin	Toxic shock syndrome
	Epidermolytic toxin	Scalded-skin syndrome
	Enterotoxins	Food poisoning
	Leucocidin	Lyses blood cells
Group A streptococcus	Streptolysins O and S	Lyses blood cells
	Erythrogenic toxins	Fever and rash
Clostridium difficile	*C. difficile* toxins A + B	Diarrhoea and colitis
Clostridium botulinum	*C. botulinum* toxin A	Prevents release of acetylcholine at neuromuscular junction, causing flaccid paralysis
Clostridium tetani	Tetanus toxin	Inhibits neurotransmitter release, causing muscle spasms
Corynebacterium diphtheriae	Diphtheria toxin	Mucosal necrosis, airway obstruction and systemic toxicity
E. coli 0157	Vero toxins VT1 and VT2	Diarrhoea, haemolytic uraemic syndrome
Shigella	Shiga toxin	Dysentery

Box 6.2   Streptokinase

Streptokinase is produced by Group A streptococcus. It activates host plasminogen to plasmin, which stops fibrin barriers building up and helps the infection to spread. This is why Group A streptococcal soft tissue infections spread so rapidly. On the plus side, streptokinase was used for many years as thrombolytic therapy for patients with myocardial infarction, because of its ability to dissolve blood clots.

---

5. a **portal of entry** into the body (respiratory tract, broken skin, etc.);
6. a **susceptible host** (infections are more likely in the immunosuppressed, very young or very elderly).

Each link must be present for infection or colonization to proceed. Breaking any of these links can prevent the infection. The easiest link to break is the mode of transmission, and infection control measures such as isolation and hand hygiene are aimed at breaking this link. Modes of transmission include the following.

Contact transmission: the commonest mode in healthcare settings. Contact transmission may be subdivided into:

Direct contact: direct person-to-person spread of microorganisms through physical contact. Examples — transfer during a dressing change or insertion of invasive device, if the healthcare worker had not cleaned their hands; or skin-to-skin contact can spread scabies or herpes simplex. Hand hygiene can prevent transmission by direct contact.

Indirect contact: occurs when a susceptible person comes in contact with a contaminated object (fomite). This may be equipment such as endoscopes, or a ward area contaminated with MRSA or *Clostridium difficile*. All healthcare settings should ensure correct cleaning of their environment and equipment, and disinfection or sterilization of equipment where appropriate, to prevent transmission by this route.

Droplet transmission: results from contact with contaminated respiratory secretions. Respiratory droplets are generated by coughing, sneezing, talking, suctioning and bronchoscopy. Droplets can travel up to 1 m and will then settle on surfaces, where the organisms may survive for some time. Examples of diseases spread by droplets include influenza and whooping cough.

Airborne transmission: occurs when small particles containing pathogens are expelled (usually from the respiratory tract), suspended in the air for a prolonged period, then spread widely by air currents and inhaled. Examples of diseases spread by the airborne route include pulmonary tuberculosis, varicella and measles.

6.2 Laboratory investigation of infection

Practical microbiology is all about the investigation of infection, and the classification and naming of microorganisms are largely based around their reactions in common tests. For this reason, laboratory investigations are described at the beginning of this chapter. It is not necessary to read this whole section at once: parts of it may make more sense if you dip in and out, and crossrefer to the section on specific organisms.

Almost any tissue or body fluid may be investigated in the laboratory. Most investigations rely on microscopy with or without stains, culture, colonial morphology on agar and biochemical tests. Organisms that are difficult to grow may be detected by nucleic acid amplification tests (NAATs), or the immune response to infection may be detected by serological methods.

6.2.1 Specimen collection

Samples must always be clearly labelled with at least two patient identifiers (such as name, date of birth, hospital or NHS number) and should also have:

- name of ward or GP practice;
- date and time of collection;
- contact details for requesting doctor;
- description of sample (culture site);
- relevant clinical details;
- antibiotic therapy details.

It is essential that the correct sample is sent for a test, otherwise the quality of the results will be affected. For example, a sputum sample is not worth investigating if it is clearly salivary, as it cannot have originated from the lungs and therefore any culture results would be uninterpretable. Another notorious example is the 'pus swab' — if there is an abscess, it is better to send a sample of pus in a sterile container rather than a swab of the pus. If in doubt, the clinician should check with the laboratory before sending a sample.

"If identification is an issue
Forget the swab and send some tissue."

6.2.2 Types of sample

Swab samples are taken from a range of body sites including eyes, ears, throat, skin and the vagina. Bacterial swabs come with the transport medium inside the accompanying tube to allow the bacteria present in the sample to survive, in the same numbers, long enough to reach the laboratory. Viral swabs come with a sponge-like section that contains a variety of antibiotics that prevent bacterial overgrowth and preserve the virus on the swab.

Sample pots are available in a variety of sizes, most commonly 30 mL, 60 ml and 250 ml. The small 30 mL pots are used for urine, semen, catheter tips and faeces collection, and occasionally for nail clippings; however, specialist dark-coloured paper envelopes are available for this sample type. The 60 mL pots are most commonly used for sputum samples, although urine is often sent in these containers as the opening is wider and more convenient for the patient. Unfortunately the urine processing racks are designed for the smaller pot and 60 mL pots can seriously unbalance a tray! The larger 250 ml containers are most commonly used for early morning urine samples (EMU), where the entire EMU is collected to investigate for TB but these containers are also used for tissue and graft samples (Figure 6.1).

Blood cultures are one of the more important samples received by the laboratory. These are used to detect the presence of microorganisms in the blood (bacteraemia or fungaemia) and are usually only taken from unwell, febrile patients. Blood culture bottles contain liquid media with suitable nutritional and environmental conditions to allow bacteria to grow, together with an indicator system to show when the bottle is positive. There are many different types of blood culture system on the market, from bottles containing liquid and solid media, which allows the user to subculture the bottle just by tilting it, to bottles loaded onto large machines (Figure 6.2). The latter bottles are continuously monitored so that when a positive is detected, by a change in the colorimetric sensor on the bottom of the bottle due to the production of CO_2 by microbial growth, the machine can alert laboratory staff to the new positive and allow it to be dealt with promptly and efficiently. The liquid in the blood culture bottles contains: complex media including amino acids, protein and growth factors; CO_2 under vacuum, to allow the correct amount of blood (usually 8–10 ml) to be drawn into the bottle

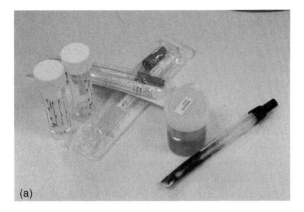

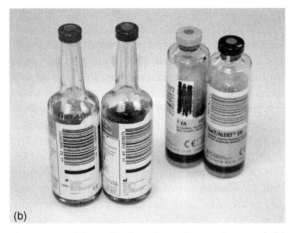

Figure 6.2 (a) Blood culture system and (b) system showing bottles inside

Figure 6.1 (a) A selection of sample containers and (b) blood culture bottles belonging to two different automated systems

when attached to blood collection systems; and occasionally beads or charcoal to counteract the effect of any antibiotics already in the patients' blood. Different bottle types are available. Most commonly, one aerobic and one anaerobic bottle are used (Box 6.3). Paediatric bottles are designed to deal with small volumes of sample and are usually sent singly. Mycobacterial and fungal blood culture bottles are also available.

6.2.3 Basic handling principles

It is important when processing a sample that nothing contaminates it whilst it is being handled in the laboratory. This is called aseptic technique.

> **Box 6.3 Blood culture pitfalls**
>
> Beware the blood culture set arriving in the laboratory with plastic covers over the septa — these are uninoculated!

There are several common methods of reducing bacterial contamination.

1. Use of correctly flamed wire loops for inoculating and spreading agar plates, or plastic disposable loops. When decontaminating a wire loop, hold it above the Bunsen at a downwards angle of approximately 30° so that the wire is red hot all the way up to the join with the handle (Figure 6.3).
2. Sterile plastic pipettes often come in packets of 10, but rarely in singles. A new packet should be opened when dealing with a sterile fluid, such as a CSF, as an

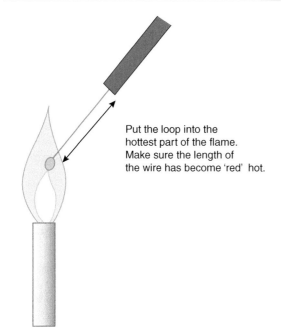

Put the loop into the hottest part of the flame. Make sure the length of the wire has become 'red' hot.

Figure 6.3   Decontaminating a wire loop

opened packet may have become contaminated with bacteria.

3. The neck of a glass bottle should be flamed before transferring any liquid into or out of it.

4. Never return any unused sample to its original container. If it is to be kept, it must be put into a fresh container so as not to contaminate the remainder.

5. When subculturing a fluid never dip the loop more than 1 cm into the sample. This avoids crosscontamination due to poor flaming.

6.2.4 Sample processing

Once all the basic checks have taken place (name, hospital number, date of birth) the samples are labelled and are ready for processing. This takes place in three stages.

1. Visual inspection of the sample for any clues to diagnosis. Examples include the presence of tapeworm segments in a faecal sample, or sulphur granules on an IUCD indicating a potential actinomyces infection. Visual inspection also gives an indication of the quality of sample that has been received. For example, a salivary specimen of sputum, if processed at all, should have its description noted so that the results may be considered accordingly. Comments are recorded of any visual clues found.

2. Microscopy — there are three commonly used microscopic techniques in routine laboratories:
   - light microscopy to look at direct preparations or stained slides;
   - phase contrast microscopy using a special condenser and objective, to look at direct preparations for organisms with flagella;
   - fluorescence microscopy using ultraviolet light to look at fluorescent stains, for example, for acid fast bacilli.

   Other microscopic techniques include polarizing microscopy, darkground microscopy and electron microscopy.

   The most common stain used in microbiology is the Gram stain. This divides bacteria into two groups, Gram-positive or Gram-negative, and is a fundamental part of bacterial identification.

3. Culture — media used to isolate bacteria may contain water, electrolytes, protein, vitamins, blood and agar.

Liquid culture media are used to enrich low bacteria numbers in some samples. It is of additional benefit in that any inhibitors present in the sample, such as antibiotics, are diluted and have less effect on bacterial growth. As liquid media often needs to be subcultured it does cause a delay in the isolation and identification of some bacteria (Table 6.5).

Solid media contains agar as a gelling agent, which does not set until it reaches 40 °C. This feature allows temperature sensitive factors, such as antibiotics, to be added as the molten agar cools (Table 6.6). Solid agar can be enriched to enhance the growth of nutritionally demanding bacteria, or it can be selective so that only certain groups of bacteria grow (Figure 6.4).

The choice of agar used varies with the different samples processed in the laboratory. This is because each site on the surface of the human body has its own mix of normal flora. The function of the media selected is to enable the biomedical scientist to find the pathogenic bacteria that may be present.

In some samples, such as CSF and joint fluids, there should be no bacteria present and any bacteria isolated will be considered significant. In contrast,

Table 6.5 Common liquid media and their uses

Common liquid media	Use
Selenite F	Enriches the growth of nonlactose fermenting coliforms by suppressing the growth of other coliforms
Brain heart infusion (BHI)	Enriches the growth of bacteria in fluids that should be sterile
Robinson's cooked meat (RCM)	As BHI and also enhances the growth of anaerobic bacteria
MRSA selective broth	Colour change to yellow indicates a potential positive that requires follow-up. No colour change indicates a negative screen and can be reported

Table 6.6 Common solid media and their uses

Common solid media	Use
Blood agar	Grows most bacteria
Fastidious anaerobe agar (FAA)	Enhances growth of anaerobic bacteria
Chocolate agar (Box 6.4)	Grows bacteria such as *Haemophilus spp.* and *Neisseria gonorrhoeae* that require extra nutrients present inside the red blood cell
MacConkey	Contains lactose and phenol red. Grows coliforms, staphylococci and some streptococci
Xylose lysine desoxycholate (XLD)	Selective media for salmonellae and shigellae
Desoxycholate citrate agar	Selective media for salmonellae and shigellae
Cystine-lactose electrolyte deficient medium (CLED)	Contains lactose and bromothymol blue. Grows coliforms, staphylococci and some streptococci

---

Box 6.4 Chocolate agar

Unfortunately, chocolate agar plates are not made from chocolate. Instead the agar is 'chocolatized' by heating at a higher temperature than usual. This lyses (bursts) the red blood cells present and releases extra nutrients — maybe it should be called gravy agar instead!

Box 6.5 Chromogenic agar

Chromogenic agar is a commercially available agar that produces specific colour changes to help the user identify bacteria grown on these plates.

Figure 6.4 Selection of culture plates used on the faeces bench

with faeces samples the laboratory is very specific about the bacteria being investigated and will often only be looking for salmonellae, shigellae, *E. coli* 0157 and *Campylobacter*, so the chosen media reflect this need.

Once the plates are inoculated, they are streaked out using wire or plastic loops and incubated. Some plates may have additional diagnostic discs added to them before incubation, such as metronidazole, used to detect anaerobes, and optochin to identify *S. pneumoniae* in a sample of sputum.

MRSA selective agar uses the phosphatase enzyme activity of the organism, together with specific antibiotic agar components, enabling the organism to be identified as blue colonies (Figure 6.5). Other

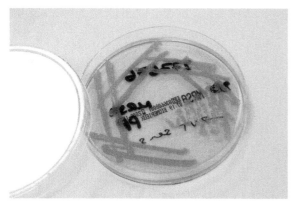

Figure 6.5   MRSA growing on a chromogenic agar plate

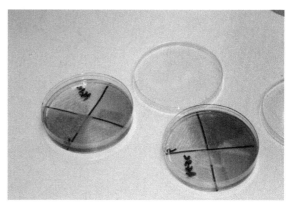

Figure 6.6   CLED plates on the urine bench

chromagar can differentiate between common coliforms isolated from urine, or species of yeasts with *Candida albicans* growing as purple colonies (Box 6.5).

6.2.5   Enumeration techniques

A link has been established between the number of organisms and the number of white blood cells (or, indeed, red blood cells) in a sample as an indicator of disease.

6.2.5.1   Enumeration of cultures

The most common sample tested for bacterial numbers is urine. Urine samples are often contaminated by flora from the external genitalia as patients are rarely advised of the correct method of specimen collection. A level of 'significant bacteriuria' of $> 10^5$ organisms/ml was established by Kass to determine the number of bacteria required to be present to indicate an infection.

There are three common methods used in the routine laboratory to enumerate bacteria in urine.

1. A 1 μL loop of urine spread onto solid agar (CLED, MacConkey or chromogenic agar); 10 colonies of growth will correspond to $10 \times 1000\,\mu L = 10^4$ orgs/mL, therefore 100 colonies corresponds to $100 \times 1000\,\mu L$, or 10^5 orgs/mL (Figure 6.6).
2. Sterile filter paper strips are dipped into the urine, allowed to soak in, and pressed onto solid agar (as above). The surface area is of a standard size and therefore the number of organisms per mL can be calculated.

3. Semiautomated machines, such as the iQ200 Automated Urine Microscopy Analyzer (from Iris Diagnostics) and the Sysmex UF Series Analyzers (developed by Biomerieux and Sysmex Corp) allow large numbers of urines to be screened. These use digital image recognition and fluorescence and forward scatter of light to detect the number of bacteria present.

6.2.5.2   Enumeration during microscopy

A significant level of white blood cells (WBC) has been established in samples of cerebrospinal fluid (CSF), ascitic fluid (ASF), peritoneal dialysis fluid (PDF) and urine (National Standard Methods, Health Protection Agency; see Table 6.7). It should be noted that immunocompromised patients may not produce WBC in these ranges.

There are three common methods of counting cells.

1. White blood cell counts in nonrepeatable samples, that is CSF, ASF and PDF, are determined using counting chambers of known surface area, such as Neubauer and Fuchs–Rosenthal. These are made of glass and are reusable. A small amount of the fluid is loaded into the chamber. Under the microscope a grid of known area can be seen and the number of WBC inside the grid can be converted into the

Table 6.7   Positive cell counts

Sample	CSF	ASF	PDF	Urine
WBC/μL	5[*]	250	100	100

[*]The normal ranges of WBC in CSF vary with the patients' age.

(a)

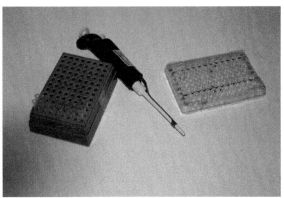

Figure 6.8   Pipette with tips and microtitre tray

(b)

Figure 6.7   (a) Fuchs—Rosenthal counting chamber and (b) inverted microscope for urine microscopy

number of WBC/μL (Figure 6.7). Plastic disposable counting chambers are available to use with samples from patients with CJD.

2. The most common method of enumerating cells is using an inverted microscope and flat bottomed wells to calculate the cells present in urine samples. A known volume (60 μL) of urine is added and allowed to settle. As the surface area of the well is known, it is possible to calculate the number of WBC and RBC/microlitre. This technique allows the operator to read 96 samples in one sitting. At least 200—300 samples of urine will be scrutinized in an average microbiology department every day (Figure 6.8).

3. Semiautomated machines, such as the iQ200 and the Sysmex UF Series, count cells using digital image processing and fluorescent and forward scatter of light.

6.2.6   Bacterial sensitivity testing

Once a bacterium has been identified as a pathogen it is necessary to ascertain the correct antibiotic treatment required to effect a cure. There are many methods of deciding whether a bacterium is sensitive or resistant to an antibiotic, but currently there are four common methods in use: Stokes' controlled sensitivity testing, BSAC (British Society for Antimicrobial Chemotherapy) sensitivity testing, MIC (minimum inhibitory concentration) testing and automated sensitivity testing.

In Stokes' controlled sensitivity testing a suspension of the test bacteria is spread onto the inner part of an agar plate (usually isosensitest or direct sensitivity test agar) and a control organism is spread around the outside edge (Figure 6.9). Filter paper discs impregnated with antibiotics are placed where the two organisms almost meet. After overnight incubation, the zone sizes are compared. If the zone size of the test organism is larger than the control organism, the antibiotic is considered sensitive. If it is more than 3 mm smaller than the control zone, then it is considered resistant. Stokes' method is no longer widely used — the main drawback was that the organism used for comparison was often not the same genus as the test organism.

More recently laboratories have adopted the BSAC method of sensitivity testing (Figure 6.10). A standardized inoculum is prepared using a densitometer. The suspension is spread over an entire agar plate, each organism having a recommended sensitivity agar to be used, temperature and atmosphere to be incubated in. Filter paper discs impregnated with antibiotics are

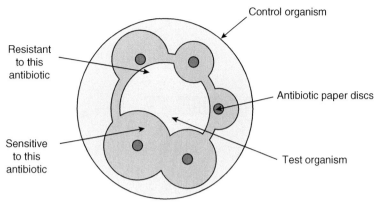

Figure 6.9   Stokes sensitivity testing

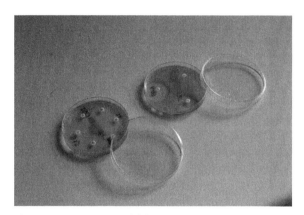

Figure 6.10   BSAC sensitivity testing

added and the zone sizes measured after overnight incubation. Each genus of bacteria (and sometimes species) has its own zone sizes allocated to it. If the zone is smaller then the antibiotic is considered resistant.

The MIC is the lowest concentration of an antibiotic required to inhibit growth of an organism. Knowledge of the MIC allows the clinician to determine the quantity of an antibiotic required to treat an infection, as well as choose one that the pathogen will be susceptible to: **the infection should respond to treatment provided that levels of antibiotic in blood are equal to or higher than the MIC.** MIC testing is performed for organisms causing endocarditis or other difficult-to-treat infections. The most common method of MIC determination is by use of E-test strips (Box 6.6 and Figure 6.11). These E-tests are small strips of 'paper' impregnated with different concentrations of antibi-

otic along its length. Growth appears, after overnight incubation, as an ellipse. The point at which the growth touches the strip is (usually) considered to be the MIC. This does vary in interpretation with yeasts, when double zones appear.

Automated systems are available in two main types — SIR systems (sensitive, intermediate, resistant) and Expert systems which use MIC ranges to determine results. The British Society for Antimicrobial Chemotherapy (BSAC) and the Clinical and Laboratory Standards Institute (CLSI) each publish MIC ranges for different organisms. These are different from each other and the rule base used by each laboratory can be fed into the automated system to produce results that follow local guidelines.

Identification of an organism takes between 5 and 18 h. A suspension of the organism is added to a test card, with the card type chosen based on the Gram stain result or on the organism type, such as yeasts or anaerobes. The wells in the card contain different concentrations of antibiotics or identification tests (Figure 6.12). These systems also include tests to determine specific resistance mechanisms, such as extended spectrum beta-lactamase production (ESBL).

6.2.7   Biochemical tests

Bacteria can be identified by the various sugar and other biochemical reactions that they perform and these reactions have been commercially produced for various bacterial groups, for example, staphylococci.

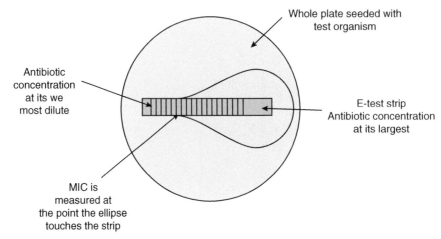

Figure 6.11   E-test

A suspension of the test organism is added to a series of small chambers, containing dehydrated substrates for several reactions, in a strip form. The reactions are converted into a numerical profile. The profile can be compared to a database to establish identity of the organism. This is more convenient than performing each test individually, and is cheaper than automated identification systems.

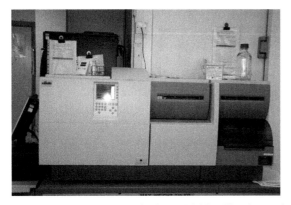

Figure 6.12   Vitek system for bacterial identification and sensitivity

Auramine stain — flood a fixed smear with auramine solution for 10 min, rinse with water then decolourize with 1% acid—alcohol for 5 min. Rinse with water then counterstain with potassium permanganate for 1 min. Wipe dry the back of the slide and air dry in a rack. **Never** blot an auramine stain as fibres transferred from the filter paper may give false positive results. Mycobacteria and cryptosporidium are auramine positive.

Beta-lactamase is an enzyme that can split beta-lactam rings. Nitrocefin discs are moistened and the test organism rubbed onto it. A red colouration indicates a positive result. N.B. Some coliforms only produce beta-lactamase in the presence of other antibiotics, such as cefoxitin and imipenem. This is known as an inducible class 1 beta-lactamase and requires disc testing to detect its presence. A seeded agar plate has cefoxitin and cefotaxime discs added. A reduced zone between the discs, characterized by a flattened edge, indicates this resistance mechanism.

Bile-aesculin is hydrolysed by enterococci and *Listeria*, producing black colouration in the media.

Bile-solubility is a method of identifying *Streptococcus pneumoniae*. A drop of 10% sodium deoxycholate solution is added to a suspicious colony and allowed to dry. The colony will dissolve if it is *S. pneumoniae*.

Catalase is an enzyme which splits hydrogen peroxide into water and oxygen. To perform a catalase test, the bacteria are brought into contact with hydrogen peroxide, and will produce oxygen bubbles if positive. The test is always performed in a capillary tube or under a cover slip, which traps oxygen bubbles and minimizes aerosol formation. Care must be taken not to pick up any agar as it may crossreact. Catalase positive — staphylococci, catalase negative — streptococci.

Coagulase is an enzyme which causes plasma to coagulate. Coagulase testing can be performed by latex agglutination on a card, or by mixing the bacteria with plasma in a tube, where a visible clot is formed. Coagulase positive — *Staphylococcus aureus*, coagulase negative — coagulase negative staphylococci (sometimes inaccurately called *Staphylococcus epidermidis*).

DNAase is also present in *S. aureus* and is often used as a confirmatory test. The test organism is inoculated onto media containing nucleic acids, and a zone of clearing indicates a positive result.

Gram stain — Flood a fixed smear with crystal violet for 1 min, rinse with tap water, then flood with Lugol's iodine for one minute. Add acetone for 2–3 s, then quickly rinse and add the counterstain, usually dilute carbol fuchsin, for 1 min, then rinse with tap water. Wipe the back of the slide, and dry the front by pressing firmly into filter paper. Allow to air dry. It is thought that Gram-positive bacteria retain the crystal violet stain because of the large amount of peptidoglycan and teichoic acid in the cell wall. The iodine step is used to fix the crystal violet stain into place.

Indole — The hydrolysis of tryptophan by tryptophanase produces indole. A solution of p-dimethyl aminocinnamaldehyde (DMACA) and hydrochloric acid is placed onto filter paper and a colony rubbed onto it. A positive result is indicated by a blue/green colour being formed. Indole positive organisms include *E. coli* and *Providentia*, indole negative organisms include *Serratia* and *Salmonella*.

Lecithinase activity is demonstrated by inoculating media containing egg yolk, such as a Nagler plate, and incubating overnight. A zone of opacity is formed around lecithinase producing colonies. A positive reaction is seen with *Clostridium perfringens* and *Bacillus anthracis*.

Oxidase — A swab or piece of filter paper soaked in oxidase reagent (p-aminodimethylaminoxalate) is applied to a colony. A positive result is a change from clear to purple. This is due to the production of iodophenol. Oxidase positive organisms include *Pseudomonas*, *Moraxella*, *Neisseria* and *Haemophilus*. Oxidase negative organisms include *Salmonella*, *Shigella* and *Proteus*.

Streptococcal grouping uses enzyme extraction, or acid extraction, to differentiate between the different groups of beta-haemolytic streptococci. Group D streptococci have protein on the cell wall, the other groups have either polysaccharide or teichoic acid. The six most common groups are screened for — A, B, C, D, F and G. Six drops of enzyme are mixed with a few colonies of the organism and incubated for 10 min at 37 °C. The solution is then mixed with latex relating to each of the groups as a slide agglutination test. All six latexes must be tested as some streptococci will react with more than one of the latexes.

Tributyrin testing is used to confirm the identification of *Moraxella catarrhalis*. It is a test looking for butyrate esterase activity. A tributyrin disc is added to a suspension of the organism in saline. Positives turn the solution yellow after 2–4 h at 37 °C.

Urease splits urea to produce ammonia. Bacteria are inoculated on to media containing urea and a pH indicator such as phenol red. Ammonia production by urease positive organisms turns the medium alkaline. Urease positive organisms include *Proteus*, *Helicobacter pylori* and *Cryptococcus neoformans*.

6.2.8 Molecular methods

The most commonly used molecular method is polymerase chain reaction (PCR). The RNA or DNA genome of the pathogen is amplified using primers and polymerase (Figure 6.13). This technique is very useful for the detection of organisms which are difficult, slow or dangerous to grow. PCR can also be used to detect antibiotic resistance genes or to quantify viral load. Methods such as PCR, which amplify nucleic acids prior to detection, are called nucleic acid amplification methods (NAATs). Examples of NAATs in

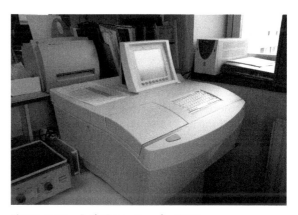

Figure 6.13   ProbeTec system for NAAT

common use include *Chlamydia trachomatis*, TB and herpesvirus PCRs. Other molecular methods include Southern blot, Western blot and 16S sequencing.

16S RNA is part of the small ribosomal subunit and is highly conserved between species. It can be amplified by reverse transcriptase PCR (RT-PCR) and then sequenced to identify the bacterium. This is a useful technique for bacteria that cannot be grown in culture. For example, a patient with empyema (chest infection with a collection of infected fluid around the lung) will often receive antibiotics before the fluid is drained. The bacteria in the pleural fluid sample are antibiotic-damaged and may not grow by conventional culture methods, but can be identified by 16S.

6.2.9   Serological tests

Serological testing detects the immune response to a pathogen. Most commonly we look for IgM, IgG or both. Serology is used to detect many viral infections and also for other difficult-to-grow organisms such as *Toxoplasma, Chlamydia, Mycoplasma, Coxiella, Helicobacter* and *Borrelia*. Methods include latex agglutination, enzyme immunoassay, complement fixation and single radial haemolysis. Most laboratories are moving away from traditional labour-intensive assays towards automated systems.

6.2.10   Antibiotic assays

Antibiotic assays involve taking a sample of blood (or rarely other body fluids) and measuring the concentration of antibiotic in it. Most antibiotics are well tolerated by patients and do not require monitoring. A few antibiotics, in which therapeutic levels are close to the toxic range at which side effects occur, have a higher risk of toxicity. The best-known examples are aminoglycoside antibiotics, such as gentamicin, and vancomycin. Possible side effects of these drugs include renal failure and deafness. Patients who are on these drugs for more than a day should have their levels monitored. If the levels are outside the desired range, the size of dose or frequency of dosing may be changed.

Traditionally, peak levels (taken 1 h after a dose) and trough levels (taken just before the next dose) were taken. Peak levels are now rarely requested unless there are concerns about underdosing and treatment failure. Trough levels give a better indication whether the

Figure 6.14   Antibiotic assays being performed on a TDX analyser

patient is likely to have problems with toxicity. For once-daily gentamicin dosing, levels are taken anywhere from 6–20 h postdose and interpreted with the aid of a graph called a nomogram.

There are two common methods of antibiotic assay. The Abbott TDX system uses fluorescence polarization immunoassay (FPIA) and can be used to measure gentamicin, amikacin and vancomycin levels (Figure 6.14). Abbott also manufacture the Axsym and the Architect i2000SR, which are immunoassay systems.

Requests for antibiotic assay should always include information about all antibiotics that the patient is on, the size and frequency of dose, and the date and time of the last dose. Without this information it is impossible to interpret and advise on the results (Box 6.7).

6.2.11   Safe working practice in the microbiology laboratory

All laboratories have working hazards to deal with. Every chemical has its own Control of Substances

Common causes of ridiculously high or low levels include:

The blood sample was taken out of the same IV line used to give the antibiotic.
The blood sample was taken at the wrong time.
The dose was not given.
The dose was given at the wrong time.
The wrong drug was given.
The wrong patient was bled.
The request form asked for the wrong assay (e.g. patient is on gentamicin but the ward staff requested vancomycin assay).

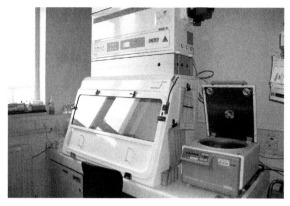

Figure 6.15   Class I safety cabinet

Hazardous to Health (COSHH) certificate. COSHH covers chemicals, products containing chemicals, fumes, dusts, vapours, mists and gases, and biological agents. If the packaging of a product has any hazard symbols then it is classed as a hazardous substance. Every procedure has its own risk assessment, as your employer has a legal responsibility to reduce risk to staff and the public. More details about COSHH and risk assessment can be found on the Health and Safety Executive website (www.hse.gov.uk).

Understanding safe working practices is part of becoming qualified as a biomedical scientist. The IBMS Certificate of Competence contains several sections on the understanding of safe working.

Every staff member, and visitor, will have safety instruction on entry to the department, with fundamental instructions on what to do if alarm bells should sound, including evacuation routes and the assembly area for the department. Every staff member will be provided with training on how to respond in the event of a fire. Every staff member, and visitor, working in the laboratory will have personal protective equipment made available to them. This includes a white coat, or protective disposable coat, disposable gloves, eye protection and the use of a Class I Safety Cabinet for handling high risk samples and organisms.

The Class I Safety Cabinet is a negative pressure, ventilated cabinet with an open front and a minimum face velocity of 75 linear feet per minute. All of the air drawn into the cabinet is filtered through an HEPA filter before being exhausted to the environment. There is a meter on the front of the cabinet showing the current air speed. The cabinet is only safe to use when this meter reading is in the safe zone (Figure 6.15).

There are set procedures for dealing with spillages, near misses and accidents. All staff should be aware of health and safety legislation and incident reporting procedures. There will be standard operating procedures (SOPs), online or provided in folders, to explain how an incident or accident must be handled.

All staff will be able to work in a safe environment, understand good laboratory practice, thus minimizing the risk to others through the understanding of any hazards present in the workplace. These hazards may include the infection risk of microorganisms, which are grouped into four categories of risk (hazard groups 1, 2, 3 and 4, Table 6.8).

Organisms in hazard groups 2–4 are considered to be pathogens. Hazard group 1 organisms would not usually cause disease, although might cause opportunistic infections under certain circumstances (see Section 6.18 on opportunistic infections).

6.2.12   Notifiable diseases

Some infections are legally notifiable to the Health Protection Agency under the Health Protection (Notification) Regulations 2010 [1]. Clinicians have a duty to notify illnesses that they suspect on clinical grounds, while the laboratory has to notify certain pathogens that it has detected. These are organisms which have implications for public health, such as food poisoning bacteria or invasive Group A streptococcal infections. Laboratory notification is usually done via an electronic link called CoSurv.

Table 6.8 Hazard groups for micro-organisms

Hazard group	Pathogenicity to humans	Example	Hazard to workforce	Hazard to community	Treatment or prophylaxis
1	Unlikely to cause human disease	—	No	No	Not usually required
2	Can cause human disease	*Staphylococcus aureus, E. coli*	Possible	Unlikely	Available
3	Can cause severe human disease	*Mycobacterium tuberculosis, Salmonella typhi*	May be serious	May spread	Usually available
4	Causes severe human disease	Lassa fever virus, Ebola virus	Serious	Likely	None

6.2.13 Waste: reduce, reuse, recycle

Waste produced by the laboratory or hospital can present a risk to staff, the community and the environment. It is necessary to treat this waste before disposal. This does not always require sterility, the complete removal of all organisms, but may require disinfection, the removal of significant organisms to the extent that the item ceases to be a potential source of infection (Table 6.9).

Methods of waste disposal are determined by the nature of the waste [2]. There is an increasing

Table 6.9 Treatment of waste

Sterilization methods	Example	Use
Dry heat	Bunsen burner	Sterilize laboratory equipment, e.g. wire loops, forceps, and scissors
	Hot air oven	Glassware, powders
Moist heat Using steam under pressure	Autoclave	Sterilize laboratory equipment that can be reused, sterilize bacterial cultures for safe disposal, sterilization of operating theatre equipment, plastic tubing for media preparation
Toxic Gas	Ethylene oxide	Sterilize laboratory equipment such as specialist instruments for re-use
Radiation	Cobalt-60 source in special facilities	Sterilize mass-produced disposable plastics, e.g. pipettes
Filtration	Hepa filters	Decontaminate air leaving safety cabinets
	Cellulose acetate membrane filters	Filter sterilize drugs for injection
Disinfection	Examples of use	
Cleaning	Hand washing to prevent transmission of infection between patients	
Pasteurization — item held at high temperature for set time period	Expressed milk from neonatal unit	
Boiling	May be used for 'medium risk' metal instruments, but rare as small autoclaves are commonly used.	
Low temperature steam	Can be used with formalin to disinfect equipment	
Chemical	Endoscope decontamination. Chemicals work well on clean objects, but not objects contaminated with bodily fluids.	

Figure 6.16   Autoclave

Figure 6.18   Yellow waste bag for clinical waste

awareness in the hospital environment of the cost of waste disposal and the staff are encouraged to dispose of waste correctly and economically.

Sharp laboratory waste will be put into rigid plastic containers before being further processed for discard (Figure 6.17). Other bench-produced waste will be put into bags to be autoclaved before disposal (Figure 6.16). Some waste, such as bacterial cultures and patient samples, are autoclaved before disposal. This often occurs on-site but can be transported off-site in rigid plastic containers to be autoclaved and incinerated elsewhere.

Secure shredding facilities are available for confidential waste, which often involves the collection of confidential waste by contractors, who then shred on-site using portable equipment. Domestic waste, such as flowers, food waste and so on, is disposed of into black plastic bags. Clinical or potentially infectious waste is discarded into orange or yellow bags, which are

sealed with bag ties, to determine their processing requirements (Figure 6.18).

6.3   Bacteria

6.3.1   Nutritional growth requirements

Bacteria chemically comprise of protein, polysaccharide, lipid, nucleic acid and peptidoglycan. This means that bacteria require hydrogen, oxygen, carbon and nitrogen in order to grow. Some bacteria can make what they need from very basic elements, others require ready-made compounds that can be broken down and used to manufacture structural components. Bacterial growth also requires an energy source, which can be as simple as sunlight, or more complicated, such as the breakdown of sugars.

Growth factors are required in very small amounts and are found in body fluids and tissues. These factors include purines and pyrimidines (to build DNA), vitamins and amino acids.

6.3.2   Environmental growth requirements

6.3.2.1   Carbon

Bacteria can be divided into two groups depending on their carbon requirement. Auxotrophs get all their carbon requirement from the CO_2 in the air and use sunlight for energy. They are free living bacteria and

Figure 6.17   Sharps bin

Table 6.10   Oxygen requirement

Group	Requirement
Obligate anaerobes	Require very low oxygen concentration gas mix and a catalyst to reduce any remaining oxygen to water
Aerotolerant anaerobes	Anaerobic bacteria that are not killed by oxygen
Facultative anaerobes	Can grow both aerobically and anaerobically
Obligate aerobes	Require oxygen to grow
Microaerophilic organisms	Grow best in low oxygen tension

nonparasitic. Heterotrophs require an organic form of carbon that can act both as a carbon source and an energy source. This group of bacteria includes human pathogenic bacteria.

6.3.2.2   Oxygen

The oxygen requirement of a bacterium allows them to be divided into five groups: obligate anaerobes, aerotolerant anaerobes, facultative anaerobes, obligate aerobes and microaerophilic organisms (Table 6.10).

6.3.2.3   Temperature

There are three groups used to describe the temperature at which bacteria grow: psychrophiles ($< 20\,°C$), mesophiles ($25 - 40\,°C$) and thermophiles ($55 - 80\,°C$). Most human pathogens are mesophiles.

6.3.3   *Bacterial growth*

Bacterial growth and division occurs by binary fission when the bacterium reaches a critical mass. The rate of growth occurs in four phases

1. The lag phase occurs when the bacteria adjust to prepare for cell division.
2. The logarithmic (or exponential) phase is when cell growth and division is at the maximum rate allowable based on the nutrients available. A bacterium can divide every 30 min during log phase.
3. The stationary phase is a levelling out period that occurs when one of the nutrients is depleted.
4. The death phase (decline) occurs when the bacteria lyse and die (Figure 6.19).

6.3.4   *Bacterial metabolism*

Many of the tests used to identify bacteria require them to metabolize various different compounds. The method in which they do that is another way of grouping bacteria through the different enzymes that they possess.

6.3.4.1   Fermentation

Fermentation is the release of energy anaerobically from carbohydrates. It occurs mainly via glycolysis, which leads to the production of pyruvic acid or pyruvate.

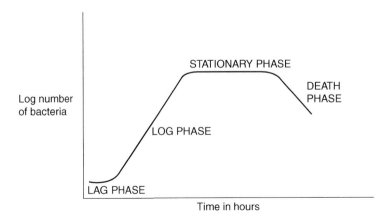

Figure 6.19   Bacterial growth curves

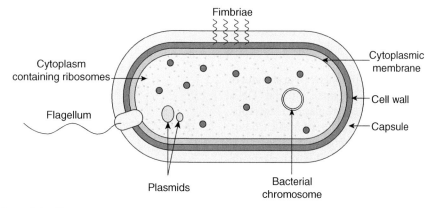

Figure 6.20   Bacterial cell structure

6.3.4.2   Oxidation

Oxidation is the release of energy in the presence of oxygen.

6.3.5   Bacterial cell structure

A typical bacterial cell is illustrated in Figure 6.20. Important cell structures include the following.

Cytoplasmic membrane: a phospholipid bilayer.

Cell wall: a rigid structure whose function is to support the cytoplasmic membrane and maintain the shape of the bacterium. Damage to the cell wall often results in lysis. The main component of the cell wall is a polymer called peptidoglycan (Figure 6.21). The staining technique known as a Gram stain will distinguish bacteria with a thick peptidoglycan layer (Gram-positive) from a thin layer (Gram-negative). Gram-positive cell walls have a simple, strong structure. Gram-negative cell walls are thinner but have a second, outer membrane which contains lipopolysaccharide (LPS), as well as many associated proteins. LPS, also known as endotoxin, may cause septic shock if it gets into the bloodstream. The Gram-negative outer membrane confers advantages such as protecting the cell from the effects of lysozyme (a host defence substance) and impeding the entry of antibiotics to the cell.

Nuclear material: the bacterial chromosome is a long molecule of double-stranded DNA which is coiled and supercoiled by the DNA gyrase enzyme system

(a)

Gram-positive cell wall

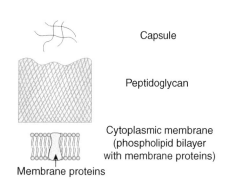

(b)

Gram-negative cell wall

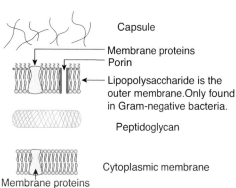

Figure 6.21   Cell wall structure

(another suitable target for antibacterial therapy). There is no nuclear membrane and no nucleus. The area where the chromosome is situated is called the nucleoid.

Many bacteria have additional DNA in structures called plasmids, which are small circular DNA elements found in bacterial cytoplasm. They replicate independently of the main bacterial chromosome and can be passed between bacteria by a process known as conjugation. They often encode genetic information for useful traits such as antibiotic resistance or toxin production.

Ribosomes: a bacterial cell contains thousands of ribosomes, whose function is protein synthesis. Bacterial ribosomes are smaller than eukaryotic ribosomes, with a sedimentation coefficient of 70S (composed of 30S and 50S subunits). This difference makes them a useful target for antibacterial therapy. Genetic study of ribosomal RNA by 16S amplification can be used to identify bacteria that cannot be grown in culture.

Fimbriae or pili: filamentous projections arising from the cell wall which mediate adhesion between the bacterium and the host cell. Sex pili are longer and initiate conjugation (transfer of DNA) between bacterial cells.

Flagella: long, thin filaments arising from the cytoplasm, used for locomotion. Flagella may be single or multiple (up to 20 per cell) and may be situated at one end of the cell (polar) or along the sides (peritrichous).

Spores: formed only by some bacterial species, notably *Bacillus* and *Clostridium*, as a response to adverse environmental conditions. The spore is a dormant form which can survive for long periods and will germinate when conditions become favourable for growth. Each cell produces only one spore, sometimes in a distinctive position within the cell which can aid us in identification of the bacteria (Figure 6.22).

Capsule or slime: many bacteria have a further polysaccharide layer external to the cell wall. This layer is often well defined and is visible on light microscopy as a halo around the bacterium, in which case it is called a capsule. If the layer is amorphous and comes away from the bacterium it is called slime; *in vitro* it causes the bacterial colonies to appear mucoid. The presence of a capsule or slime is protective against phagocytosis and complement, and reduces antibiotic penetration into the bacterial cell.

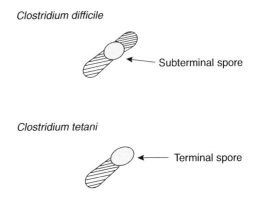

Clostridium difficile

Subterminal spore

Clostridium tetani

Terminal spore

Figure 6.22   *Clostridium difficile* with subterminal spores and *Clostridium tetani* with terminal spores

6.3.6   Common bacteria

The primary method of differentiation between groups of bacteria is through the Gram stain result. Below are some of the more common bacteria and methods of identification.

6.3.6.1   Staphylococci

Staphylococci are Gram-positive cocci growing in pairs and tetrads, resembling a bunch of grapes. They are catalase positive, which differentiates them from streptococci. Staphylococci are divided into two groups using the coagulase test. *Staphylococcus aureus* is coagulase positive, other staphylococci are not. A positive coagulase combined with a positive DNAase test confirms the identification of a *Staphylococcus aureus* which will always be followed up. *Staphylococcus aureus* is a common cause of infection of the skin and soft tissues, and can also cause deep seated infections at other sites such as endocarditis and osteomyelitis. Some *S. aureus* produce toxins which are involved in food poisoning, toxic shock syndrome and skin exfoliation. Some strains produce a toxin called Panton—Valentine leucocidin (PVL) and cause severe skin infections and necrotising pneumonia. Meticillin resistant *Staphylococcus aureus* (MRSA) causes the same infections as meticillin sensitive *S. aureus*, but is more difficult to treat and is a greater infection control problem.

The other staphylococci are usually grouped together as coagulase-negative staphylococci (CNS) and reported as such. They are usually skin flora of no real significance, but may be significant if isolated repeatedly from a sterile site.

6.3.6.2 Streptococci

Streptococci are Gram-positive cocci growing in pairs and chains. They are catalase negative. Streptococci are classified based on their ability to haemolyse blood to various degrees. Beta-haemolytic streptococci haemolyse blood completely, leaving a clear zone around colonies on a blood agar plate. Alpha-haemolytic streptococci partially haemolyse blood, producing a greenish zone around colonies on blood agar. Non-haemolytic streptococci will not haemolyse blood and have no zone around them (Figure 6.23).

Beta-haemolytic streptococci can be divided using the Lancefield grouping system into groups A, B, C, D, F and G. Group A streptococci (GAS) cause throat infections, ear infections, and skin and soft tissue infections which can progress to necrotizing fasciitis. The immunological reaction to GAS may cause rheu-

(a)

(b)

Figure 6.23 (a) Beta-haemolytic streptococci on a blood agar plate and (b) *Streptococcus pneumoniae* demonstrating alpha haemolysis on blood agar

matic fever or poststreptococcal glomerulonephritis. Group C and G streptococci cause most of the same diseases as GAS but in a milder form. Group B streptococci are carried vaginally and are linked with infections of the newborn.

Alpha-haemolytic streptococci are often nonpathogens but one species, *Streptococcus pneumoniae,* can cause serious infections such as pneumonia as well as ear and eye infections. It is differentiated from the others by looking for a zone of inhibition around an optochin disc (ethyl hydrocuprein hydrochloride).

6.3.6.3 Anaerobes

Anaerobes come in all shapes and sizes, the most common including *Clostridium sp., Bacteroides sp.* and anaerobic cocci. They are identified by adding a metronidazole disc to all anaerobic culture plates. A zone indicates the presence of an anaerobe. Anaerobes are often isolated as a mixed culture and not identified individually; however, the presence of a spreading colony sensitive to metronidazole may indicate *Clostridium sp.* and would necessitate a full identification to rule out tetanus or *Clostridium perfringens*. Some *Clostridium species* are a cause of food poisoning, others can cause gangrenous infection. Mixed cultures of anaerobic bacteria are often a cause of infection in deep tissue sites.

Clostridium difficile is a Gram-positive anaerobic rod, which is part of the normal gut flora in children aged less than 2 years but is not normal in adults. Adults may be exposed to *C. difficile* spores in the healthcare environment. Antibiotic treatment disturbs the normal bowel flora and can lead to the overgrowth of this bacterium. In these conditions it can cause antibiotic-associated diarrhoea (AAD) or pseudomembranous colitis (PMC) symptoms ranging from mild diarrhoea to life-threatening illness. The severity of the disease is related to toxin production by the bacterium. It produces two toxins: toxin A is an enterotoxin and toxin B is a cytotoxin. Diagnosis of *C. difficile* infection is based on detection of the toxins in samples of faeces. This can be performed by serological methods, such as EIA, or by cell line looking for cytopathic effect — the latter only demonstrating toxin B. Screening for this bacterium occurs daily in most hospitals, and results are available on the same day. It can be cultured in the laboratory, but is rarely requested as it takes much longer to get a result.

6.3.6.4 Coliforms

Coliforms are Gram-negative rods and encompass a wide variety of bacteria including the enterobacteriaceae. Coliforms are found in faeces but can contaminate other sites, as transient flora or as pathogens. They are the most common cause of urinary tract infections where, as treatment does not require full identification of the bacteria, the focus is on antibiotic sensitivity testing after preliminary identification.

The primary method of differentiating coliforms is to establish whether or not they can ferment lactose. Lactose-fermenting coliforms only cause disease if they are found in sterile areas. Common lactose-fermenting coliforms include *Escherichia coli* and *Klebsiella spp.* Many strains of *E. coli* have powerful toxins that cause diarrhoea and other symptoms. For example, *E. coli* O157 causes bloody diarrhoea and haemolytic uraemic syndome.

Non-lactose fermenting coliforms include diarrhoea-causing organisms, such as the salmonellae and shigellae. Salmonellae and shigellae are common causes of food poisoning. Tests to identify these pathogens initially involve oxidase to exclude pseudomonads and biochemical tests to identify which carbohydrates can be utilized by the organism. The cell wall of salmonellae contains specific lipopolysaccharides called 'O' antigens and 'H' antigens made out of flagella protein. Shigellae also possess 'O' antigens. These bacteria are identified by serological testing: a suspension of the organism is made in saline, mixed with specific antisera and rocked gently whilst looking for agglutination. Antisera are available as a mixture of antigenic types, for example polyvalent 'O' antigen, as well as singly, and so can be used for screening suspicious colonies. The results of the *Salmonella* 'serology' can be compared to a reference listing of all the antigenic types in order to identify the species; this is available in the Kauffman–White classification scheme. A copy of this can be found in every microbiology laboratory. Identification to species level is followed by phage typing by a reference laboratory (Table 6.11). It is crucial to determine identification to this level as it establishes whether or not a group of infections is due to a food poisoning outbreak from a particular event or establishment (Figure 6.24).

When performing agglutinations with antisera it is important to take great care as the risk of infection to the BMS is increased. A loop of antiserum is mixed with a saline suspension of the organism. When making the suspension and adding and mixing the

Table 6.11 Common salmonellae

Common salmonellae	Antigenic structure	
Salmonella enteritidis	'O' 1,9,12	'H' g
Salmonella typhi	'O' 9,12	'H' d Vi*
Salmonella typhimurium	'O' 1,4,5,12	'H' i
Salmonella paratyphi A	'O' 1,2,12	'H' a

*The capsule produced by some salmonellae is the 'Vi' antigen. When this is present, the organism will often not agglutinate with any 'O' antisera as the capsule gets in the way.

Figure 6.24 Probable *Salmonella* species growing as black colonies on XLD plate; lactose fermenting coliforms grow as yellow colonies on XLD

antiserum, it is important to turn the wire or plastic loop onto its side before lifting it from the mixture. This reduces the likelihood of aerosols being produced. Another common nonlactose fermenter found in faeces is *Proteus spp.*, a common cause of UTI, which is characterized by its ability to swarm on blood agar and is urease positive.

6.3.6.5 Pseudomonads and acinetobacter

Pseudomonads are Gram-negative rod-shaped bacteria that are strict aerobes. They cause infections in burns and UTI, and are very commonly found. They are nonlactose fermenters and oxidase positive. Commonly pseudomonads are further differentiated to identify *Pseudomonas aeruginosa* by a variety of test including growth at 42 °C (*Ps. aeruginosa* will grow at this temperature), and sensitivity to identification discs (e.g. C390). They are resistant to a large number of antibiotics.

Acinetobacters are short Gram-negative rod-shaped bacteria that are strict aerobes, and oxidase negative. Multiresistant strains can cause problems in hospitals with colonization and infection of patients.

6.3.6.6   Campylobacter, vibrio and helicobacter

This group are all Gram-negative curved rods. They are all oxidase positive. Campylobacter grows in micro-aerophilic conditions and causes bloody diarrhoea. There are many strains, but those pathogenic to humans are the only ones that can grow at 42 °C. Identification is done by colonial morphology on selective agar, Gram stain and oxidase.

Vibrios are found in water. *Vibrio cholerae* causes cholera. It is an aerobe and oxidase positive on nonselective agar. Thiosulphate citrate bile salt sucrose agar (TCBS) is the selective agar used for isolation of *Vibrio spp.* Antiserum is available for identification.

Helicobacter pylori is a curved Gram-negative rod, catalase positive, oxidase positive and urease positive. It is associated with chronic gastritis, peptic ulcers and gastric carcinoma. *H. pylori* strains grow in a micro-aerophilic atmosphere, at 37 °C and prefer a high humidity. It takes 3 to 5 days to grow but is notoriously difficult to isolate. For this reason, detection of *H. pylori* is done serologically. Faeces samples are tested using an EIA method, looking for bacterial antigen. A positive result indicates a current infection. Historically, antibody screening was performed, but this was less useful because antibodies remained positive after treatment and so could not be used to assess response to treatment or to detect reinfection.

6.3.6.7   Corynebacteria

Corynebacteria are Gram-positive rods, often demonstrating a pattern similar to Chinese lettering. The majority of corynebacteria are part of the normal skin flora, but *Corynebacterium diphtheriae* can cause diphtheria. This species is differentiated from other corynebacteria by isolation on selective media such as Tinsdale or Hoyle's, and being urease negative. Further biochemical testing will establish the species group from this point. There are three main biotypes of *C. diphtheriae*: gravis, intermedius and mitis (Box 6.8). Isolates are sent to the reference laboratory

> **Box 6.8   Diphtheria toxin**
>
> Diphtheria toxin production is not an intrinsic feature of *C. diphtheriae*, but is encoded on the *tox* gene carried, and inserted into the bacterial chromosome, by a bacteriophage. Only strains which contain the bacteriophage are toxigenic.

for toxigenicity testing, which takes 24–48 h. Mitis strains are less likely to be toxigenic.

6.3.6.8   Haemophilus

Small Gram-negative rods, haemophili are associated with the respiratory tract and are commonly found as a cause of eye and ear infections as well as chest infections. They require an enriched media to grow as the organism has a requirement for haemin (X-factor) and/or di- or tri-phosphopyridine (V-factor) which are present in lysed blood. The most significant isolate, *Haemophilus influenzae*, is both X and V dependent (Table 6.12). The growth of *Haemophilus* species is enhanced by an atmosphere rich in carbon dioxide (Figure 6.25).

6.3.6.9   Gram-negative cocci

Many Gram-negative cocci, such as *Neisseria gonorrhoeae* and *N. meningitidis*, are nutritionally demanding bacteria. They require a lysed blood agar (such as chocolate agar) to grow. *Moraxella catarrhalis* grows on blood agar. They are all oxidase positive but can be differentiated by tributyrin test (*Moraxella* gives a positive reaction) and sugar fermentation tests.

Neisseria gonorrhoeae is a pathogen found in the human genital tract, *N. meningitidis* can cause meningitis and is found in the upper respiratory tract, and *M. catarrhalis* can cause upper respiratory infections. All have been found as causes of eye infections.

Table 6.12   Haemophilus growth factors

Organism	Requirement
Haemophilus influenzae	X and V
Haemophilus parainfluenzae	V
Haemophilus ducreyi	X

Figure 6.25 (a) CO_2 incubator with gas cyclinder and (b) and (c) incubator with plates stacked inside

6.3.6.10 Spirochaetes

Spirochaetes are helical organisms with a similar cell wall structure to Gram-negative organisms. Spirochaetes are too slender to see on a Gram's stain so dark ground microscopy or modified staining methods, such as silver stains, are used to see them under the microscope. This group includes *Treponema*, *Borrelia* and *Leptospira*. As these organisms are very difficult to

grow, their presence is detected through serological blood testing.

Treponema pallidum causes syphilis and is detected by a range of serological tests, including TPPA (*Treponema pallidum* particle agglutination), RPR (rapid plasma reagin), EIA (enzyme immunoassay) and FTA (fluorescent treponemal antibody absorption test).

Borrelia burgdorferi is the cause of Lyme disease, a tick-borne disease that causes skin lesions, encephalitis and arthritis symptoms. The ticks are carried by deer, so this disease is found in people who have been in woodland areas. Antibodies are detected by enzyme immunoassay.

Leptospira spp. are the cause of leptospirosis or Weil's disease. It is carried in many animal hosts, most commonly the rat, and transmitted in urine. It causes fever, headache and muscle pain and can lead to jaundice and renal failure. Diagnosis is usually confirmed by serology, though it may be grown from blood cultures within the first 5 days of the illness.

6.3.7 A guide to what is normal — and what is not!

Each region of the body has its own population of bacteria that call it home (Table 6.13). This varies due to the nature of that body part but leads to a sometimes confusing mixture of bacteria and the need to decide which are in the right place, and which are unwelcome — the potential pathogens. To make this mystery a little clearer we have provided a list, region by region, of normal bacteria and common pathogens.

6.4 Fungi

Fungi are opportunistic organisms. They can cause either superficial or deep infections (mycoses).

6.4.1 Fungal cell structure

Fungi are eukaryotes and all possess membrane-bound organelles, cell membrane and cell wall composed of polysaccharides (glucan, mannan) and chitin. For practical purposes they can be divided into three main groups. Moulds or filamentous fungi are composed of branching filaments (hyphae) which intertwine to form a mass called the mycelium. They reproduce by spore production. True yeasts and yeast-like fungi are

Table 6.13   Normal bacterial flora and bacterial pathogens

	Normal flora	Potential pathogens
Mouth	Alpha haemolytic streptococci	*Candida*
	Neisseria spp.	Vincent's fusiform bacillus
	Anaerobes	
Nose and sputum		
	Alpha haemolytic streptococci	*Strep. pneumoniae*
	Neisseria spp.	*Haemophilus influenzae*
	Diphtheroids (Corynebacteria)	*Staph. aureus*
		Enterobacteriaceae
Throat	Alpha haemolytic streptococci	Beta-haemolytic streptococci (Group A)
	Neisseria spp.	*Arcanobacterium haemolyticum*
	Diphtheroids (Corynebacteria)	
Skin		
	Staphylococcus epidermidis	*Staph. aureus*
	Nonhaemolytic streptococci	Beta-haemolytic streptococci (Group A and G)
	Enterococci	*Pseudomonas* (in burns)
	Diphtheroids (Corynebacteria)	
	Propionibacteria	
Bowel (and faeces)		
	Enterobacteriacae	*Salmonella*
	Enterococci	*Shigella*
	Anaerobes	*Campylobacter*
	Candida	*E. coli O157*
Vagina		
	Lactobacilli	*Candida*
	Alpha haemolytic streptococci	*Neisseria gonorrhoeae*
	Diphtheroids (Corynebacteria)	Beta-haemolytic streptococci (Group B) (Box 6.9)
	Beta-haemolytic streptococci (Group B) (Box 6.9)	*Actinomyces*

predominantly unicellular and oval or round in shape. They may produce chains of elongated cells called pseudohyphae. Reproduction is by budding. Dimorphic fungi grow as yeasts at 37 °C (on agar and in the body) and as filamentous fungi at 28 °C. They are often Category 3 organisms, which require extra precautions during laboratory work.

6.4.2   Common fungi and laboratory identification

Direct microscopy of the sample may include potassium hydroxide to digest keratinous samples, and

> **Box 6.9   Group B streptococcus**
>
> Confusingly, Group B streptococcus appears in both columns. It is carried in the vagina of up to 40% of healthy women, and is usually regarded as a commensal organism. However, it can cause serious maternal and neonatal infections at, or soon after, birth.

lactophenol cotton blue to stain the fungal elements. All fungi stain Gram-positive. Samples are cultured onto Sabouraud's agar and incubated at 28 °C (and 37 °C). Commercial identification kits, which rely on

Box 6.10   Germ tube

Germ tube test: the fungus is incubated in serum for 2 h, then viewed down a light microscope. If it has formed protrusions known as 'germ tubes', it is almost certainly *C. Albicans.*

Box 6.11   India ink

India ink testing: india ink is added to CSF. This gives a dark background. The capsule around the cryptococcal cells stops it from taking up the stain, so the cells have a clear halo around them. This halo distinguishes them from other cells such as host lymphocytes.

biochemical tests, are available for yeasts. Indicator media such as ChromAgar, which changes colour according to the species grown, may also be used for provisional identification.

Candida albicans is a yeast that causes opportunistic infections. It is commonly isolated from the genital tract as it lives on human skin and mucous membranes. It can cause superficial disease such as thrush, or can cause systemic disease (invasive candidiasis) in the immunocompromised patient. *C. albicans* can be distinguished from other *Candida* species by the germ tube test (Box 6.10). This is important because *C. albicans* is usually sensitive to fluconazole, whereas other species of *Candida* are often resistant, so the germ tube test guides the initial choice of therapy (Figure 6.26).

Non-*albicans* species of *Candida* include *C. parapsilosis, C. tropicalis, C. krusei* and *C. dubliensis.* These are less common than *C. albicans* and are generally seen in patients who are immunocompromised, have medical devices such as IV lines, or who have previously been treated with fluconazole. Many non-*albicans* species are fluconazole resistant.

Cryptococcus neoformans is a capsulate yeast. It is found in the environment, mainly in bird droppings, but may be inhaled and act as an opportunistic pathogen. It is a cause of meningitis, particularly in immunocompromised patients. It grows as a creamy colony, is urea positive and is 'india ink' positive *in vivo* (Box 6.11 and Figure 6.27). Its presence can be detected by cryptococcal antigen testing of blood or CSF.

Malassezia furfur is a yeast that causes a superficial skin infection called pityriasis versicolor. It is lipophilic, and has been known to cause IV line sepsis in neonates receiving parenteral lipid therapy. Its lipid requirement makes it very difficult to grow in the laboratory, so diagnosis is based on its characteristic appearance on direct microscopy, known as 'spaghetti and meatballs' due to the short nonbranching hyphae and spores seen mixed together.

Aspergillus species are found in the environment; airborne spores may be inhaled to cause disease. *Aspergillus fumigatus* is the most common species found in humans. It is distinguished from other species by microscopy and colonial appearance. Other species which may cause human infection are *A. niger, A. terreus* and *A. flavus.*

The spectrum of disease caused by *Aspergillus* includes:

Allergic aspergillosis — type I or III hypersensitivity reaction causes fever, breathlessness and progressive lung fibrosis,

Aspergilloma — preexisting lung cavities become colonised to form a fungal ball which may erode blood vessels to cause haemorrhage, and

Invasive aspergillosis — in immunocompromised patients, an initial pulmonary infection spreads by the haematogenous route to other organs. This is a serious complication with high mortality.

Chronic otitis media or sinus infection may be caused by *Aspergillus* species, particularly *A. niger.*

Dermatophytes are filamentous fungi that infect the stratum corneum of skin, hair or nails to cause ringworm (tinea). There are three species:

Trichophyton spp. includes a range of dermatophytes that cause infections of skin, nail and hair. The species are differentiated by their colonial and microscopic appearance, although urea testing has some benefit.

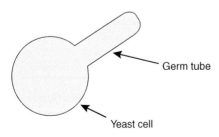

Figure 6.26   Diagram of yeast cell with germ tube

India ink preparation of CSF

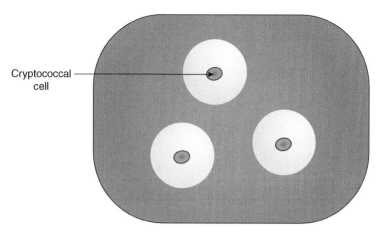

Cryptococcal cell

Figure 6.27   India ink test

Microsporum canis is a dermatophyte transmitted by animals. It causes skin infections.
Epidermophyton spp

Dermatophyte infection causes scaly red lesions on the skin (ringworm), hair loss and scarring on the scalp, or thickening and discolouration of the nails (Figure 6.28). The clinical diagnosis is given a Latin name based on the site of infection:

Tinea capitis — head and scalp.
Tinea barbae — beard area.
Tinea corporis — body.
Tinea cruris — groin.
Tinea manuum — hand.
Tinea pedis — feet (athlete's foot).
Onychomycosis — nail infection.

Other filamentous fungi such as *Mucor*, *Rhizopus* and *Absidia* are opportunistic pathogens that are usually harmless but may cause infections in severely immunocompromised patients, the elderly, diabetics or alcoholics. They may infect sites such as the eye, sinuses, lung or brain. These infections are extremely difficult to treat and have a high mortality.

Penicillium marneffei is a dimorphic fungus that causes systemic infections in AIDS patients who have visited south east Asia. It grows as a green colony with a red reverse, and is a Category 3 pathogen. Suspicious colonies should only be investigated in a safety cabinet.

Histoplasma capsulatum is a dimorphic fungus. It is found in hot climates, particularly South and Central

(a)

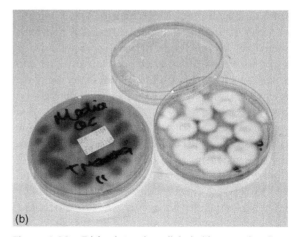

(b)

Figure 6.28   *Trichophyton interdigitale* (dermatophyte) on a Sabouraud plate

America and is rarely seen in the UK. It occasionally causes pneumonia or disseminated disease, particularly in the immunocompromised. At 28 °C it grows as a filamentous fungus with brown colonies, and is generally isolated from respiratory samples. Any brown colonies isolated from these sample types should be investigated in a safety cabinet.

6.5   Parasitology — protozoa and helminths

The study of protozoa and helminths is known as parasitology. Parasitic infections may be acquired in the UK but are often acquired during foreign travel, and are also relatively common in immunocompromised patients.

6.5.1   Parasite investigations

The most common method for the detection of parasites is through the investigation of faeces samples. Stains can be used to detect some parasites: the auramine stain is used to screen for *Cryptosporidium parvum* (and can also be used to detect *Isospora belli*) and the rapid Field's stain is good at detecting the trophozoite form of many protozoa. The biomedical scientists' treat is the discovery of a whole worm in a sample, but this is fairly unusual.

Much of parasite investigation comes from (literally) sifting through the faeces sample. The faeces is prepared for parasite examination by suspending a pea-sized amount in saline containing formalin, in order to kill off the bacteria present. This sample is vortexed to make a soup-like suspension and then filtered to remove larger debris. The remainder is mixed with 2 mL of either ethyl acetate or ether and vortexed again. This step allows the fat in the sample to be extracted. The mixture is placed in a centrifuge and spun for 2 min at 2000 rpm. The fatty plug is loosed and the entire supernatant is carefully discarded. (Due to the nature of the chemicals that have been added, this supernatant is not poured down the sink, but disposed of by collection in larger glass containers and sent for correct chemical disposal.) The remaining pellet is resuspended in one or two drops of saline and inspected under the microscope. Two drops are inspected, one in saline, another mixed with iodine to assist with the identification of cysts. The knack of

> **Box 6.12   Tapeworm motility**
>
> Occasionally faecal samples arrive in the laboratory containing tapeworm segments. Each segment can move on its own and they are fascinating to watch. They are very infectious, so they must be carefully treated, but if more than one is present it may be possible to race them up the side of the container!

telling broccoli from parasite comes only with practice. Strange as it may seem, there are some truly beautiful structures to be found in faeces (Box 6.12)!

6.5.2   Protozoal cell structure

Protozoa are eukaryotic unicellular organisms. They are relatively large, with diameters ranging from $2 - 100 \, \mu m$. Their surface membrane may be thin and flexible, as in amoebae, or relatively stiff to maintain their shape, as in ciliated protozoa. Protozoa are considered to be a low form of animal life because they do not photosynthesize: most protozoa capture and ingest food particles.

6.5.3   Common protozoa

Plasmodium species cause malaria. There are four species which infect humans: *P. falciparum, P. vivax, P. ovale* and *P. malariae. P. falciparum* causes the most severe disease. Malaria is transmitted by the bite of the female *Anopheles* mosquito. The parasites multiply within liver cells and red blood cells to cause a multi-system disease.

There are millions of cases of malaria worldwide each year, and approximately 1600 cases/year in the UK. Malaria in the UK is a travel-associated disease, usually seen in people who travelled to malaria-endemic areas without taking prophylaxis. The commonest scenario is an immigrant returning home who does not realize they need prophylaxis; sometimes this is referred to as VFR (visiting friends and relatives).

The diagnosis of malaria should be considered in any patient who presents with fever within a year of travel to an endemic area. Diagnosis is by thick and thin blood films (at least three films taken at different times) and by antigen detection dipsticks. NAATs are sometimes useful.

Malaria treatment is a specialist subject. Usually a combination of two drugs is given. Quinine plus a second agent is the treatment of choice for falciparum malaria, unless the patient has come from a region where quinine resistance is common.

Giardia lamblia is an intestinal protozoon, found worldwide, which is transmitted by contaminated water and food or by person-to-person spread (fae-cal—oral route). It is often seen in travel-associated diarrhoea or in household or nursery outbreaks. Clin-ical features include abdominal pain, diarrhoea and flatus. It has two stages: the cyst is the infective stage and survives well in the environment. The host ingests the cysts, which reach the duodenum and small intes-tine and then develop into trophozoites. Trophozoites are motile and flagellated, and attach themselves to the intestinal wall by a sucking disc. Eventually more cysts are excreted to complete the life cycle. Diagnosis is by concentration and microscopy of stool for cysts. Three stools should be examined. Trophozoites may be seen in duodenal aspirate or string test, but this test is rarely done (Figure 6.29).

Trichomonas is a flagellated protozoon which causes genital infection, usually sexually transmitted. It may be asymptomatic or cause vaginal discharge and itch (Figure 6.30). Diagnosis is by microscopy of wet preparation of vaginal discharge, or culture in liquid medium. Treatment is metronidazole. It is common knowledge in microbiology departments that, if you manage to find a *Trichomonas spp.* in a clinical sample on a Friday afternoon, you are guaranteed to have a good weekend!

Cryptosporidium parvum is an intestinal protozoon which is transmitted via contaminated water or per-son-to-person, or by direct contact with farm animals. It is highly resistant to disinfectants and may survive water purification. Infection is most commonly seen in children and in HIV-positive people. It usually causes watery diarrhoea, which is usually self-limiting but may be more severe and prolonged in the immuno-compromised. Diagnosis is by stool microscopy with modified acid-fast or immunofluorescent staining, antigen detection or NAAT. Treatment is not usually required.

Toxoplasma gondii is an intracellular parasite of macrophages. Its definitive host is the cat, but other animals and humans can become infected if they ingest cysts. It causes an acute mononucleosis-like illness. It may later reactivate during immunodeficiency, partic-ularly AIDS, most commonly as cerebral infection. It also causes congenital infection, if acquired during pregnancy, and chorioretinitis.

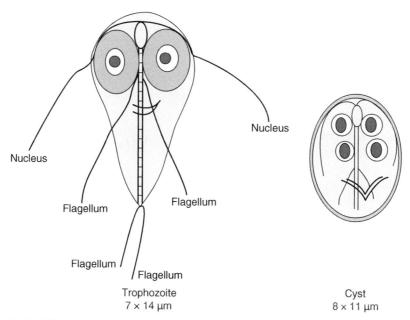

Nucleus

Nucleus

Nucleus

Flagellum

Flagellum

Flagellum

Flagellum

Trophozoite
7 × 14 μm

Cyst
8 × 11 μm

Figure 6.29   Giardia lamblia

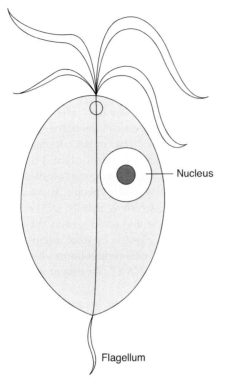

Figure 6.30 Trichomonas

Nucleus

Flagellum

Entamoeba histolytica is an intestinal protozoon which is found worldwide but is more common in developing countries (Box 6.13). It is not often seen in the UK. It is transmitted by contaminated food and water or by the faecal–oral route. The cyst is the infective stage. The host ingests the cysts, which pass to the colon and develop into trophozoites. The trophozoites adhere to colonic epithelium and damage it with protease and cytotoxin production. Clinically this produces amoebic dysentery, with bloody diarrhoea and abdominal tenderness. Complications include amoebic liver abscess and cutaneous amoebiasis. Treatment is metronidazole plus a second agent to eradicate cysts.

Box 6.13 Entamoeba cysts

E. histolytica cysts are morphologically identical to cysts of *E. dispar*, which does not cause disease. If *E. histolytica* cysts are seen, they should be reported as being '*E. histolytica* or *E. dispar*', and referred for antigen detection or NAAT to identify the species.

Diagnosis is by stool microscopy for cysts (which may be seen in fresh stools) or trophozoites. Antigen detection and NAATs are also available. Serology is useful for liver abscesses but not for intestinal disease.

Microsporidia are a group of intracellular spore-forming protozoa, which may infect the intestine or other organs. They are best known for causing infection in patients with advanced HIV. Diagnosis is by stool microscopy using modified trichrome stain or NAAT.

Isospora belli is found worldwide, most commonly in the tropics and subtropics, and causes diarrhoea with abdominal cramps. The illness is self-limiting in healthy people but may be severe and prolonged in the immunodeficient, particularly AIDS patients. Diagnosis is by stool microscopy with modified acid-fast stain.

6.5.4   Helminths

Helminths are multicellular parasitic worms. Strictly speaking they are too large to be classed as true microorganisms, but they are included here because their detection falls within the remit of the microbiology laboratory. They are a common cause of infection worldwide. Most cases in the UK are imported from abroad, with the exception of *Enterobius vermicularis* which is common in the UK. There are three major groups of helminths: nematodes (roundworms), trematodes (flukes) and cestodes (tapeworms). They can be broadly classified as intestinal or tissue parasites.

6.5.4.1   Intestinal helminths (Table 6.14)

6.5.4.2   Tissue helminths (Table 6.15)

6.6   Viruses

6.6.1   Overview of some medically important viruses (Table 6.16)

6.6.2   Virus structure

Viruses are small organisms (less than 200 nm in diameter) which consist of: nucleic acid, which may be either RNA or DNA; a protein coat called a capsid, made up of subunits called capsomeres; and an optional outer

Table 6.14   Intestinal helminths

Name	Common name	Type of helminth	Approx size of adult worm	Disease	Lab investigation
Ancylostoma duodenale	Hookworm	Nematode	1 cm	Intestinal infection, iron deficiency anaemia	Stool microscopy for eggs
Ascaris lumbricoides	Common roundworm	Nematode	15—30 cm	Intestinal infection, chest symptoms during larval migration	Stool microscopy for eggs
Enterobius vermicularis	Threadworm	Nematode	2—10 mm	Pruritis ani	Microscopy of perianal sellotape slide
Necator americanus	Hookworm	Nematode	1 cm	Intestinal infection, iron deficiency anaemia	Stool microscopy for eggs
Strongyloides stercoralis	—	Nematode	2 mm	Intestinal infection, rarely hyperinfection	Stool microscopy or culture for larvae, serology
Taenia solium	Pork tapeworm	Cestode	3—5 m	Intestinal infection, cysticercosis	Stool microscopy for eggs and proglottids. Serology (IFAT) for cysticercosis.
Taenia saginata	Beef tapeworm	Cestode	5—10 m	Intestinal infection	Stool microscopy for eggs and proglottids
Toxocara canis	Dog roundworm	Nematode	*	Visceral larva migrans	Serology
Trichuris trichiura	Whipworm	Nematode	2—5 cm	Intestinal infection	Stool microscopy for eggs

*Toxocara is a parasite of dogs and cannot complete its life cycle in humans to become an adult worm. Instead, larvae migrate around the body (visceral larva migrans) and may cause retinal damage.

Table 6.15   Tissue helminths

Name	Common name	Type of helminth	Approx size of adult worm	Disease	Lab investigation
Brugia malayi	—	Nematode	4—10 mm	Filariasis, tropical pulmonary eosinophilia	Midnight blood microscopy
Echinococcus granulosus	Dog tapeworm	Cestode	*	Hydatid disease	Serology
Onchocerca volvulus	—	Nematode	4 cm male, 50 cm female	River blindness	Microscopy of skin snips for microfilariae
Schistosoma spp.	Bilharzia	Trematode	1—2 cm	Schistosomiasis	Microscopy of terminal midday urine or stool for eggs, or serology 3 months after exposure.
Trichinella spiralis	—	Nematode	1—3 mm	Trichinosis (larvae migrate and encyst in skeletal muscle)	Serology or microscopy of muscle biopsy
Wuchereria bancrofti	—	Nematode	4—10 mm	Filariasis, tropical pulmonary eosinophilia	Midnight blood microscopy

*Echinococcus is a parasite of dogs and cannot complete its life cycle in man. It forms cysts in the liver, lungs and other organs, causing hydatid disease.

Table 6.16 Some medically important viruses

Group	Most important members	Disease caused	Genome type	Morphology	Envelope
Adenoviridae	Adenoviruses	Respiratory tract infection, conjunctivitis	dsDNA	Icosahedral	No
Herpesviridae	Herpes simplex 1 and 2	Cold sores, genital herpes, encephalitis	dsDNA	Icosahedral	Yes
	Cytomegalovirus	Glandular fever-like illness. Congenital CMV. Severe infection in immunocompromised			
	Varicella zoster	Chicken pox, shingles			
	Epstein—Barr virus	Glandular fever. Tumours (see Table 6.17)			
	HHV-6 and 7	Exanthem subitum (childhood illness)			
	HHV-8	Kaposi's sarcoma			
Poxviridae	Molluscum contagiosum	Molluscum contagiosum skin infection	dsDNA	Brick-shaped	Yes
	Orf	Orf (zoonotic skin infection)			
Parvoviridae	Parvovirus B19	Erythema infectiosum, arthralgia, aplastic crises, fetal anaemia, hydrops fetalis.	ssDNA	Icosahedral	No
Papovaviridae	Human papillomaviruses	Warts, genital warts, cervical cancer	dsDNA	Icosahedral	No
	Polyomaviruses (JC, BK)	JCV — progressive multifocal leucencephalopathy in AIDS.			
		BK — cystitis and nephritis in immunosuppressed			
Hepadnaviridae	Hepatitis B virus	Hepatitis, cirrhosis, hepatocellular carcinoma	Partial ds DNA	Spherical	Yes
Orthomyxoviridae	Influenza A,B, C	Influenza	ss(−) RNA	Spherical	Yes
Paramyxoviridae	Parainfluenza	Respiratory tract infection including bronchiolitis	ss(−) RNA	Spherical	Yes
	Mumps	Mumps			
	Measles	Measles			
	Respiratory syncytial virus	Respiratory tract infection including bronchiolitis			
Coronaviridae	Coronavirus	Common cold	ss(+) RNA	Spherical	Yes
		SARS			
Picornaviridae	Poliovirus	Polio	ss(+) RNA	Icosahedral	No
	Enteroviruses	Meningitis, encephalitis, pharyngitis, hand foot and mouth disease, myocarditis, pericarditis, conjunctivitis			
	Hepatitis A	Hepatitis			
	Rhinoviruses	Common cold			
Reoviridae	Rotavirus	Gastroenteritis	dsRNA	Icosahedral	No
	Reoviruses	Tick fever			
Retroviridae	HIV 1 and 2	HIV and AIDS	Diploid ss(+) RNA	Spherical	Yes

Table 6.16 *(Continued)*

Group	Most important members	Disease caused	Genome type	Morphology	Envelope
	HTLV-1 and 2	Leukaemia, lymphoma, tropical spastic paraparesis			
Togaviridae	Rubella	Rubella, congenital rubella	ss(+) RNA	Spherical	Yes
	Arboviruses	Encephalitis			
Flaviviridae	Yellow fever	Yellow fever	ss(+) RNA	Spherical	Yes
	Dengue	Dengue			
	Hepatitis C	Hepatitis			
	Arboviruses	Encephalitis			
Bunyaviridae	Arboviruses including hantavirus	Fever, encephalitis, renal failure	ss ambisense RNA	Spherical	Yes
Arenaviridae	Lassa	Viral haemorrhagic fever	ss ambisense RNA	Spherical	Yes
Filoviridae	Marburg, Ebola	Viral haemorrhagic fever	ss(−) RNA	Filamentous	Yes
Rhabdoviridae	Rabies	Rabies	ss(−) RNA	Bullet-shaped	Yes
Astroviridae	Astrovirus	Gastroenteritis	ss(+) RNA	Spherical	No
Caliciviridae	Caliciviruses including norovirus	Gastroenteritis	ss(+) RNA	Spherical	No
Hepevirus	Hepatitis E	Hepatitis	ss(+) RNA	Icosahedral	No

Key: AIDS — acquired immune deficiency syndrome; Arbovirus = arthropod-borne virus (numerous types exist); HHV = human herpesvirus; HIV = human immunodeficiency virus; HTLV = human T-cell lymphotropic virus.

membrane called the envelope, which is derived from host cell membranes but may be customized by the addition of viral-encoded proteins (Figure 6.31). Although viruses contain genetic material, they are unable to replicate by themselves — instead they have to 'hijack' a host cell and use its cellular machinery.

The type of nucleic acid in viruses varies enormously: it may be DNA or RNA, single or double stranded,

Viral structure (Schematic)

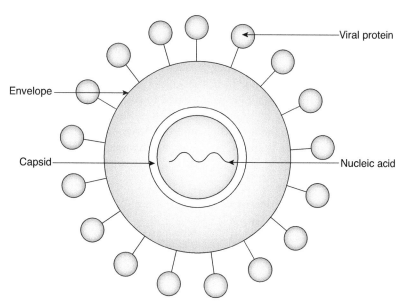

Figure 6.31   Viral structure

Positive (+) sense nucleic acid has the same polarity or sense as mRNA, so that proteins can be translated directly from it. Negative (−) sense nucleic acid has to be copied into the opposite (positive) sense before translation can start.

circular or linear, in one piece or segmented. Single-stranded nucleic acid may be described as positive sense or negative sense (Box 6.14). Viruses are classified according to their type of nucleic acid, capsid morphology, mode of replication and presence or absence of an envelope (Box 6.15).

6.6.3   Virus replication

Viruses can only replicate within living cells. They interact with host cells in a variety of ways, with different replication cycles and outcomes. Most viruses of medical importance undergo a **cytolytic cycle**, whereby the virus particle attaches to the cell, enters into it, is uncoated to reveal the viral genome, and uses the viral genome and host cell machinery to produce enzymes, viral nucleic acid and structural proteins. New virus particles are assembled and released, either through cell lysis (used by nonenveloped viruses) or through budding (used by enveloped viruses, which use the host cell membrane to make their own envelope). Completion of the cycle usually causes host cell death.

Other significant outcomes include:

Persistent infection: The virus replicates within the cell as per the cytolytic cycle but does not kill the host cell. The host cell coexists with the virus, and continues to produce some viral particles.

Latency: the viral genome persists within the host cell, but with little or no transcription of viral genes. One

RNA viruses are prone to transcription errors and therefore to frequent mutations. This is usually seen as a weakness, but their frequent changes make it difficult for us to produce effective vaccines or to sustain immunity after an infection. Examples of mutation-prone RNA-containing viruses are influenza and HIV.

example is herpes simplex virus, which can remain latent within neuronal cells for many years after an initial infection. The virus may reactivate after a period of time, particularly if host defences wane.

Transformation: the viral genome persists within the host cell, with expression of selected viral proteins which change the properties of the cell. Typically, the host cell loses contact inhibition of growth and can grow and divide indefinitely, making it behave like a tumour cell. Although transformation is a relatively uncommon event, approximately 18% of all new cancer cases are linked with viral infection.

6.6.3.1   Associations of viruses with cancer (Table 6.17)

6.6.4   Laboratory investigation of viral infection

Most microbiology laboratories will perform some routine virological investigations on site. Very few laboratories, with the exception of the largest virology centres, can offer the whole repertoire of viral investigations, so some specialized and nonroutine tests may be sent away to a reference laboratory. It is generally not financially economic to perform such tests in-house, either because the equipment required is too expensive or because the test is requested infrequently and is therefore uneconomic to provide on a day-to-day basis. Examples include nucleic acid amplification tests (NAATs) and electron microscopy.

Many viral investigations are performed in order to establish the cause of a patient's symptoms. Once diagnosis has been made, antiviral treatment may be offered to the patient if appropriate (Section 6.10.4). Contacts of the patient who are considered at risk of infection may sometimes be offered antiviral prophylaxis, for example with influenza, or immediate vaccination, for example in cases of hepatitis B (Section 6.4).

6.6.4.1   Serology

Most tests are performed on clotted blood samples, which are spun to separate the serum from the red blood cells. Methods include latex agglutination, enzyme immunoassay, complement fixation tests and single radial haemolysis. The most widely used method is enzyme immunoassay. Most laboratories are moving

Table 6.17   Associations of viruses with cancer

Virus	Cancer	% virus-positive	No. of cases worldwide/year
Human papilloma viruses (HPV)	Cervical cancer	100%	490 000
Hepatitis B virus	Liver cancer	50%	340 000
Hepatitis C virus	Liver cancer	25%	195 000
Epstein Barr virus (EBV)	Nasopharyngeal cancer; Burkitt's lymphoma; post-transplant lymphoma; Hodgkin's lymphoma	100%; > 90%; > 80%; ➤ 50%	113 000
Human herpesvirus 8	Kaposi's sarcoma	100%	66 000
Human T-cell lymphotropic virus (HTLV-1)	Adult T-cell leukaemia/lymphoma	100%	3000

away from performing these manually in 96-well plates, and towards automated systems such as the AxSym, Architect (Figure 6.32) or VIDAS.

Most virology departments will test for Hepatitis A, B and C, HIV, varicella zoster virus (VZV) and rubella. Other serology tests performed on-site may include cytomegalovirus, Epstein—Barr virus, measles and mumps.

It is fundamentally important that the requesting clinician understands that an antigen is part of the organism, and that an antibody is produced by the body. Antigen tests look for infection, IgG antibody tests look for immunity through vaccination or previous disease and IgM antibody tests look for a current infection.

6.6.4.2   Nucleic acid amplification tests (NAATs)

These may be used to detect virus in blood or other body fluids. Most viral PCR work is performed from a sample of blood in an EDTA tube, though a few PCRs give better results with serum than EDTA. For example, hepatitis C PCR on blood will distinguish patients with active infection, who have positive PCR, from those with past infection, who are PCR negative. Serology would have been positive in both scenarios. Quantitative PCR may also be done, to assess how much virus is in the bloodstream, and this result can be used to guide treatment. Another example is herpesvirus PCR, carried out by reference laboratories as a multiplex PCR which tests for herpes simplex virus (HSV) 1 and 2, CMV, VZV and EBV. This is a useful test for CSF in suspected viral meningitis or encephalitis.

6.6.4.3   Nasopharygeal aspirates

These are regularly sent to the virology laboratory in order to screen children for respiratory viruses. All laboratories screen for respiratory syncytial virus

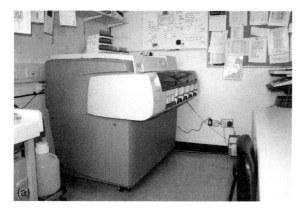

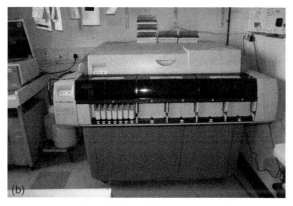

Figure 6.32   (a) and (b) Architect 200SR analyser

(RSV), and larger laboratories may also screen for influenza A and B, parainfluenzae 1, 2 and 3 and adenovirus. The results of these tests are used to reduce crossinfection within hospitals by **cohorting** (separating infectious children from noninfectious ones).

RSV is one of the most common childhood respiratory viruses, and can be detected using membrane bound immunoassay. These tests come in card or stick form and resemble pregnancy testing kits. Direct immunofluorescence can be used to look at a panel of different viruses at the same time, using fluorescently tagged monoclonal antibodies. It requires greater staff skill, more expensive equipment (including a UV microscope) and takes much longer to perform.

6.6.4.4 Viral swabs

These (together with urine and faeces samples) are investigated using tissue culture techniques. Swabs are generally used to diagnose herpes simplex virus (HSV) and varicella zoster virus (VZV). The virus grows in human embryonic lung cells. After 5—10 days a characteristic cytopathic effect can be seen. Further serotyping of positive tissue culture samples then takes place, using immunofluorescence with type-specific monoclonal antibodies.

6.6.4.5 Faeces samples

These are most often used to investigate rotavirus, using immunoassay methods. This is a common cause of gastroenteritis in children. Faeces can also be screened for enteroviruses by tissue culture and electron microscopy; this is usually done for suspected enteroviral meningitis (surprisingly, enteroviruses hardly ever cause gastrointestinal symptoms). **Urine samples** are used to screen for cytomegalovirus (CMV).

6.6.4.6 Electron microscopy

This allows visualization of a virus from clinical samples. Its great advantage is that it can be used to investigate samples even when you are not sure which virus you are looking for. For example, it was used to detect the previously unknown pathogen SARS coronavirus during the worldwide SARS (severe acute respiratory syndrome) outbreak in 2003. It is labour-intensive and requires expensive specialist equipment, so is no longer widely used.

6.6.5 Common requests in the virology laboratory

6.6.5.1 Norovirus screen

Norovirus is an important cause of diarrhoea in children and in adults. It causes an unpleasant gastroenteritis with abdominal cramps, vomiting and diarrhoea, sometimes known as "winter vomiting disease". Outbreaks can occur in hospital wards (and are also common in care homes and on cruise ships!). Investigation of a suspected norovirus outbreak will require testing of stools from several affected patients, usually coordinated by the Infection Prevention and Control Team. Norovirus PCR is available at larger virology laboratories and at reference laboratories.

6.6.5.2 Influenza testing

Influenza is usually a clinical diagnosis made in the community. Investigation may be required if there is doubt about the diagnosis or if the patient is admitted to hospital. A viral swab of the nose and throat, or a nasopharyngeal aspirate, can be sent to the reference laboratory for influenza PCR. The PCR can distinguish between different types of influenza such as influenza A, influenza B and swine flu H1N1. Influenza serology is available but does not give timely results.

6.6.5.3 The antenatal screen

All pregnant women are tested for hepatitis B, rubella and HIV, as well as syphilis, unless they opt out. This is known as the "booking bloods" and is done at their antenatal booking visit at 12—14 weeks gestation, to allow time for counselling and interventions if any of the tests are positive.

6.6.5.4 The hepatitis B screen

There are numerous different tests for hepatitis B due to the complex structure of the virus. Clinicians often find it challenging to select the correct test for hepatitis B. However, if the correct clinical details are available then the laboratory can ensure that the appropriate test is performed.

Hepatitis B surface antigen — is positive if the patient is a carrier or has had a very recent infection. This is the first line test to do 'to check a patient's hepatitis B status'.

Hepatitis B surface antibody — is positive at high titre if the patient has been successfully vaccinated. This is the test to do 'to check response to vaccine'. Anti-Hepatitis B surface antibody can also be positive **at low level** following an infection with hepatitis B, when the immune system is beginning to clear the Hepatitis B virus.

Hepatitis B core antibody — will be positive if a patient has ever had hepatitis B. This is the test to do 'to check if the patient has ever had hepatitis B'.

Hepatitis B 'e' antigen and 'e' antibody are tests performed on known carriers to assess their degree of infectivity (high infectivity carriers are 'e' antigen positive, low infectivity carriers are 'e' antibody positive). This test will eventually be superceded by quantitative PCR.

6.6.5.5 The renal screen

A laboratory covering a renal unit will receive numerous requests for hepatitis B and C screens because dialysis patients must be tested before being put on the dialysis machine to prevent cross-infection. Renal patients are retested at regular intervals, and are also screened before and after going on holiday overseas, firstly to show their current status, and secondly to check they have not become infected with anything whilst being dialysed on machinery in another country.

6.6.5.6 Checking response to vaccination

Certain occupations such as healthcare work require staff members to be vaccinated against infections such as hepatitis B, rubella, measles and mumps (Section 6.11.4). Holiday makers are advised to have pretravel vaccinations depending on the area of the world that they are travelling to. It is the job of the virology department to check that vaccine recipients have developed sufficient protective antibody levels.

6.7 Prions

Prions are proteinaceous infective agents (Box 6.16). They are composed entirely of protein, with no genetic material at all, and it is debatable whether they are actually living organisms. Prion proteins have the same amino acid sequence as a cellular protein in the host nervous system, but have a different three-dimensional conformation. Typically, the host protein is rich in

---

Box 6.16   Prions

Prions are the only known infectious agents which are able to replicate but which do not contain nucleic acid.

---

alpha helices while the prion protein is rich in beta-pleated sheets. When ingested, prions accumulate first in the peripheral lymphoid system and then reach the central nervous system, where they replicate by stimulating the host protein to adopt the abnormal conformation of the prion protein. Accumulation of these abnormal proteins leads to loss of function and spongiform degeneration of the brain.

Transmissible spongiform encephalopathies are prion diseases which cause neurological degeneration and eventually death. TSEs are seen in animals as well as humans, as shown by the epidemic of BSE in the UK in the 1980s (Box 6.17). Human TSEs are thankfully uncommon. The most important human TSE is Creutzfeld—Jakob disease (CJD), which has a variety of causes shown below (Table 6.18).

Iatrogenic CJD is very rare. There is a theoretical possibility of transmission via surgery, particularly neurosurgery, eye surgery or procedures involving lymphoid tissue. Prions are resistant to most decontamination methods including boiling, irradiation and most disinfectants. So autoclaving cannot be relied on to remove prion proteins from surgical instruments after use on a patient with CJD. As a precaution to

---

Box 6.17   Mad cow disease: an epidemic that should never have happened

Bovine spongiform encephalopathy (BSE), also known as mad cow disease, was first identified in 1985 in the UK and spread rapidly through the cattle population. At its peak the epidemic necessitated the slaughter of 200 000 diseased cattle and 4.5 million asymptomatic, but possibly exposed, cattle. It appears that the epidemic resulted from contaminated cattle feed containing sheep tissue and the sheep TSE agent scrapie. The epidemic was further fuelled by recycling contaminated cattle carcasses into cattle feed. These unsavoury practices have since been banned.

Table 6.18   Types and causes of CJD

Type of CJD	Cause
New variant (vCJD)	Infection by consuming bovine prions (BSE)
Iatrogenic (iCJD)	Infection by tissue from another infected person for example human growth hormone, dura mater graft, corneal transplant
Familial (fCJD)	Inherited mutation in prion protein gene
Sporadic (sCJD) (85% of cases)	Cause unknown

reduce risk, standards of decontamination for surgical instruments have been improved, with the options of quarantining or destroying instruments used on high-risk patients, or selective use of disposable instruments. There have also been cases in the UK of vCJD transmission via blood transfusion. This risk has now been minimized by: removing white cells from all blood used for transfusions (leucodepletion), sourcing plasma products from overseas, and banning high risk people from donating blood.

6.7.1   Practical precautions in the laboratory

If CJD is suspected, patient samples must be processed in the Class 1 safety cabinet with extra precautions that may include: lining the working area with an impermeable layer such as BenchKote, using disposable loops, counting chambers and forceps. At the end of working, all instruments are wrapped in the Bench-Kote and bagged for disposal.

6.8   Infections in the immunocompromised patient

6.8.1   Common causes of severe immunodeficiency

The immunocompromised patient is, by definition, a person whose host defences are impaired (Box 6.18). This may be due to a variety of reasons such as prematurity, cytotoxic chemotherapy, immunosuppressive drug therapy, transplantation or untreated HIV infection. There are in addition a number of inherited diseases

> **Box 6.18   Common causes of severe immunodeficiency**
>
> Neutropenia post chemotherapy.
> Post solid organ transplant.
> Post bone marrow transplant.
> HIV infection.

> **Box 6.19   Sources of infection**
>
> **Endogenous infections** arise from the body's own commensal organisms, which can proliferate to cause infection once the host is immunosuppressed.
>
> **Exogenous infections** arise from an external source, such as microorganisms transferred via the hands of healthcare workers, or *Aspergillus* from the air (a fungus found widely in air, which can cause chest infections in the immunosuppressed).

such as chronic granulomatous disorder (CGD) which affect the response to infection. The most deeply immunosuppressed patients are those who are about to have, or have just had, a bone marrow transplant.

Immunosuppressed people are prone to infection by exogenous or endogenous flora (see Box 6.19). Often these organisms are opportunistic pathogens, which are not usually very pathogenic but will cause an infection when host defences are impaired. An example is coagulase-negative staphylococci, which are normal flora on the skin but may cause line-associated infections. Remember, when reading cultures on immunosuppressed patients, that commensal organisms may cause clinical infections and may require further investigation.

6.8.2   Commonly encountered problems

6.8.2.1   Neutropenia

Neutropenia, or low neutrophil count, is most commonly seen after chemotherapy but also occurs in patients with haematological malignancy, HIV infection or bone marrow infiltration by malignant disease. Risk of infection is inversely related to the neutrophil count: there is some increased risk with a count of 1.0×10^9 cells/L and infections become very common with a count of 0.1×10^9/L or lower. Neutropenic patients who develop a fever must have cultures taken from all available sites (blood via peripheral vein and central line, urine and any other likely focus) and are immediately started on broad spectrum antibiotics.

6.8.2.2 Splenectomy

This may be performed for trauma or haematological disease, or the spleen may infarct due to sickle-cell disease. Either way, the lack of a spleen leaves patients at risk of overwhelming sepsis due to capsulated bacteria such as *Streptococcus pneumoniae, Haemophilus influenzae* and *Neisseria meningitidis,* and also at increased risk of severe malaria. Patients are warned that they are at increased risk of infection, particularly in the first 2 years after splenectomy, and to seek medical advice promptly if they become unwell. Other precautions include wearing a medical alert bracelet, lifelong antibiotic prophylaxis with penicillin V and immunization against influenza, *Streptococcus pneumoniae, Haemophilus influenzae* type b and *Neisseria meningitidis* group C.

6.8.2.3 Minimizing infections

Infections in the immunosuppressed cannot be wholly avoided but can be minimized by the following actions:

- minimize underlying problem, for example treat HIV with antiviral drugs;
- protective isolation in a side room while in hospital;
- prophylactic antibiotics are sometimes indicated, for example cotrimoxazole to prevent *Pneumocystis* (PCP) pneumonia in advanced HIV, penicillin V postsplenectomy;
- vaccination is sometimes useful, for example splenectomy patients.

6.9 Healthcare associated infections

Healthcare associated infections (HCAI) are infections acquired as a consequence of a person's treatment by a healthcare provider, or by a healthcare worker in the course of their duties. It replaces the older term hospital acquired infection (HAI) because in-

Box 6.20 Facts about HCAI in

There are over 300 000 HCAI in hospitals each year.
The cost to the NHS is at least £1 billion each year.
MRSA bacteraemia and *C. difficile* account for about 15% of HCAI.

fections may be acquired in the hospital or in the community.

HCAI is an enormous problem in modern healthcare (Box 6.20). An estimated 9% of patients develop HCAI during a hospital stay. This is obviously bad for the patient, causing anxiety, illness or even death. Furthermore, HCAIs are expensive. Recent Department of Health figures show that HCAI costs the NHS £1 billion per year [3]. This money goes towards increased length of hospital stay, extra treatment, and the cost of dealing with complaints and legal action. Many HCAIs are avoidable, as we will see below, and HCAI reduction has become a top priority for healthcare providers.

HCAIs may be usefully subdivided into:

1. community acquired infections — the infection presents in the community or within 48 h of admission to hospital (implying that it was already developing prior to admission);
2. infections contracted and developing within hospital;
3. infection contracted in hospital but not becoming clinically apparent until after discharge (e.g. postoperative wound infection).

6.9.1 Risk factors for infection

Age — elderly or very young/premature babies are most at risk. The ageing population means that there are many very elderly patients in hospital.
Surgery — wounds are prone to infection, as are prostheses such as replacement hips. Risk factors for surgical site infection include duration of surgery, surgical technique and length of stay.
Underlying illness — particularly immunosuppression, but also immobility or ischaemia.
Antibiotic use — this reduces commensal organisms, leaving a niche for resistant bacteria to fill.
Medical devices — such as intravenous cannulae, urinary catheters, endotracheal tubes for ventilation,

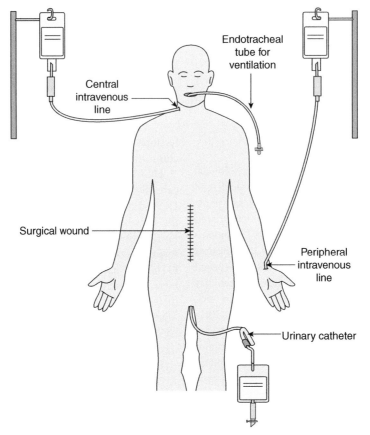

Figure 6.33   Patient with medical devices

and prosthetic devices such as artificial joints or heart valves are widely used and readily become colonized or infected (Figure 6.33).

6.9.2   Commonly occurring healthcare-associated infections, and the microorganisms that cause them (Table 6.19)

6.9.3   Antibiotic-resistant bacteria

HCAIs are often caused by multiresistant bacteria (Section 6.10.3). This is particularly likely in specialist units with heavy antibiotic use, such as intensive care or burns units, because the overuse of antibiotics selects for a population of resistant organisms. This means that broader spectrum antibiotics will be required for treatment, that a strict antibiotic policy must be implemented to reduce overprescribing, and that infection control procedures must be put in place to prevent transmission to other patients.

Examples of multiresistant bacteria implicated in HCAI include:

MRSA — commonly found on general medical and elderly care wards, and widespread in nursing homes.

VRE (vancomycin resistant enterococci) — found in intensive care units, particularly if glycopeptide antibiotics such as vancomycin are frequently prescribed.

Multiresistant coliforms (Enterobacteriaceae) — including ESBL and AmpC producers (Section 6.10.3).

Clostridium difficile

Multiresistant Acinetobacter (MRAB) — outbreaks are seen in ITUs with heavy antibiotic use. *Acinetobacter* persists for a long time in the environment and so can be difficult to eradicate from a ward.

Table 6.19   Common causes of healthcare associated infections

Urinary tract infection	*E.coli, Proteus, Klebsiella, Serratia, Enterococcus, Pseudomonas aeruginosa, Candida*
Pneumonia	*Staphylococcus aureus* (including MRSA), Enterobacteriaceae, *Candida, Aspergillus,* respiratory viruses (RSV, influenza, parainfluenza)
IV line infection	*S. aureus* (including MRSA), Group A streptococcus, *Enterococcus,* coagulase-negative staphylococci
Wound infection	*S. aureus* (including MRSA), Group A streptococcus, *Enterococcus,* anaerobes
Gastrointestinal infection	*Clostridium difficile,* norovirus

Pseudomonas survives well in moist environments (including inside bottles of aqueous antiseptic!) and can become multiply resistant.

Stenotrophomonas is notorious for being intrinsically resistant to carbapenems, and is found on ITUs with heavy carbapenem usage.

6.9.4 *The Infection Prevention and Control Team (IPCT)*

The Infection Prevention and Control Team (IPCT) is a valuable resource. Every hospital and most Primary Care Organizations will have an IPCT. The typical structure within a hospital is:

Director of Infection Prevention and Control (DIPC) — often the medical or nursing director, sometimes a consultant microbiologist;

Infection Control Doctor — a consultant microbiologist; and

Infection Control Nurse(s) — usually the only full-time team member.

The IPCT write the Infection Control Policy for the hospital, provide training for staff, and advise on day-to-day matters. They oversee HCAI surveillance and reporting, audit hand hygiene and infection control procedures, and are involved in planning hospital-wide issues such as influenza pandemic preparedness. The Infection Control Doctor or Nurse may liaise closely with the laboratory, particularly if there is an outbreak within the hospital. The DIPC leads the team and reports to the Trust Board. Management of outbreaks within the hospital is the responsibility of the IPCT. They will assess the situation, isolate patients as required and advise on appropriate samples to be sent (Box 6.21).

HCAI surveillance is a complex and difficult area. The main system within the NHS is compulsory

Figure 6.34   Handgel pump dispenser

reporting of MRSA and MSSA bacteremias and of *C. difficile* cases in the over-2s. Cases are reported by the hospital chief executive to the regional NHS authority. The figures are collated and released publicly by the Health Protection Agency every month [4]. The IPCT may carry out surveillance of other infections as part of a national scheme called Saving Lives.

Prevention of HCAI requires involvement of all staff from 'board to ward', that is. from the chief executive down to the ward and domestic staff. Basics such as thorough, regular cleaning must be put in place. Hand hygiene must be a priority for clinical staff and should be audited regularly (Figure 6.34).

The elderly care ward reports that three patients have new onset diarrhoea. The infection control nurse visits the ward and notes that all patients have loose, offensive stools. All have recently had antibiotics. This is most probably *Clostridium difficile* diarrhoea triggered by antibiotic use. *Actions*: the patients are cohorted together in a bay, which is closed to new admissions, to prevent onward transmission. The bay is cleaned twice daily with chlorine-disinfectant mix to reduce the amount of clostridial spores in the environment. Staff wear aprons and gloves. The patients' antibiotics are stopped and they are commenced on oral metronidazole. Stool samples test positive for *C. difficile* toxin. The patients remain in isolation until they have been diarrhoea-free for at least 48 h. The ward doctors are reminded not to overprescribe antibiotics.

Antibiotic overuse breeds resistance, so antibiotic stewardship by the microbiologist should restrict their use. Foreign bodies such as IV lines and catheters are a potential focus of infection, so medical devices must be inserted and looked after correctly, and should only be used if essential. Where resistant organisms are found, infection control procedures (such as single room isolation) must be followed on the advice of the IPCT.

Good infection control practices can make a noticeable difference to infection rates. The number of death certificates in England and Wales mentioning MRSA decreased from 1230 in 2008 to 781 in 2009, a drop of 37%. MRSA was recorded as the underlying cause of death in 19% of those 781 cases. Deaths involving *C. difficile* infection fell by 34% from 5931 in 2008 to 3933 in 2009 [5].

6.9.5 MRSA screening

The NHS in England requires universal MRSA screening for hospital patients. All patients admitted to hospital will be swabbed according to local policy from one or more of the following sites: nose, throat, axilla, groin, any wounds, catheter urine. MRSA positive patients are offered decolonization with topical antibacterial washes and nasal antibiotic ointment for 5–7 days, and should ideally be isolated from other patients.

6.10 Antimicrobial agents

An **antimicrobial agent** is any substance which inhibits the growth of microorganisms. An **antibiotic** is more specifically defined as a substance produced by microorganisms which is antagonistic to the life or growth of other microorganisms (Box 6.22). This definition gives away the origin of antibiotics: namely, that they occur naturally, produced by one microorganism as a weapon against others.

In the late 1930s, a team of scientists at Oxford University, led by Howard Florey and Ernst Chain, were seeking to develop antimicrobial drugs. They picked up on Alexander Fleming's observation and developed methods of growing mould and extracting penicillin on a larger scale. At this point, they were growing mould in hospital bedpans and cantaloupe melons! By 1943, penicillin went into mass production. It transformed medical practice and has saved millions of lives.

Many more antibiotics have been developed since that time. Some are derived from the original penicillin, some from other naturally occurring compounds, and some are completely synthetic. The need for new antibiotics was driven partly by the desire for better pharmacokinetic properties (such as oral bioavailability and less frequent dosing) and also by the emergence of antibiotic resistance. In everyday practice, antimicrobials which act against bacteria are referred to as antibiotics. Antiviral and antifungal agents also exist and will be covered in Sections 6.10.4 and 6.10.5.

Alexander Fleming discovered antibiotics in 1929, while working at St Mary's Hospital in London. He found fungus growing on an old Petri dish and noted that bacterial growth around the fungus was inhibited. He concluded that the mould was producing an antimicrobial substance, which he named penicillin after the *Penicillium* mould which produced it. However, he did not extract the active ingredient or develop it into a usable form.

6.10.1 Antibiotic classification

There are several different classes of antibiotics, allocated according to their chemical structure and mode of action. Numerous classes exist but the most commonly encountered are as follows.

Penicillins: derived from the original penicillin antibiotic. They all contain a beta-lactam ring in their structure (Figure 6.35). Penicillin binding proteins link the side chains in peptidoglycan, which is an integral part of the bacterial cell wall, especially in Gram-positive bacteria. Penicillins bind to these proteins, interfering with cell wall synthesis. Some penicillins have been modified to make them more stable or to give them a broader spectrum. **Examples**: benzyl penicillin, flucloxacillin, amoxicillin.

Cephalosporins: like penicillins, these contain a beta-lactam ring and inhibit cell wall synthesis. Some cephalosporins have a longer half-life meaning that they can be given just once a day. **Examples**: cefuroxime, ceftriaxone.

Carbapenems: these also contain a beta-lactam ring and inhibit cell wall synthesis. These are powerful, broad spectrum, intravenous antibiotics which are used mainly in intensive care units. **Examples**: meropenem, imipenem, ertapenem.

Glycopeptides: these large molecules also inhibit cell wall synthesis, but by a different mechanism: they block addition of new 'building blocks' to the peptidoglycan molecule. **Examples**: vancomycin or teicoplanin (Box 6.23).

Macrolides: these antibiotics inhibit bacterial protein synthesis. They stop bacteria from growing and reproducing but they do not kill them — hence they are bacteriostatic. This makes them unsuitable for use in life-threatening infections. They are used mainly to treat respiratory tract, gastrointestinal and genitourinary infections. **Examples**: erythromycin, clarithromycin.

Tetracyclines: bacteriostatic agents which inhibit protein synthesis. Not used in children under 12 years of age, or pregnant or breast-feeding women, due to

> **Box 6.23　Vancomycin**
>
> Vancomycin used to be extracted from soil, because it was naturally made there by soil-dwelling microorganisms. It was known as 'the brown powder' because it looked dirty!

adverse effects on developing tooth enamel. Used for respiratory, skin, soft tissue and genitourinary infections, and sometimes used as malaria prophylaxis. **Examples**: tetracycline, doxycycline.

Anti-folate antibiotics: these inhibit the synthesis of tetrahydrofolate, which in turn affects production of pyrimidines and purines needed for DNA synthesis. **Examples**: trimethoprim, sulphonamides.

Aminoglycosides: inhibit protein synthesis. Potent, broad spectrum agents which are given IV or IM because their oral absorption is so poor. They are active against most staphylococci and Gram-negative bacilli, but not very active against streptococci and anaerobes. They are potentially toxic to the ear and kidney, and levels should be monitored if treatment is for 48 h or more. **Examples**: gentamicin, amikacin, streptomycin.

Quinolones: these inhibit bacterial DNA gyrase and therefore kill bacterial cells by blocking DNA supercoiling. Used mainly to treat genitourinary infections, and sometimes infections at other sites caused by *Pseudomonas*.

Lipopeptides: Daptomycin is the only licensed member of this class. It has a similar spectrum of activity to vancomycin. It is used for complicated skin and soft tissue infections or endocarditis caused by resistant Gram-positive bacteria.

Oxazolidinones: Linezolid is the only licensed member of this class, used to treat pneumonia or skin/soft tissue infections caused by resistant Gram-positive bacteria.

Streptogramins: a combination of two streptogramins, quinupristin and dalfopristin, is licensed to treat serious Gram-positive infections such as pneumonia, skin and soft tissue infections which have failed to respond to other antibacterials.

Either bacteriostatic or bactericidal agents may be used (Box 6.24). In most circumstances, either type of agent is sufficient. A bactericidal agent is needed if the patient's host defences are suppressed — a common reason for this is neutropenia after cytotoxic

Figure 6.35　Beta lactam ring

Bactericidal drugs kill bacteria. Examples are the beta-lactam agents (penicillins, cephalosporins, carbapenems) and quinolones.

Bacteriostatic drugs inhibit bacterial growth, but do not kill bacteria. They rely on host defences, especially phagocytic cells, to eliminate the infection. Examples are tetracyclines and macrolides.

chemotherapy. Another example where bactericidal drugs are required is infective endocarditis. Bacteria infect the inner lining of the heart (endocardium) and reside within clumps of platelets and fibrin known as vegetations, where they are protected from phagocytic activity. It is essential to select a bactericidal drug, or combination of drugs, that can penetrate the vegetation and kill the bacteria.

6.10.2   Principles of antibiotic use

Antibiotics are given to treat infections. It sounds as simple as that. However, there are numerous factors influencing the decisions of whether to treat and, if so, which antibiotic to choose, at what dose, and for how long (Box 6.25).

1. Clinical factors. The clinical presentation is the most important factor in decision making. A good history and examination should provide a provisional diagnosis. From this, we can tell the likely site of infection and the likely organisms responsible. Specimens for culture must be taken at this stage, before starting antibiotics. Remember that antibiotics alone are not always the answer — adjuvant therapy may be needed, for example an abscess will not respond to antibiotic treatment unless it is also surgically drained.
2. Are antibiotics necessary? Many minor illnesses are selflimiting.

Clinical diagnosis.
Spectrum of activity.
Pharmacokinetics.
The need to be right.
Side effects.
Cost.

3. Spectrum of activity — choose an antibiotic that will be active against the likely pathogens. Bear in mind whether the patient has had antibiotics recently (which may have induced resistance) and any knowledge of local resistance patterns. Remember that it is a good principle to choose the narrowest spectrum option — it reduces the risk of side effects such as *Clostridium difficile* diarrhoea.
4. Pharmacokinetics — the antibiotic must achieve adequate tissue levels at the affected site. Think about how it will be given — is it absorbed orally? If not, will it be given intravenously, intramuscularly, rectally, topically? Consider also its metabolism, tissue penetration and excretion. For example, an antibiotic used to treat meningitis must be able to cross the blood—brain barrier and pass into CSF.
5. The need to be right. Put simply, can you afford for your initial choice to be wrong? If the illness is immediately life-threatening, choose a very broad spectrum antibiotic. You can swop to a narrow spectrum choice later, when the culture results are available. However, for a mild illness such as an uncomplicated UTI, it is better practice to choose a narrow spectrum option such as trimethoprim. There will be no serious harm done if your initial choice was ineffective and needs modifying.
6. Side effects. All antibiotics have possible side effects such as allergic reactions, adverse drug reactions and interactions. Glycopeptides and aminoglycosides are potentially nephrotoxic and require drug level monitoring. Antibiotics affect the normal flora, allowing superinfection, for example by *Candida* (thrush) or *Clostridium difficile.*
7. Cost — the cost of antibiotics varies widely. Newer antibiotics are generally more expensive. There is also the cost of administration (think of the nursing time involved in giving an intravenous drug several times daily) and of monitoring levels.
8. Duration of treatment. In general, short courses of treatment are preferable because they are less likely to cause side effects. For example, pneumonia is now treated for only 5 days and uncomplicated UTI in women for only 3 days. If in doubt of when to stop, be guided by 'measures of response' such as temperature, heart rate, respiratory rate, white blood cell count, C-reactive protein and culture results. There are a few conditions requiring longer antibiotic courses: these are usually deep seated infections such as endocarditis.

6.10.3 Antibiotic resistance

Resistance to antibiotics is a natural evolutionary response of bacteria to antibiotic exposure. So the introduction of antibiotics was soon followed by the advent of antibiotic resistance. Drug companies develop new classes of antibiotics to combat this, but invariably bacteria develop resistance sooner or later. Until recently, man kept ahead. From 1945 until the late 1980s, new antibiotics were produced more quickly than bacteria could develop resistance. Now, however, progress is slowing. New classes of antibiotics are hardly ever found – not many new antibiotics are being introduced, and those that do come along are often improvements within classes, with no novel mechanism of action. There is a real worry that some infections are becoming untreatable with currently available antibiotics. Also development of new antibiotics is not a priority for drug companies, as they are less lucrative than other drugs, such as statins to reduce cholesterol levels, which people take lifelong.

Intrinsic resistance occurs when the microbe does not possess a target for the drug's action or when it is impermeable to the drug. For example, the lipopolysaccharide outer envelope of Gram-negative bacteria blocks entry of many antibiotics.

Acquired resistance occurs when bacteria previously sensitive to an antibiotic become resistant to it. For example, *Staphylococcus aureus* was almost always sensitive to penicillin until the 1940s, when penicillin became widely used. It soon developed resistance by producing a beta-lactamase enzyme which breaks down the beta-lactam ring in penicillin. Now, about 90% of *S. aureus* isolates in clinical practice are resistant to penicillin.

Acquired resistance occurs because a bacterial cell mutates, or it acquires a resistance gene from another bacterium. Bacteria can divide every 20–30 min, so a single resistant cell can produce a billion cells overnight and become the predominant strain. Then resistant strains can spread between patients in hospital and in the community. This is more likely to happen if antibiotic overuse exerts selection pressure.

Mechanisms of acquired antibiotic resistance:

- inactivation of the antibiotic, for example. beta-lactamase production;
- alteration or protection of the target site within the bacterial cell, for example methylation of the 23S ribosomal RNA subunit in staphylococci confers resistance to macrolides;

- prevention of antibiotic entry into the cell, for example *Pseudomonas* species can lose porins (water-filled pores in the cell membrane) so that beta-lactam antibiotics cannot enter;
- removal of antibiotic from the cell, for example tetracycline resistance arises when bacterial cells synthesize a membrane protein which acts as a rapid efflux pump;
- metabolic bypass, for example trimethoprim resistance. Trimethoprim acts by inhibiting the dihydrofolate reductase enzyme in bacteria. Resistant bacteria synthesize a trimethoprim-insensitive variant of the dihydrofolate reductase enzyme, thus bypassing the antibiotic's mechanism of action.

Antibiotic overuse is undoubtedly a major cause of acquired antibiotic resistance. Large amounts of antibiotics are used in hospitals, particularly in intensive care areas (Section 6.9.3). General practitioners vary greatly in their prescribing habits, and 80% of antibiotics used in the UK are prescribed by GPs. Another less obvious culprit is the farming industry (see Box 6.26). A restrictive antibiotic policy can be applied in hospitals and in the community to limit antibiotic use. Such policies are drawn up by microbiologists and pharmacists, in consultation with other clinicians, and may ban use of certain antibiotics whose use is linked to problems with antibiotic resistance.

Extended spectrum beta lactamase(ESBL) simply means a beta-lactamase enzyme (one which breaks down the beta lactam ring of an antibiotic) which is very broad spectrum and can inactivate many different beta-lactam antibiotics, including extended spectrum cephalosporins such as cefotaxime. There are several

Box 6.26 Antibiotic use in animals

More than half of all antibiotics produced worldwide are used in animals, either to treat infection or as growth promoters (to increase weight gain). These antibiotic-exposed animals may become colonised with resistant bacteria and will eventually enter the food chain. Use of the glycopeptide antibiotic avoparcin as a growth promoter in Denmark led to 80% of Danish broiler chickens becoming colonized with the superbug VRE (vancomycin-resistant enterococci). The European Union subsequently banned the use of antibiotics as animal growth promoters, though they are still used elsewhere.

different types of ESBL, with names such as TEM and SHV which describe the gene which produces the enzyme. These genes are frequently carried on plasmids. ESBL producers are typically multiresistant Enterobacteriaceae. They can cause hospital outbreaks but are increasingly common in the community, often causing UTIs in the elderly [6]. Treatment options include carbapenems for severe infections, or nitrofurantoin or mecillinam for UTIs. Co-amoxiclav is sometimes active against ESBLs.

AmpC is another type of beta lactamase commonly found in multiresistant Enterobacteriaceae. It is unaffected by beta-lactamase inhibitors such as clavulanic acid, so AmpC positive bacteria are resistant to useful antibiotics such as co-amoxiclav and piperacillin-tazobactam. Treatment is usually with a carbapenem.

Carbapenem resistance is an increasing problem in Enterobacteriaceae. Carbapenem antibiotics are usually considered as a last line of defence, active even against ESBL and Amp C producers, so the emergence of carbapenemases is perceived as a real threat.

Resistance may occur as a combination of ESBL or AmpC together with porin loss, or through acquisition of carbapenemase. Acquired carbapenemases are a more serious risk. The types of carbapenemase that have been seen in the UK to date are IMP, VIM and New Delhi metallo beta lactamase NDM (all of these are metallo enzymes, with zinc at the active site), and KPC and OXA-48 (non-metallo enzymes). Most carbapenemase-producing bacteria have been imported to the UK from other countries where antibiotic overuse has led to widespread resistance [7].

Laboratory identification of carbapenemases can be difficult as carbapenem susceptibility may be only slightly reduced in vitro. Isolates with borderline or low carbapenem sensitivity should be forwarded to the reference laboratory for confirmation [8].

6.10.4 Antiviral drugs

Antiviral drugs were developed relatively recently. The intracellular location of viruses and their use of host cell machinery makes it difficult to develop a drug that can reach the virus and destroy it without damaging the host cell.

6.10.4.1 Nucleoside analogues

Aciclovir is active against herpes simplex and varicella zoster viruses. It can be taken orally and crosses the blood–brain barrier. It only becomes active in virally infected cells. **Ganciclovir** is active against cytomegalovirus. **Lamivudine** is active against hepatitis B virus and HIV. **Ribavirin** is a guanosine analogue which is active against respiratory syncytial virus and hepatitis C, and which has also been used to treat Lassa fever and hantavirus infections.

An example is the treatment of viral meningitis and encephalitis. Most cases of viral meningitis are caused by enteroviruses, which cause a selflimiting illness and do not usually require antiviral treatment. Herpes simplex meningitis or encephalitis cause a more serious illness with significant neurological symptoms, which may be provisionally diagnosed on clinical grounds and then should be confirmed by herpes PCR of CSF. This should be treated immediately with intravenous aciclovir.

6.10.4.2 Antiretroviral compounds

HIV infection is treated with a combination of antivirals known as highly active anti-retroviral therapy (HAART). Using a combination reduces the risk of resistance occurring. The exact combination can vary but is usually two nucleoside reverse transcriptase inhibitors plus either a nonnucleoside reverse transcriptase inhibitor or a boosted protease inhibitor (see below). HAART causes a rise in CD4 count and a fall in viral load, which reduces incidence of opportunistic infections and improves survival. HIV treatment is a highly complicated subject and is always overseen by a specialist.

Nucleoside reverse transcriptase inhibitors (NRTIs) are nucleoside analogues which inhibit the action of reverse transcriptase (the enzymes which converts viral RNA into a DNA copy). **Example:** zidovudine (AZT). Nonnucleoside reverse transcriptase inhibitors (NNRTIs) inhibit reverse transcriptase by a different mechanisms. **Examples:** nevirapine, efavirenz. Protease inhibitors are highly effective at reducing viral load but are prone to side effects and drug interactions. **Examples:** nelfinavir, lopinavir.

6.10.4.3 Neuraminidase inhibitors

Oseltamivir and zanamivir reduce replication of influenza A and B viruses by inhibiting viral neuraminidase. They can be used to treat influenza, when they reduce the duration of symptoms by 1–1.5 days, or to prevent it (postexposure prophylaxis). Oseltamivir was widely used to treat and to prevent swine influenza in the 2009 influenza pandemic.

6.10.4.4 Interferons

Interferons are immunomodulatory compounds rather than antiviral drugs, but they are used in combination with antivirals to treat chronic hepatitis. Such treatment must only be commenced by a specialist. Chronic hepatitis B is treated with peginterferon alfa-2a with lamivudine, for up to 48 weeks. Chronic hepatitis C is treated with peginterferon alfa and ribavirin, for 24–48 weeks. The success of hepatitis treatment can be assessed by quantitative PCR, to demonstrate a reduction in viral load, at the end of the course of treatment.

6.10.4.5 Immunoglobulins

Immunoglobulins may be given as postexposure prophylaxis to prevent or attenuate a viral infection in a susceptible person. This is known as passive immunization (Section 6.11.1).

6.10.5 *Antifungal drugs*

Fungi are eukaryotic organisms and antibacterial agents are not effective against them. Antifungal agents have been developed but are often quite toxic because of the greater similarity between fungal and human cells. Superficial infections such as dermatophytosis (infection of the skin, nail or hair) or infection of mucous membranes (thrush) are often treated with topical applications in order to minimise toxicity; deep infections require systemic treatment.

The main groups of antifungal agents are:

Azoles such as fluconazole. These interfere with the demethylation of lanosterol during the synthesis of ergosterol, which is the main sterol in the fungal cell membrane. These are the most widely used antifungal agents.
Polyenes such as amphotericin B and nystatin. These bind to sterols in the fungal cell membrane and cause damage to it.
Echinocandins such as caspofungin. These are semi-synthetic lipopeptides that interfere with the formation of glucans in the fungal cell wall. Used only for severe systemic infections.

Pyrimidine analogues such as flucytosine, which is converted intracellularly into 5-fluorouracil, which is then incorporated into fungal RNA. Used only for severe yeast infections.
Allylamines such as terbinafine. These interfere with ergosterol synthesis. It accumulates in keratin and is used to treat dermatophyte infections.

6.11 Vaccines

Immunization is used to induce immunity and control the spread of infectious diseases. It may be active or passive.

6.11.1 *Passive immunization*

Passive immunization is achieved by administration of immunoglobulin (Box 6.27). This is a short-term measure which takes effect within days but wanes within weeks. It is used to protect people with immune disorders or for postexposure management. Human normal immunoglobulin (HNIG) is derived from the pooled plasma of donors and contains antibodies to infectious agents that are prevalent in the general population. Specific immunoglobulins prepared from

---

Box 6.27 Example of passive immunization for postexposure prophylaxis

A 12-week pregnant woman spends the afternoon with her friend's child. Later that day the child develops a rash typical of chickenpox. The woman cannot recall whether she has ever had chickenpox herself. Chickenpox in pregnancy can cause congenital problems. The GP takes a blood sample for urgent varicella antibody testing. A positive result would indicate immunity. A negative result means that she is susceptible to chickenpox. She has had a significant exposure to chickenpox, so if she is susceptible she must be given IM varicella zoster immunoglobulin (VZIG) as soon as possible, provided by the local microbiology department and given by the GP or practice nurse. This will reduce the risk of her developing chickenpox.

hyperimmune donors are available for passive immunization against hepatitis B, varicella zoster, tetanus and rabies. Transfer of antibodies across the placenta from mother to child also produces a passive immunity which lasts only for weeks or months.

6.11.2   Active immunization

Active immunization or **vaccination** presents a microorganism, or antigens from a microorganism, to the host in order to stimulate the host's immune response. The idea is to produce immunity without inducing the disease or its complications. Such immunity is usually long-lasting. Vaccines may be:

- **Live attenuated** — contain whole, live microorganisms whose pathogenicity has been reduced (attenuated). To be avoided in pregnancy or immunocompromised patients. **Examples:** measles, mumps, rubella (MMR), BCG (an attenuated strain of *M. bovis* used to immunise against TB).
- **Inactivated** — contain inactivated or killed whole organisms. **Example:** hepatitis A, inactivated poliomyelitis vaccine (IPV).
- **Antigenic components** — contain the antigens from the microorganism which are required to stimulate the immune response. **Examples:** pneumococcal vaccine contains polysaccharide from the capsule, influenza vaccine contains a surface protein called haemagglutinin.
- **Toxoid** — contain toxins which have been inactivated so that they do not cause symptoms. **Examples:** tetanus and diphtheria vaccines.

6.11.3   UK vaccination programme

Universal vaccine coverage in the UK currently includes those listed in Table 6.20.

Other vaccines may be required for particular risk groups, for example over-65s are offered an annual influenza vaccination and healthcare workers are offered hepatitis B vaccination. The vaccination programme is regularly updated. The best source of information is 'The Greenbook' [9].

High levels of vaccine coverage will reduce the levels of disease in the population. This means that even unvaccinated individuals will be at less risk of catching

Table 6.20   Vaccine coverage currently in the UK

Age	Vaccine
Two months old	Diphtheria, tetanus, pertussis, polio, Hib, pneumococcus
Three months old	Diphtheria, tetanus, pertussis, polio, Hib, MenC
Four months old	Diphtheria, tetanus, pertussis, polio, Hib, MenC, pneumococcus
Twelve to thirteen months old	Hib, MenC, measles, mumps, rubella, pneumococcal
Three to five years old	Diphtheria, tetanus, pertussis, polio, measles, mumps, rubella
Twelve to thirteen years old (girls only)	Human papilloma virus
Thirteen to eighteen years old	Tetanus, diphtheria, polio

Hib = *Haemophilus influenzae* b
MenC = Meningococcus C

the disease — a concept known as **herd immunity**. A successful vaccination programme may even be able to eradicate a disease completely, as happened with smallpox in 1980. The World Health Organization aims to eradicate poliomyelitis next.

6.11.4   Vaccination of laboratory staff

The best protection against catching an infection at work is through adherence to good laboratory practice. Vaccination is never a substitute for safe working practices. All laboratory staff should be up to date with their routine immunisations. In addition, BCG and hepatitis B vaccinations are usually recommended. This can vary from laboratory to laboratory depending on what types of samples are processed.

6.11.5   Vaccine failure

Vaccine failure can occur and may be primary (when an individual fails to respond to the vaccine) or secondary (when an individual responds initially but protection wanes over time). One common example is failure to respond to hepatitis B vaccine, as demonstrated by failure to generate an hepatitis B surface antibody response. When this happens, it is worth repeating the course of vaccination. If the patient still fails to respond, check their hepatitis B core antibodies — sometimes the failure is because they have already been infected with hepatitis B.

6.11.6 Vaccine safety

Vaccine safety is a contentious topic. All vaccines are extensively tested before a license for use is granted. Most vaccines have occasional mild side effects but it is recognized that more serious reactions can rarely occur. A scheme for reporting adverse reactions exists. Most vaccine safety scares, such as the discredited 'autism link to MMR', are never proven, but have serious consequences such as reduced vaccine uptake leading to outbreaks of disease.

6.12 Conclusion

This chapter is just a taster for the speciality of microbiology. You can take away from this an overview of common pathogens and an insight into the workings of the microbiology laboratory. Microbiology is a truly fascinating field, constantly changing as new pathogens evolve and new techniques are introduced. It is also the speciality where you least want to take your work home with you! A career in microbiology may be busy but will always provide interest and variety. Thanks to: Girish Patel, John Clark, Jon Hutchinson, Clare Elwell, Cilla Hashmi, Reena Devlia, Amin Bilal, Jill Leach, Lauren Marlow, Sneha Patel, Helmut Schuster and all the staff at Kingston and St Helier Hospital Microbiology Departments, and of course our families for their support.

Bibliography

1. Health Protection (Notification) Regulations 2010. Available on the HPA website www.hpa.org.uk
2. Health Technical Memorandum 07-01: Safe management of healthcare waste., Department of Health 2006. Available at www.dh.gov.uk
3. House of Commons Public Accounts Committee "Reducing Healthcare Associated Infection in Hospitals in England" 2009 report. Available online at http://www.publications.parliament.uk
4. Healthcare-Associated Infections and Antimicrobial Resistance:, 2009/10, Health Protection Agency, report, Published online September 2010, available as free download from www.hpa.org.uk
5. Office for National, Statistics. MRSA and *Clostridium difficile* mortality rates updated annually at http://www.statistics.gov.uk. 2009 data was released on 24 August 2010.
6. Livermore DM, Canton R, Gniadkowski M *et al.* CTX-M: changing the face of ESBLs in Europe. *J Antimicrob Chemother* 2006; 59(2): 165—74.
7. Kumarasamy KK, Toleman MA, Walsh T.R. *et al.* Emergence of a new antibiotic resistance mechanism in India, Pakistan and the UK: a molecular, biological and epidemiological study. *Lancet Infect Dis* 2010; 10: 597—602.
8. Advice on Carbapenemase Producers: Recognition, infection control and, treatment. Department of Health Advisory Committee on Antimicrobial Resistance and Healthcare Associated Infection/Health Protection Agency. January 2011. Available as free download at www.hpa.org.uk
9. Immunisation against Infectious Disease ("The Green, Book"). www.dh.gov.uk/en/Publichealth/Healthprotection/Immunisation/Greenbook
10. National Health Service/Health Protection Agency/National Public Health service for Wales. National Standard Method BSOP 29 "Investigation of cerebrospinal fluid". Issued by Standards Unit, Evaluations and Standards Laboratory, Centre for Infections. Available on Health Protection Agency website www.hpa.org.uk or at www.evaluations-standards.org.uk.

Further Reading

Stokes, E.J., Ridgway, G.L. and M.W.D. Wren, M.W.D. (1993) *Clinical Microbiology*, 7th edn, Edward Arnold.

Grayson, M.L. *et al.* (2010) *Kucers' The Use of Antibiotics*, 6th edn, Hodder Arnold.

Hawkey, P. and Lewis, D. (eds) (2004) *Medical Bacteriology: A Practical Approach*, 2nd edn, Oxford University Press.

Koneman, E.W. *et al.* (1997) *Color Atlas and Textbook of Diagnostic Microbiology*, 5th edn, Lippincroft, Williams and Wilkins.

Medical, Microbiology. Edited by David Greenwood, Richard Slack, John Peutherer, Mike Barer. Elsevier. 17th edition 2007.

Notes on Medical, Microbiology. Katherine N. Ward, A. Christine McCartney, Bishan Thakker. Churchill Livingstone, 2nd edition 2009.

Oxford Handbook of Infectious Diseases and, Microbiology. Estee Torok, Ed Moran and Fiona Cooke. Oxford University Press 2009.

Problem-orientated Clinical Microbiology and, Infection. Hilary Humphreys, William Irving and C A Hart. Oxford Medical Publications 2004.

Principles and Practice of Infectious, Diseases. Edited by Gerald L. Mandell, John E. Bennett and Raphael Dolin. Churchill Livingstone, 7th edition 2009.

Chapter 7
Clinical immunology

**Professor Ray K. Iles, B.Sc., M.Sc., Ph.D., CBiol, FSB, FRSC and
Professor Ivan M. Roitt, D.Sc., FRS, Hon. FRCP**

There are huge overlaps in clinical chemistry, haematology and histology with the discipline of clinical immunology but this subarea of biomedical science covers malfunctions of the immune system in autoimmune diseases, hypersensitivities, immune deficiency and transplant rejection. Various laboratory investigations are conducted to diagnose and stage the extent of these pathological conditions using histology, clinical chemistry and haematological measurements. However, the basis of clinical Immunology is an understanding of the organs, cellular differentiation, genetics, specific proteins and biological properties of the immune system. Thus a basic understanding of the fundamentals of Immunology is essential.

Part I: The fundamentals of immunology

7.1 Overview of the immune system

The key primary lymphoid organs of the immune system are thymus and bone marrow, and secondary lymphatic tissues including both the encapsulated organs, spleen, tonsils, lymph vessels, lymph nodes, adenoids, and the unencapsulated mucosal tissue (MALT) residing in the lung, the lamina propria of the gut, genitourinary tract, salivary, lacrimal glands and lactating mammary tissue (see Figure 7.1). When health conditions warrant, immune system organs including the thymus, spleen, portions of bone marrow, lymph nodes and secondary lymphatic tissues can be surgically excised for examination while patients are still alive. The primary lymphoid organs are where cells of the immune system originate and develop into immature forms. Cells destined to be lymphocytes or myeloid cells all derive from a common progenitor 'pluripotent' stem cell population residing in the bone marrow. The secondary lymphoid tissues are where these immature forms encounter infectious challenges react and differentiate as necessary as part of the immune response. If one secondary structural tissue of critical importance to the immune system was to be singled out it would be the lymphatic system.

7.1.1 Lymphatic system

The lymphatic system is a network of very small blunt ended vessels that drains interstitial fluid which has

Biomedical Sciences: Essential Laboratory Medicine, First Edition. Edited by Ray K. Iles and Suzanne M. Docherty.
© 2012 John Wiley & Sons, Ltd. Published 2012 by John Wiley & Sons, Ltd.

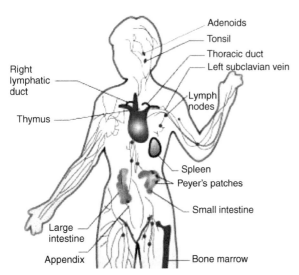

Figure 7.1 Diagrammatic representations of the primary and secondary immune tissues

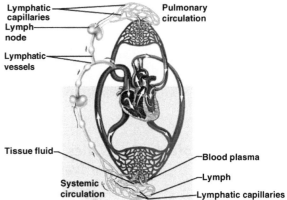

Figure 7.2 Diagrammatic representation of the parallel relationship of the lymphatic system to the pressurized blood system, illustrating how leaked blood plasma is collected and returned as lymph

been extruded from the blood capillaries (which are under pressure) as part of their nourishment of tissues or produced by the tissues themselves. Called lymph the fluid is clear straw-coloured and contains protein molecules, salts, glucose, urea and other potent biochemical substances. This drainage network is essential to keep the bodily fluid level in balance. However, it is not a completely acellular fluid as immune system white cells constantly migrate from the blood system through tissue into the lymphatic drainage, hence the term lymphocytes.

Although the lymph fluid is returned to the blood system via openings in major vessels such as the left subclavian vein (drained from most of the lower body) and right lymphatic duct (draining the right neck, chest and arms) to the right subclavian vein; it is constantly filtered by lymph nodes and other immunological tissue including the spleen (see Figure 7.2). It should be noted that the blood is also filtered by the spleen. Since foreign and invading organisms and substances will enter tissue, this drainage network is the primary site in which to first detect and then defend the body against infections that are or will happen.

Blockage of a lymph channel or draining lymph gland (e.g. due to infection) will lead to swelling due to the build up of localized lymph fluid in the interstitial tissue spaces. However, this swelling (interstitial tissue buildup of lymph fluids) is also a feature of immobility as the fluid in lymph vessels can only flow due to muscle contraction and movement. For example feet and legs swell when sitting too long on trains and planes because the lymph fluid is not being made to flow by muscle contractions as a result of walking.

Lymph nodes (see Figure 7.3) are bean-shaped and covered with a thick fibrous capsule subdivided into compartments by trabeculae consisting of functional cortex and medullary cells. The cortex is populated mainly with lymphocytes and germinal centres, the primary resting place for B Cells which

Structure of a lymph node

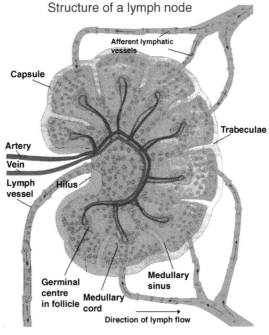

Figure 7.3 Typical lymph node

will divide in response to antigen stimulation and produce specific antibodies. The paracortex contains T- cells that have circulated through the lymph nodes, blood stream and lymphatic ducts monitoring the tissues for infection. The medulla of the lymph nodes is primarily made up of macrophages attached to reticular fibres. Subcapsular macrophages process afferent lymphatic material and present antigens to B cells in the cortex.

7.1.2   T-Cells

Immature cells destined to become T-lymphocytes migrate from the bone marrow and enter the thymus as thymocytes. These cells express the T-cell receptor. Akin to the antigen binding domain of antibodies the T-cell receptors, specifically complement the antigenic shape they recognize and are the product of random T-cell receptor domain gene rearrangement. In this nursery the T-cell receptor of an individual thymocyte will be tested for reaction against selfantigens and the thymocytes with autoreactive T-cell receptors ruthlessly deleted from the population. The next maturation step is the designation of the response pathway which activation of the T-cell receptor will elicit. Those which differentiate to express a CD4 antigen (cluster differentiation antigen 4) will be hardwired to be T-helper cells and express cytokines; or CD8 (cluster differentiation antigen 8) cells which once activated will kill any cell which their T-receptor binds to.

These antigen naive T-cells migrate from the thymus and populate the secondary lymphatic tissue awaiting encounter with foreign antigens. On binding a foreign antigen the primary function of the T-cell receptor is to drive this antigen naive first generation effector T—cell into a replication frenzy generating thousands of clones which will be execution CD4+ or CD8+ cells respectively either expressing cytokines to aid the immune response or kill the antigen expressing cells. A few of the daughter cells will not be execution T-cells and will migrate into the spleen and lymph nodes as CD4 and CD8 memory cells for the antigen.

7.1.3   B-Cells

The progenitor B lymphocytes undergo maturation and differentiation within the bone marrow. Genetic rearranging occurs within the component domains that make up an antibody, both antigen binding Fab portions and the structural Fc domains that determine whether the immunoglobulin is as an IgM, IgA, IgG, IgE or IgD subtype, produced by this gene rearrangement. Like T-cells those first generation B-cells expressing an antibody (usually surface expressed in these immature B-cells) reacting against selfantigens are deleted by apoptotic mechanisms. The antigen-specific naive B-cells migrate from the bone marrow and populate the follicular centres of lymph nodes and other lymphoid tissues waiting to react to passing antigens which bind to their surface membrane bound antibody. As with naive T-cells, the reaction of naive B-cells to the antibody binding of its specific target antigen is to drive the primary B-cell into frenzied replication of genetic clones most of which will be antibody factories called plasma cells churning out billions of copies of the antibody - monoclonal antibodies. A few of the daughter cells will not be plasma cells and will migrate to the spleen and lymph nodes as memory B-cells.

The memory T- and B-cells form the basis of immunity and why vaccination is so successful: a large population of memory cells means that a very rapid and overwhelming immune response can be mounted if a subsequent challenge with even the tiniest contact with the foreign antigen/organism occurs. Clinical illness resulting from the foreign organism is therefore much less serious and often not noticed as the immune response is so efficient at eliminating the foreign organism.

7.1.4   Immunoglobulins

The product of specific plasma cell clones of activated naive B-cells, there are five main classes of immunoglobulin IgM, IgG, IgE, IgA and IgD (see Figure 7.4 and Table 7.1).

The basic structure of the immunoglobulins is illustrated in Figure 7.4(a). Although different immunoglobulins can differ structurally, they all are built from the same four chain structure as their basic unit. They are composed of two identical light chains (23 kD) and two identical heavy chains (50−70 kD). The heavy and light chains and the two heavy chains are held together by interchain disulphide bonds and by noncovalent interactions. The number of interchain disulphide bonds varies among

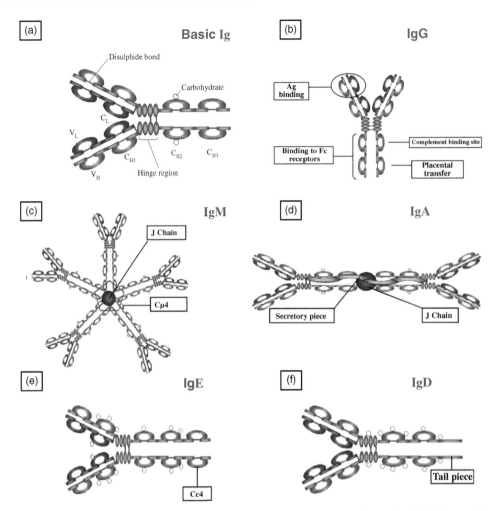

Figure 7.4   Generalized structure of immunoglobulins demonstrating sites of glycosylation disulphide bonds, hinge region and the variable and constant domains (labelled C and V respectively) is illustrated in (a). The functional regions of the Ag binding site, Fc complement and placental transfer receptors are indicated in (b). The IgM pentameric structure is shown as is the position of an addition Fc domain — $C\mu 4$ — and the J chain in (c). The dimeric structure of IgA with its J Chain and secretory signalling protein is indicated in (d). The generalized structure of IgE with its additional $C\epsilon 4$ Fc domain is shown in (e), whilst IgD's structure with its additional membrane insertion tail piece to the Fc is indicated in (f)

different immunoglobulin molecules. Within each of the polypeptide chains there are also intrachain disulphide bonds.

When the amino acid sequences of many different heavy chains and light chains were compared, it became clear that both the heavy and light chain could be divided into two regions based on variability in the amino acid sequences. These are the light chain — V_L (110 amino acids) and C_L (110 amino acids) — and the heavy chain — V_H (110 amino acids) and C_H (330—440 amino acids) — domains. The region where the arms of

the antibody molecule form a Y is called the hinge region because there is some flexibility in the molecule at this point. Carbohydrates are attached to the C_{H2} domain in most immunoglobulins. However, in some cases carbohydrates may also be attached at other locations.

Immunoglobulin fragments produced by proteolytic digestion have proven very useful in elucidating structure/function relationships in immunoglobulins. Digestion with papain breaks the immunoglobulin molecule in the hinge region on the N-terminal side

Table 7.1 Characteristics of the main immunoglobulin isotypes

	IgG	Ig M	IgA	IgE	IgD
Clinical feature	Dominant class of antibody	Produced first to response to a new immune antigen	Crosses into mucous membrane secretions	Allergy specific Ig, but designed to work in defence of nematode infections	Found predominately on lymphocyte membranes
Mean serum level (adults)	8–16 g/L	0.5–2 g/L	1.4–4 g/L	17–450 mg/L	0–0.4 g/L
Half-life (days)	21	10	6	2	3
Complement fixation: classic/alternative pathways	YES ++	YES +++	YES +	NO	NO
	Classic	Classic	Alternative	NA	NA
Mast cell binding	No	No	No	YES +++	No
Traverses placenta	Yes +++	No	No	No	No

of the H-H interchain disulphide bond. This results in the formation of two identical fragments that contain the light chain and the V_H and C_{H1} domains of the heavy chain. These fragments were called the Fab fragments because they contained the antigen binding sites of the antibody. Each Fab fragment is monovalent whereas the original molecule was divalent. The combining site of the antibody is created by both V_H and V_L. An antibody is able to bind a particular antigenic determinant because it has a particular combination of V_H and V_L. Different combinations of a V_H and V_L result in antibodies that can bind different antigenic determinants.

Digestion with papain also produces a fragment that contains the remainder of the two heavy chains each containing a C_{H2} and C_{H3} domain. This fragment was called Fc because it was easily crystallized. The effector functions of immunoglobulins are mediated by this part of the molecule. Different functions are mediated by the different domains in this fragment. Normally the ability of an antibody to carry out an effector function requires the prior binding of an antigen;

The clinical essential about immunoglobulin production is that IgM is the first type to be produced in a primary immune response and measurement of an IgM specific isotype is usually diagnostic of a primary infection. Higher affinity IgG dominates at second exposure to antigen is diagnostic of an established immune response to the infection. Both IgM and IgG are efficient activator of complement when bound to

an antigen in an immune complex. IgG is the only maternal antibody that can cross the placenta and therefore can transfer immunity (and sometimes a foreign antigen/pathogen) to the fetus. IgA are specific to secretion and are carried across mucosal cells to give protection within the mucosae. The IgE class is designed to activate mast cells and IgD is essentially the only class found in lymphocyte membranes (as receptors).

As B-cells mature they undergo instruction from T-cells to switch the class of their heavy chain (i.e. μ to γ, α or ϵ) but never the specificity for the target antigen.

7.1.5 T-cell receptors

The T-cell receptor exists as a transmembrane heterodimer with peptide domain similar to the Fab fragment of the immunoglobulin (see Figure 7.5) with variable and constant domains. There are two types of T-cell receptor (TCR) and $\alpha\beta$ chain TCR found on all CD4 T cells and the majority of CD8 T-cells and a $\gamma\lambda$ chain TCR found on 20% CD8 T-cells. The TCR works in conjunction with the CD3 receptor found on all T-cells. Another crucial ligand on CD4 T cells is CD154; this is a surface ligand for the CD40 receptor found on antigen presenting cells. The cell-to-cell interaction via antigen presenting cells CD40 receptor with CD154L on CD4 T helper cells 'licences' the antigen presenting cell to activate other CD8 cytotoxic T- and B-cells.

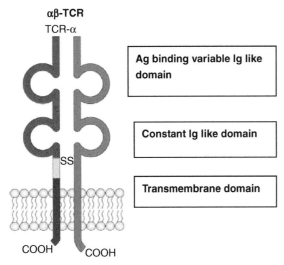

αβ-TCR

TCR-α

| Ag binding variable Ig like domain |
| Constant Ig like domain |
| Transmembrane domain |

SS

COOH   COOH

Figure 7.5   Diagrammatic representation of the αβ chain T-cell receptor (TCR) illustrating its transmembrane domain and immunoglobulin (Ig) like Ag binding and constant domains

7.2   Overview of the immune response

The immune response is a complex interplay of innate immunity, driven by cells which recognize set patterns of molecules found on invading organisms, and adaptive immunity driven by cells that can mount a more precise response to a specific immune challenge.

7.2.1   Orchestration of the immune response by CD4 T-cell subsets

Although delineation of function is primarily determined by expression of CD4 helper or CD8 cytotoxic markers, various subsets of these T-cells exist; there are at least three classes of CD4 Pro-inflammatory T-helper cells; the so-called Th1, Th2 and Th17 response regulators. There is a further class of CD4 antiinflammatory T-cells or regulatory T-cells (Tregs) of which three subsets are clinically identified. Various cytokines can polarize the immune response based of the CD4 T-cell subset they promote. These subsets are described in Table 7.2.

7.2.2   Components of the innate immune system

So far we have described the underlying cellular nature of the adaptive immune system which has been divided into humoral B-cell and cellular T-cell components; however, a more primitive innate immune system,

Table 7.2   Functionality of CD4 T-cell subsets

T-cell subset	Function	Action	Main polarizing cytokine	Main cytokines produced
Th1	Proinflammatory	Orchestrate killing of bacteria, fungi and viruses by activating macrophages and cytotoxic T-cells	IL-12	IFNγ, IL-2, TNFα
Th2	Proinflammatory	Orchestrate killing of parasites by recruiting Eosinophils promote antibody response specifically switching to IgE	IL-4	IL-4, IL-5, IL-13
Th17	Proinflammatory	Recruiting cells to immune response more resistant to Treg cell dampening of responses	IL-6, IL-23	IL-17
Antigen induced Treg — Tr1 CD25+ve Foxp3+ve	Antiinflammatory	Suppression of immune responses by IL-10	IL-10	IL-10
Antigen-induced Treg (Th3) CD25 −ve, Foxp3-ve	Antiinflammatory	Down regulates proinflammatory responses	TGFβ	TGFβ
Naturally occurring Treg CD25+ve Foxp3 -ve	Antiinflammatory	Cell-to-cell down regulation of immune response	IL-2, TGFβ	TGFβ

Table 7.3   Cells on the innate immune respons

Class of cell	Subtype	Function	Primary target
Granulocyte	Neutrophils	Phagocytose and kill bacteria, release antibacterial molecules	Bacteria, fungi
	Eosinophils	Release cellular toxins and proinflammatory factors	Multicellular parasites
	Basophils	Release proinflammatory factors including histamine	
	Mast cells	Release soluble proinflammatory histamine within tissues	Parasites
Mononuclear	Monocytes (blood) Macrophages (tissue)	Ingest and kill bacteria present processed antigens to T-cells release proinflammatory molecules IL-1 -, TNF-α, IFN-γ, etc.	Intracellular pathogens recruit adaptive immune response
Dendritic	Myeloid dentricytes	Ingest pathogen igrate to lymph nodes and present antigens to T-cells	Recruit adaptive immune response
	Plasmactoid dentrictic	Migrate to inflammation sites and reactive lymph nodes, release IFN's and present antigens to T-cells	Recruit adaptive immune response to viral infections
Lymphoid	Natural killer cells	Lyse viral infected cells	Viruses

which does not improve with repeated exposure to antigen, preexists. Many components of this immune system are also cellular in nature, not associated with any specific organ but are embedded in or circulating through, various body tissues.

The first line of defence to infection is provided by an array of cells which engulf or otherwise destroy foreign organisms (see Table 7.3). Many of the white blood cells are part of this innate immune system:

Neutrophils (alternative name polymorphs due to their distinctive multilobed nucleus) contain neutral staining granules. More often found in the blood and lymphatics than tissue, except in acute inflammatory responses, they phagocytose bacteria either triggered by recognizing key antigens of the foreign organism or because they have been coated by immune reactive molecules such as antibodies (*opsonization*).

Eosinophils are much more prevalent in tissue than neutrophils and less prevalent in blood. They have dense eosin-staining granules packed with cationic proteins enzymes and toxins. Their main function is to target large muticellular parasites such as helminths (worms) causing destruction of the organism and recruiting other immune cells to clean up the debris in a localized inflammatory response.

Unfortunately they can become sensitized to generally innocuous substances and play a major causative role in allergies like asthma. Indeed, in addition to toxic substances eosinophils release leukotrienes and platelet derived activating factor (PAF) which have an effect on smooth muscles and vasculature — in the case of asthma constriction of the air passages. Eosinophils are activated by a variety of set chemicals such as complement factors, chemokines (eotaxin−1 and eotaxin−2) and cytokines such as IL-3 and IL-5.

Mast cells and basophils are almost exclusively found in skin and mucosal tissue and are characterized by the expression of high affinity IgE receptors which trigger degranulations (an antibody subtype therefore associated with allergies) and a high content of histamine-containing granules. Histamine is a simple 111 Dalton amine which is rapidly metabolized (blood half-life 5 min) but is the major product of mast cells. Its localized biological action, constituting the acute inflammatory reaction is threefold:

1. erythema (reddening) of the skin due to arteriole dilation and venule contraction;
2. wheal, swelling due to increased vascular permeability leaking blood plasma into the tissue; and
3. flare, direct action on local nerve axons causing widespread and distant vascular changes.

Table 7.4   Circulating and humeral innate immune response molecules

Class of molecule	Example	Primary target	Source	Function
Complement	(40 plus circulating proteins)	Bacteria, viruses	Blood and lymph	Lyse bacteria, opsonize, recruit and activate immune cells and promote inflammation
Collectins	Mannose binding protein	Bacteria	Blood and lymph	Binds bacteria and activates complement
Pentraxins	C-reactive protein	Bacteria	Blood and lymph	Opsonize bacteria
Enzymes	Secretary lysozyme	Bacteria	Secretions and neutrophil released	Cleave bacterial cell wall
	Defensins	Bacteria and Fungi	Neutrophil released	Antibacterial and antifungal polypeptides
	Cathepsins	Bacteria	Neutrophil released	Cellular digestion of phagocytized pathogen
	Elastase	Bacteria	Neutrophil released	Intracellular digestion of phagocytised pathogen, but also breakdown of extracellular matrix to aid migration
	Collagenase	Migration	Neutrophil released	Breakdown of extracellular matrix to aid migration
	Myeloperoxidase and cytochrome B_{558}	Toxic damage to pathogens	Neutrophil released	Halide and respiratory burst — free radical damage
	Eosinophil peroxidase	Toxic damage to pathogens	Eosinophils released	Reactive oxygen free radical damage
Proteins	Bacterial permeability increase protein	Bacteria	Neutrophil released	Cellular digestion of phagocytized pathogen
	Major basic protein	Helminths	Eosinophils released	Toxic damage to helminths
	Eosinophil cationic protein	Helminths	Eosinophils released	Toxic damage to helminths
Reactive amines	Histamine	Pathogens and allergens	Basophils/mast cell release	Vasodilatation, increased permeability smooth muscle contraction.
Leukotrienes	C4 and D4	Pathogens and allergens	Eosinophils released	Alter smooth muscle and vascular responses.
Prostaglandins	PGD_2	Pathogens and allergens	Basophils/mast cell release	Vasodilatation, increased permeability, bronchoconstriction
Proteoglycan	Heparin	Pathogens and allergens	Basophils/mast cell release	Anticoagulation

In addition to these cellular elements of the innate immune response, a variety of soluble factors contribute to innate antimicrobial defences of the body (see Table 7.4).

The innate cellular response is more limited and preprogrammed in its repertoire than the adaptive immune system. However, it can not only kill infectious organisms and infected cells in its own right, but is also crucial in controlling the adaptive immune response of antibodies and T-cell mediated cell killing. Immunology itself rests on an understanding of the properties of these two biological entities and how both systems are highly interdependent. This is no better

Table 7.5 Toll receptor and their target pathogen associated molecular pattern (PAMP) molecules

Pathogenic organism	PAMP	Toll receptor	Antigen presenting cell type
Gram-positive bacteria	Peptidoglycan	TLR2	Myeloid dendritic cells
Viruses	Double-stranded RNA	TLR3	Myeloid dendritic cells
Gram-negative bacteria		TLR4	Myeloid dendritic cells
Viruses	Single-stranded RNA	TLR7	Plasmacytoid dendritic cells
Viruses	Double-stranded DNA	TLR9	Plasmacytoid dendritic cells

illustrated than by the functioning of antigen presentation functions of the monocyte/macrophage dendritic cell system.

7.2.3 Antigen presenting cells

Monocytes and macrophages derive from a sophisticated lineage of mononuclear phagocytic cells. Monocytes are blood circulating cells that subsequently migrate to tissues as macrophages and some forms of dendritic cell, particularly in response to an inflammatory event. Monocytes are divided in to two subtypes: one positive for CD14 which is a receptor for bacterial lipopolysaccharide, and the other CD14/CD16 positive monocytes, CD16 being a receptor for IgG antibodies.

In response to proinflammatory stimuli these cells are activated and engulf bacteria, fungi and tissue debris at the site of inflammation. Monocytes differentiate into dendritic cells and macrophages in response to the cytokine IL-4 and GM-CSF. At all stages of differentiation they produce proinflammatory cytokine in particular TNF-α, IL-1 and IFNγ as part of the mechanism of presenting processed fragments — antigenic portions — of the pathogen to T-cells.

Dendritic cells have a distinctive morphology of projecting tentacles (dendrites) from their ovoid cell shape and very large distorted, but not lobular, nuclei. They have the receptors to sense pathogens (TLR - Toll-like receptors see Table 7.5) to sense pathogens, phagocytose, process and present antigens to CD4 and CD8 T-cells and are unique in their ability to activate naive T-cells. What is more they determine the future differentiation and function of the T-cells they recruit. Activated dendritic cells secrete enormous quantities of IFNα and migrate from tissues to local lymph nodes to achieve T-cell activation while the direction of T-cell functional differentiation they orchestrate is tailored to the antigenic determinants of the pathogen.

There are various subtypes of dendritic cells, such as myeloid, plasmacytoid and Langerhans cells of skin, as well as immature and mature counterparts.

7.2.4 Histocompatibility and regulation of antigen presentation

A feature of all cells is that they will express antigens which are specific to an individual and identify cells of one person as being different from that of another. These are termed histocompatability antigens and are coded on chromosome 6 as part of a major histocompatability complex (MHC) of genes (see Figure 7.6). This MHC gene cluster codes for three classes of protein: two are concerned with how foreign antigens are presented to T- and B-cells to elicit an adaptive immune response — MHC class I and class II — and a third cluster of genes codes for complement genes, heat shock proteins (HSPs) and the cytokine TNF α and TNF β. MHC class I are still referred to as HLA genes, reflecting their historical discovery as antigens present on leucocytes (human leucocyte antigens) which determined whether transplant organs are matched or induced transplantation rejection. There are three major types termed HLA-A, HLA-B and HLA-C and along with the class II MHC are highly polymorphic. Equally expressed by both maternal and paternal derived copies of chromosomes 6 all our nucleated cells will express two variants of HLA-A, HLA-B, and so on (except when the antigen in both parents is identical). The reason they induced such a potent adaptive immune response is because they are in fact functional

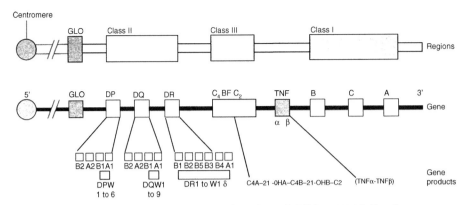

Figure 7.6   Human HLA complex. From Male *et al, Immunology, Seventh Edition*, 2006© Elsevier

receptor molecules which present small antigenic fragments of digested proteins to immune T-cells, to induce a specific adaptive response. Thus, in addition to the polymorphic differences which make them appear 'foreign' to the immune system, the grafted MHC molecules will differ from autologous MHC in the structures of their receptor grooves which will accommodate a different extensive set of antigen fragments, even those derived from 'self' proteins to which the recipient's T-cells are not tolerant. Thus the graft induces a highly potent immunological rejection mechanism, just as though it were an invading pathogen.

7.2.5   Antigen presentation

HLA class I molecules have the role of presenting short 8—10 amino acids antigenic peptides to the T-cell receptor of those T-cells also expressing the CD8 receptor. The HLA molecule consists of a transmembrane protein in which the large extra cellular domain

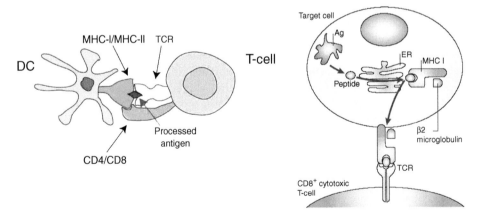

Figure 7.7   (a) Schematic representation of dentritic cell (DC)—T-cell interaction. This interaction, is mediated mostly by the class II major histocompatibility complex presentation molecule presenting a foreign pathogens antigenic molecule to the T-cell receptor (TCR) and CD 4 molecule on the DC and T-cell surface, respectively, leading to T-cell activation and proliferation. In so-called 'cross-priming' the processed antigen associates with MHC class I and interacts with CD8 cytotoxic T-cell precursors. Costimulatory molecules B7 on the DC and CD28 on the T-cell are obligatory for successful activation. (b) Foreign antigenic peptides processed by an infected cell are presented on the cells as a complex with class I/HLA molecules which presents and interacts with the TCR on activated cytotoxic immune cells like CD8 T-cell, stimulating the immune cell to kill and recruit disposal of the infected cell

noncovalently binds a common serum protein β2 Microglobulin. HLA-A, HLA-B and HLA-C are constituently expressed by all nucleated cells; however, HLA − E, F and G are only expressed by specialized cells or in special circumstance, such as by fetal tissue *in utero*, and can alter − even downregulate − the immune response. A subgroup of MHC class I related (MICA) antigens is expressed by stressed/virally infected epithelial cells and is targeted by natural killer cells of the innate immune system as the recognition receptor for cell−cell killing (Figure 7.7).

MHC class II genes have three major subgroups termed DP, DQ and DR of which each contain an A and B gene. The mature MHC class II antigen is a hetro dimeric αβ surface protein complex. These molecules present slightly longer 12−15 amino acid antigenic fragments of processed pathogens to T-cells that are also expressing the CD4 receptor. MHC class II antigens are not constituently expressed by all nucleated cells and were originally thought restricted to cells of the immune systems. However, cytokines − and in particular IFN-γ − can induce MHC class II antigen expression by all cell types.

There are three major routes to antigen processing and presentation:

Endogenous route A property of all nucleated cells whereby peptides processed as part of intracellular turnover of peptides are incorporated and displayed by HLA class I antigen complexes. Self proteins are ignored, but if virally infected fragments of the viral protein made by the cell are presented they are recognized as foreign. The HLA-class I viral peptide complex is surveyed by T-cells bearing CD8 and induce a cytotoxic response.

Exogenous route A differentiated function of antigen presenting cells (APC) whereby phagocytized pathogens and other antigenic complexes are broken down and the resultant peptides complexed and presented by MHC class II antigens. This MHC class II−pathogen peptide complex interacts with CD4 T-cells. These in turn can facilitate and enhance B-cell response (antibody production) as well as T-cell responses.

Cross presentation Dendritic cells in particular are able to process foreign antigens so that both HLA class I activation of CD8 T-cells (endogenous route short 8−10 amino acid fragments) and MHC class II activation of CD4 T-C cells (exogenous route 12−15 amino acid fragments) occurs simultaneously.

7.3 MHC genotyping, autoimmunity and susceptibility to disease

Although it is necessary to match HLA and MHC antigens for donor recipient compatibility in transplantation (except corneal grafts as this tissue is not surveyed by T-cells due to the lack of a blood supply to the eye's surface) particular MHC genotypes are associated with susceptibility to various diseases.

Genetic polymorphism and the fact that both paternal and maternal HLA/MHC complex genes are expressed equally, results in huge phenotypic variation in the antigen presenting HLA/MHC complex molecules in a population. The structural molecular variability is predominantly in the peptide binding groove, the region that recognizes the foreign antigen. Thus the ability to hold on to − and therefore present − any given pathogenic antigen to induce a specific adaptive immune response can be dependent on the binding ability and repertoire of HLA/MHC genes one inherits. This variability is a good thing in gross population terms as it ensures that at least some of the population will be immune and be able to mount an effective adaptive immune response to an epidemic/pandemic infectious agent. However, on an individual basis this may mean that a certain foreign antigens are never effectively presented or indeed that one presents degraded selfantigens, which may closely resemble a pathogen's antigen and induce autoimmunity.

7.3.1 Conditions associated with particular genotypes

Genotyping has involved identifying the specific variants of the HLA and MHC class II genes. Labelling is denoted first by the gene locus from which they originate, that is HLA-A, HLA-B, HLA-C, DP, DR, DQ (see Figure 7.4). For the class I genes this is followed by an asterisk and a two−four digit number that defines the variant (e.g. HLA-B*02). The same system applies for MHC class II molecules but both the A and B genes are both listed variants, with the exception of the DR locus where all of our A genes are identical and only the B gene variable. A lower case 'w' after the gene locus refers to a classification whereby the international

Table 7.6   HLA and MHC class II genotypes associated with disease phenotypes

HLA Class I or MHC Class II genotype	Condition	Disease classification
HLA—A*02, HLA—A*24, HLA—B*18, HLA—B*35	Susceptibility to type 1 diabetes	Autoimmunity
HLA—B*27	Susceptibility to ankylosing spondylitis	Inflammation
HLA—B*27, HLA-B*51, HLA—B*57	Human immunodeficiency virus infection — associated with slow progressive disease	Infection
HLA-B*35	Human immunodeficiency virus infection — associated with rapid progressive disease	Infection
HLA—Cw*0602	Susceptibility to psoriasis	Inflammation
DQA1*0102/DQB1*0602	Protection from type 1 diabetes	Autoimmunity
DQA1*0301/DQB1*0302	Susceptibility to type 1 diabetes	Autoimmunity
DQA1*0501/DQB1*0201	Susceptibility to type 1 diabetes	Autoimmunity
DQA1*0501/DQB1*0201	Susceptibility to coeliac disease	Autoimmunity
DQB1*0201 & DRB1*0402	Susceptibility to pemphigus vulgaris	Autoimmunity
DRB1*03 & DRB*04	Susceptibility to autoimmune- hepatitis	Autoimmunity
DRB1*0404	Susceptibility to rheumatoid arthritis	Autoimmunity
DRB1*1501	Susceptibility to multiple sclerosis	Autoimmunity
DRB1*1501	Susceptibility to antiglomerular basement membrane disease	Autoimmunity

tissue typing association are agreeing the unique genotype and have assigned a preliminary '*workshop*' numeric designation. A person's HLA—MHC geneotype may look like:

HLA—A*02/A*24, HLA—B*18/B*35, HLA—C0001/Cw*0602,

DPA1*0111/DPB1*0112 DQA1*0102/DQB1*0602, DRA/DRB1*0402.

Certain alleles are associated with disease process and are illustrated in Table 7.6.

7.4   Physical age and immunocompetency

The body's ability to react to antigen varies according to the age of the person, antigen type, maternal factors and the area where the antigen is presented. Neonatal fetus is considered to be in a state of physiological immunodeficiency, because both the innate and adaptive immunological responses are greatly suppressed. Once born, a child's immune system responds favourably to protein antigens while not as well to glycoproteins and polysaccharides. Consequently many of the infections acquired by neonates are caused by low virulence organisms like Staphylococcus and Pseudomonas. In neonates, opsonic activity and the ability to activate the complement cascade is very limited; the mean level of C3 in a newborn is approximately 65% of that found in the adult. Phagocytic activity is also greatly impaired in newborns. By 6—9 months after birth, a child's immune system begins to respond more strongly to glycoproteins. Not until 12—24 months of age is there a marked improvement in the body's response to polysaccharides. This can be the reason for the specific time frames found in vaccination schedules.

Maternal factors also play a role in the body's immune response. At birth most of the immunoglobulin present is maternal IgG. Since IgM, IgD, IgE and IgA do not cross the placenta, they are almost undetectable at birth; although some IgA is provided in breast milk. These passively acquired antibodies can protect the newborn for up to 18 months. However, these antibodies can also produce a negative response: if a child is exposed to the antibody for a particular antigen before being exposed to the antigen itself then the child will produce a dampened response. Similarly

Table 7.7   Common clinical laboratory tests evaluating the immune system

Parameter	Interpretation	Diagnostic method
C-reactive protein	Raised levels indicative of infection or inflammation	Immunoassay
Complement C3 and C4	Low levels indicate consumption in immune complex disease	Immunoassay
Immunoglobulins	Low levels indicate antibody deficiency, usually due to underlying disease or immunodeficiency Elevated levels, particularly IgM, indicate acute infection	Colorimetric protein and specific immunoassay
IgE	Raised levels in autoimmunity, can pinpoint allergen specific IgE responses	Immunoassays
Neutrophil respiratory burst	Absent in immune deficiency chronic granulomatous disease	Functional assay
T-cell proliferation	Low in primary T-cell immunodeficiencies	Functional assay
Neutrophils	High levels in bacterial infections low in secondary immunodeficiency	Haematological cell count
Eosinophils	High levels in parasite infections or allergic reactions	Haematological cell count
CD4 T cells	Low levels in HIV infections	Haematological cell count
Rheumatoid factor and antiCCP autoantibodies	Rheumatoid arthritis	Immunoassay
Double-stranded DNA antibodies	Systemic lupus erythematosus (SLE)	Immunoassay
Acetylcholine receptor antibodies	Myasthaemia gravis	Immunoassay

the response of T-cells to vaccination differs in children compared with adults, and vaccines that induce Th1 responses in adults do not readily elicit these same responses in neonates.

During adolescence the human body undergoes several physical, physiological and immunological changes. These changes are started and mediated by different hormones. Depending on the sex, at 12 and 10 years either testosterone or 17-β-oestradiol, start acting. However, these steroids act not only on the primary and secondary sexual characteristics, but also have an effect on the development and regulation of the immune system: the female sex hormone 17-β-oestradiol has been shown to regulate the level of immunological response. Similarly, some androgens, like testosterone, seem to suppress the stress response to infection, but other androgens like DHEA have the opposite effect. There is an increased risk in developing autoimmunity for pubescent and postpubescent females and males. It has been suggested that cell surface receptors on B-cells and macrophages may respond to various reproductive peptide hormones. In addition to hormonal changes physical changes such as the involution of the Thymus during puberty also affect the immunological response.

Part II: Laboratory investigations and immune assessments (Table 7.7)

7.5 Inflammation and chronic infection

Acute inflammation sees activation of the innate immune system: direct activation of complement, mast cell degranulation, swelling, changes in blood flow, pain and redness. As a result of cytokine release such as IL-1, IL-6, TNF-α, and so on, fever often results as these reach the hypothalamus. C-reactive protein is also produced by the liver due to increasing circulatory levels, particularly of IL-6. Neutrophils and granulocytes migrate to the site(s) of infection due to changes in adhesion molecule expression by local blood vessel endothelial cells and interstitial spaces may fill with pus — a mixture of pathogen and cellular debris and abundant granulocytes. This will be walled in by fibrin and reactive fibroblasts forming a containment sphere. Lymphatic traffic of the foreign antigens, either free or carried by antigen presenting cells to the lymph nodes, may also induce localized inflammation – red tracks in the tissue – lymphangitis. This triggers the adaptive immune response activating T- and B-cells and the first response is to produce IgM antibodies within 7–14 days postinfection which will change to high affinity IgG antibodies and IgA antibodies if it is a mucosal tissue response. Scavenging macrophages help resolution of the infection.

Chronic infection is where the immune response is not as dramatic nor are infections resolved so quickly. Immune responses may be triggered – or only partially triggered – as the infectious agent is mostly hidden intracellularly such as latent hepatitis B and C and mycobacteria. Even environmental toxins such as asbsestos and silicon are partially hidden but slowly leach giving a burst of immune response. Intermittent exposure as in hay fever and even autoimmune stimuli catches the immune system in an insidious trap of not properly clearing the supposed infection and the tissue damage caused by chronic repeated activation of lymphoid and mononuclear cells.

Chronic inflammation can lead to permanent organ and vascular damage and the rapid recall inflammatory response, evident within 24–48 h postexposure is both a feature and test for chronic infection/inflammatory conditions – the so-called type IV hypersensitivity tests of contact dermatitis and the Mantoux test for tuberculosis immunity. The immunological event is the presences of proinflammatory antigen presenting cells – particularly macrophages, B- and T-cells. Indeed these cells can be present in such organization that it resembles the germinal centres of lymph nodes misplaced in other tissue such as joints in rheumatoid arthritis. Both Th1 and Th2 responses are seen but are often polarized such as Th1 response in cases of chronic mycobacterium and viral infections but Th2 responses in cases of chronic allergic reactions.

7.6 Autoimmune diseases

Autoimmune disease is estimated to affect 5% of the population at some stage of their life. This is a situation in which the immune system responds to specific self molecules as if they were foreign particles. Given the vast array of possible antibodies and T-cell receptors it is highly likely that some will be generated that are autoreactive. Indeed a degree of autoreactivity exists as the T-cell receptor is designed to react with the selfHLA and MHC class II molecules but recognize the tiny fragment held within the binding groove is foreign.

In order to prevent autoimmune responses resulting from the inevitable generation of novel naive autoreactive T- and B-cells three major (and other minor) checkpoints are in place:

- First, during the early stages of development in the thymus, T-cell clones with strong affinity antiself T-cell receptors are deleted.
- Secondly in any immune reaction naturally arising T regulatory cells (Tregs) are generated to damp down or even stop a reaction particularly if it is to – or crossreacts significantly with – abundant selfantigens.
- Thirdly, dendritic cells have to overcome selfawareness blockage before they can activate CD4 T-cells.

Autoimmunity therefore can occur in cases of failures in either complete Thymic development or the inability of the given selfantigen to be expressed/presented within the thymus. Similarly genetic defects can and do effect Treg cells and subtle failure of Tregs has been implicated in diseases such as type 1 diabetes. Linked to this is failure of antigen presenting cells to distinguish activation of an autoreactive T-cell. Dendritic cells being particularly important in recruiting the entire gambit of immune responses are therefore key in escalating autoreactivity to a clinically florid autoimmune disease. This is proposed to occur in two major ways:

- Tissue damage – the release of hidden selfantigens, which were never presented by the thymus and are now actively presented by dendritic cells, particularly in circumstance of an infection as a reactive adjuvant allows activation of autoreactive CD4 T-cells along with the pathogen specific ones – bystander activation.
- Pathogen mimics of selfantigens – an entirely appropriate immune response to a pathogen results in T- and B-cells that react with similar (not necessarily identical) selfantigens – molecular mimicry.

We know more of the molecular mechanism of tissue damage due to stimulation of autoimmune

Table 7.8   Examples of common autoimmune disease

Disease	Target antigen	Pathology
Addison's disease	Adrenal cortical cells, 21α-hydroxylase	Adrenal failure as cortical cells are destroyed, antibodies target corticosteroid synthesis enzyme
Goodpasture syndrome	Glomerular basement membrane (GBM)Type IV collagen fragment	Antibody reaction to Goodpasture antigens on the basement membrane of the glomerulus of the kidneys and the pulmonary alveolus
Graves' thyroiditis	Thyroid stimulating hormone receptor (TSHr)	Autoantibodies bind TSHr continually stimulating production of thyroid hormones — hyperthroidism
Hashimoto's thyroiditis	Thyroid peroxidize, thyroglobulin	Antibodies cause gradual destruction of follicles in the thyroid gland leading to hypothroidism
Vitiligo	Pigment cell antigen	Cellular destruction mediated rather than antibody direct killing of pigment cells
Systemic lupus erythematosus (SLE)	? Antinuclear antigens nRNP A and nRNP C including, in particular, DNA plus a variety of other antigens	Widespread symptoms of immunological destruction of tissues and deposition of antinuclear protein complexes thought to be triggered by failure to clear apoptotic cells by scavenger macrophages
Multiple sclerosis (MS)	Myelin basic protein, myelin oligodendritic glycoproteins	Immune cell mediated progressive demyelination of nerves leading to muscle fatigue and nervous system collapse
Type I diabetes	? Antiproinsulin and antiglutamic acid decarboxylase.	Autoimmune cell mediated destruction of insulin-producing beta cells of the pancreas
Primary biliary cirrhosis	? anti mitochondrial pyruvate dehydrogenase and anti nuclear antigen	Cell mediated slow progressive destruction of the small bile ducts (bile canaliculi) within the liver
Rheumatoid arthritis	Citrullinated protein/peptide, rheumatoid factor (antibody to the Fc portion of IgG !)	Inflammatory response of the synovium (synovitis) secondary to hyperplasia of synovial cells, excess synovial fluid and the development of pannus in the synovium
Scleroderma	? Anti-topoisomerase, anti-centromere anti-U3 and anti-RNA polymerase antibodies	Chronic CD4 T cell mediated autoimmune disease characterized by fibrosis (or hardening) of tissue, vascular alterations and autoantibodies
Sjögren's syndrome	Alpha-fodrin	Immune cells attack and destroy the exocrine glands that produce tears and saliva
Myasthenia gravis	Acetylcholine receptors	Antibodies block the muscle cells from receiving neurotransmitters — voluntary muscle weakness
Pernicious anaemia	Gastric parietal cells and intrinsic factor	Destruction of cells responsible for absorption of B12 or antibody blockade of the B12 carrier protein — intrinsic factor
Pemphigus vulgaris	Desmoglein	Antibodies attack desmosmal proteins of the skin causing blistering

responses: direct CD8 T cell cytotoxity, cytokine damage such as IL1 and TNF; complement fixation as a result of autoantibodies binding to target tissue; autoantibodies binding soluble selfantigens and precipitating as complexes in tissue (and the kidneys' nephrons) then recruiting complements to produce a damaging chronic inflammatory response which nonspecifically kills all adjacent cells in the proximity; autoantibodies binding a surface receptor activating or blocking it to cause pathology (see Table 7.8).

7.7   Transplant rejection

Immunological rejection is the inevitable consequence of a graft of tissue between genetically nonidentical individuals in the absence of adequate immune suppression. Rejection occurs in waves, the first is hyperacute which is a response that is seen within minutes and hours when preformed antibodies predominantly antiblood group (or HLA if this foreign tissue type has been encountered before) affix to the grafted tissues. An acute response occurs 7—14 days later when T-cells join the attack (this can be accelerated acute if the tissue type has prior sensitized the recipient) and if this is controlled by steroid and immune suppressive drugs like ciclosporin A, a chronic phase ensues. Months to years later reactive B-cells are producing new tissue-specific antibodies that affix complement bringing destruction of the graft. The T-cell response is against the whole foreign HLA/MHC molecule — direct allorecognition — whilst the chronic response is a processed one against small peptides of the foreign HLA/MHC antigens — indirect allorecognition.

Aside from matching a suitable blood group, some HLA and MHC antigens are more important than others and so partial matching in addition to appropriate immunosuppressive therapy is common. HLA-A, HLA- B and DR genotypes are regarded as the most important to match.

7.8   Hypersensitivities

Under some circumstances a harmless molecule can initiate an immune response that can lead to critical tissue damage and death. In type 1 (immediate) hypersensitivity the binding of an antigen to specific IgE bound by IgE receptors on mast cells results in immediate and massive degranulation; the consequential inflammatory response is ferocious and very damaging. Yet the antigens (allergens) involved are typically inert molecules present in the environment.

The immediate effects of the allergen are often very florid (early response) and are seen within a few hours. Of even greater concern are the late phase responses which are mediated by Th2 cells recognizing peptide epitopes of the allergen. They recruit eosinophils and the late phase response is chronic inflammation which is much more difficult to control. In asthma the late phase response gives rise to wheezing due to the eosinophil leukotryene release causing constrictions of airways. The immediate hypersensitivity is normally responsive to antihistamines but late-phase responses will not be and

require much more potent, but slower acting, corticosteroids to control the immune cells.

The tendency to develop allergic responses (atopy) display strong inheritance tendencies and 20—30% of the UK population are atopic. Two thirds of atopic individuals, who have been characterized due to circulating levels of specific antiallergen IgE, have clinical allergic disease. This ranges from hay fever (allergic rhinitis), allergic eczema, uritcara, angioedema, to asthma, food, bee and wasp venom allergies. Diagnosis is usually made from clinical history and skin prick testing — introducing a tiny amount of the allergen under the skin and measuring the size of the weal — or measurement of the levels of allergen specific IgE.

The term anaphylaxis describes an allergic reaction that is so rapid on onset and serious that it may cause death. Due to an acute generalized (not local site) IgE mediated immune reaction of mast cells and basophils it is characterized by an allergen priming followed by reexposure. The allergen has to be systematically absorbed either by ingestion or parenteral injection in order to provoke this whole body response. Of the foods that often do this nuts (peanut, Brazil, cashew), shellfish (shrimp, lobster), dairy products and eggs are the most noted. The less common food anaphylactic agents are citrus fruits, mango, strawberries and tomatoes. Of the directly injected known anaphylactic agents venoms (bees, wasps, etc.) are the best recognized, whilst of the medications antisera and antibiotics are the most recognized. However, dextrans and latex (from surgical gloves and condoms) can induce anaphylaxis and are absorbed into the systemic circulation through the skin.

7.9   Immune deficiency

Infections associated with immunodeficiency have several typical clinical features:

- The causative infecting organism is often unusual being either atypical or opportunistic.
- The infection is often chronic, severe or recurrent.
- Following antibiotic treatment they return quickly and may only ever partially resolve during therapy.

Most forms of clinical immunodeficiency are secondary and result from infection (e.g. HIV) or therapy (e.g. cytotoxic drugs, corticosteroids, antiTNFα antibody therapy, radiation damage or indeed bone marrow ablation prior to stem cell transplantation). Disease or drug induced myelosuppression, increased hypersplenic destruction and autoimmune reactivity to neutrophils makes acquired neutopenias very common and as blood neutrophil counts drop below 5×10^8/L infection rapidly overwhelms innate immunity.

Hypogammaglobulinaemias are seen in patients with myeloma and chronic lymphocytic leukaemia and lymphoma. Spleenectomy, although not fatal will lead to an inability to mount an effective immune response against pneumococcal, meningococcal and other capsulated bacteria.

Primary immunodeficiency is rare and is the result of either a congenital developmental defect of immune cells or inherited genetic defects (invariably autosomal or X-linked recessive). They manifest after birth anywhere from infancy to adulthood depending on exposure to different pathogens. Examples of primary immunodeficiency are given in Table 7.9.

Table 7.9   Examples of primary immunodeficiency

Disease	Phenotypic defect in the immune system
Chronic granulomatous disease (GCD) — failure in generating antibacterial respiratory burst metabolites	Neutrophil functional deficiency
Leucocyte adhesion deficiency (LAD) — failure of components of cell adhesion responsible for leucocyte rolling and passage from blood stream at sites of infection	Neutrophil and other leucocyte deficiency in tissue penetration
Hyper Ig E syndrome — failure in STAT3 secondary messenger signalling, high serum IgE, cysts, dermatitis as neutrophils fail to respond to IgE binding to pathogens	Neutrophil functional deficiency
Mutation in compliment proteins — depending on the protein the alternative, classic or indeed both pathways may not function impairing innate or adaptive immune responses	Complement deficiency
DiGeorge Syndrome — The thymus (and parathyroids and aortic arch) fails to develop. Affected children have reduced/absent T-cells and immune proliferation responses	T-cell deficiency
CD3 and IL-2 gene mutations — Deficiencies in CD3 receptor and IL-2 mean that the necessary growth/proliferation signal to activated T-cells is absent	T-cell deficiency
X-linked agammaglobinopathy (XLA) — IgG levels are low usually less than 2 g/L, B-cell development is arrested at the preB-cell stage and B-cells are absent from the blood. (After maternal IgG levels fall newborns succumb to recurrent infection)	B-cell deficiency
Selective IgA deficiency — Most common immunodeficiency (~1/600 North Europeans); normal IgG and IgM but very low IgA — recurrent mucosal tissue infections	B-cell deficiency
Severe combined immunodeficiencies (SCID) — This is a collection of diseases with various molecular causes (including adenosine deaminase deficiency) whereby T-cell, NK-cell and B-cell development is impaired. The most common is an X-linked defect of the IL-2 receptor γ chain but as this is a shared component of IL-4, IL-7, IL-9 and IL-15 receptors its affects are widespread	T- and B-cell deficiency
Hyper IgM syndrome (HIGM) — Only effectively produce an IgM response, an X-linked defect on the CD40 ligand: signalling between CD40L on T-cells with CD40 on B-cells is necessary to effect class switch from low-affinity IgM to high-affinity immunoglobulin G, A, etc.	T- and B-cell deficiency
Wiskott—Aldrich syndrome — An X-linked defect associated with eczema and thrombocytopenia an unexplained fall in immunoglobulin levels which is cell mediated and considered autoimmune	T- and B-cell deficiency

Bibliography

Delves, P.J., Martin, S.J., Burton, D.R. and Roitt, I.M. (2011) *Roitt's Essential Immunology*, Wiley-Blackwell.

Hall, A. and Yates, C. (2010) *Immunology: Fundamentals of Biomedical Science*, Oxford University Press.

Male, D., Brostoff, J., Roth, D. and Roitt, I. (2006) *Immunology*, Mosby.

Sompayrac, L. (2008) *How the Immune System Works*, Wiley-Blackwell.

Chapter 8

Haematology and transfusion science

Dr Suzanne M. Docherty, BMedSci, MBBS, Ph.D.

8.1 Introduction and components of blood

Haematology is the study of blood. Blood is a connective tissue comprising cells (red blood cells and white blood cells, plus cellular fragments; platelets) suspended in a straw-coloured fluid, plasma, to circulate through the arteries, capillaries and veins of the body. The average adult has around 5 L of blood. Haematology also considers the disorders of blood that can arise, while transfusion science focuses on the different products that are made from donated blood, and how these are best used for different patients with different markers on their own red blood cells, and in a variety of haematological disorders.

8.1.1 Red blood cells

Red blood cells — or erythrocytes — are the most numerous cells found in blood, and have a lifespan of around 120 days. There are some 5 million red blood cells/µL of blood, giving an overall average red blood cell count of around 5×10^{12}/L in adults. Red blood cells are a biconcave disc shape (i.e. they look like a flat disc that someone has squeezed between finger and thumb in the middle!) and about 7 µm in diameter. The structure of the red blood cell makes it uniquely flexible; red blood cells need to be able to squeeze through capillaries that are only 3 µm in diameter, despite their 7 µm size, for which they obviously need to be able to bend considerably to fit. The flattened disc shape also gives the red blood cell a large surface area/volume ratio, which is ideal for efficient gas exchange. The structure of the red blood cell is shown in Figure 8.1.

8.1.2 White blood cells

The normal range for white blood cells (also known as leucocytes) is around $4.0-11.0 \times 10^9$/L in humans. There are five types of white blood cells, given in descending order of frequency in blood: neutrophils, lymphocytes, monocytes, eosinophils and basophils (many people use the mnemonic 'Never Let Monkeys Eat Bananas' to remember this order!). The characteristics of each are described below.

Biomedical Sciences: Essential Laboratory Medicine, First Edition. Edited by Ray K. Iles and Suzanne M. Docherty.
© 2012 John Wiley & Sons, Ltd. Published 2012 by John Wiley & Sons, Ltd.

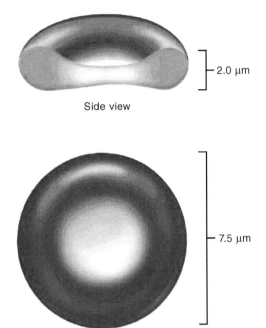

Side view

2.0 μm

Top view

7.5 μm

Figure 8.1   The red blood cell

8.1.2.1   Neutrophils

Neutrophils are the most abundant of the white blood cells; the normal range in blood is $1.5-7.5 \times 10^9$/L (Figure 8.2). Neutrophils are 9–15 μm in diameter, and contain numerous small granules and a nucleus with a variable number of segments (usually 2–5 segments). They are a pale pink/purple in a Romanowsky stain. The function of neutrophils is chemotaxis into areas of

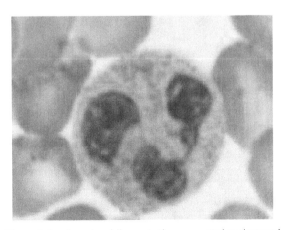

Figure 8.2   A neutrophil — note the segmented nucleus and fine granules

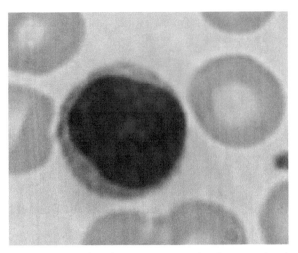

Figure 8.3   A lymphocyte — note the large nucleus/cytoplasm ratio

inflammation, then phagocytosis, and the killing of phagocytozed bacteria. The majority of neutrophils in blood are circulating in plasma, but a minority form a marginated eutrophils pool, attached to the endothelium of small blood vessels. Neutrophils have a half-life in blood of around 7 h.

8.1.2.2   Lymphocytes

The normal range for lymphocytes in blood is $1.2-3.5 \times 10^9$/L, making them the second most frequent white blood cells in plasma (Figure 8.3). Lymphocytes are 7–16 μm in diameter, with a large round nucleus that has large chromatin clumps and almost fills the cell. They have a few fine granules in their small rim of cytoplasm, and appear pale blue in Romanowsky stains. B-cells comprise 10–30% of lymphocytes, T-cells form 65–80% of lymphocytes, while 2–10% are nonB-, nonT- (null) cells. Lymphocytes are involved in specific immune responses; the production of antibodies by B-lymphocytes, and cell-mediated immunity by T-lymphocytes. They have a half-life of around 70 h, though some B-lymphocytes are immortal.

8.1.2.3   Monocytes

The normal range for monocytes in blood is $0.2-0.8 \times 10^9$/L (Figure 8.4). Monocytes are 15–30 μm in diameter, with a moderately low nucleus/cytoplasmic ratio. The nucleus is of variable shape, but often C- or U-shaped, with some lacy chromatin. With a Romanowsky stain, monocytes have a variable number of fine

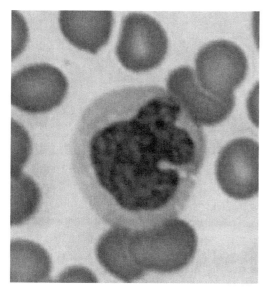

Figure 8.4   A monocyte

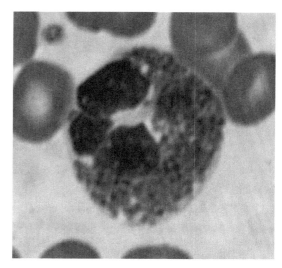

Figure 8.5   An eosinophil – note the multiple red granules, and nucleus with multiple segments

purplish granules amid a pale blue cytoplasm, in which vacuoles are often visible. Around half of all monocytes are found in the spleen, while the rest typically circulate in the bloodstream for 1–3 days before migrating into tissues and differentiating into larger tissue macrophages. Monocytes and macrophages are phagocytes capable of killing some microorganisms.

8.1.2.4   Eosinophils

The normal range for eosinophils in blood is $0.02 - 0.6 \times 10^9$/L (Figure 8.5). Eosinophils are 12–17 μm in diameter, with a low nucleus/cytoplasm ratio. The nucleus usually has two segments and eosinophils contain numerous large, round reddish granules (since they readily take up eosin in the Romanowsky stain), giving them a distinctive appearance. They have a half-life of around 6 h in blood. Eosinophils' granules contain histamine, peroxidase, ribonuclease, deoxyribonuclease, lipase and plasminogen among other substances, and the function of eosinophils in defence against parasite, viruses and mediation of allergy often results in release of the toxic contents of their granules (degranulation).

8.1.2.5   Basophils

The normal range for basophils in blood is $0.01 - 0.15 \times 10^9$/L (Figure 8.6). Basophils are 10–14 μm in diameter, with a low nucleus/cytoplasm ratio. The nucleus usually has two segments, often seen with overlying

granules, and a basophil's several granules stain purplish-black on a Romanowsky stain. The granules contain histamine, proteolytic enzymes, cytokines and heparin. Basophils mediate immediate-type hypersensitivity with histamine release, and modulate inflammatory responses using substances like heparin to increase blood flow to tissues.

8.1.3   Platelets

Although platelets appear to be tiny, simple fragments of a larger cell (the megakaryocyte), their complex composite structure underlines their important function

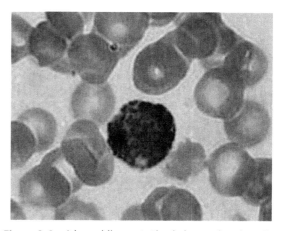

Figure 8.6   A basophil – note the dark granules obscuring the nucleus

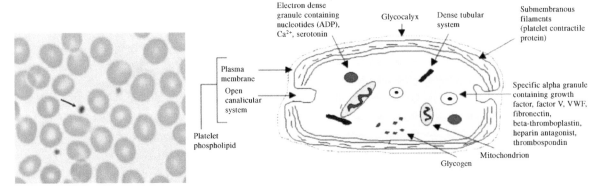

Figure 8.7   Blood film with platelets arrowed (left). Structure of a platelet

in haemostasis. As shown in Figure 8.7, they contain mitochondria, a glycogen store and different types of granules, the contents of which can be released using an intricate tubular network within the platelet.

8.1.4   Plasma

Plasma is a pale yellow, clear fluid in which the cellular components of blood are suspended, and it comprises around 55% of blood volume (Box 8.1). Around 90% of plasma is water, and dissolved in this are numerous proteins (including clotting factors and fibrinogen), glucose, hormones, ions and carbon dioxide (Figure 8.8).

8.2   Routine laboratory blood tests

8.2.1   Full blood count

The full blood count or FBC (known as the 'complete blood count', CBC, in the USA) is the most frequently

> **Box 8.1   Serum v. plasma**
>
> 'Serum' is not an interchangeable word for 'plasma'. Serum does not contain clotting factors and fibrinogen, and is thus found in blood that has been allowed to clot. Plasma is prepared from blood with an added anticoagulant to prevent clotting.

requested haematological investigation on patients (Box 8.2). Blood is collected into tubes that contain a compound that will prevent clotting (e.g. potassium citrate), and in the laboratory, samples are drawn into an automated analyser which dilutes and then runs a set volume of the blood through an electrical current to enable all of the cells in the blood to be counted as they pass through the current and individually

Figure 8.8   Centrifuged blood sample showing plasma and cellular components of blood; a thin 'buffy coat' layer (platelets and white blood cells) sits between the red cells and plasma

Note that males and females differ in terms of their red blood cell count, and the haemoglobin concentration packed into each red blood cell. After puberty, males have higher red cell counts (which in turn gives them a higher haematocrit than females) and haemoglobin concentrations. This is mainly due to the anabolic hormone testosterone, of which males have considerably higher levels after puberty.

8.2.1.1   Indices measured in a full blood count and normal ranges

Haemoglobin level

Haemoglobin level is the quantity of haemoglobin in a given volume of blood, and is given in g/dL (dL = one tenth of a litre, i.e. 100 ml) or g/L.

Normal ranges for haemoglobin

Adult males: 13.5—17.5 g/dL (135—175 g/L).

Adult females: 11.5—16 g/dL (115—160 g/L).

Red blood cell count(RBC).

Red blood cell count (RBC) is simply the number of red blood cells in 1 L of blood, as counted by automated analysers.

RBC normal ranges

Adult males: $4.5-6.0 \times 10^{12}$/L.

Adult females: $3.9-5.0 \times 10^{12}$/L.

interrupt it. The red blood cells are then burst with a lytic solution, and white blood cells and platelets are counted again using the same method (platelets are easily distinguished from white blood cells using this method as they interfere less with the electrical current by virtue of being so much smaller than white cells). The white blood cell counts and platelet counts obtained are subtracted from the initial total cell count to enable the red blood cell count to be calculated (Figure 8.9). The different types of white blood cells are estimated by running the dilute sample through a very narrow capillary tube one by one, and a laser beam directed onto each cell gives a characteristic pattern of reflection, scattering or transmission according to which type of white blood cell it is. In this way, white cell populations are estimated.

8.2.2   Red blood cell distribution width (RDW)

Red blood cell distribution width (RDW) is a measurement of how widely red blood cells vary in volume in one blood sample. In normal circumstances, red blood cells vary little around their normal volume, but in some forms of anaemia, they can vary widely in volume and shape — this is known as anisocytosis. RDW is calculated in automated analysers by dividing the standard deviation of the mean red blood cell

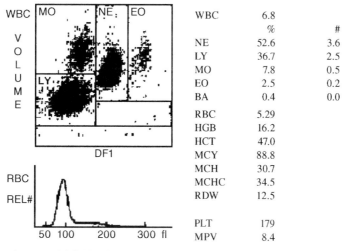

		%	#
WBC	6.8		
NE		52.6	3.6
LY		36.7	2.5
MO		7.8	0.5
EO		2.5	0.2
BA		0.4	0.0
RBC	5.29		
HGB	16.2		
HCT	47.0		
MCY	88.8		
MCH	30.7		
MCHC	34.5		
RDW	12.5		
PLT	179		
MPV	8.4		

Figure 8.9   An example of a normal full blood count produced by an automated analyser

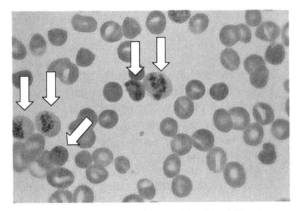

Figure 8.10   Reticulocytes (arrowed) in a blood film. Source: www.hematologyatlas.com

volume by red cell count (this latter method has the added advantage of being slightly more accurate than directly measuring PCV, as PCV will also include some serum between cells, while calculation of haematocrit excludes this). The normal ranges (NB: normal ranges vary from lab to lab, and those given here are average values) are:

adult males:	43–49%
adult females:	37–43%.

volume, by the mean cell volume (MCV: see Section 8.2.5). The RDW normal range is 11–42%.

8.2.3   Reticulocytes

The reticulocyte count is a measurement of the proportion of red cells that are immature reticulocytes, and gives an insight into bone marrow activity. It is expressed as a percentage of the total number of red cells. Reticulocytes are slightly larger than mature red blood cells and still contain some ribosomal RNA (see Figure 8.10). After bleeding or haemolysis, the reticulocyte count will be increased; if the reticulocyte count is low despite anaemia, it may be due to bone marrow failure, or a lack of iron, B12 or folate for red blood cell formation. The reticulocyte normal range is < 2%.

8.2.4   Haematocrit

The term 'haematocrit' refers to the total volume of blood that is comprised of cells. It is also known as the packed cell volume (PCV); in practice, the vast majority of cells in blood are red blood cells, so haematocrit is considered equivalent to the total red blood cell volume. Haematocrit is expressed either as a percentage (e.g. HCT 38% – this means that cells comprise 38% of blood volume), or in L/L (e.g. HCT 0.38 L/L – so 0.38 L out of 1 L of blood is the cellular component). It can be measured by centrifuging heparinized blood in microcapillary tubes and simply measuring the cellular and serum components in a specialized haematocrit reader, but it is more usually calculated in automated analysers in haematology laboratories by multiplying red cell

8.2.5   Mean cell volume (MCV)

Mean cell volume (MCV) is the average volume of a single red blood cell. It is an extremely useful index in patients with anaemia, as different causes of anaemia are associated with small, normal-sized or large red blood cells. MCV is measured in femtolitres (abbreviated as fL, femto is 10^{-15}), and can be calculated by dividing the haematocrit by the red blood cell count; that is, the total volume of red blood cells divided by the number present will give the volume of one red blood cell. The normal range for MCV is 80–96 fL.

8.2.6   Mean corpuscular haemoglobin (MCH)

Mean corpuscular haemoglobin (MCH) is a calculated value for the amount of haemoglobin in each red blood cell, and is measured in pg (picograms, where pico is 10^{-12}). It is calculated by dividing the haematocrit by the red blood cell count. The normal range for MCH is 27–32 pg.

8.2.7   Mean corpuscular haemoglobin concentration (MCHC)

Mean corpuscular haemoglobin concentration (MCHC) is the amount of haemoglobin in a given volume of packed cells and as such is measured in g/dL. It is calculated by dividing the haemoglobin level by the haematocrit. The normal range for MCHC is 32–36 g/dL.

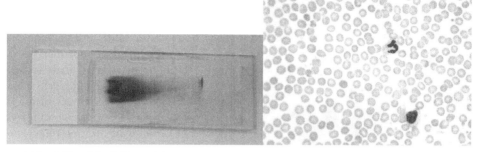

Figure 8.11   A blood film slide (left), and its appearance under a light microscope (right)

8.2.8   White blood cell count (WBC)

The white blood cell count (WBC) is the number of white blood cells in 1 L of blood, and may also include a differential white cell count (in which the numbers of each type of white blood cell are also given). The normal range for WBC is $4.0 - 11.0 \times 10^9/L$.

8.2.9   Platelets

The platelet count (often abbreviated as 'Plt') gives the number of platelets in 1 L of blood. The normal range for platelets is $150 - 400 \times 10^9/L$.

8.2.10   Blood films

Blood films (see Figure 8.11) are microscope slides with a drop of a patient's blood smeared over the length using a second, spreader slide (although automated blood film preparation is now used in many hospital laboratories). Slides are left to air dry, fixed with methanol and then stained, typically with a Romanowsky stain, though others (e.g. Giemsa or Wright stain) may be used (Figure 8.11). Not every sample sent in to the haematology laboratory will have a blood film prepared; this is reserved for those with significant abnormalities in the full blood count, or with a problem performing an automated full blood count that may suggest the presence of unusual sized or shaped cells (e.g. blast cells in leukaemia, or the aberrant lymphocytes of glandular fever).

8.2.11   ESR

The erythrocyte sedimentation rate (ESR) is the rate of precipitation of red blood cells in a vertical column in 1 h. Anticoagulated blood is placed in a Westergren tube and the height of the column of red blood cells is measured after a specific period of time, and reported in mm/h (Figure 8.12).

ESR gives a measure of inflammatory processes in the body nonspecifically, and an ESR greater than that expected for age and gender suggests the presence of an inflammatory process in the body, for example chronic renal disease or rheumatoid arthritis. This is because of higher levels of fibrinogen in the blood during

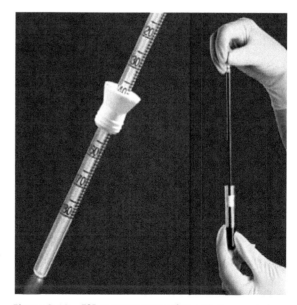

Figure 8.12   ESR measurement using a Westergren tube

inflammatory diseases, which causes red cells to stick together in stacks ('rouleaux'), which pile up faster and create an elevated ESR. However, ESR increases with age, is higher in females than in males, and is dependent on Hb concentration (it is increased in anaemia, and low where Hb is high). The normal ESR is calculated as follows:

$</ =$ (age in years/2) for male adults.
$</ = ((age$ in years $+10)/2)$ for female adults.

8.2.12   Plasma viscosity

Some laboratories report plasma viscosity instead of ESR, which is similarly dependent on the concentration of large molecules – such as fibrinogen and immunoglobulins – in blood. The advantages of using plasma viscosity instead of ESR are that the normal range is independent of Hb level, it is the same in males and females, and only slightly higher in older adults. Plasma viscosity results can also be produced rapidly, in around 15 min.

8.2.13   Bone marrow biopsy

Samples of a patient's bone marrow may be required to assist in the diagnosis of some haematological conditions, and are usually acquired from the posterior iliac crest of the pelvis using either trephine biopsy, or bone marrow aspiration (occasionally, both are needed). For a bone marrow aspirate, a special aspirate needle is inserted through the skin over the pelvis under local anaesthetic and then twisted through the bone into the marrow cavity. A syringe is attached, and a sample of liquid bone marrow aspirated into it. For a trephine biopsy, a larger needle with a hollow cavity is inserted into the pelvic bone and a solid core of bone marrow obtained. For both forms of bone marrow sampling, a film of the bone marrow is made to be examined under the microscope, and the cells obtained may also be analysed using techniques such as flow cytometry to look at surface markers on the stem cells, or fluorescent *in situ* hybridization to analyse chromosomes (Figure 8.13).

8.3   Haemopoiesis

The bone marrow is one of the most fertile tissues of the body and is responsible for the production of more than 10^{13} new blood cells each day – an incredible feat. Haemopoiesis is the term for the production of new blood cells. The first primitive blood cells are detectable from around the fourth week of gestation, where they are produced in the fetal yolk sac and migrate to the liver and spleen. The latter are the major sites of haemopoiesis until the third trimester of gestation, when the bone marrow becomes the main source of blood cells. Throughout postnatal life, the bone marrow is the only normal site of haemopoiesis and almost all bones of the body contain red marrow, producing blood cells, at birth. However, by adulthood, most of these red marrow cavities have been replaced by yellow marrow (composed predominantly of fat), and only the long bones and some of the cavities within the axial skeleton still contain red marrow for haemopoiesis. The liver and spleen may still be pressed into service as sites of extramedullary haemopoiesis if disease processes demand more haemopoiesis than the red marrow can produce, but this is always abnormal.

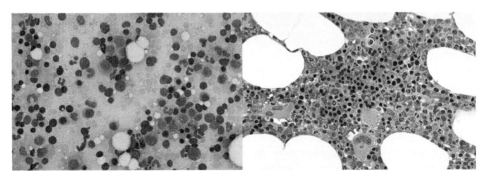

Figure 8.13   Normal bone marrow aspirate film (left) and trephine biopsy film (right)

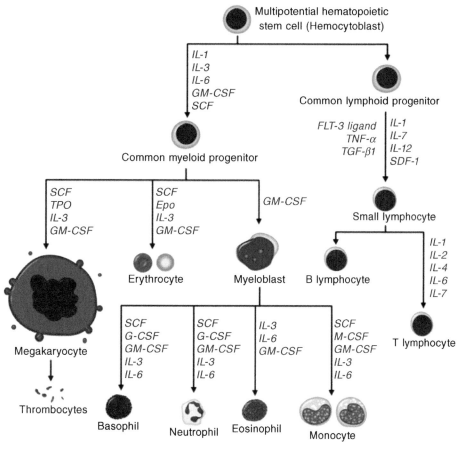

Figure 8.14   Normal haemopoiesis and associated growth factors. Source: http://en.wikipedia.org/wiki/File:Hematopoietic_growth_factors.png

Every type of blood cell is ultimately derived from pluripotent stem cells of the red marrow. Such cells are capable of both self and renewal and also of dividing along different lineage pathways to give rise to a range of possible cells. The two major pathways that pluripotent stem cells of bone marrow divide along initially are into either lymphoid stem cells (from which lymphocytes are produced), or myeloid stem cells (from which all other types of blood cell derive). The possible pathways of haemopoiesis, along with the main growth factors involved in the control of each arm, are shown here in Figure 8.14.

The differentiation of the pluripotent haemopoietic stem cell into different blood cells is controlled by a wide range of growth factors, which act on receptors expressed by blood cells at their various stages of development as illustrated in Figure 8.14. Some of these growth factors have been developed into clinical treatments; erythropoietin (EPO) is normally produced by the kidney and causes increased production of erythrocytes, and this can be given to patients who are anaemic due to renal disease, with a consequent lack of EPO. Granulocyte colony stimulating factor (G-CSF) can be used in patients who have received chemotherapy to improve their full blood count.

8.4   Red blood cell structure, disorders and metabolism

8.4.1   The red blood cell membrane

The plasma membrane of the red blood cell comprises the lipid bilayer with associated membrane proteins found on all cells of the body, but it has a

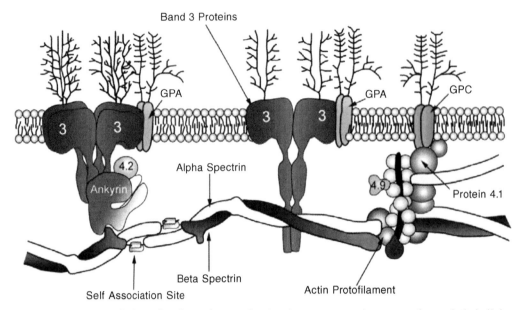

Figure 8.15 The structure of the red cell membrane, showing important membrane proteins and their links to the cytoskeleton. From Hoffbrand et al, *Postgraduate Haematology*, Sixth Edition 2010, reproduced by permission of John Wiley & Sons Ltd

number of special features necessary to fulfil its multiple functions and retain its inherit strength and flexibility while travelling the equivalent of hundreds of miles around the circulation in its life-span. In addition to the lipid bilayer, the red cell membrane also has a glycocalyx on its external surface; this is rich in carbohydrates. The lipid bilayer contains a large number of associated membrane proteins and within the red cell is an extensive cytoskeleton linked to these — it is these membrane proteins and their connection to the cytoskeleton within the red cell that enables the red cell to combine flexibility with robustness. Integral proteins that provide the links across the entire plasma membrane and cytoskeleton are known as vertical connections, while the proteins of the cytoskeleton present just on the internal face of the plasma membrane are called horizontal connections (Figure 8.15).

8.4.2 The glycocalyx

Many of the proteins found in the plasma membrane of the red blood cell are profoundly glycosylated; that is, they have a network of polysaccharide carbohydrate groups associated with them along the external face of the plasma membrane. This glycocalyx has numerous functions on the red cell, including assisting with cell adhesion, immune system recognition and carbohydrate antigens that are important in human blood grouping.

8.4.3 Major membrane proteins

More than 50 proteins have been described in the red cell membrane in various quantities. Some are rare, while others have several hundred thousand copies per red cell. Around half carry blood group antigens, while others act as ion channels, transport proteins or cell adhesion molecules. The most important structurally are the glycophorins, spectrin, ankyrin, Band 3, proteins 4.1 and 4.2.

8.4.3.1 Glycophorins

The glycophorins are heavily glycosylated, major integral proteins that cross the entire red cell plasma membrane. Glycophorin A is part of the MN blood grouping system, glycophorin B bears the Ss blood group proteins and glycophorin C carries antigens of the Gerbich blood groups.

8.4.3.2   Spectrin, ankyrin, Band 3 and proteins 4.1 and 4.2

The α and β subunits of the spectrin form a hetero-dimeric molecule. Ankyrin is a large molecule that attaches spectrin to Band 3, a major integral protein that spans the red cell membrane, along with protein 4.2. Together, these molecules — spectrin, Band 3, protein 4.2 and ankyrin — form a network to maintain red cell conformation and flexibility.

8.4.4   Hereditary disorders of the red cell membrane

The shape of the red blood cell is intimately dependent on the presence of membrane proteins, and mutations in the genes that encode these proteins can adversely affect red cell conformation, resulting in characteristic clinical syndromes. The most important of these are hereditary spherocytosis, hereditary elliptocytosis, hereditary pyropoikilocytosis and hereditary stomatocytosis.

8.4.4.1   Hereditary spherocytosis

Hereditary spherocytosis is an inherited form of haemolytic anaemia in which affected red blood cells take on a characteristic small, spherical shape — spherocytes — rather than the usual biconcave disc shape. It is relatively common in Caucasians (with an incidence of 1 in ∼ 2000 in Northern Europe), and usually inherited in an autosomal dominant pattern. It is usually caused by the inheritance of a defective gene for ankyrin, but can also be due to defects in spectrin, Band 3 or protein 4.2, any of which can lead to cytoskeletal dysfunction in the red cell due to the adverse effects on vertical connections. The presence of small spherocytes in the circulation triggers their haemolysis, with resultant mild anaemia and jaundice, and a small degree of splenomegaly. Full blood count shows mild anaemia and the blood film will show the presence of spherocytes (Figure 8.16).

The red cells in hereditary spherocytosis are more rigid than normal red cells and tend to prove relatively resistant to haemolysis if incubated with 0.5% NaCl (the osmotic fragility test). The clinical course of hereditary spherocytosis is variable in different individuals, even within the same family. Most will experience a mild anaemia but occasionally there is severe

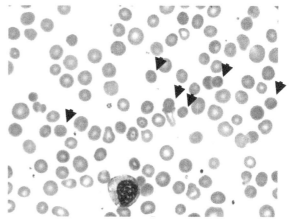

Figure 8.16   Blood film in hereditary spherocytosis, with spherocytes arrowed. From Hoffbrand et al, *Postgraduate Haematology*, Sixth Edition 2010, reproduced by permission of John Wiley & Sons Ltd

anaemia in some individuals, who may be treated with removal of the spleen (splenectomy — this works as removing the spleen removes a substantial source of premature red cell removal from the circulation). Aplastic crises (bone marrow failure, usually transient) can develop after certain infections such as with Parvovirus 19, requiring transfusion and support. Gallstones are a common complication, as in other forms of haemolytic anaemia. Folic acid supplements are usually recommended to allow for increased requirements due to the shortened lifespan and rapid turnover of red cells.

8.4.4.2   Hereditary elliptocytosis and hereditary pyropoikilocytosis

Inherited mutations of the α or β spectrin genes causes a dysfunction in the horizontal connections along the red cell membrane and gives rise to a spectrum of red cell morphological disorders. These range from the mild common hereditary elliptocytosis (HE) to the more distorted red cell forms seen in hereditary pyropoikilocytosis (HPP). Most forms of HE are inherited in an autosomal dominant fashion, while HPP is of autosomal recessive inheritance. HE and HPP are relatively common in malarial regions of the world, due to the relative resistance to malaria conferred by these conditions. Both tend to cause haemolytic anaemia, though HE may be clinically silent in the majority of cases. The blood film appearances vary according to the mutation inherited and typical appearances of HE and HPP are given in Figure 8.17. In HE elliptocytes are

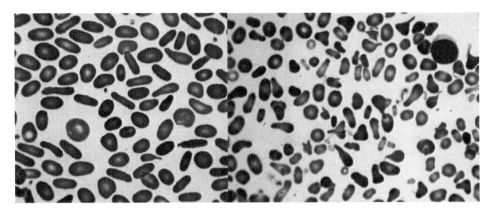

Figure 8.17 Blood film appearances of mild common HE (left) and HPP (right). From Hoffbrand et al, *Postgraduate Haematology*, Sixth Edition 2010, reproduced by permission of John Wiley & Sons Ltd

seen, while in HPP, cells are highly variable in size (anisocytosis) and shape (poikilocytosis).

The clinical picture in HE varies from silent carriers who are asymptomatic through to mildly shortened red cell life span, and to full blown haemolytic anaemia. No treatment is usually necessary for HE. However, in HPP there is moderate to severe haemolytic anaemia in most cases, and folic acid supplements are usually used, with splenectomy where there is severe anaemia.

8.4.4.3 Hereditary stomatocytosis

Stomatocytes are unusual looking red blood cells, named after the slit that appears to look like a mouth ('stoma') that is seen on the red cells in blood films in this condition. This disorder is inherited in an autosomal dominant fashion and is caused by a leak of Na^+ and K^+ ions from the red cell membrane. Most individuals who inherit this disorder have a moderate haemolytic anaemia associated with macrocytosis (large red cells) and there will be an increased serum K^+ level caused by the leaks from red blood cells. The blood film generally shows the presence of stomatocytes, as shown in Figure 8.18. Most cases of hereditary stomatocytosis require no treatment, but in some cases, splenectomy may be recommended.

8.4.5 Red cell metabolism

In order to maintain its shape and perform gas exchange effectively using its full complement of haemoglobin molecules, the red blood cell needs a source of energy, and also reducing power to reduce methaemo-

globin (see Section 8.5.5) back into deoxyhaemoglobin. Since it has no nucleus or mitochondria, the red cell is not capable of protein synthesis or oxidative phosphorylation. It thus relies on anaerobic glycolysis, the hexose monophosphate shunt (also known as the pentose phosphate pathway) and the glutathione cycle for generating ATP and providing reducing power. However, as no protein synthesis is possible in the red cell, no enzymes can be replaced in these metabolic pathways during its lifespan — enzyme activity thus declines over the red cell's life, contributing to its ageing.

8.4.5.1 Glycolysis in the red blood cell: the Embden–Meyerhof pathway

During glycolysis, glucose molecules in the red cell are converted into pyruvate molecules, with two ATP molecules generated for each molecule of glucose

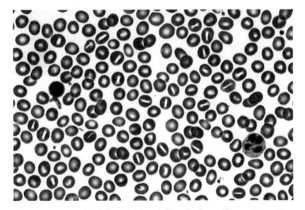

Figure 8.18 Peripheral blood film in hereditary stomatocytosis. From Hoffbrand et al, *Postgraduate Haematology*, Sixth Edition 2010, reproduced by permission of John Wiley & Sons Ltd

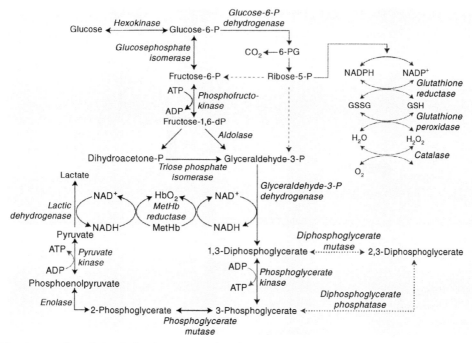

Figure 8.19   An overview of glycolysis, the pentose phosphate pathway and the glutathione cycle in the red cell. From Hoffbrand et al, *Postgraduate Haematology*, Sixth Edition 2010, reproduced by permission of John Wiley & Sons Ltd

metabolized (Figure 8.19). This pathway also serves the dual purpose of providing the reduction reaction to convert methaemoglobin back into deoxyhaemoglobin by using NADH that is produced from NAD^+ during glycolysis.

A sidebranch off glycolysis is the Rapoport—Luebering shunt, which generates 2,3-diphosphoglycerate (2,3-DPG). 2,3-DPG is essential to decrease the affinity of Hb for oxygen, which makes Hb give up oxygen to tissues more readily when 2,3-DPG levels are high (as occurs in red cells during hypoxia or anaemia, making this an important survival mechanism, see also Section 8.5.3).

The glycolytic pathway utilizes enzymes at each stage and mutations in many of these have been described which result in haemolytic anaemia due to failure to generate sufficient ATP (Figure 8.20). The commonest by far is pyruvate kinase deficiency.

Pyruvate kinase deficiency

Pyruvate kinase (PK) catalyses the conversion of pyruvate into phosphoenolpyruvate, one of the last stages of the glycolytic pathway and the step that produces phosphorylation of ADP to ATP. PK deficiency is the most frequent enzyme defect in the glycolytic pathway, and is inherited as an autosomal recessive disorder. However, over 100 different mutations of the PK gene

have been described, leading to huge variability in the clinical effects of inheriting two defective alleles. PK deficiency may give rise to a spectrum of disease, from very mild, symptomless haemolytic anaemia through to severe anaemia and jaundice. Gallstones due to the haemolysis are common in PK deficiency and spleen enlargement may occur. A full blood count shows a normocytic anaemia and reticulocytosis, and enzyme assay of red cell pyruvate kinase activity will show a decrease in the maximal activation of PK from that of normal controls. Treatment consists of removal of the spleen for severe haemolytic anaemia.

8.4.5.2   The pentose phosphate pathway

The pentose phosphate pathway in the red blood cell gives it reducing power via the production of NADPH from $NADP^+$. The first step of this pathway converts glucose-6-phosphate (G6P) into 6-phosphogluconate (6-PG), which is catalysed by the enzyme glucose-6-phosphate dehydrogenase (G6PD). G6P is also a substrate in the glycolytic pathway, and the final products of the pentose phosphate pathway reenter the glycolytic pathway — there is overlap between the two pathways, and the proportions of glucose metabolized

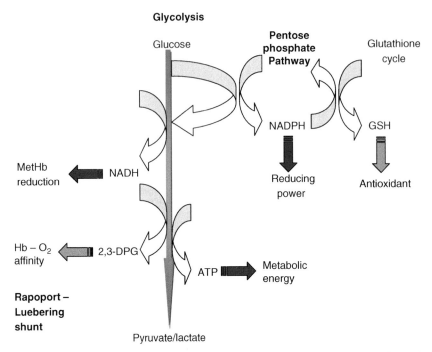

Figure 8.20   Glycolytic pathway showing enzymes. From Hoffbrand et al, *Postgraduate Haematology*, Sixth Edition 2010, reproduced by permission of John Wiley & Sons Ltd

via each pathway vary according to the cell's need for reducing power.

Different steps of the pentose phosphate pathway can be affected by inherited and acquired disorders, but the most important haematological disorder arising from it is glucose-6-phosphate dehydrogenase deficiency.

Glucose-(6)-phosphate   dehydrogenase   deficiency (G6PDD)

G6PDD is the commonest inherited enzyme defect in the world and some 130 variants of the gene have been described. The gene for the red cell isoenzyme of G6P is found on the X chromosome, and thus G6PDD is inherited as an X-linked disorder (however, the other cells of the body will have normal G6PD enzyme). Since the mutated genes for G6PD are common in certain regions of the world, female homozygotes for G6PDD are not uncommon, and female heterozygotes may experience symptoms of G6PDD where X-chromosome inactivation has randomly inactivated relatively more X chromosomes with the normal gene.

In individuals with G6PDD, haemolysis is increased by exposure to oxidative stress, which can be caused by ingestion of oxidative foods or drugs. Haemolytic anaemia can be precipitated by exposure to broad beans ('favism', as broad beans are known as fava beans in some regions of the world) which can produce free oxygen radicals. Antimalarial drugs can also produce a dose-related haemolysis. Neonatal jaundice can also be seen with G6PD.

Investigations in G6PDD show variable levels of anaemia in the full blood count. If G6PDD is suspected, screening tests can be used to distinguish between deficient and nondeficient individuals, for example using a rapid fluorescent spot test which looks for the generation of fluorescent NADPH from $NADP^+$ by G6PD — a deficiency of G6PD is indicated by a lack of fluorescence, though this may not detect heterozygous females. A quantitative assay can then be performed to measure G6PD activity.

Management of G6PDD depends upon the severity of the condition in an individual, though all should be advised to avoid oxidizing agents. Some episodes of severe haemolysis may require blood transfusion for support, and removal of the spleen may be advised for some cases.

8.5   Haemoglobin

Haemoglobin (abbreviated as 'Hb') is the major protein found within red blood cells, and is responsible for

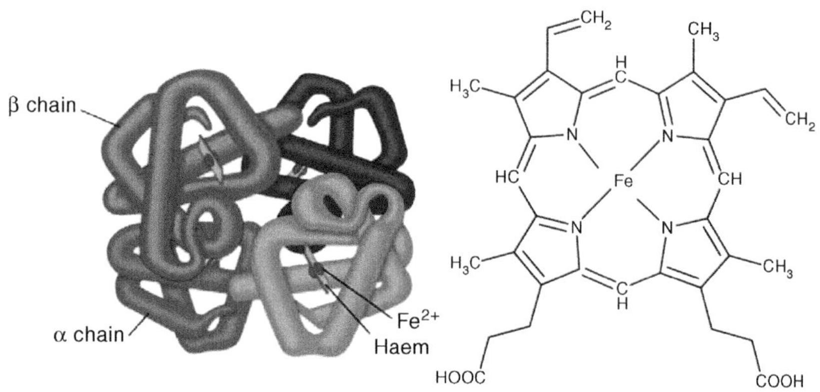

Figure 8.21   Structure of the haemoglobin A molecule (left) and a haem group (right)

carrying oxygen from the lungs to tissues of the body (and to an extent, returning carbon dioxide from tissues to the lungs). Haemoglobin is a 68kD tetrameric molecule, meaning that it consists of four globin subunits. Each globin subunit in a haemoglobin molecule contains a haem group, which has a porphyrin ring with a ferrous atom that is capable of reversibly binding one oxygen molecule.

The majority of haemoglobin in the normal adult is HbA, comprising two α subunits and two β subunits. The globin structure of HbA (and of a haem group) is shown in Figure 8.21.

About 1.5–3% of adult haemoglobin is HbA_2, which comprises two α globin chains and two δ (delta) globin chains. The major haemoglobin expressed by the term baby is HbF, the structure of which contains two α globins and two γ (gamma) globins. Normal adults express <2% HbF.

8.5.1   Haemoglobin synthesis

Haemoglobin synthesis occurs largely within the mitochondria of the 'newborn' red blood cell. The enzyme erythropoietin stimulates conversion of glycine and succinyl CoA into δ-aminolaevulinic acid (ALA) in the rate-limiting step of the reaction. This uses the enzyme ALA synthetase, along with vitamin B6 acting as a cofactor; completed haem molecules inhibit this initial stage of haemoglobin synthesis in classical negative feedback. Two molecules of δ-ALA then form porphobilinogen, which go on to form protoporphrin via uroporphyrinogen and coproporphyrinogen.

Iron atoms are imported into the cell, bound to transferrin receptors on the cell surface. Iron combines with protoporphyrin to form haem, which is incorporated into globin chains. These combine to form haemoglobin. The entire scheme is illustrated in Figure 8.22.

8.5.2   Function of haemoglobin

The function of haemoglobin is to bind oxygen readily — and reversibly — in the capillaries of the lungs (forming oxyhaemoglobin) and carry it to the tissues, where Hb needs to give up oxygen. Some oxygen is also carried in solution in plasma, but the majority is bound to Hb. The percentage of Hb that is saturated with oxygen at any one time depends on the partial pressure of oxygen in the surrounding environment. Haemoglobin arriving in the capillaries of the lungs encounters an environment of high partial pressure of oxygen and picks up oxygen molecules readily. However, in the tissues, oxyhaemoglobin encounters an environment of low partial pressure of oxygen and gives up its oxygen to diffuse into those tissues. Since each Hb molecule is a tetramer with four haem groups, each molecule of Hb can carry four oxygen molecules. Binding of the first molecule of oxygen to an Hb molecule is the most difficult as the Hb molecule undergoes a shape change to accommodate the first oxygen molecule; binding of the second and third molecules of oxygen is then easier, but binding of the final, fourth oxygen molecule is harder again due to 'steric hindrance'; that is, the first three oxygen molecules get in the way! The relationship between the

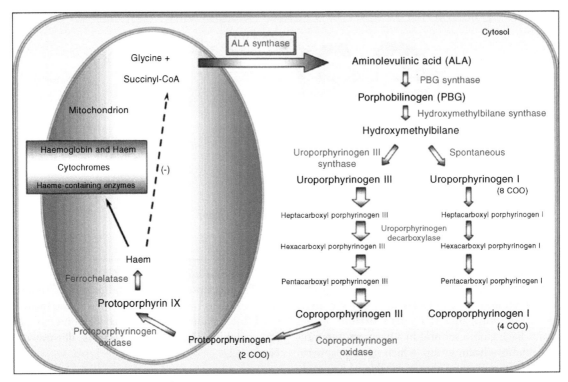

Figure 8.22   Haemoglobin synthesis

partial pressure of oxygen in an environment and the saturation of haemoglobin that results can be plotted, and the oxygen–haemoglobin dissociation curve that results has a sigmoidal shape due to the relative difficulty of binding the first and fourth oxygen molecules to haemoglobin, as compared to the ease of binding the second and third. This is shown in Figure 8.23.

A number of factors can decrease the affinity of haemoglobin for oxygen. In the tissues, high levels of CO_2 and the presence of H^+ ions produce this effect, causing Hb to give up oxygen readily there (this is known as the 'Bohr effect'), and this shifts the oxygen–haemoglobin dissociation curve to the right. The presence of high levels of 2,3-DPG in the red blood cell (see Section 8.4.5.) also decreases the affinity of haemoglobin for oxygen, as does hyperthermia. However, fetal haemoglobin, HbF, has a higher affinity for oxygen than HbA; this enables the fetal circulation to attain higher levels of O_2 from the maternal circulation, and causes the oxygen–haemoglobin curve in the fetus to shift to the left.

Carbon monoxide (CO) has a much higher affinity for haemoglobin than oxygen does and CO binds irreversibly to haemoglobin — once this occurs with high enough levels of CO there is a lack of spaces on haemoglobin molecules to bind oxygen molecules and CO poisoning results, often with fatal consequences.

8.5.3   Carbon dioxide transport in blood

Carbon dioxide is transported from the tissues to the lungs for excretion by exhalation. It is carried in blood in three ways:

1. About 5% of CO_2 is transported dissolved in plasma
2. 10% of CO_2 is carried reversibly bound to the amino groups of haemoglobin (as carbaminohaemoglobin) and to the amino groups of plasma proteins
3. The majority of carbon dioxide is transported inside red blood cells as bicarbonate ions (HCO_3^-), formed by the reversible reaction of CO_2 with water inside the red blood cell. This reaction is catalysed by the enzyme carbonic anhydrase, and is shown here in Figure 8.24.

This reaction does also occur to an extent in plasma, but is very slow due to the lack of carbonic

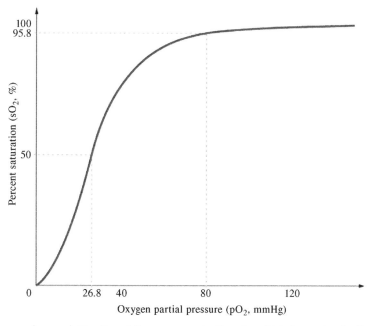

Figure 8.23   The oxygen–haemoglobin dissociation curve: note the sigmoidal shape, due to the relative difficulty of binding the first and fourth oxygen molecules to haemoglobin

anhydrase there. The HCO_3^- ions that are produced within the red cell diffuse out of it, and Cl^- ions diffuse in to balance the charges; this is known as the chloride shift.

Once blood reaches the lungs – an environment of low partial pressure of CO_2 – the reverse reaction occurs, again catalysed by carbonic anhydrase, and CO_2 is reformed within the red blood cell. This diffuses out of the red cells, into the plasma and then out of the alveolar capillaries and into the alveoli of the lung for exhalation. This process is illustrated in Figure 8.25.

8.5.4   Globin genes

The production of the globin chains found in haemoglobin is from genes at two loci. Two genes for α globins are found on each copy of chromosome 16 (giving a total of four α globin genes), along with the ζ (zeta) globin gene that is only expressed in early

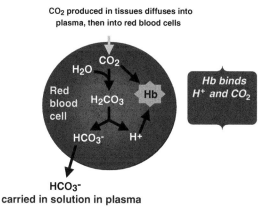

Figure 8.24   Carriage of CO_2 in the blood as bicarbonate ions – reaction in tissue capillaries

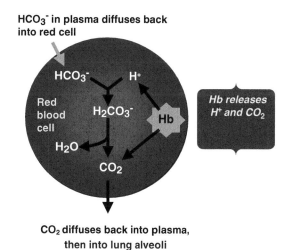

Figure 8.25   Excretion of carbon dioxide – reaction in alveolar capillaries

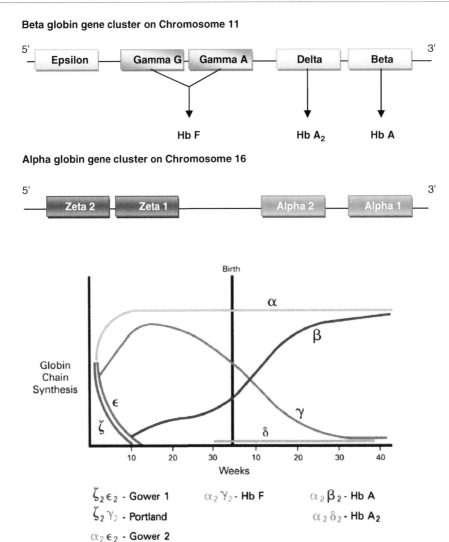

Beta globin gene cluster on Chromosome 11

Alpha globin gene cluster on Chromosome 16

$\zeta_2\epsilon_2$ - Gower 1 $\alpha_2\gamma_2$ - Hb F $\alpha_2\beta_2$ - Hb A

$\zeta_2\gamma_2$ - Portland $\alpha_2\delta_2$ - Hb A$_2$

$\alpha_2\epsilon_2$ - Gower 2

Figure 8.26   Gene loci for globins and haemoglobins produced at different stages of life

embryonic life. On chromosome 11, genes coding for the production ε (epsilon), γ, δ and β globins are present in a gene cluster. The gene loci for globins and the haemoglobins they produce at different life stages are shown in Figure 8.26.

The first globins produced in embryonic life are Hb Gower 1 ($\zeta_2\epsilon_2$), Hb Gower 2 ($\alpha_2\epsilon_2$) and Hb Portland ($\zeta_2\gamma_2$), though these are rapidly superceded by HbF (Box 8.3). HbF ($\alpha_2\gamma_2$) comprises 80% of the haemoglobin of a human baby at term; the switchover to HbA occurs after birth, when β globin chain synthesis increases rapidly due to mechanisms as yet unknown. By the age of 6 months, there is little HbF remaining.

Expression of the δ globin chain commences just before birth and ~2% of total haemoglobin is HbA2 ($\alpha_2\delta_2$).

8.5.5   Haemoglobinopathies

Haemoglobin disorders may be related to:

- abnormal globin chain structure;
- reduced or absent production of globin; or
- a combination of abnormal globin chain structure and reduced production.

Hb Gower (1 and 2) were named by Huehns and colleagues in 1961 after Gower Street, where University College Hospital in London is situated, and where the Hb Gower variants were first described. Hb Portland was named after its discovery at the University of Oregon in Portland, USA.

8.5.5.1   Abnormal globin chain structure

Haemoglobin S (HbS) is the best-known and commonest example of a structurally abnormal haemoglobin molecule. It is caused by a mutation of the β globin gene in which a substituted base results in the insertion of a valine residue at the sixth position in the β globin chain, where a glutamic acid would normally be present. This results in an abnormal haemoglobin molecule, HbS, which polymerizes at low oxygen tension, resulting in a rigid, sickle-shaped erythrocyte with membrane damage from the sharp crystals of polymerized HbS, giving it a shortened lifespan.

The gene for HbS is inherited in an autosomal recessive pattern; in the patient who is homozygous for the HbS gene this gives rise to sickle cell anaemia (SCA, in which the phenotype is HbSS). Those individuals who inherit a single copy of the defective gene are said to have sickle cell trait (HbAS). It is of note that, as the defect affects the β globin chain which is present in HbA, sickle cell anaemia does not become apparent before the age of around 6 months, as HbF ($\alpha_2\gamma_2$) is the prevalent Hb before this age. The gene for HbS is widespread in tropical and subtropical regions of the world, since individuals who are carriers had a good degree of protection against malaria infection, and is most often found in individuals of Black African or Black Caribbean descent. However, individuals with ancestry in the Middle East, Asia and the Mediterranean may also be affected.

8.5.5.2   Clinical features of sickle cell anaemia

Individuals with SCA typically suffer repeated episodes of vaso-occlusive crisis, where large proportions of erythrocytes sickle and occlude small blood vessels, giving rise to the clinical picture of pain in the hands and feet, and in bones. Fever usually occurs with severe pain, and occlusion of small vessels in the spleen gives painful splenomegaly (an enlarged spleen). After repeated episodes, normal spleen is replaced by nonfunctional scar tissue (fibrosis). Massive pooling of cells in the spleen can be fatal, as can bone marrow aplasia, which may be seen following infection in individuals with SCA (usually with parvovirus B19).

8.5.5.3   Laboratory diagnosis of sickle cell anaemia

The only single test that can diagnose a haemoglobin variant with certainty is DNA analysis but a combination of laboratory tests can diagnose the most common Hb variants with confidence. In practice, a combination of full blood count and film, with a solubility test to determine the presence or absence of a sickling molecule, is used alongside high performance liquid chromatography (HPLC) and/or haemoglobin electrophoresis. If doubt still exists as to the identity of a variant haemoglobin, DNA analysis to look at a precise globin gene mutation can then be performed.

8.5.5.4   Full blood count and blood film

Hb levels are typically 6–8 g/dL in individuals with SCA, and reticulocyte count is high (~15%). The blood film will show sickle cells, as in Figure 8.27.

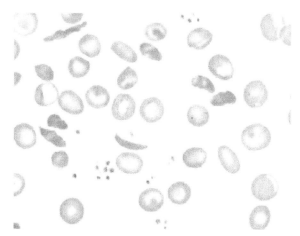

Figure 8.27 Blood film in sickle cell anaemia. From Hoffbrand et al, *Postgraduate Haematology*, Sixth Edition 2010, reproduced by permission of John Wiley & Sons Ltd

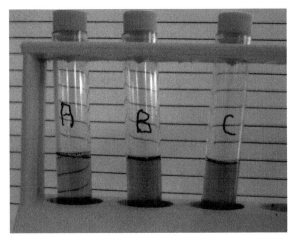

Figure 8.28   Sickle solubility test result — positive result ((b) and (c)) and negative result (a). Lines are not visible through the turbid samples positive for a sickling Hb

8.5.5.5   The sickle solubility test

In the hospital setting, patients admitted in emergency circumstances may need urgent surgery under anaesthesia and the most important piece of information the clinician needs about a potential haemoglobinopathy is whether it will cause a patient's blood to sickle at low oxygen partial pressure (a situation which may occur transiently in a patient under general anaesthesia). Nonsickling variants of haemoglobin are of less interest in the emergency situation and a sickle solubility test can be performed in minutes if required. The sickle solubility test utilizes the relative insolubility of HbS compared to HbA in a concentrated phosphate buffer solution, and the addition of red cells to such a solution produces a cloudy (turbid) appearance if HbS is present, while HbA gives a clear solution (Figure 8.28). The test should be read by holding the tube against a background of white card with black lines.

The sickle solubility test cannot distinguish between individuals with SCA and those with sickle cell trait — both will produce a positive sickle solubility test result — but allows a rapid assessment of whether or not a patient has a sickling Hb variant (Box 8.4).

8.5.5.6   HPLC of Haemoglobins

Automated separation and detection of different haemoglobins using high performance liquid chromatography (HPLC) is becoming increasingly popular in diagnostic laboratories. The principles of HPLC are explained in more detail in Chapter 5, but in essence, HPLC of haemoglobin is separation on the basis of charge difference (as haemoglobin variants with amino acid substitutions often have a net charge difference from HbA) using ion-exchange column matrices. Very little sample is required and resolution takes just a few minutes. Identification of Hb variants is dependent on the time it takes for them to elute from the column, which in turn depends on how strongly they interact with the charged column matrix (Figure 8.29).

8.5.5.7   Hb electrophoresis

Electrophoresis of haemoglobins, like HPLC, is also based on the charge of variant haemoglobin molecules (see Chapter 5 for further details of electrophoresis). Most laboratories use a gel matrix, as these are easier to handle, and stain to see where the haemoglobin molecules have migrated to. Acid or alkali conditions are employed to maximize the difference in charge of variant haemoglobins due to amino acid substitutions; extremes of pH are needed to enhance separation of subtle charge differences in the variant haemoglobin molecules (Figure 8.30).

8.5.5.8   DNA analysis of globin genes

In some situations the usual combination of tests for a haemoglobinopathy may fail to identify the haemoglobin form present; in such situations analysis of the DNA sequence of globin genes may be required. A mutation may be found in any of the major globin genes (α, β, γ and δ), giving rise to variants of the relevant globin chain, and the production of abnormal forms of HbA, HbF or HbA_2. This enables the sequence determined to be compared with a world database of

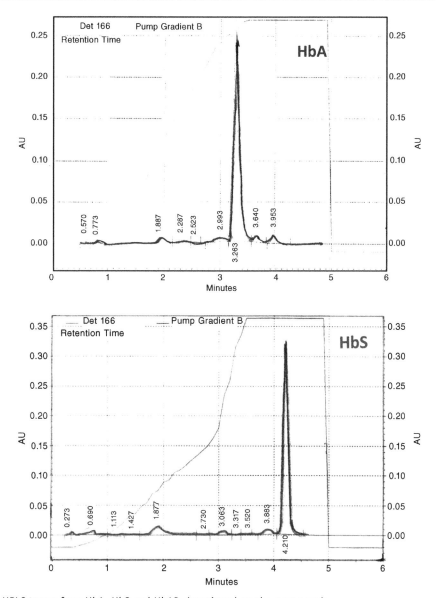

Figure 8.29 HPLC traces from HbA, HbS and HbAS (continued on the next page)

known variant globin genes and gives a definitive diagnosis. However, new globin gene mutations are constantly being published and a novel gene may be identified.

8.5.5.9 Treatment of sickle cell anaemia

As the clinical picture of sickle cell anaemia is highly variable, the treatment is tailored to the individual's pattern of illness. Some people with SCA will suffer frequent painful crises, with hospital admission for pain control and hydration. They may require regular blood transfusions to maintain an acceptable Hb level and along with regular blood transfusions goes the requirement for near daily iron chelation therapy with desferrioxamine to remove the excess iron from frequent transfusion that might otherwise accumulate in tissues and cause harm. Some will require splenectomy (removal of the spleen) to lessen the likelihood of frequent haemolytic episodes. Still others will lead a relatively normal life, with few — if any — crisis episodes.

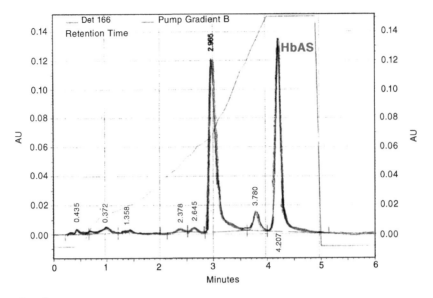

Figure 8.29   Continued

8.5.5.10   Other structural haemoglobin variants

More than 1000 haemoglobin variants have been described at the time of writing. Many are clinically insignificant in that they have no effect on the haemoglobin molecule's function and are 'silent'

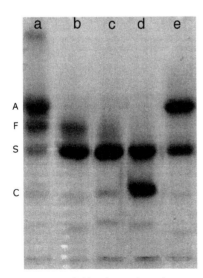

Figure 8.30   Haemoglobin electrophoresis — standards in lane A, HbSS in lanes B and C, HbSC in lane D and HbAS in lane E. From Hoffbrand et al, *Postgraduate Haematology*, Sixth Edition 2010, reproduced by permission of John Wiley & Sons Ltd

variants which are incidentally discovered, perhaps when a patient is screened for haemoglobinopathies preoperatively or during antenatal care. HbC is much like HbS in terms of the population it is found in, its inheritance and effect on the Hb molecule, and is produced when a lysine residue is substituted for a glutamic acid residue in the β globin chain (rather than valine in HbS). The coinheritance of HbC along with HbS (HbSC disease) gives rise to a clinical picture much like that of SCA, albeit a somewhat milder version in many individuals. HbE is a further variant of Hb that is common in Asia, and caused by a mutation in the β globin gene that causes lysine to be substituted for a glutamic acid residue in the β globin chain. HbE trait (HbAE) is generally asymptomatic, while HbE disease (HbEE) usually gives rise to a mild haemolytic anaemia, but the combination of an HbE gene with a β^0 thalassaemia gene can produce very severe anaemia with serious morbidity and mortality. Hb Lepore is caused by a fusion of the δ and β globin genes, giving rise to an Hb molecule with normal α globin chains coupled to δ-β fusion chains. Hb Lepore trait is asymptomatic and Hb Lepore homozygotes are fortunately rare, but the combination of an Hb Lepore gene with a β thalassaemia gene gives rise to anaemia of variable severity where few/no normal β globin chains can be produced.

8.5.5.11 Methaemoglobins

Methaemoglobin is haemoglobin in which the normal ferrous (Fe^{2+}) ion of the haem group is oxidized to the ferric form (Fe^{3+}) and rendered unable to bind oxygen. Normal individuals should have <1% of all haemoglobin in the methaemoglobin form, as the NADH and NADPH methaemoglobin reductase pathways are protective against oxidation of Fe^{2+}. However, larger proportions of methaemoglobin are seen in congenital and acquired methaemoglobinaemia.

Congenital methaemoglobinaemia may be due to a hereditary defect in the protective enzyme pathway, an autosomal recessive condition in which two defective copies of the gene for the enzyme diaphorase I (NADH methaemoglobin reductase) are inherited, compromising the major protective system against oxidation of the ferrous atoms in haemoglobin. It may also be seen when a variant haemoglobin molecule is inherited in which an amino acid substitution on either the α or β globin chain alters the haem—globin bond in a way that makes the molecule more stable in the oxidized form. Several variants of such 'haemoglobin M' molecules have been described, including Haemoglobin M Saskatoon and Haemoglobin M Hyde Park, and most are inherited in an autosomal dominant pattern.

Methaemoglobinaemia is, however, more commonly acquired by exposure to oxidizing agents in drugs, which can swamp the protective enzyme systems and cause an acute rise in methaemoglobin levels. Drugs that can trigger this include antibiotics (e.g. dapsone, trimethoprim), local anaesthetics, metaclopramide and nitrate drugs. Infants under 6 months of age are more prone to acquired methaemoglobinaemia as HbF is more readily oxidized than HbA, and levels of the protective NADH reductase are lower in the early months of life.

Affected individuals have methaemoglobin levels >1% and are usually asymptomatic until levels are above 10—20%. Above such levels, symptoms of hypoxia (low oxygen levels) are seen, that is shortness of breath, cyanosis (blue-tinged skin), dizziness and poor exercise tolerance (Box 8.5). Once methaemoglobin levels reach >50%, arrhthymias, unconsciousness and ultimately coma and death may ensue.

Diagnosis of methaemoglobinaemia is usually based on cooximetry, a form of spectrophotometry which can identify the different signature wavelengths of light absorption by normal oxyhaemoglobin, deox-

Box 8.5 The blue men of Lurgan

The 'Blue Men of Lurgan' were described in the County Armagh town of Lurgan, Northern Ireland, by local General Practitioner Dr James Deeny in 1942, who diagnosed 'familial idiopathic methaemoglobinaemia', and made the discovery that the reducing agent ascorbic acid (vitamin C) was an effective treatment for methaemoglobinaemia.

yhaemoglobin, carboxyhaemoglobin and methaemoglobin. The variant haemoglobin M molecules can often be identified using haemoglobin electrophoresis, but DNA sequencing of the globin chain genes may be necessary to identify them. Enzyme assays will allow diagnosis of methaemoglobinaemia due to defects in the protective pathway enzymes. Methaemoglobinaemia is treated with supplementary oxygen and by administering methylene blue, which accepts electrons from NADPH methaemoglobin reductase and increases its activity.

8.5.5.12 Unstable haemoglobins

More than 140 variants of haemoglobin have now been reported in which a mutation in a globin gene produces a haemoglobin molecule that is relatively insoluble and comes out of solution readily, with Heinz bodies (clusters of denatured Hb) seen in the red cells. Inheritance is usually autosomal dominant and haemolytic anaemia or altered affinity for oxygen results in some (but not all) cases. Hb Köln is the commonest unstable haemoglobin encountered and often arises as a new mutation in a family with no history of the disorder.

8.5.5.13 Reduced or absent production of haemoglobin: the thalassaemias

In contrast to the structural haemoglobin variants, where an abnormal globin molecule is generally produced at a normal rate, the thalassaemias are caused by a decreased or absent production of a normal globin molecule. This is usually due to gene deletion, or mutations to the gene or its regulatory elements that caused reduced production of a globin. The thalassaemias are classified according to the affected globin — α thalassaemia or β thalassaemia — and also according to

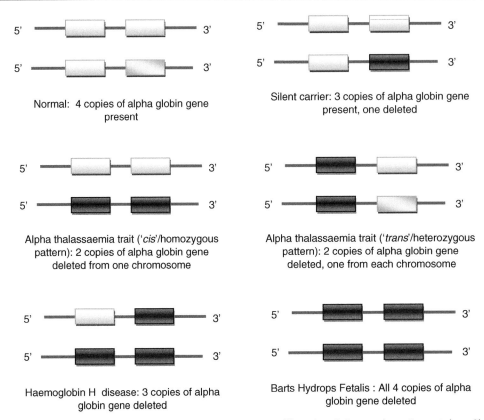

Figure 8.31   Possible gene deletions in alpha thalassaemia. From Hoffbrand et al, *Postgraduate Haematology*, Sixth Edition 2010, reproduced by permission of John Wiley & Sons Ltd

clinical picture. Patients may thus have thalassaemia minor, major or intermedia.

Alpha thalassaemia

Alpha thalassaemia is due to a reduced or absent production of α globin molecules. Since there are four α globin genes, it is possible to have one, two, three or four α globin genes deleted or mutated, as illustrated in Figure 8.31.

Defect of one α globin gene; deletion or mutation of one of the four α globin genes is clinically silent, with no apparent effect on full blood count, or blood film. It is usually only discovered when a family is screened for alpha thalassaemia after the diagnosis of a more serious form of the condition in other family members.

Defects of two α globin genes – α thalassaemia trait: deletion or mutation of two α globin genes gives rise to alpha thalassaemia trait. Alpha thalassaemia trait can be inherited in one of two ways: patients may have two missing genes on one allele of chromosome 16, or one missing gene from each copy of chromosome 16. These patterns are found in different ethnic groups; people of

Southeast Asian origin with α thalassaemia trait tend to have the homozygous ('*cis*') pattern of gene deletions, while people of Black ethnic origin with α thalassaemia trait tend to inherit the heterozygous ('*trans*') pattern of gene deletion. This pattern of gene inheritance has very important implications for the children of parents with α thalassaemia trait. The combination of a '*cis*' chromosome 16 with two missing α globin genes with a second chromosome 16 with one deleted gene will give rise to the severe haemoglobin H disease, while the combination of two '*cis*' chromosomes 16 would give rise to the fatal condition Bart's Hydrops fetalis, where an infant has no functional α globin genes and thus no functional Hb molecules. This is not possible in populations where α thalassaemia trait is caused by the '*trans*' pattern of gene deletion across both copies of chromosome 16. This is illustrated in Figure 8.32.

Haemoglobin H disease

In this variant of alpha thalassaemia, individuals inherit three defective α globin genes; two gene deletions are seen on one allele of chromosome 16 and one gene

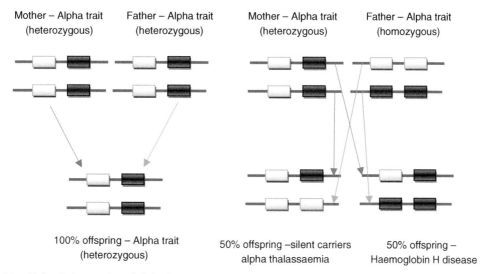

Figure 8.32 Alpha thalassaemia trait inheritance patterns. From Hoffbrand et al, *Postgraduate Haematology*, Sixth Edition 2010, reproduced by permission of John Wiley & Sons Ltd

deletion on the second allele, as illustrated in Figure 8.33.

Individuals with haemoglobin H disease are affected by a chronic haemolytic anaemia, though the degree of anaemia is variable. They may be transfusion-dependent and will require iron chelation therapy to counteract the effects of iron loading from multiple transfusions if so. Infections may trigger bouts of increased haemolysis. The lack of α globin causes a gross imbalance between α and β globin such that tetramers of β globins form into β_4 'haemoglobin H' molecules; this comprises $\sim$20% of total Hb in people with haemoglobin H disease. The β_4 HbH molecule has very high affinity for oxygen and is unaffected by the Bohr effect such that it does not relinquish O_2 to the tissues; this causes a worse O_2 carrying capacity than would be expected from overall Hb levels. It is an unstable molecule which precipitates within erythrocytes, giving rise to the appearance of Heinz bodies in a blood film (as shown in Figure 8.33).

Barts Hydrops fetalis

Barts Hydrops fetalis (named after the London teaching hospital, St Bartholomew's, where the associated haemoglobin was first described) is the variant of α thalassaemia with all four genes deleted — individuals with this condition inherit two copies of chromosome 16 with the '*cis*' double gene deletion, meaning that they have no functional α globin genes. Haemoglobin Barts, comprising four γ globin chains, γ_4, is expressed in the fetus, a molecule which has such high O_2 affinity

that it does not relinquish it to tissues and is thus effectively nonfunctional. This condition is incompatible with life and infants with it are grossly oedematous and are either stillborn or will die shortly after birth.

8.5.5.14 Combinations of haemoglobinopathies

Contrary to what might instinctively be expected, inheriting combinations of haemoglobinopathies often gives rise to an improved clinical picture! Coinheritance of genes for α and β thalassaemia, for example, can bring the production of the α and β globin chains back into balance with less serious clinical effect than would be expected from either alone.

8.6 Anaemia

Anaemia is defined as a haemoglobin level below the normal range for age and gender. It is not a condition in its own right, but rather is a symptom of another problem, and haematological indices can give important clues as to the underlying cause. Apparent anaemia can also be seen due to changes in plasma volume — an increase in plasma volume will result in greater dilution of Hb, as occurs during pregnancy. Anaemia is classified according to mean cell volume (MCV) and divided into microcytic anaemia (small cells — low MCV), normocytic anaemia (cells of normal size, with a normal MCV) and macrocytic anaemia (large cells — high MCV). Patients with anaemia may be asymptomatic,

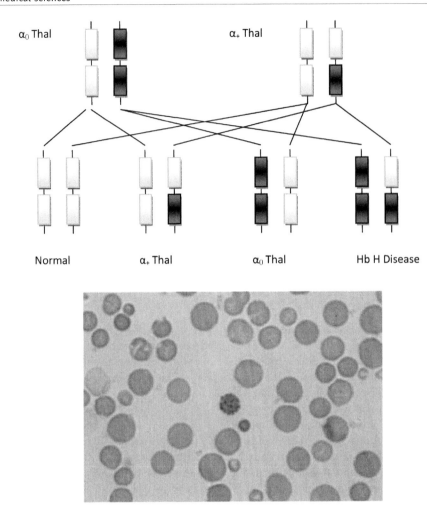

Figure 8.33   Haemoglobin H disease genes and blood film showing Heinz bodies

particularly where the fall in Hb has been gradual. In patients with symptoms of anaemia, tiredness, faintness and shortness of breath are common, while some may additionally experience angina and palpitations (the latter particularly in the elderly). They will often look pale, with a rapid heart rate and sometimes with signs of heart failure too.

Investigations for suspected anaemia include a full blood count to assess Hb levels alongside red cell count, white cell count, platelet count and reticulocyte count. A blood film to look at red cell shape should usually be performed too. Some patients will need a bone marrow biopsy to supplement findings from the FBC and blood film; this reveals details of how cellular the marrow is, whether erythroblasts are normoblastic (normal-sized blasts) or megaloblastic (large blasts). Further investigations should be based on findings in

the FBC and blood film — for example, iron store assessment where iron deficiency anaemia is a possible diagnosis (Table 8.1).

8.6.1   Microcytic anaemia

Iron deficiency is the commonest cause of microcytic anaemia, though thalassaemia (see Section 8.5.5), sideroblastic anaemia and the anaemia of chronic disease are also relatively common possible causes.

8.6.1.1   Iron

Humans have a number of mechanisms to absorb iron and regulate this, but no physiological mechanism to excrete iron from the body. Most iron in the

Table 8.1 Common causes of microcytic, normocytic and macrocytic anaemia

	MCV	Causes
Microcytic anaemia	< 80 fL	Iron deficiency
		Thalassaemia
		Sideroblastic anaemia
		Chronic disease
Normocytic anaemia	Normal	Acute haemorrhage
		Haemolytic anaemia
		Chronic disease
		Bone marrow infiltration
		Renal failure
Macrocytic anaemia	> 96 fL	Vitamin B12/folate deficiency (megaloblasts in marrow)
		Alcoholism
		Drugs e.g. methotrexate, trimethoprim
		Liver disease
		Hypothyroidism

body is found in haemoglobin ($\sim$3 g), while the rest is found in macrophages of the reticuloendothelial system, skeletal muscle and hepatocytes ($\sim$1 g), mostly in the storage form of iron, ferritin, a soluble complex of iron and protein, and haemosiderin, an insoluble iron—protein complex found particularly in macrophages.

The average diet contains around 20 mg of iron per day but under normal circumstances only around 10% of this is absorbed. Each day, around 1 mg of iron is lost in urine, faeces and sweat, though in menstruating women, the average daily loss of iron is around 1.7 mg, though it can be considerably higher. Haem iron — the iron found in red meat — is the best absorbed form of iron, but most dietary iron is in the nonhaem form found in vegetables and grains. Ferrous iron (Fe^{2+}) is also better absorbed than ferric iron (Fe^{3+}) and gastric acid helps to keep iron in the ferrous format. Iron absorption from food can be increased up to around 20—30% in iron deficiency, and by the action of erythropoietin (e.g. at high altitudes, or after bleeding or haemolysis), but decreased where there is iron overload. Nonhaem iron is absorbed in the small intestine after gastric acid in the stomach — and the enzyme ferriductase in the duodenum — have reduced iron from the ferric to the ferrous form. Haem iron can

be absorbed by intestinal haem transporters; some passes intact into the bloodstream. Iron absorption is regulated by the cells of duodenal villi, which absorb iron in proportion to recent dietary iron intake, to body iron stores and according to bone marrow erythropoiesis via different pathways. Iron that has been absorbed by duodenal epithelial cells may be stored as ferritin (and later shed with the cell into the gut at the end of its lifespan, thus regulating iron balance), or passed across the cell into the plasma via the iron transporter protein ferroportin. The activity of ferroportin can be reduced by hepcidin, a peptide made in the liver that binds ferroportin and triggers its destruction, thus preventing the movement of iron into the plasma and its loss when the duodenal epithelial cells are shed.

Once in plasma, iron is transported bound to transferrin. Each transferrin molecule binds two iron atoms and at the bone marrow this binds to receptors on reticulocytes and erythroblasts to deliver iron to them. Normal serum iron range is 13—32 μmol/L.

8.6.1.2 Defective iron absorption — hereditary haemochromatosis

Hereditary haemochromatosis is an inherited disorder of iron absorption that results in excessive iron deposition in various tissues of the body. It is inherited as an autosomal recessive disorder, and the gene is common in Caucasian populations where around one in 10 people are carriers. The mechanism for excessive iron absorption in the intestine is unclear, but may result from absence of hepcidin — or defective forms — resulting in a lack of regulation of dietary iron absorption. Iron levels exceed the plasma carriage capacity of transferrin and free iron is taken up by most tissues, including the liver, pancreas, heart and skin. The iron causes fibrosis (the production of scar tissue), with ensuing liver cirrhosis. People with hereditary haemochromatosis develop bronzed skin, diabetes, heart arrhythmias and an enlarged liver, typically around the age of 45. Some may develop hepatocellular carcinoma after liver cirrhosis. Investigations show elevated serum iron and ferritin and genetic testing is performed to test for the gene for hereditary haemochromatosis. Treatment is with venesection — 500 mL of blood is taken up to twice a week initially, then three or four times per year until serum ferritin is within the normal range. Chelation therapy with the iron-binding desferrioxamine may be used in some patients who cannot tolerate venesection. Desferrioxamine binds free iron in plasma, thus enhancing its excretion in urine.

8.6.1.3 Iron deficiency

Iron deficiency is very common worldwide and is caused by a lack of sufficient iron for haemoglobin synthesis. Normal haemoglobin levels are maintained initially while the body's iron stores are depleted; once this occurs Hb levels begin to fall. Iron deficiency can result from inadequate iron in the diet, increased iron demands (e.g. during pregnancy), decreased iron absorption (e.g. coeliac disease) or blood loss, and the latter is the most common cause of iron deficiency. Menstruating women are particularly prone to iron deficiency but blood loss can also be from the gastrointestinal tract (e.g. where there are gastric or duodenal ulcers present, or bleeding tumours in any part of the gastrointestinal tract, or from hookworm infestation in developing countries).

Once haemoglobin levels begin to fall after iron stores have been depleted, iron deficiency anaemia is said to be present. Patients with this condition may experience the classical symptoms of anaemia — fatigue, shortness of breath and palpitations on exertion — as well as the effects of tissue iron deficiency such as brittle nails and hair and altered shape of nails. A full blood count will show a lowered Hb with microcytic, hypochromic red cells of varied sizes (anisocytosis) and shapes (poikilocytosis, see Figure 8.34). Serum iron levels and serum ferritin will be low, with increased iron-binding capacity (i.e. transferrin levels).

The treatment of iron deficiency anaemia relies on identifying the underlying cause and treating that (e.g. treating heavy periods, gastric ulcers or gastrointestinal tumours), and give oral iron supplementation with a preparation such as ferrous sulphate.

8.6.2 Normocytic anaemia

Acute haemorrhage, chronic haemolysis, chronic disease (such as rheumatoid arthritis, cancer or renal failure) and bone marrow infiltration can all cause a normocytic anaemia, in which there is a decreased haemoglobin level with a normal MCV. Treatment of the cause where possible will treat the anaemia, and treatment with erythropoietin where the cause cannot be treated may also be suggested.

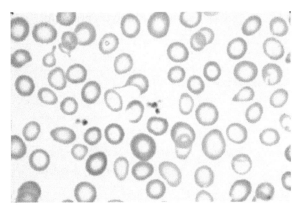

Figure 8.34 Peripheral blood film in iron deficiency anaemia. From Hoffbrand et al, *Postgraduate Haematology*, Sixth Edition 2010, reproduced by permission of John Wiley & Sons Ltd

8.6.3 Haemolytic anaemia

The normal lifespan of the red cell is around 4 months (approximately 120 days). In haemolytic anaemia, red cells are destroyed at an unusually high rate by macrophages of the spleen, bone marrow and liver, and survival times for red cells fall. Red cell production will increase to compensate and some circumstances this may suffice to prevent anaemia developing. However, once red cell survival is down to around 15 days the bone marrow will no longer be able to compensate, and anaemia will ensue.

Investigations in haemolytic anaemia will show a decrease in Hb level, with increase in reticulocytes and a raised serum bilirubin level (indicating breakdown of red cells is occurring), along with an increase in urobilinogen and urinary haemosiderin. A blood film will give further clues as to the likely cause: abnormal red cells may be seen in sickle cell disease or thalassamia (Section 8.5.5), Heinz bodies in red cell enzyme defects (Section 8.4.5) or spherocytes in autoimmune haemolytic disease, malaria or hereditary spherocytosis (Section 8.4.4).

8.6.3.1 Acquired causes of haemolytic anaemia

The cause of destruction of red cells can be divided into immune destruction, nonimmune destruction and other causes.

1. Immune destruction of red cells may be due to the presence of autoantibodies, alloantibodies (antibodies from another individual), or drug-induced antibodies in the circulation. The presence of

antibodies coating red cells gives a positive direct antiglobulin (Coombs) test (see Section 8.13.6). The antibodies that cause this autoimmune haemolytic anaemia (AIHA) can be classified into 'warm' and 'cold' antibodies based on whether they attach best to red cells at 37°C (warm) or <37°C (cold). In warm AIHA, the antibodies are usually IgG, whilst cold AIHA is usually caused by IgM.

The presence of autoantibodies in AIHA which attach to the red cells triggers the complement cascade and then intravascular haemolysis ensues. Some IgG antibodies do not activate complement, and these are removed in the spleen via extravascular haemolysis. 'Warm' autoimmune haemolytic anaemia occurs most frequently in middle-aged females, usually as an episode of anaemia with jaundice that may remit and relapse. Investigations reveal haemolytic anaemia, spherocytosis and a positive direct antiglobulin test with autoantibodies against red cells present. Around a third of cases have no identifiable underlying cause, whilst others are associated with rheumatoid arthritis, drugs or cancer of the lymphoid system. Treatment with steroids or other immunosuppressive drugs is used, possibly with blood transfusion for severe anaemia.

'Cold' autoimmune haemolytic anaemia may be seen after infections such as cytomegalovirus or the Epstein–Barr virus, or in the elderly. Investigations show red cell agglutination at room temperature or cooler, as well as a positive direct antiglobulin test and the presence of monoclonal antibodies against the Ii blood group system. Patients should avoid exposure to the cold and blood transfusion may be required for severe anaemia.

Alloimmune haemolytic anaemia is seen where antibodies from one individual react against red cell antigens on another individual, for example in haemolytic transfusion reactions or haemolytic disease of the newborn (see Section 8.13.5).

2. Nonimmune destruction of red cells may be due to red cell membrane defects (e.g. hereditary spherocytosis), mechanical red cell destruction (e.g. due to a prosthetic heart valve) or due to chronic disease, such as liver or kidney disease.

3. Other causes for acquired haemolytic anaemia include malaria, extensive burns, drugs and chemicals that can induce oxidative haemolysis (e.g. dapsone and certain weedkillers) and toxins such as arsenic and hypersplenism.

8.6.4 Macrocytic anaemia

Macrocytic anaemia, in which the MCV is >96 fL, can be subcategorized according to the appearance of bone marrow into megaloblastic and nonmegaloblastic macrocytic anaemia. Megaloblastic anaemia is defined by the presence in bone marrow of large erythroblasts with immature nuclei (megaloblasts). The most important causes of megaloblastic macrocytic anaemia are vitamin B12 deficiency and folate deficiency. Nonmegaloblastic macrocytic anaemia, where normal erythroblasts are seen in bone marrow, is seen in normal pregnancy, but also in alcoholism, liver disease, hypothyroidism, aplastic anaemia and following treatment with chemotherapy drugs.

8.6.5 Megaloblastic macrocytic anaemia

8.6.5.1 Vitamin B12 and haemoglobin

Humans are dependent on animal sources for vitamin B12, which is found in eggs, milk, meat and fish (and not in vegetables). The average diet contains 5–30 μg per day of vitamin B12, and around 2 μg is absorbed. Body stores — found mainly in the liver — of around 3 mg of vitamin B12 will last for approximately 2 years if absorption failure occurs, after which deficiency will develop.

The B12 molecule is a cobalamin, consisting of planar (flat) group made up of four pyrrole rings with a central cobalt atom bound to them, and a ribonucleotide set at 90° to this (Figure 8.35).

Vitamin B12 is necessary for the metabolism of almost every cell in the body. It is required for the successful methylation of homocysteine to convert it to methionine during DNA synthesis, and as a coenzyme for conversion of methylmalonyl CoA to succinyl CoA, a key molecule of the citric acid cycle. After ingestion in food, the action of gastric enzymes releases it from proteins and it is bound to an 'R' binder protein. Pancreatic enzymes later separate the two and at this point B12 is bound to intrinsic factor in the small intestine. Intrinsic factor is a glycoprotein excreted by parietal cells of the stomach and binds B12 to carry it to receptors on the mucosa in the ileum — B12 is transported into ileal cells while intrinsic factor stays behind. B12 is then taken to the bone marrow via the bloodstream on the transport protein transcobalamin

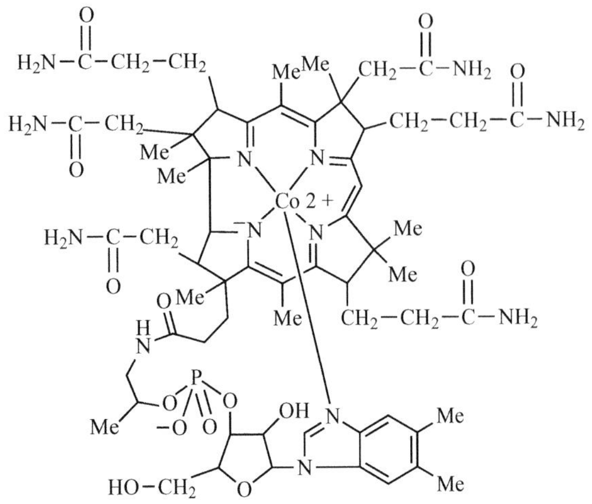

Figure 8.35   Structure of vitamin B12

II (TCII), though most B12 is bound to transcobalamin I, of which the function is unknown.

8.6.5.2   Vitamin B12 deficiency

Deficiency of vitamin B12 is most commonly due to pernicious anaemia (see Section 8.6.5), but may also be seen where dietary intake is poor (e.g. in vegans), with congenital deficiency of transcobalamin II or intrinsic factor, following removal of the stomach (gastrectomy) or small intestine and due to infection with fish tapeworm.

8.6.5.3   Pernicious anaemia

Pernicious anaemia is an autoimmune condition in which the gastric parietal cells are destroyed, leading to a loss of intrinsic factor and consequent vitamin B12 malabsorption. It is relatively common in the elderly, particularly in females, people of blood group A and those with fair hair and pale eye colour. More than 90% will have antibodies against parietal cells in serum and 50% will have antibodies against intrinsic factor. The onset of pernicious anaemia is gradual, with progressive symptoms of anaemia and often a mild jaundice due to breakdown of Hb. Neurological problems also occur with B12 deficiency, causing parasthesia (tingling) in fingers and toes, weakness and loss of vibration and spatial sense, which can develop into paraplegia, dementia and psychiatric problems.

A full blood count will show a decreased Hb with a raised MCV — red cells are large (megaloblastic) and hypersegmented neutrophils are seen on a blood film.

Serum bilirubin is often raised, and serum B12 levels are well below the normal range (below 160 ng/L).

The treatment of pernicious anaemia is usually with regular intramuscular injections of vitamin B12 (as hydroxycobalamin) over the rest of the patient's life, though oral B12 is now being used since 2% of the dose is absorbed by diffusion into the blood without intrinsic factor.

8.6.5.4   Folate

Folate (folic acid) is present in green vegetables and in offal (liver and kidney). The average diet contains around 400 µg of folate per day, though cooking markedly reduces the amount present.

The folate molecule comprises a pteridine nucleus, p-aminobenzoic acid and glutamic acid. Folate is inactive until it is reduced in the body to dihydrofolic acid and then to tetrahydrofolate (THF) (Figure 8.36). These two latter forms are coenzymes used in amino acid metabolism and in the synthesis of purines and pyrimidines for DNA and RNA synthesis.

8.6.5.5   Folate deficiency

The body's folate stores are relatively low (~10 mg) and folate deficiency can develop after about 4 months of a folic acid deficient diet; this may be more rapid when folate utilization by the body is high. The major cause of folate deficiency is nutritional due to poor intake and absorption (e.g. in starvation, following gastrectomy, coeliac disease, alcoholism), though anti-folate drugs such as phenytoin, trimethoprim or methotrexate may also be a cause. Increased utilization of

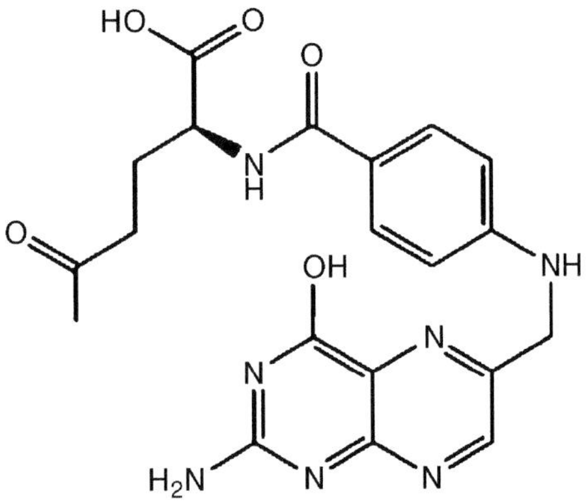

Figure 8.36   Structure of the folic acid molecule

folate during pregnancy and lactation may result in its deficiency, as can haemolysis, malignancy and inflammatory diseases. During pregnancy, folate deficiency is known to be associated with neural tube defects such as spina bifida in the fetus, though the mechanism for this is not entirely clear. Folic acid supplementation for 3 months prior to conception, and during the first 12 weeks of pregnancy, is recommended to decrease the risk of this occurring.

In folate deficiency symptoms of anaemia may be present, or patients may be asymptomatic. Full blood count and blood film features are very similar to those for vitamin B12 deficiency; a megaloblastic anaemia is found and levels of serum folate are low (normal range 5–63 nmol/L), as are levels and red cell folate (normal range 160–640 µg/L; this is a more sensitive indicator of tissue folate levels).

Folate deficiency is treated by treating the underlying cause (e.g. coeliac disease) if necessary, and folic acid supplements are given orally for at least 4 months to replace body stores. Folic acid supplementation is also used in haematological disease with chronic haemolysis to prevent deficiency.

8.6.6   Aplastic anaemia

Aplastic anaemia is a deficiency of all of the cellular components of blood and is due to failure of the bone marrow. It is, fortunately, rare, and may be congenital or acquired during life. There are many different causes of aplastic anaemia. Primary aplastic anaemia may be genetic (e.g. Fanconi anaemia, an autosomal recessive condition) or idiopathic (in which the cause cannot be determined, although most cases are probably autoimmune in origin). Secondary aplastic anaemia may follow exposure to drugs (e.g. chemotherapy, phenytoin, chloramphenicol), chemicals (e.g. benzene), insect killers, radiation and infections (e.g. Epstein–Barr virus, hepatitis, HIV). Patients with aplastic anaemia will experience symptoms of anaemia, leucopenia and thrombocytopenia, such as pallor, shortness of breath, infections, bleeding and bruising. A full blood count will show a decrease in red cell count, white cell count and platelet count (pancytopenia) with absence of reticulocytes; a bone marrow biopsy will show a bone marrow with a greatly decreased number of cells and large fat spaces

Aplastic anaemia is treated by removing exposure to any causative agent where relevant, and providing red cells and platelet transfusions to support the patient while the bone marrow recovers. Antibiotic cover is used to protect against infection. Some patients will require a bone marrow transplant, particularly for the congenital forms of aplastic anaemia.

8.7   Benign white blood cell disorders

In a number of pathological conditions, white cells display alterations in their number, shape or function and provide important diagnostic information that will help identify the underlying cause.

8.7.1   Leucopenia

Leucopenia refers to a reduction in the total number of white blood cells found in the circulation and a differential white cell count will show whether this is a generalized leucopenia (which may be part of a pancytopenia, affecting red blood cells and platelets too) or a selective leucopenia predominantly affecting one type of white blood cell.

8.7.2   Neutropenia

A selective fall in the neutrophil count below the normal range is termed neutropenia and may be caused by a wide range of conditions, particularly the use of certain drugs. Once the neutrophils count is $<0.5 \times 10^9$/L, there is a serious risk of infection. Table 8.2 gives some of the common causes of neutropenia.

8.7.3   Lymphopenia (lymphocytopenia)

A selective reduction in lymphocytes below the normal range is called lymphopenia and may be found following trauma, surgery, in people taking corticosteroid drugs, in some acute infections and after chemotherapy or radiotherapy. It is also sometimes seen in individuals with HIV infection, sarcoidosis, Hodgkin's disease and systemic lupus erythematosus (SLE).

Table 8.2   Common causes of neutropenia

Physiological	Neutrophil counts lower in Black populations
Infections	Overwhelming bacterial infection
	Some viral, fungal and protozoal infections
Drugs	Antibiotics: chloramphenicol, cotrimoxazole
	Antiinflammatories: indomethacin, phenylbutazone
	Some anticonvulsant drugs, oral hypoglycaemic drugs, antithyroid drugs, antimalarial drugs, antihistamines, antidepressants
Immune	Autoimmune neutropenia, systemic lupus erythematosus (SLE)
Other causes	Familial benign chronic neutropenia, hypothyroidism, hypopituitarism

Table 8.3   Some causes of neutrophilia

Physiological	Neonates, pregnancy, lactation, exercise
Infections	Acute infections with pyogenic bacteria
Inflammation	Burns, surgery, myocardial infarction, rheumatoid arthritis
Metabolic	Diabetic ketoacidosis, gout, thyrotoxicosis
Drugs	Adrenalin (epinephrine), steroids
Nonhaematological malignancies	Lymphoma, malignant melanoma, carcinoma
Myeloproliferative disorders	Myelofibrosis, polycythaemia, leukaemia
Other causes	Electric shock, convulsions, following removal of the spleen

8.7.4 Leucocytosis

Leucocytosis is an increase in the absolute white cell count. A differential white cell count will reveal which type of white cell predominates.

8.7.5 Neutrophilia

A selective increase in the neutrophils count above the normal range is called neutrophilia, and this is the commonest abnormality of white cell count (Figure 8.37). The most important cause of neutrophilia is bacterial infection and neutrophils may show additional features on a blood film such as toxic granulation (abnormal, coarse reddish granules) and the presence of Döhle bodies (small, pale blue peripheral cytoplasmic inclusions). These features indicate accelerated neutrophils production in the bone marrow. Some of the causes of neutrophilia are given in Table 8.3.

8.7.6 Eosinophilia

An increase in eosinophil count above the normal range is called eosinophilia, and is usually due to allergy in developed countries from conditions such as asthma, hay fever and drug sensitivity (Box 8.6 and Figure 8.38). It is sometimes found in people with eczema, psoriasis and with malignancies such as lymphoma and metastatic carcinoma. However, in other regions of the world eosinophilia is most often due to parasitic infections such as hookworm, filariasis, hydatid cysts and strongyloidiasis.

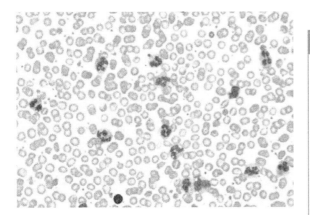

Figure 8.37   Peripheral blood film showing neutrophilia

Box 8.6   Spanish Toxic Oil Syndrome

In 1981, some 20 000 people in Spain became acutely unwell with intense muscle pain and investigations revealed that they had a marked eosinophilia too. This was eventually discovered to be due to ingestion of a toxic oil that had been fraudulently marketed as olive oil; it contained fatty acid esters of 3-(N-phenylamino)-1,2-propanediol (PAP). Some 300 people died and many have since developed chronic tissue and neurological disorders.

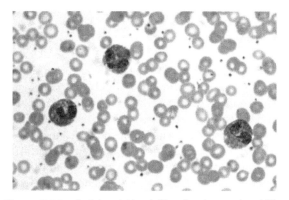

Figure 8.38   Peripheral blood film showing eosinophilia. Source: http://en.wikipedia.org/wiki/File:Eosinophils_in_ peripheral_blood.jpg©Ed Uthman, MD, Houston, Texas, USA

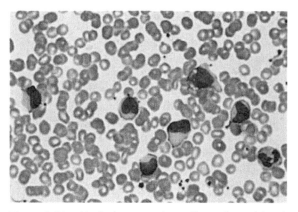

Figure 8.39   Atypical mononuclear cells on the peripheral blood film from a patient with infectious mononucleosis. Source: www.hematologyatlas.com

8.7.7   Lymphocytosis

An increase in lymphocyte count above the normal range for one's age is called lymphocytosis; however, a relative lymphocytosis is normal in young children below the age of 2, hence the need to correct for age. Lymphocytosis is most often due to viral infection, for example in glandular fever, hepatitis, cytomegalovirus, rubella or pertussis, and may also be seen in thyrotoxicosis and some connective tissue diseases. However, it may also be seen in chronic lymphocytic leukaemia.

8.7.8   Infectious mononucleosis (glandular fever)

Infectious mononucleosis (commonly known as 'mono' in the USA, and as 'glandular fever' in the UK) is an infection of the nasopharyngeal epithelial cells and B-lymphocytes with the Epstein—Barr virus (EBV). Most people infected have a mild illness, with sore throat, tiredness, headaches and fever, and swollen lymph nodes (lymphadenopathy) in the neck. Around half also experience swelling of the spleen and a very small proportion may have splenic rupture, thrombocytopenia, haemolytic anaemia, aplastic anaemia or liver failure. A full blood count will show an increase in the absolute lymphocyte count and a blood film shows the appearance of lymphocytes with abnormal morphology ('atypical mononuclear cells') — they are large, with abundant basophilic cytoplasm and an indented nucleus with prominent nucleoli and fine chromatin patterns (Figure 8.39). The abnormal lymphocytes typically show scalloping of their plasma membranes at the point of contact with other blood cells on the film, and appear to be almost wrapped around them. These atypical cells are T-lymphocytes which are reacting against infected B-lymphocytes; the latter appear morphologically normal on a blood film. The diagnosis of mononucleosis can be made by detection of the characteristic heterophilic antibodies present in serum from 2—3 weeks after the onset of illness to around 3 months later — these antibodies crossreact against the red cells of different species and will agglutinate them ('monospot' test). IgM antibodies (indicating recent/current infection) and IgG antibodies (indicating infection in the past) against EBV can also be used in the diagnosis if necessary.

8.7.9   Monocytosis

An increase in monocyte count above the normal range is monocytosis; it may be caused by a range of bacterial or viral infections, including glandular fever, tuberculosis, brucellosis, subacute bacterial endocarditis, syphilis and malaria. It can also be seen in autoimmune conditions such as rheumatoid arthritis, ulcerative colitis and systemic lupus erythematosus, as well as in malignancies such as Hodgkin's disease and some forms of leukaemia.

8.8   Haemostasis

Blood faces an interesting dilemma in survival terms. To function as a transport medium for oxygen,

Virchow's triad- The nineteenth century German pathologist Rudolph Virchow made a number of important discoveries in his lifetime and several clinical signs now bear his name (e.g. Virchow's law, Virchow's node). He discerned that changes in a blood vessel, changes in blood flow and changes in the content of blood could all cause thrombosis, or the formation of abnormal clots within blood vessels — this is known as **Virchow's triad.**

nutrients and carbon dioxide, blood needs to be a fluid that can be pumped readily around the vascular network. However, this also means that large quantities of blood could be readily lost from broken blood vessels if blood remained a fluid in that circumstance. Happily we have evolved haemostasis — the ability of blood to clot in the event of injury. Blood is usually in a state of equilibrium such that it is neither forming pathological clots (thrombi) within blood vessels nor haemorrhaging abnormally from damaged blood vessels. Several factors contribute to this haemostatic homeostasis. Blood vessels themselves are involved, as are platelets and the clotting factors and inhibitors of the clotting cascade, all of which are detailed here (Box 8.7).

8.8.1   Blood vessels and haemostasis

While capillaries are simple structures comprising a single layer of endothelium, veins and arteries are more complex affairs with three layers. The outer layer, the tunica adventitia, is a tough connective tissue 'coat'. The middle layer, the tunica media, has different amounts of muscle and elastin in arteries — relatively more elastin in elastic arteries and more muscle in muscular arteries, though this layer is thin in veins. The inner layer of veins and arteries is the tunica intima, and this layer comprises a single layer of endothelial cells which are critical to successful haemostasis. Endothelial cells contain Weibel-Palade bodies, which produce the glycoprotein von Willebrand factor (vWF). vWF functions as a binding protein; it acts as a bridging protein for platelet adhesion when blood vessels are damaged. vWF is also produced by megakaryocytes, and can thus be found in platelets too (Figure 8.40). It can also be found in solution in plasma, where it binds clotting factor VIII and increases its half-life five times.

8.8.2   The role of platelets in haemostasis

Platelets need to be activated before they can aggregate into a blood clot, and activation requires platelets

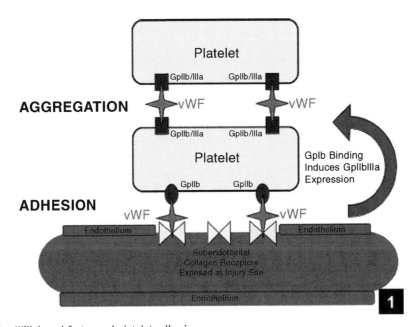

Figure 8.40   Von Willebrand factor and platelet adhesion

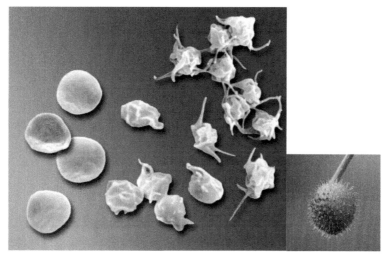

Figure 8.41　Inactivated platelets on left; activated platelets on right (left). Litvinov et al in *The Journal of Biological Chemistry* Vol 278 2003 © The American Society for Biochemistry and Molecular Biology. Goosegrass fruit (right). Goosegrass fruit photograph©www.english-country-garden.com

coming into contact with one of a range of surfaces. The most obvious one is a damaged blood vessel; platelets immediately adhere to, and are activated by, contact with the tissues beneath vascular endothelial cells, and they have specific receptors on their plasma membrane for vWF to enable this adhesion. Other substances are also capable of activating platelets, such as ADP and bacterial endotoxins. Once platelets have adhered, they undergo a shape change from their small, discoid form to become a highly irregular shape with multiple projections from the plasma membrane. This makes the platelet extremely 'sticky', rather like the barbed fruit burrs from the goosegrass weed, which are almost impossible to comb out of a pet's fur (see Figure 8.41)!

As the external shape of the platelet changes upon activation, so its internal structure is altered to enable the platelet's granules to fuse with its inner network of tubules, allowing the contents of the granules to be released along the tubule tunnels to the surface of the platelet. The released contents of the granules cause the adherence of further platelets, which are themselves activated and degranulate, recruiting more and more platelets to the damaged area.

8.8.3　The clotting cascade

The ultimate goal of the clotting cascade is to convert soluble fibrinogen in plasma into insoluble fibrin, a mesh-like structure that converts relatively weak platelet aggregates into a strong, stable clot. To achieve this, a cascade of enzyme reactions takes place. Although the clotting cascade has been traditionally delineated into 'extrinsic' and 'intrinsic' branches, this is not the case in reality (though it remains a useful concept when interpreting the results of some clotting tests). The cascade is one unified pathway, and proceeds as follows.

The cascade starts with the exposure of tissue factor (TF) when tissue damage occurs. TF immediately binds to factor VII in plasma and the clotting cascade is initiated. The TF—factor VII complex turns inactive factor X into active factor Xa ('a' denotes the active form of clotting factors). Once factor Xa is produced, inhibition of further conversion of factor X to Xa is achieved by tissue factor pathway inhibitor (TFPI), and further Xa must be generated via factors IX and VIII (see Figure 8.42). Factor Xa converts prothrombin into thrombin, and thrombin serves the purposes of activating factors IX, VIII and XI (producing further Xa — again, refer to Figure 8.42), and factor V, which hugely amplifies the conversion of prothrombin to thrombin. Finally, thrombin also triggers polymerization of fibrinogen monomers into the net-like molecule fibrin, all whilst enhancing its own production via activating clotting factors. Thrombin requires Ca^{2+} ions to achieve this, and also activates factor XIII to assist with the fibrin/fibrinogen conversion.

von Willebrand factor also plays an important role in the clotting cascade, in that it associates with factor

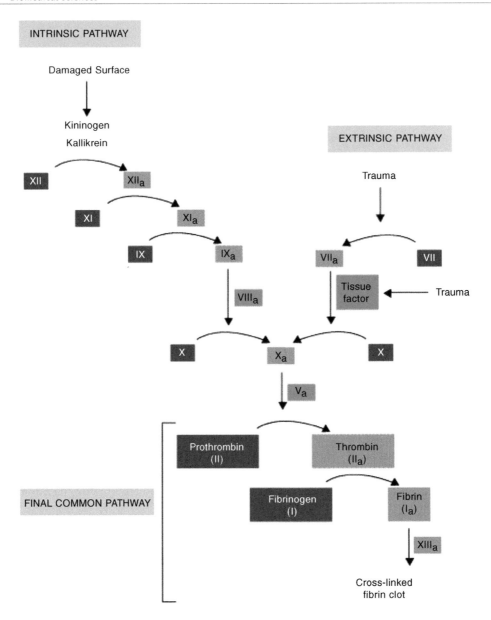

Figure 8.42 The clotting cascade

VIII in plasma, forming VIII:C, and protects it from being broken down in plasma.

8.8.4 Control of the clotting cascade

Haemostasis must be held in check in the normal person — it would plainly be disastrous if blood clotted too readily in the absence of tissue damage. Antic-

oagulants are present to keep the clotting cascade from activating inappropriately, or extending too widely once the cascade is initiated.

8.8.4.1 Protein C and protein S

Protein C is activated by thrombin, and activated protein C inactivates factors V and VIII to reduce further production of thrombin. Protein S acts as a cofactor for protein C; for reasons that are as yet unclear, the majority of protein S is bound in the circulation (by C4b), but it is the free, unbound fraction of protein S in plasma that is active.

8.8.4.2 Antithrombin

Antithrombin achieves its anticoagulant activity by binding a range of activated clotting factors (factors Xa, IXa, XIa, XIIa, VIIa and thrombin) and in so doing 'hides' their active site, preventing their participation in the cascade with their substrates. Antithrombin is a serpin (serine protease inhibitor) molecule and its inactivation of its target clotting factors is greatly increased by heparin.

8.8.4.3 Fibrinolysis

After the clotting cascade has been triggered and resulted in the formation of fibrin and a stable clot, fibrinolysis will be initiated by the presence of fibrin. This process aims to restore blood flow in the damaged vessel. The enzyme plasmin is fundamental to breaking down fibrin; it is produced from inactive plasminogen by the action of tissue plasminogen activator (tPA), which is derived from endothelial cells, and fibrin itself has binding sites for plasminogen activators. Plasmin degrades both fibrin and fibrinogen into fibrin degradation products (FDPs), fragments known as X, Y, D and E. D-dimer levels can be measured to establish that the clotting cascade has been initiated in patients who may have a thrombus, for example in suspected deep venous thrombosis or pulmonary embolus.

8.8.5 Laboratory clotting tests

There is an enormous range of specialized tests that can be undertaken to investigate disorders of blood coagulation, including assays of the various clotting factors. Here, the commonest laboratory coagulation tests are considered.

8.8.5.1 Activated partial thromboplastin time (aPTT)

In the aPTT assay, a patient's plasma is combined with a platelet membrane substitute, calcium (which reverses the effect of the anticoagulant in the tube the blood is collected into) and an activator such as silica, after which the time taken to form a clot is measured in

seconds. This test assesses the intrinsic and common branches of the clotting cascade pathways. A prolonged aPTT is most often caused by deficiency of factors VIII, IX, XI or XII, by treatment with the anticoagulant heparin, or by the presence of inhibitors to clotting factors. If aPTT is prolonged, the test is repeated with a 50:50 mixture of the patient's plasma and plasma that is known to be normal; if the problem is corrected, it is likely that the patient has a clotting factor deficiency and clotting factor assays can be undertaken. If it does not correct, the presence of an inhibitor to clotting is likely.

The normal range will be produced by the laboratory for each batch of reagents used in the test; it can thus vary between laboratories, as well as at different times in the same laboratory. A typical normal range for aPTT is 25–39 s.

8.8.5.2 Prothrombin time (PT)

The prothrombin time measures the extrinsic branch of the clotting cascade, looking at the activity of factors I, II, VII and X. Calcium is added to a patient's plasma at 37 °C to reverse the effects of the citrate anticoagulant in the tube the blood was collected in; then tissue factor (factor III) is added and the time taken to form a clot is measured. A typical reference range for PT is 11–16 s but again this is dependent on the laboratory and the reagents used.

A prolonged PT is most commonly caused by a deficiency of vitamin K (which is essential to produce factors II, VII, IX and X), by treatment with the anticoagulant warfarin, by liver failure (as the liver synthesizes clotting factors) and by deficiencies of factors V or X.

8.8.5.3 The international normalized ratio (INR)

The INR compares the patient's PT time to a standardized normal PT time for a particular batch of tissue factor and produces a ratio according to this formula:

$$INR = \frac{PT_{patient}}{PT_{normal}}.$$

The normal INR for a healthy person is 0.9–1.3; an increase in the INR indicates that a patient has a prolonged PT and can be caused by all the causes of a prolonged PT. In patients treated with the anticoagulant warfarin, the INR is measured on a regular basis

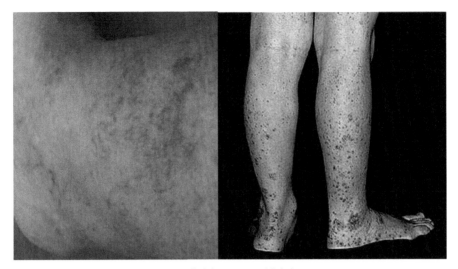

Figure 8.43   Telangiectasia (left) and Henoch—Schönlein purpura (right)

and the dose of warfarin adjusted as necessary to achieve the target INR. A target INR of 2.0—3.0 is often used for patients taking warfarin after a deep vein thrombosis, while a higher target INR (e.g. 3.0—4.0) may be required in patients with artificial heart valves.

8.9   Coagulation disorders

8.9.1   Vascular disorders

Structural blood vessel disorders can cause easy bruising in people with normal platelet counts, and coagulation and bleeding test results. Connective tissue disorders such as Ehlers—Danlos syndrome (in which abnormal collagen is found in blood vessel walls) can cause this, as can the autosomal dominantly inherited Hereditary Haemorrhagic Telangiectasia, with dilated capillaries present in different areas of the body. Easy bruising can also be seen in people taking steroids, in the elderly, and due to certain infections. Henoch—Schönlein purpura may be seen following chest infections in children and is due to the formation of IgA complexes triggering vasculitis (inflammation of blood vessels), with subsequent bleeding (Figure 8.43).

8.9.2   Platelet disorders

The normal range for platelets is $150-400 \times 10^9$/L, and abnormal bleeding may occur in individuals with a lowered platelet count (thrombocytopenia), or who have a normal platelet count but functionally abnormal platelets.

8.9.2.1   Thrombocytopenia

A low platelet count can be caused by destruction of platelets in the circulation, by reduced production of platelets in bone marrow or because an enlarged spleen is sequestering platelets. Some instances of reduced platelet count may be caused by the dilutional effect of a massive transfusion of red blood cells. Bone marrow biopsy is a key tool in identifying the cause of thrombocytopenia as it will show the level of megakaryocytes present in bone marrow, as well as their morphology. The risk of bleeding in thrombocytopenia is directly proportional to platelet count: a platelet count $<50 \times 10^9$/L carries a risk of haemorrhage after trauma and will be closely monitored with repeated full blood counts but at $<20 \times 10^9$/L, the risk of spontaneous bleeding and even bleeding into the brain is real and such patients will often require platelet transfusion.

8.9.2.2   Reduced platelet production in the bone marrow

There are numerous causes of decreased megakaryocyte production in the bone marrow leading to thrombocytopenia, including bone marrow failure, leukaemia, myelodysplasia, aplastic anaemia, some rare inherited syndromes (e.g. inherited amegakaryocytic thrombocytopenia), numerous drugs (including antibiotics, proton pump inhibitors and chemotherapy drugs) and systemic bacterial and viral infections (e. g. HIV). Bone marrow biopsy will show decreased numbers of megakaryocytes, and may show morphological abnormality. Treatment is that of the underlying cause or stopping any drugs where this may be a factor.

8.9.2.3 Excessive destruction of platelets in the peripheral circulation

Platelet destruction in the peripheral circulation is mainly mediated by immune mechanisms, of which the most clinically important is the autoimmune condition immune thrombocytopenia purpura. Other causes of excessive destruction are disseminated intravascular coagulation, systemic lupus erythematosus and thrombotic thrombocytopenia purpura (TTP). Drugs may also lead to immune-mediated platelet destruction. Heparin-induced thrombocytopenia (HIT) is one example where antibodies are produced against heparin-platelet factor 4 complexes in a small proportion of patients receiving heparin for anticoagulation, and alongside thrombocytopenia, HIT paradoxically predisposes to severe thrombosis, as the antibody triggers platelet aggregation.

8.9.2.4 Immune thrombocytopenia purpura (ITP)

Immune thrombocytopenia purpura (ITP) occurs where macrophages remove antibody-coated platelets from the peripheral circulation, causing a fall in platelet count and leading to bruising, nose bleeds and petechiae (pin-prick sized bruises on the skin). It may appear rapidly in younger children, but usually has a more insidious onset in adults who develop ITP. Platelet autoantibodies will be found in ~70% of people with ITP and in adults women are most often affected. Full blood count shows an isolated thrombocytopenia and bone marrow biopsy will show normal or increased levels of megakaryocytes. Most people with ITP require no treatment, as it is usually a mild, selflimiting condition. High dose oral steroids may be used in both adults and children where necessary, and immunoglobulin can be given intravenously in adults where a rapid rise in platelets is needed (e.g. prior to surgery). Removal of the spleen may be required in some adults and this has high success rates in improving platelet count; however, a few go on to require additional immunosuppressive drugs. Platelet transfu-

sions are generally reserved for urgent clinical situations only in ITP, and are not part of routine management – it is usually pointless to transfuse platelets in ITP as transfused platelets will also be rapidly destroyed!

8.9.2.5 Thrombotic thrombocytopenia purpura (TTP)

Thrombotic thrombocytopenia purpura (TTP) is uncommon; it occurs where endothelial damage triggers widespread platelet aggregation and extensive microscopic thrombosis throughout the body, with ultimate platelet consumption and severe thrombocytopenia. Red blood cells that are forced through the tiny thromboses are damaged too, with haemolytic anaemia ensuing, and numerous schistocytes (damaged, fragmented red cells) seen in the circulation. This all causes widespread bruising, fever, neurological problems such as headache, hallucinations and often with renal failure too. Primary TTP is caused by autoantibodies inhibiting an enzyme called ADAMTS13, which is normally responsible for cleaving the vWF molecules – the presence of large vWF molecules (ultra large vWF multimers) is known to trigger platelet aggregation and vWF is a presence in both platelets and endothelium for this purpose. Secondary TTP may occur due to the presence of tumours, in pregnancy, with some infections and certain drugs, but the mechanism of action is not yet well understood for secondary TTP.

Treatment of TTP relies on plasmapheresis, which is removing the patient's own plasma – and thus their autoantibodies - from one vein, and exchanging it for donated plasma containing a fresh source of uninhibited ADAMTS13 via another vein.

8.9.2.6 Splenic sequestration of platelets

Splenomegaly is associated with a range of haematological conditions and the presence of an enlarged spleen can lead to sequestration of platelets there, causing a decreased platelet count in the peripheral circulation.

8.9.3 Platelet function disorders

People may inherit or acquire dysfunctional platelets. This will usually result in bruising and haemorrhage in the presence of a normal (or high) platelet count, with a prolonged bleeding time due to those platelets being functionally abnormal. Inherited platelet dysfunction disorders encompass a range of anomalies, including a lack of dense bodies within platelets (storage pool deficiency), a lack of the binding site for vWF on platelets (Bernard—Soulier syndrome), or a defective fibrinogen binding site (Glanzmann's thrombasthenia). Acquired platelet dysfunction may occur in a setting of renal and liver disease, myeloproliferative disorders or simply via the action of NSAIDs which act as platelet inhibitors.

8.9.4 Thrombocytosis

A platelet count above the normal range — a count greater than $400 \times 10^9/L$ — is thrombocytosis; it is said to be an 'acute phase reaction', meaning something that is a physiological response to disease. It can be seen in the setting of iron deficiency, cancer, myeleoproliferative disease and inflammatory disease. It is also found after splenectomy since the site of the majority of platelet elimination has been removed. The underlying cause of thrombocytosis should be treated where possible, but low dose aspirin may be given to alleviate some of the risk of thrombosis in the presence of a chronically elevated platelet count.

8.9.5 Inherited disorders of blood clotting

8.9.5.1 Haemophilia A

Haemophilia A is a haemorrhagic tendency caused by insufficient production of factor VIII. It is an X-linked condition, and thus almost all sufferers are male. It is possible to be a female with haemophilia, but this requires inheritance of two defective genes on both X chromosomes, one from a father with haemophilia and the other from a carrier mother; as such this is far rarer in females than males, who only need to inherit one defective factor VIII gene from a carrier mother. The incidence is about one in 5000 of the male UK population and new mutations of the factor VIII gene that produce sporadic cases of haemophilia are relatively common, since it is a very large gene, which is thus statistically more prone to mutation.

Haemophilia A varies in severity in different individuals and to a large extent this is due to the levels of factor VIII they produce. Severe haemophilia is experienced by haemophilia sufferers with factor VIII levels of <1 IU/dL (normal range 50—150 IU/dL), with spontaneous bleeding, particularly into joints, occurring regularly from infancy if untreated. Moderate haemophilia is experienced if factor VIII levels are 1—5 IU/dL, with rare spontaneous haemorrhage, and most bleeding occurring where injury has preceded it. Mild haemophilia is defined by factor VIII levels >5 IU/dL, and diagnosis is often made late in this group of individuals. Haemorrhage usually only occurs after surgery or significant injury.

Laboratory testing in haemophilia
Clotting tests on individuals with haemophilia will show a normal bleeding time and prothrombin time (PT), but a prolonged activated partial thromboplastin time (APTT). Levels of factor VIII:C will be low.

Management of haemophilia
Recombinant preparations of factor VIII can be infused intravenously to treat haemophilia A, either at the time of haemorrhage, for planned surgery or prophylactically in people with severe haemophilia to prevent recurrent joint bleeds (haemarthroses) and the risk of ensuing deformity and disability from these (Box 8.8). Since the half-life of factor VIII is only 12 h, infusions are given twice daily where there is bleeding

---

Box 8.8 The legacy of plasma-derived use in haemophilia

In most developed countries, recombinant factor VIII and IX is now produced in the laboratory for use in people with haemophilia. However, until relatively recently, these factors were manufactured from pooled plasma from blood donors, and this came with the risk of viral transmission in the era prior to exclusion of high-risk blood donors and screening for viruses. Infection with HIV, Hepatitis B and Hepatitis C was sadly not uncommon among people with haemophilia who received plasma-derived factor, and lawsuits continue in this area. The risk of vCJD from blood products remains, and plasma products in the UK are currently obtained from countries with a low incidence of BSE.

or planned surgery, and three times a week in young people with severe haemophilia who are receiving it prophylactically (usually from around the age of 2 into early adulthood). However, a significant proportion of people with haemophilia will develop antibodies against factor VIII and such individuals may require alternative therapy with recombinant factor VIIa (which effectively bypasses factor IX and VIII activity in the clotting cascade – see Figure 8.42).

Vasopressin (often given as a nasal spray) may also be useful in people with mild haemophilia, in whom it is capable of producing a rise in factor VIII levels that may suffice to treat minor haemorrhage.

Female carriers of haemophilia

Female carriers of haemophilia A may have reduced levels of factor VIII, but this is highly variable between individual females due to the random nature of lyonization (in which one X chromosome is inactivated in each cell). DNA studies to look for factor VIII gene mutations are often required to confidently identify female carriers where a family history is present and antenatal testing for haemophiliac status of any male fetuses may be offered if a pregnancy ensues. Obstetricians should be aware of a woman's haemophilia carrier status since the delivery of a male baby with potential haemophilia should be as atraumatic as possible – babies with haemophilia have a higher risk of cerebral haemorrhage at the time of delivery.

8.9.5.2   Haemophilia B

Haemophilia B is often known as Christmas disease after the first patient identified with it, Stephen Christmas, in 1952, and it is due to factor IX deficiency. It is also X-linked, but far less common than haemophilia A, with an incidence of one in about 30 000 males in the UK population. This is probably because the gene for factor IX is smaller than that for factor VIII and thus less prone to mutation.

Haemophilia B is managed in a similar manner to haemophilia A, using recombinant factor IX. However, vasopressin will not increase levels of factor IX.

8.9.5.3   Von Willebrand disease

Von Willebrand disease (vWD) is the most common bleeding disorder, thought to be present in around 1% of the population although only symptomatic in a

---

Box 8.9   vWF and blood group

Interestingly, levels of vWF have been found to be related to ABO blood group. Group O individuals have the least vWF, while group AB individuals have the highest levels. This is related to the frequency of symptomatic vWD, which is commonest in people who are blood group O.

---

small proportion of these. It is usually caused by a qualitative or quantitative lack of vWF, that is vWF may be present at very low levels in vWD, or present at normal levels but as an abnormal molecule which does not function well (Box 8.9). The congenital forms of vWD may be inherited in autosomal dominant or recessive forms. Some cases of vWD are acquired, and due to the production of autoantibodies against a normal vWF molecule.

vWD is diagnosed in the laboratory by measuring the amount present in plasma and also how well it is functioning using a glycoprotein binding assay. Factor VIII levels can also be measured; these can be low in vWF deficiency, as vWF binds factor VIII in plasma and increases its half-life.

vWD is generally mild enough to need no treatment, though additional factor VIII may be given at times of increased bleeding (e.g. for elective surgery) for people badly affected. For milder cases, vasopressin nasal spray can be used as it increases the release of vWF from the Weibel–Palade bodies in endothelial cells.

8.9.6   Acquired clotting disorders

8.9.6.1   Disseminated intravascular coagulation (DIC)

Disseminated intravascular coagulation (DIC) occurs where there is universal triggering of the clotting cascade throughout the body, with generalized platelet aggregation, ultimately with consumption of all clotting factors and platelets. This in turn causes widespread fibrinolysis and then inhibition of further fibrin formation, so the clinical picture that was initially one of thrombosis may become one of profound haemorrhage. DIC is seen in some people after large burns, in the presence of eclampsia, septicaemia and falciparum malaria, and with malignant disease. In serious cases, PT and APTT are prolonged, with depletion of fibrinogen levels and severe thrombocytopenia. Treatment is of the underlying condition and this will often suffice for mild DIC. In worse cases of DIC, replacing platelets and clotting factors and maintaining blood volume, is the treatment of choice.

8.9.6.2   Liver disease

Clotting factors are made in the liver, and liver disease will often result in deranged coagulation tests as a consequence. Platelet dysfunction and defective fibrinogen may also be sequelae to liver disease for some individuals.

8.9.6.3   Vitamin K deficiency

Without vitamin K, factors II, VII, IX and X (and proteins C and S) are unable to bind Ca^{2+} ions, rendering them inactive and giving a prolonged PT and APTT. This can cause bruising, haemorrhage and haematuria (blood in urine). Vitamin K deficiency can be seen in newborn babies, who are routinely prescribed vitamin K shortly after birth, and also in people who malabsorb vitamin K or are severely malnourished. Warfarin is a vitamin K antagonist, which is the basis of its anticoagulant activity.

Treatment of vitamin K deficiency is to treat the cause and replace vitamin K if necessary.

8.9.6.4   Thrombotic disorders

Thrombosis

A thrombus is defined as an abnormal clot formed within an intact blood vessel in a living person and it can be caused by any of the three components of Virchow's triad: changes in the blood vessel, changes in blood flow and changes in the constituents of blood. Deep vein thrombosis (DVT) is the most common type of thrombus and forms in the deep veins of the calf. The danger of having a thrombus is the possibility of a section of it breaking off (an embolus) and travelling to another site in the body, where it may block other blood vessels. Venous thrombi usually occur in normal blood vessels, but in areas of stasis of blood, or where alteration to components of blood make thrombus formation more likely.

Thrombi may occur in the arteries, usually in the presence of preexisting atherosclerosis of the vessel. The rough plaque of atheroma and associated endothelial damage creates a site where platelets aggregate; if clotting is triggered an arterial thrombus may result. This is commonly seen in the coronary arteries during myocardial infarction (MI) or within the ventricles of the heart.

Risk factors for thrombosis include increasing age, obesity, long-haul travel, pregnancy, recent surgery, the oral contraceptive pill, cancer, hormone replacement therapy, varicose veins, sickle cell anaemia and the thrombophilias.

Thrombophilias

The tendency to form thrombi is known as thrombophilia, and this may be inherited or acquired. This may be the cause of recurrent deep vein thrombosis and of recurrent miscarriages. One inherited cause that has been discovered in recent years is factor V Leiden, a variant form of clotting factor V that is not easily cleaved by protein C, making it far more likely to proceed in the clotting cascade and cause the formation of a thrombus. It is found in around 30% of patients with deep vein thrombosis. The risk of thrombosis from factor V Leiden is compounded by the addition of other risk factors given above, particularly the oral contraceptive pill.

Other less common forms of thrombophilia include antithrombin deficiency (which may be inherited, or acquired after trauma or major surgery), an inherited variant of prothrombin giving rise to elevated levels of it, deficiencies of Protein S and Protein C, and antiphospholipid antibody (see Section 8.8.4).

Investigation of thrombosis and thrombophilia

In the laboratory setting, a platelet count should be performed as part of FBC, along with a fibrinogen level and coagulation screen. If a thrombophilia is suspected after recurrent instances of thrombosis (or miscarriage), assays can be performed for the specific causes, for example levels of protein C and protein S.

Treatment of thrombosis

Arterial thrombosis treatment is tailored to the precise clinical situation. It may include the use of thrombolytic ('clot busting') drugs to break down a thrombus. Streptokinase or recombinant tissue plasminogen activator (rTPA) are thrombolytic drugs: both promote the formation of plasmin, and thus the degradation of fibrin and breakdown of a thrombus. Antiplatelet drugs such as aspirin or clopidogrel may be given to help prevent platelet activation at the site of atheroma in patients known to be at risk. The treatment of arterial thrombosis must, however, include management of risk factors for atheroma, such as improving diabetic control or losing weight as needed to attain a healthy BMI.

Venous thromboembolism is usually managed with anticoagulants such as heparin and warfarin. Heparin is given either as an intravenous infusion or via subcutaneous injection and it binds antithrombin to have an immediate anticoagulant effect. Direct thrombin inhibitors such as hirudin may be preferred in some patients (e.g. those suffering from heparin-induced thrombocytopenia). Many patients will be started on the oral anticoagulant warfarin at the same time as commencing treatment with heparin; warfarin takes several days to establish its antivitamin K activity and diminish levels of factors II, VII, IX and X, but once it is working well, heparin can be discontinued. The efficacy of warfarin in an individual patient is established by measuring the international normalized ratio (INR), where the patient's PT is compared to a normal control, and expressing the results as a ratio — an INR of 2.5 is the usual target in most patients, but this will be higher (3.5) for patients who have had recurrent thromboembolism while on warfarin.

Subcutaneous heparin injections can also be used for prophylaxis (prevention) of thrombosis in high-risk patients, for example patients in hospital around the time of surgery. They are usually used along with thromboembolic disease (TED) compression stockings, worn on the legs to prevent stasis of blood in the deep veins there. Up to 10% of all hospital deaths may be due to pulmonary embolism, so this is a large-scale problem.

8.10   Myeloproliferative disorders

Myeloproliferative disorders are characterized by clonal proliferations of specific cell lines in the bone marrow; polycythaemia, thrombocythaemia and myelofibrosis are considered here, while chronic myeloid leukaemia (CML) is described in Section 8.11.1.3.

8.10.1   Polycythaemia

Polycythaemia comprises an increase in Hb, haematocrit and red cell count. It may be a primary disorder (where the changes in full blood count are due to clonal stem cell expansion) or where red cell count is raised, or apparently raised, secondary to other conditions (e.g. dehydration, cardiac or lung disease, or travelling to high altitude).

8.10.1.1   Primary polycythaemia

Primary polycythaemia (also known as polycythaemia vera (PV) or polycythaemia rubra vera) is due to a clonal expansion of a pluripotent stem cell leading to an increase in the lineages that give rise to erythroid, megakaryocyte and myeloid cells. More than 90% are now known to be due to mutations in the Janus Kinase 2 gene (JAK2), which codes for a cytoplasmic tyrosine kinase. The effect of this clonal expansion is to cause an increase in Hb levels (>18.5 g/dL in males; >16.5 g/dL in females) and bone marrow biopsy in such patients will show hypercellularity with characteristic increase in the cell lines detailed above (Figure 8.44). Serum

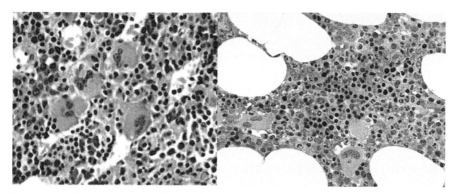

Figure 8.44   Bone marrow film in polycythaemia vera showing marked hypercellularity (left) Pardanani et al in Leukemia, 21, 2007 © Nature Publishing Group, a division of Macmillan Publishers Limited; normal bone marrow for comparison (right)

erythropoietin levels will be low or normal. Patients with PV are usually over 60 years old and experience a gradual onset of vague symptoms such as dizziness, tiredness and visual problems. They may also have hypertension, angina and gout and often look very pink!

Treatment of PV tries to bring the haematocrit below 45% and venesection is the mainstay in most patients, with ~500 mL blood removed each week in most patients. Intermittent chemotherapy with hydroxyurea may also be used to control platelet counts, with low dose aspirin if blood clotting episodes are experienced. PV will develop into myelofibrosis in ~30% of cases, and into acute leukaemia in ~5%.

8.10.1.2 Secondary polycythaemia

Secondary polycythaemia is, like PV, an increase in Hb, haematocrit and red cell count, but is not due to stem cell mutation. It may be relative, where plasma volume is reduced relative to the cellular component of blood, due to dehydration or burns. Secondary polycythaemia may also be seen due to an increase in erythropoietin, due to high altitude, cardiovascular disease, lung disease, the inheritance of Hb variants with high O_2 affinity, or tumours that produce erythropoietin (such as renal cell carcinoma, hepatocellular carcinoma). Investigations will show a normal or raised erythropoietin level and treatment is to tackle the cause.

8.10.2 Myelofibrosis

Myelofibrosis (sometimes known as myelosclerosis) is due to clonal expansion of a myeloid stem cell lineage in the bone marrow, with additional fibrosis of the marrow due to the production of fibroblast-stimulating factors by abnormal megakaryocytes. A quarter of all cases arise following polycythaemia vera. Patients with myelofibrosis typically have a gradual onset of tiredness and weakness and will have marked splenomegaly — the spleen may undergo infarction, causing abdominal pain. Patients with myelofibrosis may also experience bruising and bleeding due to abnormal platelets. Investigations show an increase in platelet and white blood cell counts with anaemia, and bone marrow trephine biopsy shows increased fibrosis of the bone marrow (bone marrow aspiration is usually unsuccessful — the so-called 'dry tap' — due to the amount of fibrous tissue in the marrow making it

difficult to aspirate cells). Myelofibrosis can be hard to distinguish from CML, but the Philadelphia chromosome (see Section 8.11.1) is absent in myelofibrosis.

Treatment of myelofibrosis is difficult; drugs can be used to suppress bone marrow activity (e.g. busulfan), and blood transfusion and folic acid to support the patient. Radiotherapy/chemotherapy may be employed to reduce the size of the spleen, though splenectomy (removal of the spleen) may ultimately be required. The prognosis of myelofibrosis is poor in most cases, with median survival of 3 years (20% will transform into acute leukaemia). Bone marrow transplantation can be curative if available for a fit enough patient.

8.10.3 Myelodysplasia

Myelodysplasia is a group of conditions that cause gradual bone marrow failure and is typically seen in the elderly. It is caused by stem cell disorders in the myeloid lineages. Patients with myelodysplasia develop a gradual pancytopenia; low Hb, red cell, white cell and platelet count are seen, singly or in various combinations, but alongside increased bone marrow cellularity. Myelodysplasia is classified according to the percentage of blast cells seen in peripheral blood and bone marrow; the prognosis worsens with increasing numbers of blasts seen in the bone marrow, and patients with >5% blasts are treated with chemotherapy if fit enough to tolerate it. Bone marrow transplantation may be curative if a match is available, in patients who are less than 50 years old.

8.10.4 Thrombocythaemia

Thrombocythaemia (also known as essential thrombocythaemia (ET)) is related to PV, and another disorder caused by clonal stem cell expansion in the bone marrow, but patients with ET have normal Hb and WBC counts, and increased platelet count to $>600 \times 10^9$/L, often into the low thousands. Patients with ET will usually experience a thrombosis (+/− embolism), but some may have bleeding problems due to abnormal platelet function. Many are diagnosed during routine tests for other reasons. Diagnosis can prove difficult as there is no single diagnostic test for ET, and patients may need to be observed over a period of time, with regular FBC, before a firm diagnosis is established. Treatment aims to bring the platelet count below 400×10^9/L, and uses hydroxyurea, busulphan or α-interferon to achieve this. ET can progress into PV or even acute leukaemia, but many cases follow an indolent course for decades.

8.11   Haematological malignancies

Malignant disorders of blood are relatively uncommon and all are potentially life-threatening. The cause is unknown in most cases, but exposure to radiation, previous chemotherapy and certain viruses have all been found to increase the risk of haematological malignancy.

8.11.1   Leukaemia

Leukaemia is uncommon, with an incidence of one in 10 000 per year across all age groups. The subclassification of all types of leukaemia is becoming increasingly complex, but leukaemias can be divided into acute and chronic types based on what the natural history would be if untreated. They may be further divided according to whether the affected cell type is of lymphoid or myeloid origin. Thus the acute leukaemias are subdivided into acute lymphoblastic leukaemia (ALL) and acute myeloid leukaemia (AML), while the chronic leukaemias are divided into chronic lymphocytic leukaemia (CLL) and chronic myeloid leukaemia (CML).

8.11.1.1   Diagnosis of leukaemia

Patients with leukaemia will classically present with symptoms of anaemia (pallor, breathlessness), leucopenia (infections), thrombocytopenia (bruising) and general symptoms such as weight loss, and bone pain from marrow infiltration by malignant cells. A full blood count will show a low Hb, low platelets and usually high WBC count (though paradoxically, this can sometimes be low). A blood film will show the presence of immature white blood cells (blasts) and a bone marrow biopsy will show increased cellularity; a lumbar puncture may also be performed to assess for the presence of blasts in the central nervous system. The precise subtype of a leukaemia depends upon the appearance of the cells on a blood film, with an assessment of enzymes in the cytosol of the blasts, establishing which cell surface markers are present on the blasts with monoclonal antibodies (e.g. using flow cytometry) and testing for chromosomal rearrangements (by cytogenetic analysis, fluorescent *in situ* hybridization (FISH) or PCR for specific fusion genes). Most cases of leukaemia have an acquired cytogenetic abnormality, of which the most well known is the Philadelphia chromosome, which is present in 97% of cases of CML.

8.11.1.2   Acute leukaemias

Acute lymphoblastic leukaemia (ALL) is the commonest malignancy in childhood, while AML is found mainly in older adults (the median age at presentation is 65). It can be variously subclassified according to the French—American—British (FAB) scheme, or more recently according to the World Health Organization's (WHO) scheme. A simplified version of the WHO's classification is given in Table 8.4.

The diagnosis of the specific subtype of acute leukaemia is arrived at with a combination of blood film and bone marrow aspirate appearances, as well as confirmation of blast lineage using cytogenetic analysis, immunophenotyping and molecular genetics (see Figures 8.45 and 8.46). Most patients with ALL will have the common ALL antigen, CD10, on their blasts,

Table 8.4   WHO classification of acute leukaemia

AML with genetic abnormalities (the best prognosis category of AML)	AML with translocations between chromosomes 8 and 21
	AML with translocations between chromosomes 15 and 17
	AML with inversions in chromosome 16
	AML with abnormalities in 11q23
AML with multilineage dysplasia	AML following myelodysplasia or myeloproliferative disease
AML and MDS, therapy-related	AML in patients who have had prior chemotherapy or radiotherapy
AML not otherwise categorized	Other categories of AML
ALL	Precursor B-cell acute lymphoblastic leukaemia
	Precursor T-cell acute lymphoblastic leukaemia
	Burkitt cell leukaemia
	Biphenotypic acute leukaemia

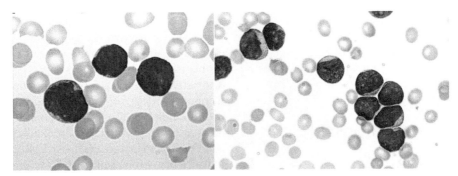

Figure 8.45   Blood film appearance in ALL (left) and AML (right). From Hoffbrand et al, *Postgraduate Haematology*, Sixth Edition 2010, reproduced by permission of John Wiley & Sons Ltd

often with CD19, while immature myeloid cells of AML express CD33 or CD19.

Acute leukaemia would always prove fatal without treatment, and the different subtypes have different prognoses — the age and health of the patient otherwise are also factors to take into account when deciding on a treatment regime. Childhood ALL is usually curable, for example, while some types of AML are usually incurable. Treatment will involve supportive therapy with transfusions of red cells and platelets, antibiotic prophylaxis to counter the high risk of infection and antibiotics to treat active infections; then different regimes of chemotherapy may be used to try to induce complete remission (i.e. the return of normal blood and bone marrow). This is tailored to the patient, their precise leukaemia subtype and their other risk factors, and should be carried out in isolation facilities in hospital to reduce the risk of infection. Risk factors include increasing age, high initial white cell counts at

the time of diagnosis, the presence of certain cytogenetic abnormalities and the presence of minimal residual disease (MRD: low levels of persistent leukaemic cells in apparently normal bone marrow during remission, detected using flow cytometry with monoclonal antibodies, or PCR for chromosome/gene rearrangements). If remission is successfully achieved, consolidation therapy will follow to reduce the risk of a recurrence of the leukaemia — this may entail an early bone marrow transplant, or further cycles of chemotherapy.

8.11.1.3   Chronic leukaemia

Chronic lymphocytic leukaemia (CLL) is the commonest form of leukaemia and is usually seen in older adults (median age at presentation is 66). Some 95% are of B-cell origin. Chronic myeloid leukaemia (CML) also occurs almost entirely in adults, but with a younger medial age at presentation of around 50. Most

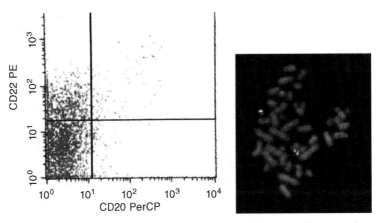

Figure 8.46   Flow cytometry immunophenotyping of leukaemia (from Hoffbrand et al, *Postgraduate Haematology*, Sixth Edition 2010, reproduced by permission of John Wiley & Sons Ltd) and fluorescent *in situ* hybridization (FISH) of fusion gene products in chromosomal rearrangement in leukaemia

patients who present with chronic leukaemia have no symptoms, and the condition is found incidentally during a full blood count for other reasons.

Chronic myeloid leukaemia

Many patients with CML will be asymptomatic but some may have symptoms of anaemia, a swollen spleen, weight loss, bruising or fevers. Diagnosis of chronic leukaemia includes a full blood count, which typically shows a normocytic anaemia, a raised WBC count and variable platelet numbers. A blood film will show myeloid precursors (with some blasts) and a neutrophilia, while a bone marrow aspirate will have increased cellularity with myeloid precursor cells (Figure 8.47).

Cytogenetic analysis will show the presence of a Philadelphia (Ph) chromosome in 97% of cases of CML. This is a reciprocal translocation of part of the long arms (q) of chromosome 22 to chromosome 9. This translocation results in part of an oncogene, *c-ABL*, on chromosome 9, becoming fused with a 'breakpoint cluster region' (*BCR*) from chromosome 22, creating a novel hybrid gene, *BCR-ABL*. This new gene is transcribed into a fusion protein that has tyrosine kinase activity resulting in altered growth and apoptosis (see Figure 8.48)

Treatment of CML thus uses tyrosine kinase inhibition to block the activity of the BCR-ABL fusion protein with imatinib, and this will produce remission in 95% of patients. However, this response is often only temporary, and chemotherapy is then used as for acute leukaemia to produce further remission. Stem cell

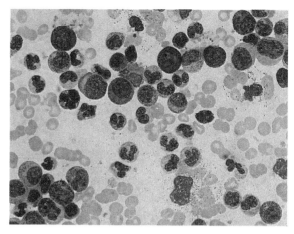

Figure 8.47　Blood film in CML. From Hoffbrand et al, *Postgraduate Haematology*, Sixth Edition 2010, reproduced by permission of John Wiley & Sons Ltd

transplantation is offered to those who do not continue to respond well to imatinib, and are otherwise well enough. Cure is achieved in around 70% of patients with CML.

Chronic lymphocytic leukaemia

CLL patients are usually asymptomatic at diagnosis, though some will present with symptoms of bone marrow failure (anaemia, infections, bruising), or from abdominal discomfort due to a swollen spleen. Full blood count shows a normal or low Hb, a raised WBC count (with lymphocytosis) and usually normal platelets. A blood film will typically show small

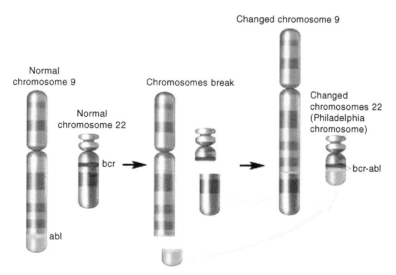

Figure 8.48　The Philadelphia chromosome and formation of the *BCR-ABL* hybrid gene in CML

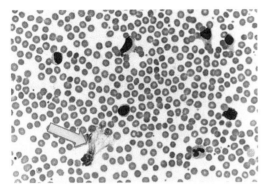

Figure 8.49 Blood film in CLL showing characteristic smear cell. From Hoffbrand et al, *Postgraduate Haematology*, Sixth Edition 2010, reproduced by permission of John Wiley & Sons Ltd

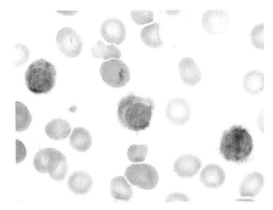

Figure 8.50 Hairy cells seen on blood film in HCL. From Hoffbrand et al, *Postgraduate Haematology*, Sixth Edition 2010, reproduced by permission of John Wiley & Sons Ltd

lymphocytes and 'smudge cells' where the fragile malignant lymphocytes are fragmented during the spreading of the film of blood (Figure 8.49).

Treatment options are variable in CLL and 30% of all patients will not require treatment. A further proportion will only be treated when progression of their condition occurs: treatment with chemotherapy and antibody therapy may be undertaken. Supportive treatment — red cell transfusions, platelet transfusions and antibiotics — should also be used as necessary. The benefits of stem cell transplantation (particularly in younger patients) in CLL are still being studied.

Hairy cell leukaemia

Hairy cell leukaemia (HCL) is a rare form of chronic leukaemia found predominately in males in the sixth decade of life. It is due to a clonal proliferation of abnormal B cells which have cytoplasmic projections that look hair-like! Patients with HCL usually present with anaemia, weight loss and fever, and most will have a swollen spleen. The FBC will show anaemia, low platelet count and low neutrophil and monocyte counts; immunophenotyping will show a range of cellular markers including CD19, CD20 and CD103 (Figure 8.50).

HCL is treated with chemotherapy and this is generally very successful with one cycle of chemotherapy producing remission in 90%.

8.11.2 Lymphoma

The lymphomas are a further type of haematological malignancy that arise due to an abnormal proliferation of lymphoid tissue. Lymphoma is commoner than leukaemia and its incidence is increasing. The lymphomas usually start as enlarged lymph nodes (lymphadenopathy) at single or multiple sites in the body and the prognosis is related to how many sites are involved, where in the body they are, and what subtype of lymphoma this represents. Lymphomas are initially divided into two major groups; Hodgkin lymphoma and non-Hodgkin lymphoma.

8.11.2.1 Hodgkin's lymphoma (HL)

The great majority of cases of HL occur in adults, with a peak in diagnoses in the third decade of life, and the UK incidence is around three per 100 000. The Epstein—Barr virus (EBV) is implicated in the aetiology of HL; EBV has been found in tissue from HL patients suggesting it plays a role in the development of the condition. Patients with HL typically present with enlarged, rubbery lymph nodes, usually in the neck. They may have a swollen spleen and liver, and some will have symptoms of systemic illness such as fever, night sweats and weight loss ('B symptoms' in HL, as these affect the classification and prognosis of HL). Diagnosis is made on biopsy of an enlarged lymph node, in which the characteristic malignant Reed—Sternberg cell is seen — these cells have two nuclei and are said to have an 'owl's eye' appearance. They are derived from B cells and usually express CD30 and CD15 markers (Figure 8.51).

A full blood count in a patient with HL may be normal or show a normocytic anaemia, and ESR is usually raised. CT scans, a chest X-ray and a positron emission tomography (PET) scan will look for the involvement of other lymph node groups in the body.

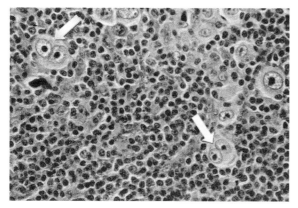

Figure 8.51   Reed—Sternberg cells seen in a lymph node biopsy in Hodgkin lymphoma. From Hoffbrand et al, *Postgraduate Haematology*, Sixth Edition 2010, reproduced by permission of John Wiley & Sons Ltd

Treatment for HL usually aims for a cure, and is tailored according to how far advanced the disease is and whether 'B' symptoms are present. It generally employs one or more cycles of chemotherapy, followed by radiotherapy of involved lymphoid tissue, to induce remission. This is successful for about 75% of HL patients and further remissions may be achieved after recurrences with more chemotherapy.

8.11.2.2   Non-hodgkin lymphoma (NHL)

The incidence of NHL is around 15 per 100 000 per year and, like HL, this has been increasing in recent years. In terms of aetiology, the EBV is also implicated in NHL, particularly in Burkitt lymphoma (which occurs mainly in Africa). HIV infection confers an increased risk of NHL, as does immunosuppression (e.g. after organ transplantation) and the human T-cell lymphotropic virus (found mainly in Japan) is a major risk factor. Patients with NHL usually present with painless enlargement of superficial lymph nodes and often with systemic symptoms (weight loss, fevers and sweating), or sometimes from the appearance of NHL in other tissues such as the skin, lung, brain or gastrointestinal tract. Full blood count usually shows a normocytic anaemia, raised WBC count, and low platelet count — some will have renal impairment with increased urea levels and bone marrow biopsy will show infiltration of the marrow with malignant cells in a proportion of patients. Lymph node biopsy with immunophenotyping and cytogenetic analysis will allow the subtype of NHL to be diagnosed. All NHL

cells express CD20, while the subtypes vary, and chromosomal translocations are frequent in NHL.

There are some 23 types of NHL according to the WHO classification (as shown here in Table 8.5) and the treatment and prognosis for NHL depends on the accurate classification of each case using biopsy appearances, immunophenotyping and cytogenetic analysis.

The treatment of NHL depends on the disease subtype, and the age and condition of the patient. The antiCD20 antibody rituximab may be used alongside chemotherapy ('chemoimmunotherapy'), or either

Table 8.5   Modified WHO classification of lymphoid neoplasms

B-cell lymphomas

Precursor B-cell lymphoma	Precursor B lymphoblastic leukaemia/lymphoma
Mature B-cell lymphoma	Chronic lymphocytic leukaemia/lymphoma
	Lymphoplasmacytic lymphoma
	Splenic marginal zone lymphoma
	Extranodal marginal zone B-cell lymphoma of mucosa-associated lymphoid tissue
	Nodal marginal zone B-cell lymphoma
	Follicular lymphoma
	Mantle cell lymphoma
	Diffuse large B-cell lymphoma
	Burkitt lymphoma

T-cell lymphomas

Precursor T-cell lymphoma	Precursor T-cell lymphoblastic leukaemia/lymphoma
	Blastic natural killer lymphoma
Mature T-cell/natural killer cell lymphoma	Adult T-cell leukaemia/lymphoma
	Extranodal natural killer cell/T-cell lymphoma
	Enteropathy-type T-cell lymphoma
	Hepatosplenic T cell lymphoma
	Subcutaneous panniculitis-like T-cell lymphoma
	Mycosis fungoides
	Sézary syndrome
	Primary cutaneous anaplastic large cell lymphoma
	Peripheral T-cell lymphoma
	Angioimmunoblastic T-cell lymphoma
	Anaplastic large cell lymphoma

alone, with supportive measures such as red cell and platelet transfusions, erythropoietin to increase red cell count and antibiotics as needed. For some cases of NHL, supportive treatment may be used in the first instance and plasmapheresis in some patients where a B-cell lymphoma is producing large quantities of immunoglobulin molecules.

8.11.3 Myeloma

Myeloma is a rare haematological malignancy and it is a malignant proliferation of plasma cells which produces monoclonal paraproteins, usually IgG or IgA. It is usually seen in the elderly, with a medial presentation age of 60. It is slightly commoner in males and the incidence is four per 100 000. A major characteristic of myeloma is bone destruction, which is due to increased osteoclastic (bone breakdown) activity, as myeloma cells produce interleukin-6 (IL-6) and other factors to increase osteoclast numbers and activity. The huge number of paraproteins also results in deposition of light chains of antibodies in the renal tubules, with renal impairment resulting. The presentation of patients with myeloma is thus often due to bone pain from destruction of bones (with hypercalcaemia symptoms due to the release of calcium from bone), symptoms of renal failure and symptoms of bone marrow infiltration (anaemia, infection, sometimes bleeding). A full blood count may be normal, or show decreased Hb, WBC count and platelets. ESR is high and a blood film may show rouleaux formation – stacked red blood cells – due to the sticky paraprotein molecules sticking red cells together. Urea and creatinine levels may be raised due to renal impairment, possibly with raised serum calcium too. Bone marrow aspirate will show infiltration with plasma cells and X-rays will show characteristic areas of bone breakdown (see Figure 8.52). Serum protein electrophoresis and immunofixation will show a monoclonal band.

Treatment of myeloma involves supportive measures such as red cell transfusion and erythropoietin for anaemia, antibiotics for infections, bisphosphonate drugs to reduce osteoclast activity and pinning areas of thin bone before a fracture occurs. Myeloma is currently incurable and specific treatment regimes depend on the age and health of the patient; combinations of cytotoxic drugs with steroids and thalidomide derivatives, or using thalidomide derivatives alone, can improve survival. In younger patients, autologous stem cell transplants may also help prolong life. The potential benefit of allogeneic stem cell transplant is still unclear.

8.12 Complement

Complement is not one entity, but a group of around 25 different proteins found in plasma that 'complement' antibody activity. They are made in the liver, but can also be produced by macrophages, monocytes and some epithelial cells. Complement proteins are generally in an inactive precursor form in plasma, but once activated they interact in cascades that ultimately stimulate lysis of antibody-coated red blood cells (e.g. in haemolytic syndromes, or transfusion reactions), to help attract phagocytes into inflamed regions and stimulate phagocytosis, and to clear immune complexes. The two major pathways through which this can happen are the classical path-

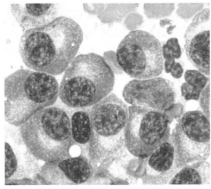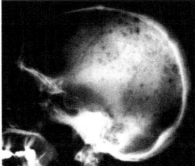

Figure 8.52 Bone marrow biopsy showing abnormal plasma cells (left); skull X-ray showing 'pepper pot skull' appearance in myeloma, caused by bone breakdown. From Hoffbrand et al, *Postgraduate Haematology*, Sixth Edition 2010, reproduced by permission of John Wiley & Sons Ltd

way and the alternative pathway, and both result in lysis of a target cell and the induction of opsonization, inflammation and phagocytosis.

8.12.1 The classical complement pathway

This pathway is triggered by IgG or IgM antibodies binding antigen, for example in a transfusion reaction with antibody binding a red cell antigen (Figure 8.53). It relies on there being a large enough number of these happening in close proximity to each other, which

results in two Fc regions of the antibodies being able to bind to one molecule of C1q. Since IgM is a complex of five antibodies this is more effective than IgG at triggering the classical pathway. Once the large C1q molecule has bound two Fc regions on Ig, it undergoes a conformational change that activates two C1r serine protease molecules and these in turn cleave another serine protease, C1s. Activated C1s cleaves C4 into two molecules, C4a and C4b, and C4b binds briefly to the membrane of red blood cells via a hydrophobic binding site. C1s also cleaves C2 into C2a and C2b, and C2a binds to C4b (a process requiring Mg^{2+} ions) on the red blood cell membrane

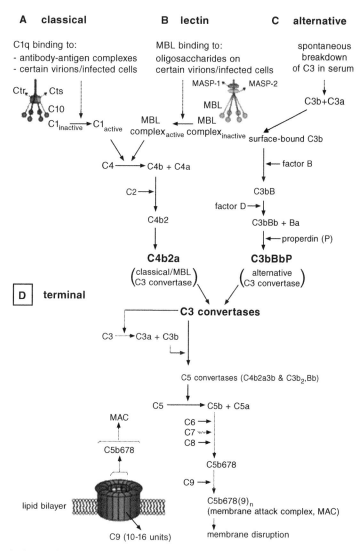

Figure 8.53 The classical complement pathway, with link to the alternative pathway shown. From Hoffbrand et al, *Postgraduate Haematology*, Sixth Edition 2010, reproduced by permission of John Wiley & Sons Ltd

to form the novel enzyme C4b2a. This new molecule splits C3 into C3a and C3b, and the larger C3b molecule can also bind to the red cell membrane or to C4b2a. If it binds to the latter, it produces the complex C4b2a3b which cleaves C5 into C5a and C5b. C5b can also bind to the red cell membrane and C5b molecules then undergo the addition of C6, C7 and C8 molecules to their structure, resulting in a very large molecule known as the membrane attack complex (MAC). The MAC is built up into a cylindrical structure at the surface of the target cell (a red blood cell here) and creates a pore at the surface of the red cell. These pores enable the movement of ions and small molecules between the red cell and its external environment resulting in the loss of the usual concentration gradients and ultimate equilibration. The red cell starts to swell as water moves in to dilute the large molecules that are unable to leave the red blood cell and ultimately it will burst (haemolysis).

The number of components of the complement pathway that can bind red cells once antibody has bound them means that there is considerable amplification of the original number of molecules bound to the cell after the antibody binding occurred and, moreover, it is much easier to test for the presence of complement than antibodies on the surface of red blood cells as there is so much more complement present.

Apart from lysis, the complement pathways also stimulate opsonization, phagocytosis and inflammation. C3a and C5a are chemotactic molecules for neutrophils, drawing them into areas of complement activation. Phagocytic cells (e.g. macrophages, neutrophils) have receptors for C3b and will bind cells coated with C3b — this binding triggers very efficient phagocytosis of the C3b-coated cell, for example the phagocytosis of of C3b-coated red cells by phagocytes in the spleen. C3b can also bind immune complexes, stimulating their phagocytosis. C3a, C4a and C5a can trigger degranulation of mast cells, with resultant histamine-induced vasodilation and plasma arriving in the area of complement activation; this results in more complement and immunoglobulin in the region too. Complement can also enhance platelet aggregation in blood clotting and the breakdown of fibrin in the fibrinolytic system.

8.12.2 The alternative complement pathway

The alternative complement pathway is misnamed, as it is simply a loop off the classical pathway in which

> **Box 8.10 Paroxysmal nocturnal haemoglobinuria**
>
> This results from an acquired deficiency of the enzyme that allows linkage of factors that protect red blood cells from complement to the red blood cell membrane, for example decay-accelerating factor (DAF), which decreases formation of C3 convertase. A lack of these proteins causes complement-mediated red cell haemolysis and the appearance of dark urine (due to haemoglobin) predominantly at night.

C3b production enhances further cleavage of C3 into more C3a and C3b (Box 8.10 and Figure 8.54). It is not dependent on the presence of antibody for activation but may be initiated by natural breakdown of C3 in plasma into C3a and C3b. Factor B (found in plasma) binds C3b to form C3bB. In the presence of factor D, this complex will be cleaved into Ba and Bb, and Bb will bind the C3bB molecule to form C3bBb. This complex catalyses the cleavage of C3 into C3a and C3b, and the cycle continues. Ultimately, all C3 would be exhausted if this continued unchecked, but it is kept under control by the instability of the C3bBb complex (which rapidly breaks down once formed) and by the conversion of C3b into an inactive form (C3bi) by the action of factors H and I in plasma.

8.12.3 Complement disorders

Absence or suboptimal function of most types of complement have been described in humans, and may be inherited. Deficiency of C1, C2 and C4 are associated with systemic lupus erythematosus (SLE), while deficiencies of C3, factor H and factor I are associated with an increased risk of bacterial infection.

8.13 Blood transfusion

Human blood groups are determined by the antigens present on the surface of an individual's red blood cells. Although some 30 blood grouping systems are recognized, it is the ABO and Rhesus systems which are of greatest clinical importance (Box 8.11).

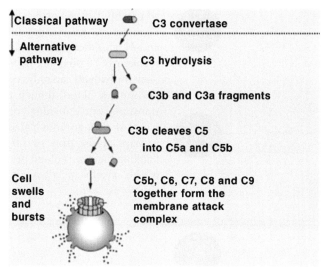

Figure 8.54   Focus on the alternative complement pathway

8.13.1 The ABO system

The great importance of the ABO blood grouping system in blood transfusion arises from the fact that humans develop natural antibodies against the ABO antigens early in life, before ever apparently being exposed to other ABO antigens — a phenomenon sometimes called Landsteiner's Law. (Contrast this to other types of antibodies in humans, which we acquire during life after exposure to infection or foreign antigens.) These naturally occurring anti-A and anti-B antibodies can produce severe haemolytic reactions to transfusions of incompatible red blood cells — such reactions can be severe enough to prove fatal. ABO antibodies start to appear at around 18 months of age

and it is thought that this is due to exposure to similar carbohydrates in food after weaning of infants.

An individual's ABO blood group is determined by codominant genes A, B and O, which encode the expression of specific carbohydrate antigens on the surface of the red blood cell. The A gene codes for expression of the A antigen and the B gene for the B antigen but the O gene codes for the H substance, which is not generally antigenic. These three molecules are nonetheless very similar: the H substance is the 'basic building block', while the A and B antigens comprise additional N-acetylgalactosamine and D-galactose respectively, added onto the end of the H substance structure by specific enzymes. These antigens are illustrated in Figure 8.55.

The antigens, antibodies and possible genotypes for each ABO blood group are shown in Table 8.6.

There are subtypes of A and B antigen within the ABO system; most individuals who are blood group A have A_1 antigens and express large numbers of these. A_3, A_4 (sometimes designated A_X), A_Y and A_{end} antigens can also be found in the population, and are associated with less antigen presence on the red cell surface than for the A_1 antigen. The B antigen similarly has other forms that are more poorly expressed, including B_3, B_w, B_x and B_m. These antigen variations are unimportant in terms of transfusion medicine.

The ABO system antigens are also expressed in body fluids in the 80% of the population who have the

Box 8.11 Landsteiner and blood discovery

Karl Landsteiner was a scientist in Vienna and in 1900 he took red cells suspensions and serum from six men in his laboratory (including himself); by mixing them in different combinations and noting the agglutination patterns, he first described the A, B and O blood groups — sheer luck awarded him two men of each group, against all the odds in a random selection of six men! Landsteiner was awarded the Nobel Prize for Medicine in 1930.

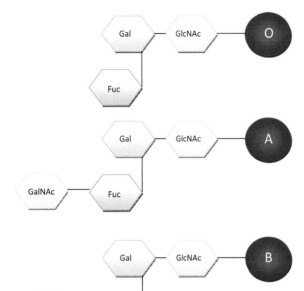

Figure 8.55   The antigens of the ABO system

'secretor' gene — they can be found in saliva, for example, or in gastric juices.

8.13.2   The hh blood group

A very rare blood group known as the hh or Bombay blood group is found in around 0.0004% of the world population, though more frequently in the Mumbai (formerly Bombay) region of India. Individuals with this blood group inherit two copies of a recessive h gene which results in no expression of the H antigen and thus no expression of any ABO antigens, regardless of which ABO blood group genes they have inherited. Their serum will contain anti-A and anti-B, but also anti-H, and their blood will appear to be Group O in regular ABO typing. The hh blood type is written as O_h. Such individuals can only receive blood from the same hh/Bombay blood, though their blood can safely be transfused into individuals of all ABO groups. Since hh blood is rare, an anticipated need for transfusion — perhaps at the time of surgery — will often entail banking their own blood prior to the event, but blood of hh group is often unavailable in emergency situations.

8.13.3   The Rhesus system

Individuals may be either Rhesus positive or Rhesus negative and this is determined by whether or not they express the Rhesus D antigen on the surface of erythrocytes — people who possess at least one D antigen are said to be Rhesus positive (Box 8.12). The Rhesus (Rh) system comprises some 45 different antigens but C and c, E and e, and D and absence of D (denoted d) are the most important clinically. Of these, the D antigen (or its absence) is the most critical to establish, as RhD negative individuals have a high incidence of generating anti-RhD antibodies if exposed to RhD positive blood. Rh antigens are strongly expressed in the fetus too, which has important implications during pregnancy (see Section 8.13.5).

The Rh antigens are large transmembrane polypeptide, with six loops of the polypeptide exposed on the external surface of the red cell where the Rh antigens are expressed (Figure 8.56). The polypeptides are combined with an Rh glycoprotein into tetrameric

Table 8.6   The ABO system antigens, antibodies and genotypes, and frequency of each ABO group in the UK population[*]

	Antigens on red blood cell surface	Antibodies in plasm	Genotype	Frequency in UK population 2009
Group A	A	Anti-B	AA or AO	42%
Group B	B	Anti-A	BB or BO	10%
Group AB	A + B	None	AB	4%
Group O	None	Anti-A + Anti-B	OO	44%

[*]Note that an individual can only possess antibodies against antigens they do not themselves possess (e.g. Group A individuals only have anti-B antibodies) and the H antigen found in Group O individuals is not generally antigenic (so Group O individuals are considered to have no antigens, and there is no 'anti-O' antibody). It is also of interest to note that individuals who are heterozygous for Group A or Group B genes (genotypes AO or BO) will have half as many A or B antigens as homozygotes (AA or BB).

The Rhesus system was also discovered by Karl Landsteiner, along with Alexander Wiener, in 1940. It is named after the Macaque rhesus monkeys, whose red cells were injected into rabbits and gave rise to the production of an 'antiRhesus' antibody that reacted against 85% of the red cells of blood donors in New York. They named the antigen that caused this reaction the, 'Rhesus factor', and blood that reacted to the antibody was said to be 'Rhesus positive'.

Table 8.7   Fisher's model of Rhesus gene combinations on one chromosome

Haplotype	Symbol	Rh status
DCe	R_1	Rh positive
DcE	R_2	Rh positive
Dce	R_0	Rh positive
DCE	R_z	Rh positive
dCE	r^1	Rh negative
dcE	r^n	Rh negative
dce	r	Rh negative
dCE	r_y	Rh negative

structures to form a Rhesus core complex; some individuals inherit a defective gene for this essential glycoprotein and are said to have a Rh_{null} phenotype with no Rh genes expressed. The function of the Rh complex is unknown, but its structure strongly suggests it is an ion channel.

8.13.3.1   Inheritance of rhesus genes

Combinations of genes that can be inherited in the Rh system are complex and a lot of investigation went into elucidating this following the discovery of Rh antigens. Fisher and Race coined a model in which there are three pairs of closely-linked alleles (D/d, C/c and E/e), allowing for eight different possible combinations of Rh genes on one chromosome and resulting in seven common combinations of Rh genes

overall in the population. These are given below in Tables 8.7 and 8.8.

Fisher's model is now known not to be entirely correct, but nonetheless it works well in clinical practice and continues to be used. In 1986, Tippet proposed that the Rh system is actually based on two closely linked gene loci on chromosome 1, with one ten exon gene for the *RHD* gene, and another separate ten exon gene for *RHCcEe* which encodes the alleles (C or c and E or e) that an individual expresses; this is now known to be the correct model for inheritance of Rh genes. Interestingly, the *RHD* gene is thought to have arisen from duplication of the *RHCcEe* gene during evolution. Rh negative individuals lack a functioning *RHD* gene (Box 8.13).

Some Rh positive individuals may express fewer of their D antigens than normal for unknown reasons, a

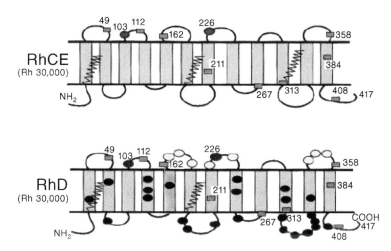

Figure 8.56   The structure of the Rhesus antigen

Table 8.8 Fisher's model of Rh genotypes given in order of approximate frequency

Geno-type	Fisher shorthand	Frequency	Rh status
DCe/dce	R_1r	34%	Rh positive
DCe/DCe	R_1R_1	17%	Rh positive
dce/dce	rr	15%	Rh negative
DCe/DcE	R_1R_2	14%	Rh positive
DcE/dce	R_2r	11%	Rh positive
Dce/dce	R_0r	2%	Rh positive
DcE/DcE	R_2R_2	2%	Rh positive
DCe/Dce	R_1R_0	2%	Rh positive
Other combinations	Uncommon (< 1%)		

condition designated D^U — this may make their blood harder to correctly group (historically, such donors were sometimes mistakenly grouped as Rh negative), but blood from these donors would stimulate the production of anti-D if transfused into an Rh negative recipient.

8.13.3.2 Transfusion and Rhesus groups

At its simplest, RhD positive patients should have RhD positive blood (it is safe for RhD positive individuals to have RhD negative blood but, since this is in short supply, it is generally reserved for RhD negative patients). RhD negative patients should usually have RhD negative blood to prevent the formation of the highly potent anti-D antibody. However, at times blood may need to be provided that is also matched to a patient's Ce/Ee alleles, since patients who have previously been pregnant, or transfused with blood matched solely to their RhD status, may have formed anti-C, anti-c,

anti-E, or anti-e antibodies. For example, recipients with the genotype DCe/dce (R_1r) often develop anti-E antibodies if transfused with blood that is RhE positive and future blood will need to be provided that is RhE negative.

8.13.4 Laboratory blood grouping

In the modern laboratory, blood grouping is mostly done using automated systems which use monoclonal antibodies against the ABO and Rh antigens to arrive at a full blood group, and an antibody screen will be performed to look for preformed antibodies that will affect the choice of units of blood for transfusion (or may cause a problem for the fetus during pregnancy). These automated systems commonly use microcolumns of gel with antibodies or plasma lying on top of a gel column. A solution of a patient's is added to an antibody (anti-A, anti-B or anti-D) and their serum is added to known Group A and Group B cells as a cross-check. The mixtures are allowed time for any agglutination to occur, then the tubes are centrifuged. Where there is no agglutination (e.g. a combination of Group B cells with anti-A antibody will not cause agglutination), red cells will fall through the gel and appear at the bottom of the column to give a negative result for that column. Where there is agglutination (e.g. Group B cells with anti-B antibody), the agglutinated cells will remain on the top of the gel column as the clusters of cells are too large to pass through the pores of the column, giving a positive result for that column. The results can be read automatically if they are clear. This is illustrated in Figure 8.57.

8.13.5 Other important human blood grouping systems

There are more than 600 different antigens identified on the surface of human red blood cells and, following transfusion or pregnancy, antibodies may be produced against a wide variety of red cell epitopes. The ABO and Rhesus blood group systems are the most important clinically' but several other major systems must also be considered.

Blood group antigens may be divided into carbohydrate and protein structures. Protein antigens are determined by the inheritance of genes for different

> **Box 8.13 Rhesus factor advantage**
>
> Around 15% of the European population is Rh negative, implying that the r allele must have conferred a survival advantage at one time, particularly since Rh negativity was associated with poorer pregnancy outcomes. The Rh complex is thought to be an ion channel or pump and some studies have suggested that carriers of the r allele (i.e. Rh positive heterozygotes) have a degree of protection against toxoplasma infection.

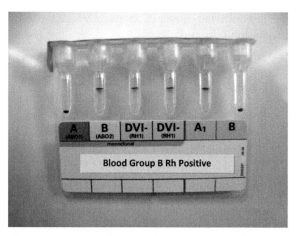

Figure 8.57 Use of microcolumn technology for blood grouping

membrane protein molecules on red cells. They are may be further divided into:

1. membrane channels, for example Rhesus, Kidd;
2. structural proteins, for example MNS;
3. membrane-bound enzymes, for example Kell;
4. cell adhesion molecules, for example Lutheran;
5. chemokine receptors, for example Duffy;
6. complement regulatory proteins, for example Knops.

The carbohydrate antigens are the product of genetically inherited glycosyl transferase enzymes that attach specific carbohydrate molecules to the surface of red blood cells. The ABO antigens are an example of carbohydrate red cell antigens and the ABO blood group is determined by which carbohydrates are present, according to which genes for the glycosyl transferase enzymes have been inherited.

The naming of the blood group systems historically lacked standardization and in 1980 the International Society of Blood Transfusion (ISBT)'s Working Party on Red Cell Surface Antigens gave each known red cell antigen a unique six digit number. In this system, the first three digits represent the blood group system and the second three represent the antigen.

8.13.5.1 Protein antigens

Kell system

The Kell system is represented by the symbol KEL; ISBT number is 006. Some 24 Kell antigens are

associated with the highly polymorphic Kell system and they are epitopes on the Kell protein which is a transmembrane endopeptidase molecule. After the ABO and Rh systems, the Kell system is the most likely to cause transfusion problems following transfusion or pregnancy and is thus of considerable clinical importance. Six major alleles of the Kell system are most clinically important: these are the K (Kell) antigen, k (also known as Cellano), Kp^a (Penney), Kp^b (Rautenberg), Js^a (Sutter) and Js^b (Matthews). Antibodies may be produced against any of these Kell antigens following transfusion or pregnancy, and transfusion reactions or haemolytic disease of the newborn may result.

The Kell protein is present on the red cell membrane bound to another protein, K_x, via a disulphide bond; absence of the K_x protein thus greatly diminishes expression of the Kell protein, though absence of the Kell protein (due to homozygosity for the rare K_0 genes) has no effect on K_x expression. Individuals who are homozygous for K_0 and express no kell protein may produce an antiKell antibody (called anti-K_U) if exposed to Kell antigens.

MNSs system

The MNSs system is represented by the symbol MNSs; ISBT number is 002. Some 45 antigens of the MNS system have now been identified, of which the major ones are M, N, S and s. The antigens of this system are glycophorins A and B, structural components which span the red cell membrane and render it hydrophilic; the M and N genes are expressed on glycophorin A and the S and s genes on glycophorin B. The genes are found in codominant pairs at two closely-linked loci, with the major combinations for MN in the population being M+N−, M+n+, M−N+ and, rarely, M−N−. (note that since glycophorin A (MN) acts as a receptor for the malarial parasite *Plasmodium falciparum*, red cells that lack glycophorin A [M−N−] have resistance to this strain of malaria.) Common population combinations for Ss are S+s−, S+s+, S−s+ and, very rarely, S−s−.

Antibodies against M, N, S or s may be acquired by individuals without the relevant antigens, and can cause both haemolytic transfusion reactions and haemolytic disease of the newborn.

Duffy system

The Duffy system is represented by the symbol FY; ISBT number is 008. The Duffy protein is a nonspecific chemokine receptor that is found on the surface of the red blood cell and was discovered after an antibody was found in a haemophiliac patient who had received multiple transfusions. The two major alleles of the FY system are Fy^a and Fy^b, with four common phenotypes of the Duffy system described: Fy(a+b−), Fy(a+b+), Fy(a−b+) and Fy(a−b−). This last phenotype, Fy (a−b−) (the 'minus minus' phenotype) is almost unknown in Caucasians but is common in Black populations of African descent, and reflects the Duffy antigen's role as the site of attachment of the malarial parasites, *Plasmodium vivax* and *Plasmodium knowlesi*. Individuals who have the Fy(a−b−) genotype are thus at a survival advantage in malarial regions, hence the frequency of this genotype in Africa.

Anti-Fy^a and anti-Fy^b antibodies may be acquired following transfusion or pregnancy, and can cause transfusion reactions and haemolytic disease of the newborn. However, the Duffy antigen is only moderately immunogenic.

Kidd system

The Kidd system is represented by the symbol JK; ISBT number is 009. The Kidd (Jk) antigen is found on a urea transport protein on the red cell membrane and the two major allelic forms are Jk^a and Jk^b, which are codominant; thus the major phenotypes found in the population are Jk(a+b−), Jk (a+b+), Jk(a−b−) and Jk(a−b+). Rarely, individuals may be homozygous for the Jk null gene and express no Jk antigen. Antibodies against Jka and Jkb may be acquired following exposure during pregnancy or via transfusion − for example, a Jk(a+b−) individual may form anti-Jk^b antibodies. However, anti-Jk antibodies tend to be unstable and levels often decline rapidly. They can cause mild haemolytic disease of the newborn or haemolytic transfusion reactions.

Lutheran system

The Lutheran system is represented by the symbol LU; ISBT number is 005. Lutheran antigens are expressed on glycoproteins on the red blood cell membrane that function as cell adhesion molecules and the major antigens are Lu^a and Lu^b, coded for by codominant alleles. The major phenotypes in the population are Lu (a+b−), Lu(a+b+) and Lu(a−b+), though rarely some individuals inherit a null Lu phenotype, Lu (a−b−). Antibodies against Lu^a and Lu^b may be acquired after pregnancy or transfusion; they are generally weak and may cause mild haemolytic disease of the newborn but they are an uncommon cause of transfusion reactions.

8.13.5.2 Carbohydrate antigens

Lewis system

The Lewis system is represented by the symbol LE; ISBT number is 007. The major antigens of the Lewis system are Le^a and Le^b although Lewis antigens are not strictly red cell antigens. They are actually secreted by exocrine epithelial tissue in the body and then reversibly adsorbed onto the surface of red cells from plasma. The Le^a antigen is encoded by the presence of at least one *Le* gene and this antigen interacts with the ABO secretor gene (see Section 8.13.1) where this is present. This results in the conversion of Le^a into Le^b and gives Le(b+) red cells, the most common phenotype (interestingly, Le^b acts as a receptor for *Helicobacter pylori* infection, which causes gastritis). If no secretor gene is present, only Le^a remains in plasma with just Le(a+) positive red cells. Individuals in a population can thus have the phenotypes Le(a+b−) and Le(a−b+) but not Le (a+b+). Individuals who lack an *Le* gene (i.e. who are homozygous for *lele*) will not express any Le antigens on their red cells and have the phenotype Le(a−b−).

Antibodies against Le^a and Le^b may be found naturally in people without the relevant antigen; however, they do not cause haemolytic disease of the newborn although anti-Le^a can cause some haemolytic transfusion reactions. Lewis antigens may be lost from stored red cells so it is always advisable to use the freshest possible samples for Lewis grouping of red cells.

P system

The P system is represented by the symbol P1; ISBT number is 003. The major antigens of the P system are P$_1$, P and P^k, although only P$_1$ is considered a true antigen. Individuals may express just P, P$_1$ and P, P$_1$ and P^k, just P^k or, rarely, no P antigens at all. Naturally-occurring antibodies may be found in serum against antigens an individual does not possess and these may be anti-P$_1$, anti-P or anti-PP$_1$P^k in those rare individuals who express no P antigens. These antibodies may cause haemolytic transfusion reactions but typically do not cause haemolytic disease of the newborn.

Haemolytic disease of the newborn (HDNB)

Haemolytic disease of the newborn (HDNB) is caused by the presence of IgG antibodies in the mother that can cross the placenta and coat the red cells of the fetus to bring about their destruction. The most clinically important form of HDNB is Rhesus incompatibility. Many babies were stillborn several decades ago, and in 1939 Levine and Stetson published the first paper, 'An unusual case of intra-group agglutination', in which they first described Rhesus haemolytic disease of the newborn. Since approximately 15% of pregnant mothers are Rhesus negative — while 85% of their male partners will be Rhesus positive — most of these women's pregnancies will be with Rhesus positive fetuses; before effective prophylaxis was available, one in 200 mothers developed anti-D antibodies following exposure to their baby's Rhesus positive red cells during the pregnancy, or around the time of the birth (fetal-maternal haemorrhage, or FMH). Once anti-D was made by the mother in response to exposure to the baby's RhD antigens, an increasing proportion of the babies of Rhesus negative mothers were stillborn with each subsequent pregnancy. The principles of Rh sensitization during pregnancy are shown in Figure 8.58.

Once anti-red cell antigen IgG has crossed the placenta and coated fetal red cells, these will agglutinate and be removed by macrophages in the circulation or spleen, causing a rise in fetal bilirubin levels. This is not usually a problem in itself before birth, as the placenta passes excess bilirubin to the maternal circulation, where it is safely processed by the maternal liver. However, after birth bilirubin levels can rise sufficiently high to cause early onset jaundice (jaundice within the first 24 h or so of birth is always pathological). Very high levels of bilirubin are toxic to the central nervous system and can cross the blood–brain barrier once at sufficient concentration, so treatment should be instigated to protect the newborn baby from brain damage due to bilirubin (kernicterus). Using ultraviolet light to break down bilirubin in the skin may be sufficient (see Figure 8.59), but for higher bilirubin concentrations, an exchange transfusion may be necessary. In an exchange transfusion, the two veins in the umbilical cord are used to remove the baby's blood and replace it with blood that is compatible with the maternal antibodies in the baby's circulation — so, for example, a RhD positive baby whose red cells were being destroyed by anti-D would have that RhD positive blood drawn off and receive a transfusion of RhD negative blood to protect him/her until maternal antibodies are broken down.

If there are high levels of a potent IgG during the pregnancy, ultimately the fetus will become severely anaemic, pale and swollen, and will often die in the uterus or soon after birth. This condition is known as hydrops fetalis, and the commonest cause of this condition is the highly potent anti-D. Rhesus haemolytic disease of the newborn has thankfully been very

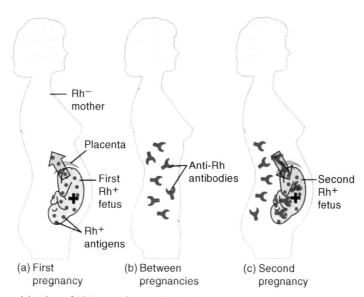

Figure 8.58   Rhesus sensitization of RhD negative mother during pregnancy

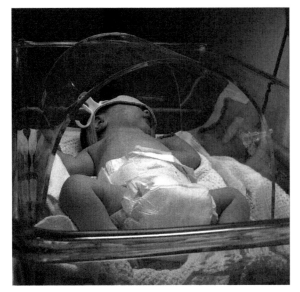

Figure 8.59   A baby undergoing ultraviolet light treatment for moderately severe haemolytic disease of the newborn (this baby was suffering from ABO incompatibility at birth — the mother is blood group O and her anti-A antibodies caused haemolysis of her newborn son's Group A cells)

rare since the advent of prophylactic treatment for RhD negative pregnant women.

8.13.6   Laboratory detection of maternal antibodies

During pregnancy, all women are tested for ABO and RhD blood group and screened at several points during the pregnancy for the presence and amount (titre) of any antibodies against red cell antigens that could cross the placenta, using an indirect antiglobulin test (IAT — also known as an indirect Coombs' test). The fetus is assumed to be at risk from high titre red cell antibodies and the fetus's condition can be assessed using ultra-sound scans, amniocentesis (with measurement of bilirubin in the amniotic fluid), or using fetal blood sampling via the umbilicus in the process of cordo-centesis to assess fetal haemoglobin and bilirubin levels. Early in gestation the baby's blood group can be confirmed by obtaining fetal cells with chorionic villus biopsy (which removes early placental tissues and is possible from 8 weeks of gestation). Exchange transfusion is also possible while the fetus is still in the uterus using the umbilical veins and the blood is crossmatched with maternal serum in this situation to check its compatibility with maternal antibodies. A direct antiglobulin test (DAT — also known as a direct Coombs' test) can be performed on fetal blood to confirm that fetal cells are coated in antibody.

8.13.6.1   Indirect antiglobulin test (IAT)/indirect Coombs test

The indirect antiglobulin test (IAT) looks for the presence of anti-red cell antibodies in serum (Figure 8.60). In the IAT, a panel of washed red cells containing a variety of common red cell antigens is incubated with maternal serum in separate tubes. If the maternal serum contains antibodies against any red cell antigen, the maternal antibodies will bind to the

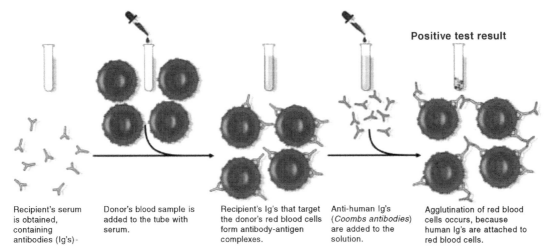

Positive test result

Recipient's serum is obtained, containing antibodies (Ig's)-

Donor's blood sample is added to the tube with serum.

Recipient's Ig's that target the donor's red blood cells form antibody-antigen complexes.

Anti-human Ig's (*Coombs antibodies*) are added to the solution.

Agglutination of red blood cells occurs, because human Ig's are attached to red blood cells.

Figure 8.60   Indirect antiglobulin test

relevant red cell antigen in that tube. The red cells are then washed with saline and incubated with antihuman globulin (also known as Coombs' reagent). If any red cells are coated with maternal antibodies, they will be agglutinated by the antihuman globulin, and this will be apparent.

8.13.6.2 Direct antiglobulin test (DAT)/direct Coombs test

The direct antiglobulin test (DAT) looks for the presence of antibody on red cells (Figure 8.61). In the DAT, the patient's red cells are washed with saline and then incubated with antihuman globulin (Coombs reagent). If anti-red cell antibodies are present on the surface of the red cells, they will be visibly agglutinated by the antihuman globulin.

8.13.6.3 Prevention of HDNB in RhD negative pregnant women

Early researchers working on Rhesus HDNB noticed that RhD negative women were less likely to be sensitized to the D antigen during pregnancies with RhD positive babies if mother and baby were ABO incompatible. They realized that this effect was likely to be because preformed ABO antibodies were destroying incompatible fetal red cells as soon as they got into the maternal circulation and before RhD sensitization could occur. In 1960, Finn suggested that RhD negative women could receive injected anti-D to destroy RhD positive fetal cells in the maternal circulation before they could trigger RhD sensitization in the mother (with subsequent seriously problematic production of her own anti-D). Clinical trials were set up in several countries in the 1960s, which showed the success of giving exogenous anti-D to RhD negative women immediately after the birth of an RhD positive baby in preventing RhD sensitization; in 1968, anti-D was licensed for use.

The use of anti-D was recently extended so that pregnant women also receive two injections of anti-D during pregnancy (at 28 and 34 weeks of gestation) as well as immediately after the birth, and this further prevents the small number of RhD sensitisation events due to feto-maternal haemorrhage during pregnancy. A dose of 500IU of anti-D is used during pregnancy, and following delivery, and this is sufficient to clear less than 4 mL of RhD positive fetal blood in the maternal circulation. However, sometimes larger FMH can occur at the time of delivery, and for this reason, quantification of FMH is performed after delivery in RhD negative women whose baby is RhD positive so that they can be given further anti-D if necessary.

8.13.6.4 Assessment of feto-maternal haemorrhage

Immediately after birth, blood samples are taken from the babies of RhD negative women to have the baby's blood group ascertained as well as from the mother to

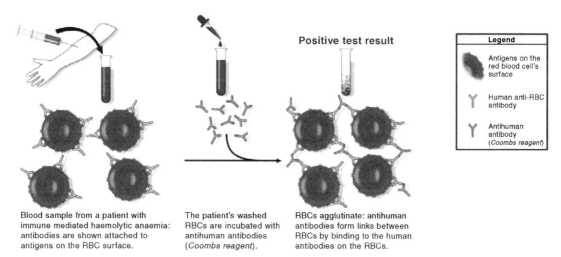

Positive test result

Legend
Antigens on the red blood cell's surface
Human anti-RBC antibody
Antihuman antibody (*Coombs reagent*)

Blood sample from a patient with immune mediated haemolytic anaemia: antibodies are shown attached to antigens on the RBC surface.

The patient's washed RBCs are incubated with antihuman antibodies (*Coombs reagent*).

RBCs agglutinate: antihuman antibodies form links between RBCs by binding to the human antibodies on the RBCs.

Figure 8.61   Direct antiglobulin test

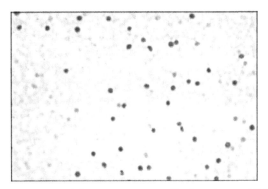

Figure 8.62   Kleihauer test to quantify the volume of FMH

quantify the volume of fetal blood that is present in the maternal circulation. If the baby is also found to be RhD negative after birth, then no further action is necessary. However, if the baby is RhD positive (as the majority will be), the blood sample from the mother is tested to see how many fetal cells are present in a given volume of maternal blood. This is done using the Kleihauer test, in which a blood film is made from the maternal blood sample following delivery and treated with haemotoxylin and hydrochloric acid before counterstaining with eosin (Figure 8.62). Adult red cells contain predominantly HbA, which is removed by acid such that maternal cells appear as 'ghost cells'. Fetal cells contain predominantly HbF, which is resistant to acid – thus the different populations of red cells from mother and baby can easily be distinguished. The ratio of fetal/maternal cells is counted under a microscope and thus the extent of FMH can be calculated. Large FMH will require a further dose of anti-D.

8.13.6.5   Other causes of HDNB

Anti-D is the most clinically important cause of HDNB since it has the most potential to cause serious morbidity and mortality in the fetus, but the commonest cause of HDNB is actually ABO incompatibility. A large number of pregnancies will be ABO incompatible but this is not usually a problem as ABO antigens are not strongly expressed by the fetus in most pregnancies. However, in some pregnancies the mother may produce high titre anti-A or anti-B IgG in response to antigens on the fetus's red cells and in this situation they will cross the placenta and a degree of haemolysis of the fetal red cells is seen. This is not usually serious enough to cause harm to the fetus, though a mild degree of early onset jaundice may be experienced after

the birth. ABO incompatibility rarely requires treatment.

Other IgG antibodies can also cause HDNB, including anti-K, anti-C, anti-c, anti-E, anti-e and anti-Fy[a]. Of these, anti-K and anti-c are the most likely to cause clinical problems but these are rare; interestingly, anti-K antibodies cause suppression of red cell haemopoiesis rather than haemolysis of fetal red cells.

8.14   Blood products

In the UK, the NHS Blood Service collects blood in units of around 470 mL from volunteer donors at donation sessions. Whole blood is collected from donors into sterile blood bags containing an anticoagulant. It is tested for ABO and Rhesus blood groups and for the presence of transmissible infections before being processed into blood components and blood products – note that whole blood is rarely transfused back into patients.

8.14.1   Red cell concentrates

Red cell concentrates are manufactured from units of donated blood by extraction of leucocytes ('leucocyte depletion') and plasma and replacement of the latter with 100 mL of an additive solution. They may be further washed in saline for patients who have previously suffered anaphylactic reactions from transfusion. Red cell packs last for 5 weeks and are stored at 4 °C until required. Red cells are used for patients with low Hb who require blood transfusion and typically they are infused with sufficient units to bring Hb to 10 g/dL – the rule of thumb commonly used is that one unit of red cells equates to an Hb rise of 1 g/dL, so that a patient with an Hb level of 7 g/dL would typically receive three units of red cells where clinically necessary. Note that a low Hb is not an automatic reason for transfusion. In an otherwise well patient who is not at risk of bleeding for any reason, iron supplementation and FBC monitoring will often suffice.

8.14.2   Platelet concentrates

Cell separation on whole blood allows separate preparation of packs of concentrated platelets, which last around 5 days. They need to be stored at 22 °C and constantly agitated to prevent platelet aggregation. Platelet concentrates are used for patients with thrombocytopenia and bone marrow failure.

8.14.3 Fresh frozen plasma

Around 200 mL of plasma is removed from a unit of whole blood in creating red cell concentrates and this plasma is frozen down ('fresh frozen plasma', or FFP) to be used in patients with acquired clotting factor deficiencies since it contains all clotting factors.

8.14.4 Cryoprecipitate

Cryoprecipitate comprises the ~20 mL supernatant produced from allowing frozen FFP to thaw to 4 °C. It contains fibrinogen along with von Willebrand factor (vWF) and factor VIII:C and is thus particularly useful in conditions where fibrinogen is depleted, such as disseminated intravascular coagulation (DIC).

8.14.5 Immunoglobulins

Normal immunoglobulin is a preparation made from human plasma and encompasses a range of Ig, which is useful in patients with hypogammaglobulinaemias. Single immunoglobulin preparations are made from the plasma of blood donors who have high titre antibodies against specific antigens, for example anti-Hepatitis B Ig, anti-D.

8.14.6 Clotting factors

Patients with haemophilia generally lack factor VIII or factor IX and recombinant concentrates of these factors are now available to treat haemophilia. In areas where these are unavailable, concentrates may be manufactured from pooled plasma of blood donors.

8.14.7 Albumin

Two different human albumin solutions are available, at either 4.5% or 20% albumin concentration. They are indicated in some patients with hypoalbuminaemia and in plasma exchange procedures.

8.14.8 Autologous blood transfusion

In some situations, a patient's blood may be collected and reinfused into them at the time of surgery. This may be done as a predeposit of blood before planned surgery, such that that a patient donates a few units of blood at weekly intervals prior to their surgery and then receives it back. Patients may also have one or two units of blood removed just prior to surgery (effectively diluting the blood that remains in their circulation so that less red cells are lost during surgery: this is preoperative haemodilution) and receive it back during or after their surgery to replace lost blood. Cell salvage is also increasingly being used to collect blood lost during planned surgery for retransfusion into the patient, though not all operations are suited to this (e.g. bowel surgery releases bacteria that make cell salvage from the site unsafe).

8.14.9 Adverse reactions to blood transfusion

Most reactions to blood products are minor and easily managed. However, in some rare situations, serious, life-threatening reactions to transfusion of blood components can develop. Any new symptoms that appear while a patient is being transfused must be taken seriously as they may be the first warnings of a serious reaction. Symptoms of a reaction include fever, rash, high or low blood pressure developing, shortness of breath, chest or abdominal pain, nausea and generally feeling unwell. Transfusion reactions may be divided into:

- febrile nonhaemolytic transfusion reactions;
- acute haemolytic transfusion reactions;
- reaction to infusion of a bacterially contaminated unit;
- transfusion associated circulatory overload (TACO);
- transfusion related acute lung injury (TRALI);
- severe allergic reaction;
- anaphylaxis.

Where symptoms develop that may suggest an adverse reaction to a transfusion, the transfusion should be stopped immediately and the patient's basic observations done (i.e. pulse, blood pressure, temperature, respiratory rate and O_2 saturation). The details of the unit of blood should again be checked against the patient's details on their wristband to check that the right unit is being given to the right patient. If the patient is otherwise well and a rise in temperature of $<1.5\,°C$ has occurred in isolation, paracetamol can be administered and the transfusion continued at a slower rate — this usually means a febrile nonhaemolytic transfusion reaction. If an urticarial rash is present an antihistamine may be administered and the transfusion continued at a slower rate, as this would indicate a mild allergic reaction. However, more serious reactions may be due to ABO incompatibility of the unit, bacterial infection of the unit, fluid overload, transfusion-related acute lung injury or serious allergic reactions, and these require more specialized medical management to reverse associated symptoms. The transfusion must be stopped immediately in these situations and the unit returned to the transfusion laboratory for tests on the component of blood.

8.14.10 Safety of blood products

Ensuring the safety of blood products begins with donor selection and, in the UK, the National Health Service Blood and Transplant (NHSBT) service has stringent guidelines for who is eligible to donate blood. A donor health check survey is taken by each donor every time they donate blood. Some guidelines within this (e.g. minimum age and weight, and minimum time between donations of 16 weeks) aim to keep the donor safe and ensure they do not come to harm through donating blood, but others (e.g. ineligibility to donate after recent travel to certain regions of the world, for 6 months after ear-piercing or tattoos, for 12 months after having hepatitis or jaundice, or where any member of their family has had Creutzfeld—Jacob disease) aim to ensure that infections are not present in any units of blood collected. The donor is also asked to give an initial drop of blood which is dropped into standardized concentrations of copper sulphate solution to check their haemoglobin level is above the minimum acceptable limit for their gender (currently 13.5 g/dL for males and 12.5 g/dL for females). Donated units of blood undergo rigor-

ous testing for a range of blood-borne infections, including HIV, Hepatitis B and C and syphilis. Some units may be screened for cytomegalovirus too, as CMV-negative units are required for transfusion in immunosuppressed patients.

8.14.11 Ensuring the right patient receives the right blood product

In the UK, patients receiving blood products in the hospital setting is a common aspect of patient care, but the pathway for providing the correct product for the correct patient is complex and has occasionally led to fatal errors where a patient has been given the wrong blood product — ABO incompatible red cell transfusions are one example. Several steps are involved in the pathway of ensuring the correct blood product reaches a patient: a sample for blood grouping and crossmatching must be correctly labelled with full details of the patient and with barcoding of the sample at the laboratory to track the specimen through the lab. Blood grouping and crossmatching is performed in the laboratory that has appropriate quality assurance and quality control in place. Once blood is crossmatched and one or more units for the patient is issued to a blood fridge, the correct units must be collected for the patient it has been ordered for and then units must be crosschecked against the patient's identification details (usually a hospital wristband, ideally with verbal checking with the patient of their identity too) before they can be administered to the patient. Any of these stages can be error-prone and a lot of effort has gone into ensuring that procedures are streamlined and clear protocols are followed to keep the patient safe. The use of barcoding of specimens in the laboratory, with automated blood grouping and electronic crossmatching in well-defined situations, has seen the laboratory error rate in transfusion adverse events fall to very low levels. Most errors now relate to mislabelling of the original blood sample for grouping and crossmatching, and to the wrong unit being given to a patient on the ward.

The UK was the first country in the world to set up a professionally led 'haemovigilance' scheme, Serious Hazards of Transfusion (SHOT), in which anonymized data on serious adverse events and reactions to blood transfusion is collected via the Serious Adverse Blood Reactions and Events (SABRE) reporting website and analysed. SHOT produces an annual report and recommendations which are circulated to all relevant bodies (including blood and transplantation services, Departments of Health and all reporting hospitals in the UK). Analysis of where adverse events occur in the transfusion pathway has led to several recommendations of change, including trials of electronic patient identification at the bedside to match unit to patient (e.g. using barcodes on patient wristbands and using patient identification cards for patients having regular transfusions in the outpatient setting).

8.15   Haemopoetic stem cell transplantation

The first successful human tissue transplant occurred in 1954 when a kidney transplant was carried out by Joseph Murray. Since that time, successful transplants of other tissues as various as heart, lung, liver, cornea, pancreas and bone have been achieved and here we will consider transplantation of bone marrow, peripheral blood stem cells and umbilical stem cells in patients with serious haematological and immunological disorders.

With most types of transplantation, the danger arises from recognition by the recipient's immune system that a foreign tissue has been implanted, with ensuing rejection. In the case of allogeneic bone marrow and blood stem cell transplantation into a deliberately immunosuppressed recipient, this danger is turned on its head; healthy white blood cells in the donated tissue can recognize the recipient's tissues as foreign, resulting in 'graft versus host disease' (GvHD). This can prove serious or even fatal for the recipient and its prevention relies on a close match of HLA markers between donor and recipient.

8.15.1   Matching donor and recipient for bone marrow and blood stem cell transplants

In the ideal blood stem cell transplant scenario, donor and recipient would be genetically identical to completely eliminate the risk of GvHD. This is the case where a prospective recipient has a healthy iden-

tical twin or, in the case of autologous transplants, in which diseased marrow is collected from a patient and treated (e.g. with chemotherapy to attempt to eradicate tumour cells) before infusion back into the patient. However, autologous transplantation often has a high disease recurrence rate and most people who require a transplant will not have an identical twin to donate. Most transplants of bone marrow and blood stem cells are thus allogeneic transplants, where the donor is matched as closely as possible with the recipient for their major histocompatibility (MHC) molecules. In most cases, a sibling has the greatest chance of a good MHC match with the patient.

The MHC molecules are cell surface markers that present foreign antigens to T lymphocytes. They are divided into Class I and Class II MHC; Class I MHC molecules are present on all nucleated cells and present antigens to cytotoxic T-cells, while Class II MHC molecules are only found on antigen-presenting cells of the immune system, and present antigen to helper T-cells. The genes that encode the MHC molecules are often called human leucocyte antigen (HLA) genes, and there are billions of different possible combinations of MHC that can be inherited. The success of a bone marrow or blood stem cell transplant depends on a good match of MHC between donor and recipient. The enormous variability of MHC makes this difficult when attempting to match unrelated donors with recipients in need of a transplant where matched siblings are unavailable. Large international registers have been created detailing the MHC molecules of healthy people who are willing to volunteer to donate bone marrow or peripheral blood stem cells (in the UK, the Anthony Nolan Bone Marrow Register is one example). Some differences in MHC — particularly in Class I MHC — can be tolerated and immunosuppressive drugs are given following transplantation to enhance this tolerance. However, the degree of GvHD experienced following transplantion is known to be directly related to the extent of MHC mismatch between donor and recipient.

8.15.2   Performing a bone marrow or stem cell transplant

If a patient is in need of bone marrow or stem cells (e.g. a young adult with acute leukaemia), a search for a donor with a good MHC match to the patient is undertaken. The patient's MHC molecules are identified using monoclonal antibodies, analysis of restriction fragment length polymorphisms (RFLPs) or DNA gene sequence identification with DNA probes, after which an initial assessment of the patient's family will be undertaken to identify any likely donors among siblings (and possibly parents, if no siblings are available — however, each parent is likely to be no better than a 50% MHC match unless parents are related). If this is unsuccessful, the medical team will turn to the international bone marrow registers to identify a possible MHC-matched donor. For many patients in need of a transplant no donors who are a suitable MHC match will be found, sadly, and in this situation an autologous transplant may be considered. However, for patients where a possible MHC-matched donor is identified, further testing will be conducted on the possible donor and recipient to establish full MHC profiles and ultimately crossmatching between recipient serum and peripheral blood lymphocytes from the potential donor to look for recipient antibodies against donor cells. Note that the ABO and Rh blood group of donor and recipient are unimportant in matching them for this type of transplant.

Once a suitable donor has been identified and has consented to donate bone marrow or blood stem cells, the recipient must be 'conditioned' before the transplant can take place to destroy their own bone marrow. This is usually achieved using high doses of chemotherapy, possibly with radiotherapy too. Following this, the recipient will be completely immunosup-

pressed and is kept in a special isolation unit to minimize the risk of infection. Bone marrow or stem cells may be collected from the donor using either marrow harvesting under general anaesthetic (in which marrow is extracted from multiple sites on the pelvic bone — a minimum of around 3×10^8 marrow cells per kg of recipient body weight is required), or using peripheral blood stem cell (PBSC) collection. For PBSC, the donor is given injections of granulocyte colony-stimulating factor (G-CSF) over several days prior to their collection to stimulate increased production and mobilization of the CD34-bearing haemopoietic stem cells, from marrow into the peripheral circulation. After around 5 days of these injections, PBSCs are collected from the donor using apheresis. Blood from the donor is removed from a vein in one arm and PBSC are removed from it using centrifugation, with the rest of the blood returned to the donor into a vein in their other arm. This involves minimal discomfort for the donor and no general anaesthesia; however, it does not always result in the collection of sufficient stem cells for larger recipients and the older bone marrow harvest technique may still sometimes be preferred (Figure 8.63).

Other sources of stem cells have been identified in recent years, with the umbilical cord of newborn infants identified in recent years as a rich source of haemopoietic stem cells. Cord blood banks are now being established in several countries and research is still underway in this area.

Following bone marrow harvesting or apheresis, stem cells are concentrated and can then be infused into the conditioned recipient. The recipient is kept in

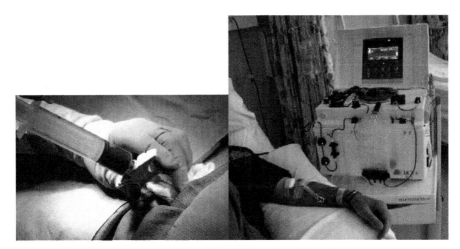

Figure 8.63   Stem cell harvesting using marrow harvesting (left) and peripheral blood stem cell collection (right)

isolation until blood tests show that the new stem cells have engrafted, with a sufficiently high white cell count to offer sufficient protection against infection. Engraftment usually takes around 15–30 days. Before this occurs, other blood components such as red cells and platelets may be needed to support the recipient, and these should be cytomegalovirus (CMV) negative and irradiated before being given to the immunosuppressed recipient. Antibiotics may also be given to decrease the risk of infection.

8.15.3 Complications of bone marrow and PBSC transplantation

Bone marrow and PBSC transplants carry a considerable mortality risk which varies according to the disease process being treated and the MHC match achieved between donor and recipient, and it is not undertaken lightly.

Infection is a common and sometimes fatal complication of bone marrow and PBSC transplantation and occurs due to the immunosuppression of the recipient around the time of transplantation, both from chemotherapy +/− radiotherapy before the transplant, and from immunosuppressive drugs given following it.

Graft versus host disease (GvHD) may occur. This is seen where immunocompetent cells are infused into an immunosuppressed recipient and can occur following allogeneic transplantation of bone marrow or stem cells. T lymphocytes in the donated bone marrow recognize the recipient's MHC as foreign and launch an immunological attack. Acute GvHD occurs within the first 30 days, and is characterized by a skin rash, which rapidly becomes extensive and often blistered. Epithelial cells of the skin and gastrointestinal tract may become necrotic and slough off; the patient typically experiences severe diarrhoea, sometimes with jaundice and splenomegaly as the liver and spleen are affected. Chronic GvHD may evolve from acute GvHD beyond the first 3 months after transplant, with a dry rash, dry skin and eyes and hair loss. GvHD is treated with immunosuppressive drugs and steroids but may be fatal if serious enough.

Bibliography

Green, A. (2010) *Postgraduate Haematology* (eds A.V. Hoffbrand, D. Catovsky and E.G.D Tuttenham), Wiley-Blackwell.

Hoffbrand, V. and Moss, P. (2011) *Essential Haematology*, Wiley-Blackwell.

Lewis, S.M., Bain, B.J. and Bates, I. (2006) *Dacie and Lewis Practical Haematology*, Churchill Livingstone.

Overfield, J., Dawson, M.M. and Hamer, D. (2007) *Transfusion Science*, Scion Publishing.

Pallister, C.J. and Watson, M.S. (2010) *Haematology*, Scion Publishing.

Chapter 9

Professional practice and biomedical science

Dr David Ricketts CSci, DBMS, FIBMS

9.1 What is a biomedical scientist?

The term biomedical scientist is a protected title in law (Health Professions Council 2003). This means that if a person has a test performed by a biomedical scientist they can be assured that the person is registered with the Health Professions Council (HPC) as competent to perform that analysis. In order to be registered with the HPC, students need to have been deemed competent to practice by their peers and met the HPC Standards of proficiency for Biomedical Scientists.

Professionalism is not just being paid to do a job, but to perform duties while adhering to a set of values, conforming to a code of conduct and being trusted by those who you are providing a service for. Results provided by biomedical scientists are trusted as they are provided as part of a professional service to other healthcare practitioners. Quality is a given by the users of the service and the results and advice given impact directly on the patient pathway. The HPC provides the peer assessed framework against which biomedical scientists can be judged.

The term biomedical scientist was confirmed as the legally protected title when the HPC replaced the Council for Professions Supplementary to Medicine (CPSM). Under the CPSM the term Medical Laboratory Scientific Officers (MLSO) was used to describe the profession, although ministers used the term biomedical scientist in 2000. MLSOs were not graduate entry and developed through a junior A, junior B grade before registering with the CPSM on completion of a Higher National Diploma and an oral examination. To qualify as a biomedical scientist a Bachelor of Science (B.Sc.) (Hons.) degree is required and a certificate of competence awarded by the Institute of Biomedical Science (IBMS).

9.2 The IBMS

The IBMS is the professional body for practitioners of biomedical science. It was formed in 1912 as the Pathological and Bacteriological Laboratory Assistants Association (PBLAA) and then it became the Institute of Medical Laboratory Technology (IMLT) in 1942 changing its name with the role of its members to the Institute of Medical Laboratory Science (IMLS) in 1975. Unusually for a professional body, membership is not compulsory for biomedical scientists.

The IBMS provides education and support for biomedical scientists including assessing the competence

Biomedical Sciences: Essential Laboratory Medicine, First Edition. Edited by Ray K. Iles and Suzanne M. Docherty.
© 2012 John Wiley & Sons, Ltd. Published 2012 by John Wiley & Sons, Ltd.

of a prospective registrant onto the HPC register. The IBMS's primary role is to develop biomedical scientists within healthcare and support their role in providing patient care.

The IBMS provides an education stream along with a professional support network for all levels of education from student to senior professional qualifications at career stage 9 of the healthcare science career pathway. This career escalator was published by the Department of Health in 2005. The IBMS qualification structure (Table 9.1) commences with an accredited biomedical science degree. The combination of the degree and a period of training in an Institute approved laboratory, during which a Registration Portfolio is completed, leads to the issue of a certificate of competence. This award is dependent upon external verification of the training process — as evidenced by the portfolio and a guided laboratory tour conducted by the trainee in the presence of the external verifier. Together with the academic degree, this demonstrates the learning outcomes needed to register as a biomedical scientist have been met. As well as being able to apply for registration with the HPC, the student also becomes eligible to apply for Corporate Membership of the Institute as a Licentiate. To become a Member

Table 9.1 IBMS qualifications taken from the IBMS website at www.ibms.org

Career framework stage	Membership class	Additional optional qualifications
9	Professional Doctorate	
8	**Advanced Specialist Diploma** ↑	Fellow	
7	**Higher Specialist Diploma*** ↑ MSc	Member	Diploma of Expert Practice and Certificate of Expert Practice
6	**Specialist Diploma*** ↑	Licentiate	
5	**Certificate of Competence** ↑ B.Sc. (Hons.) Biomedical Science		

Source: The Institute of Biomedical Sciences, UK www.ibms.org

(the next class of Institute membership) the Licentiate is required to pass the specialist diploma in the subject area of their specialty. This is a portfolio-based professional qualification and is assessed through a presentation and an accompanied tour of the laboratory, where their expert knowledge is tested by the external examiner.

The higher specialist diploma (HSD) was developed to fill the void left by the special and fellowship examinations. These examinations were used to award the Fellow membership class of the IBMS. With the advent of Masters Degrees in biomedical science, they became less popular and were no longer offered. The Masters program offers an excellent academic qualification but a vocational qualification was lacking for experienced practitioners to demonstrate their level of practice. The HSD is designed to test the application of a practitioner's knowledge at a high level of practice. A portfolio of evidence is submitted and, if it is of a sufficiently high quality, is followed by a written examination. Success in this examination leads to eligibility to apply for Fellowship of the Institute. There are now Advanced Specialist Diplomas in discrete areas of high level practice developed with the Royal College of Pathologists to provide an academic and clinical framework to support advanced clinical roles in biomedical science.

Supporting this spine of qualifications are the diplomas and certificate of expert practice. These qualifications are used to support the education and development of members in specialist areas such as diabetes, histological dissection, quality management and point of care in the community.

The IBMS also supports a mature continual professional development (CPD) programme to support lifelong learning in the profession and evidence the requirements of both the HPC and the Science Council for continued learning. The IBMS is an awarding body for Chartered Science status. This is a nationally recognized mark of excellence that is now available to scientists and requires the applicant to have a Masters level qualification and CPD.

9.3 Professional practice and the role of the HPC

Codes of practice are in place to ensure patient safety, which is the primary role of the regulator. Registrants falling short of these standards can be referred to the

HPC for investigation and be liable to action — the ultimate sanction including being struck off from the register. The HPC provides advice for practitioners and aspiring practitioners in the form of brochures outlining what is expected from them as they develop in their career as a biomedical scientist.

The HPC produces a document entitled *Guidance on conduct and ethics for students* (HPC, December 2009) aimed at students, practitioners in training and those who are delivering the training for a career in a profession regulated by the HPC, including biomedical science. The brochure covers information on the purpose of the HPC as a regulator and an overview of what you can expect as a registered healthcare practitioner. Importantly it highlights that conduct leading up to registration can impact the ability of the potential candidate to be accepted onto the register. Guidance on conduct and ethics is covered with statements provided on the expectation of the practitioner (Table 9.2). What the document does not provide are examples of good and bad practice.

The information given on acting in the patient's best interest is fairly straight forward to understand, but does raise a dilemma that professionals often have to face in its last bullet point. Even at the beginning of your career as a healthcare professional you need to be able to exhibit a mindset that looks for risks to the patient and have the courage to raise them with either your placement team or education provider. This is an ongoing theme for biomedical scientists who need to develop critical reflective skills to enable them to grow and develop as a healthcare provider. Not all guidance is as straight forward as the example given (Table 9.3).

Table 9.2   Guidance on conduct and ethics taken from the HPC pamphlet for best interest

1 You should always act in the best interests of your service users:

- You should respect a person's right to have their interventions carried out by a professional not a student
- You should not exploit or abuse your relationships with service users
- You should treat everyone equally
- You should not do anything that you think will put someone in danger
- If you are worried about a situation which might put someone at risk, you should speak to a member of the placement team or your education provider

Source: The Health Professions Council, UK www.hpc-uk.org

Table 9.3   Guidance on conduct and ethics taken from the HPC pamphlet for personal conduct

3 You should keep high standards of personal conduct:

- You should be aware that conduct outside your programme may affect whether or not you are allowed to complete your programme or register with us
- You should be polite with service users, your colleagues and the programme team
- You should make sure that your personal appearance is appropriate for your placement environment
- You should follow your education provider's or placement provider's policy on attendance

Source: The Health Professions Council, UK www.hpc-uk.org

Bullet points two to four are fairly self explanatory but the first point is one that is important to all registrants. Your conduct is important in work as well as away from the workplace. It is possible to be referred to the HPC for conduct unfitting for a professional due to issues arising from non work-related situations. This does not mean that you will be struck off for these situations, but the HPC will investigate and agencies such as the police force will refer any criminal activity to the HPC for investigation.

The student guidance covers key areas that a biomedical scientist will need throughout their career such as competence, health, operating within the limits of knowledge, development, consent and communication. All of these will be covered later in this chapter.

The HPC covers a diverse spectrum of healthcare professions from biomedical scientists to arts therapists. They all need to meet the same standards of proficiency, known as generic standards. There are profession-specific standards to underpin these, which can be found in the Standards of Proficiency — Biomedical Scientists, produced by the HPC in 2003 and updated in November 2007.

9.4 Standards of proficiency — biomedical scientists

These Standards — coupled with the Standards of Conduct, Performance and Ethics — are the backbone of a biomedical scientist's duties. As stated they provide the framework against which a biomedical scientist's professionalism is judged and therefore they require some scrutiny to help define how they can be used to improve practice.

The HPC states in its literature that:

'Your scope of practice is the area or areas of your profession in which you have the knowledge, skills and experience to practice lawfully, safely and effectively, in a way that meets our standards and does not pose any danger to the public or yourself.'

The HPC's primary concern is that of safety to the public, which is why a practitioner who is struck off from the register cannot lawfully practice biomedical science. The register is open to the public so they can check if the people they entrust their samples to are fit to practice. This is not the case in voluntary registers where a practitioner may be able to practice in another part of the country and not declare previous short comings in his or her performance.

9.5   Expectations of a health professional

The expectations for a biomedical scientist's scope of practice are clearly laid out by the HPC. The first guidance provided is in the area of professional autonomy and accountability. Under the old CPSM standards, biomedical scientists could not offer advice — this rule seems to be upheld in some workplaces even though the HPC view differs. The HPC standards ask practitioners to perform their duties within the legal and ethical boundaries of there profession. There is a need to look after the best interests of those who you offer the service to as well as a need to understand current standards and legislation. Confidentiality and seeking appropriate consent are key areas of responsibility, as are the ability to exercise a duty of care. One of the most important areas this standard of proficiency covers is in understanding the limits of practice and when to seek advice. A biomedical scientist is permitted to practice as an autonomous practitioner using professional judgment. It is essential that a biomedical scientist understands any advice they provide can have an impact on the patient care pathway, especially as pathology training in junior doctors is not as extensive as that which was previously delivered.

The issue of consent is particularly important due to the Human Tissue Act (2004) (HTA) clarifying the issue and setting it in a legal framework following some high-profile cases with human organs. Biomedical scientists are increasingly asked for expert opinion on consent regarding this Act; therefore it is important that the knowledge of the latest advice is up to date. This Act is covered in the expectation of the HPC that practitioners should be aware of standards, be they British or more global, that impact on the pathology laboratory.

Self management of workload and the effective use of resources are expected of a biomedical scientist. This can be a challenging standard to meet when resources are tight due to global financial issues. It is the duty of a professional to make full use of resources available and also to highlight to senior management if the lack of resources could have a detrimental impact on patient care.

Personal conduct has been mentioned in the student booklet and is as important when registered as a professional. The HPC has a public record of any disciplinary hearings on its website. Reviewing these hearings shows that often offences outside of the workplace are brought to their attention. It is also the duty of a registered professional to report poor conduct in others if it can impact patient care. This can be an issue as there are local policies and procedures in place to deal with conduct. The golden rule is: if in doubt, discuss this locally and obtain advice from the HPC. An employer would not like to know there is an issue from the HPC if it had not first been raised locally. It is also important that a patient safety issue concerning a registrant is not kept from the HPC. Failure of a registered practitioner to notify HPC of a fellow registrant acting in such a way as to contravene HPC standards may place their own registration at jeopardy.

The issue of personal health is important to the HPC. Registrants are required to alert the HPC if their health can jeopardise the service they offer. This covers all forms of health issues including mental health and potential substance dependency — if in doubt check with the HPC. It is important that if there is a short-term issue that may impact judgment, such as a heavy cold, care must be taken as to whether it is safe to be on duty or not.

The need for life-long learning is ingrained in the Standard. No one would wish to have their results provided by someone who does not have access to the skills and training of an up-to-date healthcare professional. Medicine continually evolves and those providing diagnostics need to be in a position to respond. As a result, the HPC randomly samples a number of registrants to check their CPD records to ensure they are keeping their skills current. CPD will be discussed later in this chapter, but it is an essential part of a professional's career.

9.6 Professional relationships

There have always been some tensions between various groups delivering healthcare and science in healthcare. Some of these tensions are due to historical issues, whilst others are due to service development and the roles of healthcare professions changing. Point of Care Testing (POCT) can be performed by numerous staff of different grades and abilities. This tests the skills of a biomedical scientist to be able to help support and develop these services. The training and support will need to be provided to a range of healthcare professionals meaning that the information given needs to be set at an appropriate level and that the biomedical scientist can ensure the testing is performed safely.

Increasingly, the laboratory is viewed as a key part of the patient pathway. This requires biomedical scientists to link with clinical teams and provide expert input. Traditionally, most biomedical scientists have the view that their role is laboratory based. The advent of advanced roles, such as in cytology and the involvement of biomedical scientists in histological dissection means that biomedical scientists now input into the multidisciplinary team (MDT) meetings to review cancer cases. This role is recognised as key by the HPC who make it one of their Standards of professional relationships — Standard 1b.2.

Standard 1b.3 relates to the appropriate communication of advice, instruction and professional opinion to colleagues, service users, their relatives and carers. As previously stated in the chapter, this is a culture change for some in pathology and there remains some confusion as to when and where this is appropriate. The skill and knowledge of the individual should be the deciding factor in this issue. This of course should be underpinned by supporting evidence from academic and vocational training and CPD.

The last area covered in this section is about effective communication throughout the care of the service users. Biomedical scientists usually do not treat patients directly as part of their normally duties, although they can do as part of clinics, such as warfarin dosing.

However it is important that the reports they produce are clearly understood by those using the service. This links to the need to communicate at the level of the recipient of information's understanding.

9.7 The skills required for the application of practice

The HPC has included many profession-specific standards in this section. The section concentrates on assessment of need, delivery of service, review of service and audit as its key themes. It encourages biomedical scientists to move from the delivery of a routine service to be part of the team planning future services. The ability of the biomedical scientist to know which is the most appropriate specimen and procedure for analysis forms the first standard; it is a key skill for a professional to have, as often this is the sort of basic advice needed by other healthcare professionals.

There is a detailed list in the Standards, where the registrant must be able to perform and demonstrate key laboratory functions, such as to 'be able to validate scientific and technical data and observations according to predetermined quality standards'. These are important skills for the biomedical scientist and require evidence to support that these have been met. The Clinical Pathology Accreditation Scheme (CPA), which is aligned to the International Standard ISO 15189 for medical laboratories, sets a standard that requires evidence of competencies for staff performing laboratory investigations. Signing a Standard Operating Procedure (SOP) to indicate it has been read is now insufficient as evidence. It is important that the person undergoing the testing has not only read the SOP but understands what they are doing and can demonstrate this competence to their peers. There are many ways of doing this, a simple one being including a series of questions as part of the SOP that require being answered and assessed before the task is performed for the first time. This assessment must be recorded as evidence.

The HPC also uses this section of the Standards to alert the biomedical scientist to the importance of using their knowledge to identify risk and try to remove opportunity for error from their systems. Data quality and data retention are important issues as the rules on personal information change with the advances in information technology. These advances prove an ongoing challenge due to the regulations, such as those of retention of blood transfusion records requiring increasingly long periods of storage. The move from paper-based, to tape, to floppy disk, to CD, to DVD, to memory stick has happened over the last 15 years. The challenge is not in the keeping of records, but keeping them in a format that will be future-proof in terms of access as different technologies become obsolete: even the ink used in modern printers fades over time.

The HPC requires professionals to be able evaluate their own impact on the users of the service and the actions as a result of their decisions. Critical reflection is an import tool, one traditionally associated with nurses. Reflection is now a recognised activity for CPD and forms an important part of a registrant's practice. The use of qualitative as well as quantitative data is important to registrants, especially as scientists are more often used to dealing with numerical data. Understanding feedback from users and staff is becoming increasingly more important in the management of resources and being able to respond to the changes in healthcare delivery.

The use of auditing is an important tool to help identify areas for service improvement and to provide a rich source of data for research and development. Auditing is a skill that is required by both the HPC and CPA. The HPC includes the audit of internal quality systems such as quality control and external quality assessment. The use of audit and review as part of a quality management system, and the application of the findings to the service delivery, are expected qualities of a biomedical scientist. This is one of the areas the biomedical scientist can offer an important skill set to help other healthcare professionals. All doctors are expected to perform a clinical audit; many of these audits require data from the laboratory and therefore it is an ideal opportunity to become involved in the clinical interface with colleagues from other departments. The training that biomedical scientists have in quality control and quality assurance is transferable to those performing analysis near the patient, and is required to ensure good governance and safe outcomes for patients.

9.8 Knowledge, understanding and skills

The final set of Standards for biomedical scientists examines the knowledge that a practitioner must have.

The understanding of the anatomy and function of the human body are the basic building blocks of this knowledge. What is 'normal' and how to identify disease and disorder is expected for a biomedical scientist. There must be an awareness of current research and scientific enquiry to help influence appropriate investigation of the diagnostic questions posed by requesting clinicians.

Biomedical scientists must understand the limitations of the molecules being examined and convey important scientific information to users. The move towards greater patient choice, where a test may be carried out in different laboratories or non-laboratory environments, can lead to confusion and error due to a lack of standardisation in methodologies and terms, which is being address by the professional bodies in the Harmony Project (2010). Similar tests are described by different names, some profiles are called the same but have different analytes prescribed to them and even the same test can have a different result depending on how the reaction is performed or the epitope being measured. This level of detail is not understood by the patient or the requesting clinician who would, rightly, expect the test to be the same no matter where and when ordered.

Pathology laboratories have many sub-specialties, the most common being clinical biochemistry, haematology, microbiology, histopathology, cytology and blood transfusion. In some centres these may even be combined into larger multi-disciplinary departments such as blood sciences. Biomedical scientists should be aware of the function of all of these and be able to direct service users to the most appropriate area to answer the clinical question needed. There are more specialised areas of pathology such as virology, immunology and departments offering esoteric testing which may offer a changing menu and more complex testing, so keeping up-to-date requires good communication and documentation between departments. The advent of molecular techniques into routine diagnostics offers a challenge as the same technology can cross traditional diagnostic departmental boundaries and there is a new challenge for learning as the data produced requires advanced bioinformatics skills.

The adaptation of the testing and information for specific clinical and cultural challenges is part of the knowledge and understanding expected from a biomedical scientist. An example of this is biomedical scientists who help with advice to set up and provide a cell salvage service for those who will not allow blood transfusion due to religious belief.

There are many aspects of health and safety that a biomedical scientist has to be aware of and implement. The medical laboratory contains many hazards governed by diverse legislation such as Control of Substances Hazardous to Health (COSHH) which was revised in 2009, ISO 15190 (2003) and the Health and Safety Law. There is a need to understand the biological, chemical and physical aspects of working in pathology; the risks to staff and visitors as well as the need to ensure proper precautions are taken to ensure the same vigor is used when placing equipment in a POCT setting.

9.9 Standards of conduct, performance and ethics

The HPC provides this document for all registrants as an overarching guide to inform the public and the professionals on what to expect and what is expected of them. Most of the Standards have been covered earlier in the chapter so only a brief summary of this document will be given.

There are 14 Standards covered which are set out as one list (Table 9.4). These Standards all start with the descriptor of 'You must... ', leaving no room for ambiguity as to the need for compliance. The Standards listed are taken from the HPC website and are free to download.

Often the statements contain phrases that are an expectation, such as keeping a high standard of personal conduct. It does not define what a high standard is, but it is up to the registrant to prove that if referred they meet this expectation. A high standard may be hard to define, but a low standard is easily spotted by your peers.

The emphasis is on communication between the registrant and the HPC, and between the registrant and the service users. If you train someone to do a task, you must ensure that adequate support is there should there be an issue; this is particularly relevant as diagnostics are being moved away from the traditional laboratory setting and closer to the patient.

Operating within the scope of knowledge and skills is important. The HPC guides the registrant to refer issues to another practitioner for a second opinion if unsure on how to proceed. It also clearly states that

Table 9.4   Duties of a registrant

Duties as a registrant

1. You must act in the best interest of service users

2. You must respect the confidentiality of service users

3. You must keep high standards of personal conduct

4. You must provide (to us and any relevant regulators) any important information about your conduct and competence

5. You must keep your professional knowledge and skills up-to-date

6. You must act within the limits of your knowledge, skills and experience and, if necessary, refer the matter to another practitioner

7. You must communicate properly and effectively with service users and other practitioners

8. You must effectively supervise tasks that you have asked other people to carry out

9. You must get informed consent to give treatment (except in an emergency)

10. You must keep accurate records

11. You must deal fairly and safely with the risks of infection

12. You must limit your work or stop practicing if your performance or judgment is affected by your health

13. You must behave with honesty and integrity and make sure that your behaviour does not damage the public's confidence in you or your profession

14. You must make sure that any advertising you do is accurate

registrants are responsible for ensuring that they are able to identify when they need to moderate or stop practice if unwell, a point alluded to earlier in this chapter.

The HPC relies on the professional biomedical scientist to make judgment calls and to back these up with evidence. As a professional you are expected to identify practice in yourselves and your colleagues that may impact on the patient, and improve on it — or refer it — if patient safety is put at risk. Keeping up-to-date and aware of service change is incumbent on all biomedical scientists, coming in daily and just doing a job is no longer sufficient for a professional practitioner. The appropriate use of CPD is an essential tool in maintaining registration.

9.10 CPD

In 2006 the HPC produced another helpful brochure entitled *Your guide to our standards for continuing professional development.* CPD is defined in this document as:

a range of learning activities through which
health professionals maintain and develop throughout their
career to ensure that they retain their capacity to practise safely,
effectively and legally within their evolving scope of practice.

(This definition is taken from the Allied Health Professions project, *Demonstrating competence through CPD,* 2002.)

Until 2005 CPD was an optional exercise using voluntary schemes such as the one offered by the IBMS. It is now a requirement of registration and a detailed record needs to be kept of CPD activities. These activities need to be relevant to current practice and focus on service delivery offered now and in the future. The HPC allows practitioners to define their own learning but provides a list of suggested activities to help. These can be done as part of a group, such as journal clubs, or as individual focused learning such as critical reflection on one's own performance.

9.11   Critical reflection

Critical reflection has been used by nurses to improve their practice for a number of years (Rolfe *et al.,* 2001) and was incorporated into the IBMS CPD scheme in 2003. Critical reflection looks at learning from an experience and disseminating and changing practice as a result. This technique is suited to the HPC CPD evidence and uses the Gibbs model of reflection as describe in Rolfe's book.

The Gibbs model requires that you describe what happened and then you describe your feelings about this. The next stage is to evaluate what was good and bad about this and to analyse this data, to make sense of it. In this case the data is thoughts and feelings which is something of an unfamiliar approach to learning for scientists. Using this data you elucidate what else could be done and then finally come up with an action plan as to what to do in the future to prevent this from happening again. This model can be taken further and used positively to change the culture of the laboratory. With agreement of all staff, thoughts on the input of

colleagues and management can be added and kept private at the time of the incident or process being reflected upon. After a while these thoughts can be revisited and reviewed to see if, when emotions are no longer running high, the feelings as to how the interaction went can be revisited and then reviewed in a constructive way to see if a service or professional relationship improvement can result.

Good practice as a result of reflection should be shared with colleagues and should be encouraged to input into the Knowledge and Skills Framework (KSF). This can then lead to improved training needs identified as part of the appraisal process.

Reflective learning logs should be performed as a result of attending training courses and seminars so the information gained from them can be used to improve the service and to inform your colleagues. These should be documents that all can see rather than an intimated reflection of one's self which are more the result of reflection after an incident or issue. The information should convey which areas are: suitable for immediate implementation, require more research, a detailed business plan or what is there for general interest and information.

9.12   IBMS CPD scheme

This scheme complements the requirements needed for the HPC evidence but is not sufficient to provide this as the only evidence for the HPC. The scheme came about due to the vision of one of the Institute's past Presidents, the late Russ Allison, and was launched in 1992. This scheme is free to members of the IBMS and consists of the accreditation of events. Accumulation of credits leads to the award of a CPD certificate. Online journal-based learning, structured essays and credits awarded to meetings and seminars all go towards the credit total. Credits are awarded for work-based journal clubs, for professional activity and for attending events such as the IBMS congress, held every 2 years. With the advent of reflective learning as part of the role of a biomedical scientist, CPD credits can now be given for this, although no formal assessment is made of CPD learning sheets.

The IBMS CPD scheme is designed to reward activity that encourages biomedical scientist to share their professional knowledge. Credits are awarded for educational activity such as teaching on the degree programs, running workshops and acting as an external examiner for the degree courses. Publishing books and chapters of books attracts CPD credits as does reviewing articles and books. Lecturing to share knowledge is encouraged and is also recognised by the scheme; this can be as a presenter of research at an international or national meeting or as a speaker on science or management.

Getting involved in the professions as an elected member of Council or inputting scientifically or academically in one of the specialist advisory bodies attracts credits. At a local level this may mean organising local meetings through the region or branch network of the IBMS or even at laboratory level running discussion groups. Case presentations to fellow biomedical scientists or other healthcare professionals are also a good way of gaining credits and establishing professional recognition.

One of the criticisms of the IBMS scheme is that credits are awarded for attending activities in nondiscipline-related subject areas, a biochemist gaining CPD points for attending a haematology seminar, for example. This has, in hindsight, proved to be a strength of the scheme as technology is increasingly blurring the boundaries of traditional pathology subject areas. As well as a drive from technology to combine disciplines, workforce reconfiguration due to budgetary restraints is also causing departments such as clinical biochemistry and haematology to combine as a blood science department. The HPC also requires a broad knowledge of other disciplines' functions as part of the evidence they wish to see from their registrants.

9.13 The professional biomedical scientist as an agent for change in the wider healthcare setting

The knowledge and skills of biomedical scientists can offer a huge amount of value to the wider healthcare community should they look beyond the doors of the laboratory. Biomedical scientists, through their accreditation scheme — CPA, have developed internal systems which others in the healthcare professional can learn from.

The delivery of training for biomedical scientists is a mature structured scheme, supported by education tools from the IBMS. Each laboratory that is approved for training by the IBMS has a person responsible for delivering and coordinating training. Training is agreed in a structured way to benefit the individual and the service. These systems can be used to help structure wider delivery of education and often biomedical scientists can be found at the heart of Trust training.

CPA has also brought structure to the quality management systems used in pathology. These systems have allowed the development of competency based SOPs and good document control systems to be put in place. It is possible to link the KSF dimension to the SOPs, making the appraisal target setting and evidence gathering easier. These systems seem second nature to many in pathology and can greatly benefit other areas by sharing the learning and developing similar systems across the health system. Quality management systems are not just about complying with paperwork, but using the system to look for opportunities for improvement rather than being reactive to problems when the system fails. The three strands of a good quality management system are: quality control, quality assurance and quality improvement. These skills are well developed in to biomedical scientists. Controlling quality is about systems being in place rather than just running a known sample through an analyser and getting a result. It looks for systems to be in place to look for consistent quality as well as focusing on the quality of the service and/or product.

Recently process management tools such as LEAN and Six Sigma have become very popular with Health Service Managers in both primary and secondary care. Schemes such as the productive ward and looking to bring LEAN into nursing have been set up to reduce error and to improve the experience of patients by making the ward more efficient and safer. These techniques have been used in pathology since 2003 and there is a considerable body of knowledge that can be used to help shape these projects. The danger is that these resources will be kept in pathology due to the fear of losing staff without backfill and the pathology service suffering as a result. The true cost will be the host organisation not realising the benefit biomedical scientists can offer in terms of quality and innovation. If this is not realised then there is a risk that the pathology department will be seen as an overhead only, which can be disposed of or downgraded to help meet short-term financial expediency.

Biomedical scientists have been in existence, under various titles, for almost 100 years. They have developed as an essential part of the healthcare team through good training and adaptation of new ideas and systems. The HPC regulates their conduct to ensure they operate at the highest level of professionalism needed to fulfill their role in delivery of service. This is supported by a professional body that provides a structure to training, education and CPD. One of the big challenges facing biomedical scientists is the low profile they seem to have with the public and their peers. This challenge can only be met by leaving the confines of the laboratory and getting into the wider healthcare environment offering their unique skills to help develop systems throughout the wider service.

The NHS has devised a scheme known as Modernising Scientific Careers (MSC). This looks to replace the established B.Sc. (Hons.) degree with a practitioner training programme and a voluntary register. The impact of this is unclear and comes at a time when pathology is under pressure to save 20% of its budget. Whatever the outcome of these challenges, there has never been such a time when the biomedical scientists' skills have been needed as much as now.

Appendix

Index of diseases, disorders and syndromes

Biomedical Sciences: Essential Laboratory Medicine, First Edition. Edited by Ray K. Iles and Suzanne M. Docherty.
© 2012 John Wiley & Sons, Ltd. Published 2012 by John Wiley & Sons, Ltd.

Index

Biomedical Sciences: Essential Laboratory Medicine, First Edition. Edited by Ray K. Iles and Suzanne M. Docherty.
© 2012 John Wiley & Sons, Ltd. Published 2012 by John Wiley & Sons, Ltd.

Printed and bound by CPI Group (UK) Ltd, Croydon, CR0 4YY

27/10/2024

14580371-0002